Smart Flow Control Processes in Micro Scale

Smart Flow Control Processes in Micro Scale

Volume 1

Special Issue Editors

Bengt Sunden
Jin-yuan Qian
Junhui Zhang
Zan Wu

MDPI • Basel • Beijing • Wuhan • Barcelona • Belgrade • Manchester • Tokyo • Cluj • Tianjin

Special Issue Editors

Bengt Sunden
Lund University
Sweden

Jin-yuan Qian
Zhejiang University
China

Junhui Zhang
Zhejiang University
China

Zan Wu
Lund University
Sweden

Editorial Office
MDPI
St. Alban-Anlage 66
4052 Basel, Switzerland

This is a reprint of articles from the Special Issue published online in the open access journal *Processes* (ISSN 2227-9717) (available at: https://www.mdpi.com/journal/processes/special_issues/Flow_Micro_Scale).

For citation purposes, cite each article independently as indicated on the article page online and as indicated below:

LastName, A.A.; LastName, B.B.; LastName, C.C. Article Title. *Journal Name* **Year**, *Article Number*, Page Range.

Volume 1
ISBN 978-3-03936-493-0 (Hbk)
ISBN 978-3-03936-494-7 (PDF)

Volume 1-2
ISBN 978-3-03936-513-5 (Hbk)
ISBN 978-3-03936-514-2 (PDF)

Contents

About the Special Issue Editors

Bengt Sunden received his M.Sc. in 1973, Ph.D. in 1979, and was appointed Docent in 1980, all at Chalmers University of Technology, Gothenburg, Sweden. He was appointed Professor of Heat Transfer at Lund University, Lund, Sweden, in 1992. He has served as Professor Emeritus and Senior Professor since 2016. His main research interests include heat transfer enhancement techniques, gas turbine heat transfer, and computational modeling and analysis of multiphysics and multiscale transport phenomena for fuel cells. He serves as Guest Professor of numerous prestigious universities. He is a Fellow of ASME, regional editor for *Journal of Enhanced Heat Transfer* since 2007, and associate editor of *Heat Transfer Research* since 2011, the ASME J. Thermal Science, Engineering and Applications (2010–2016), and ASME *Journal of Electrochemical Energy Conversion and Storage* since 2017. He is a recipient of the ASME Heat Transfer Memorial Award 2011 and Donald Q. Kern Award 2016. He received the ASME HTD 75th Anniversary Medal 2013. He has edited 30 books and authored three textbooks. He has published over 400 papers in numerous journals, with a h-index of 39 and over 6400 citations.

Jin-yuan Qian is presently Lecturer at the Institute of Process Equipment, College of Energy Engineering, Zhejiang University, China, a position he has held since his appointment in 2018. He received his B.Sc. and Ph.D. degrees, both in Chemical Process Equipment, from Zhejiang University, China, in 2011 and 2016, respectively. He was a joint Ph.D. student at TU Bergakademie Freiberg, Germany, from 2013 to 2014 and a Postdoc Researcher in the Department of Energy Sciences, Lund University, Sweden, from 2016 to 2017. His research interests include heat transfer, multiphase flow, flow control, and computational fluid dynamics. He has co-authored around 50 papers in international journals and conference proceedings.

Junhui Zhang is presently Research Professor at College of Mechanical Engineering, Zhejiang University, China, a position he has held since 2013, and has been Deputy Director of the State Key Laboratory of Fluid Power and Mechatronic Systems since 2019. He received his B.Sc. and Ph.D. degrees, both in Mechatronic Engineering, from Zhejiang University, China, in 2007 and 2012, respectively, and won the Outstanding Youth Science Foundation in 2019. His research interests are focused on the design and measurement of hydraulic components, especially the high-power-density axial piston pump. He has co-authored about 60 papers in international journals and conference proceedings.

Zan Wu is presently Senior Lecturer in the Department of Energy Sciences, Lund University, Lund, Sweden. He received his B.Sc. in Energy and Environmental System Engineering in 2008, and Ph.D. in Energy Engineering, both from Zhejiang University, Hangzhou, China. His research interests include multiphase flow, phase-change heat transfer enhancement techniques, microfluidics, surface modification, nanofluids, thermophysical properties, compact heat exchangers, and proton exchange membrane fuel cells. He has co-authored around 70 papers in international journals and conference proceedings as well as four book chapters.

Preface to "Smart Flow Control Processes in Micro Scale"

In recent years, microfluidic devices with a large surface-to-volume ratio have witnessed rapid development, allowing them to be successfully utilized in many engineering applications. Within microfluidic devices, the fluid flow at microscale shows obvious differences and unique flow characteristics compared to that at the common macroscale. Thus, the flow behaviors at microscale have attracted many researchers for the purpose of innovative heat and mass transfer enhancement. A smart control process has been proposed for many years, while many new innovations and enabling technologies have been developed for smart flow control, especially concerning "smart flow control" at the microscale. This Special Issue aims to highlight the current research trends related to this topic, presenting a collection of 33 papers from leading scholars in this field. Among these include studies and demonstrations of flow characteristics in pumps or valves as well as dynamic performance in roiling mill systems or jet systems to the optimal design of special components in smart control systems. We do think smart flow control at the microscale will continue to become more and more useful in the near future. To end, we would like to express our heartful gratitude to all the scientific contributors of the papers submitted to this Special Issue.

Bengt Sunden, Jin-yuan Qian, Junhui Zhang, Zan Wu

Special Issue Editors

 *processes*

Editorial

Special Issue: Smart Flow Control in Micro Scale

Jin-yuan Qian [1,2,3], **Junhui Zhang** [2], **Zan Wu** [3] **and Bengt Sunden** [3,*]

[1] Institute of Process Equipment, College of Energy Engineering, Zhejiang University, Hangzhou 310027, China; qianjy@zju.edu.cn
[2] State Key Laboratory of Fluid Power and Mechatronic Systems, Zhejiang University, Hangzhou 310027, China; benzjh@zju.edu.cn
[3] Department of Energy Sciences, Lund University, P.O. Box 118, SE-22100 Lund, Sweden; Zan.Wu@energy.lth.se
* Correspondence: bengt.sunden@energy.lth.se; Tel.: +46-462228605

Received: 28 April 2020; Accepted: 28 April 2020; Published: 8 May 2020

1. Introduction

Smart control processes have been proposed for many years, while for smart flow control—especially when "smart flow control" comes at the microscale—it turns out that many new innovations and enabling technologies are possible. For instance, precise flow rate in a microreactor means high reaction efficiency. Similarly, for micromixers, smart fluid control can improve the precise distribution of every constituent. Such systems are made up of micropumps, microchannels, and microvalves, etc.

In this Special Issue on "Smart Flow Control Processes at the Microscale", 33 papers have been published, ranging from studies of flow characteristics in pumps or valves, dynamic performances in roiling mill systems or jet systems, to optimal design of special components in smart control systems. The Special Issue is available online at the following link:

https://www.mdpi.com/journal/processes/special_issues/Flow_Micro_Scale.

The contributions are summarized in four parts as follows:

2. Smart Flow Control in Pumps

A pump is one of the most important devices in fluid transportation systems. Research aiming at the flow field and pressure characteristics is of great importance for improvement of the operating performance.

Bai et al. [1] numerically studied the influence of pressure fluctuations and unsteady flow patterns in a pump flow channel with different diffuser vane numbers. The results indicate that the lower number of diffuser vanes was beneficial to obtain weaker pressure fluctuation intensity.

Cao et al. [2] investigated the whole flow field of a low specific speed centrifugal pump with five blades at different flow rates in order to study the near-wall region flow characteristics in a low-specific-speed centrifugal impeller. The main contribution of this work is the illustrations of pressure distribution and relative velocity distribution profiles on the pressure side of different blades.

Si et al. [3] studied the mechanism of radiated noise and its relationship with hydraulics in centrifugal pumps via a numerical method combined with an experimental approach. The results reveal that the radiated noise exhibits a typical dipole characteristic behavior and its directivity varies with the flow rate. In addition, Si et al. [4] carried out an experimental and numerical study aiming at the internal flow characteristics under gas-liquid two-phase flow in a miniature drainage pump. The pump performance and emitted noise measurements were monitored at various conditions. The study is a good reference for low noise design of drainage pumps.

Wang et al. [5] also analyzed the gas-water two-phase flow in a self-priming centrifugal pump. The results illustrate the three stages in a self-priming process. The effect of the middle stage is highlighted, which further determines the length of the self-priming time.

Jiao et al. [6] studied the three-dimensional cavitation flow in a waterjet propulsion pump based on Zwart–Gerber–Belamri cavitation model and the RNG (Renormalization Group) k-ε model. The study demonstrates that the potential dangerous regions of cavitation are the lip of the inlet passage and the upper and lower connecting curved section of the inlet passage.

Luo et al. [7] analyzed the unsteady flow process in waterjet pumps to improve the overall performance and optimization of the structure design. The surface vortex of the blade and the unsteady flow process of the propulsion pump at different times of the same period were demonstrated.

Cao et al. [8] studied the evolution of vortex structures in a laminar boundary layer over a flat plate by the Fourier spectral hybrid method. Results show that the maximum amplitudes of the vortex structures experience a process of linear growth and nonlinear rapid growth. The change in the mean flow profile further induces or promotes the growth or formation of vortex structures.

Jin et al. [9] investigated the external flow characteristics and pressure fluctuation in a submersible tubular pumping system. Results indicate that the pressure pulsation is less affected by the blade frequency with an increase of the measuring point from the impeller.

Xue et al. [10] proposed a design method based on Amesim and a Python script for the purpose of multi-objective optimization in static and dynamic performances of a pump-driven actuator. The mapping between the design parameters and the relations between the objectives are plotted. The results highlight the feasibility of the proposed method in achieving the multi-objective optimization.

Zhang et al. [11] investigated the structural characteristics of an ultra-high pressure axial piston pump. Via an analysis of the oil film pressure and thickness in different rotating angles of the piston–cylinder pair, it was found that the oil film pressure achieves the maximum value when the rotating angle increases to 90°, while the film thickness reaches the minimum at the same time.

Zheng et al. [12] provided a fluid pressure signal method for hydraulic pumps based on Autogram for solving the fluid pressure fluctuations caused by the center spring wear faults. The results highlight the superiority of standard Autogram on the extraction of fault feature information on center spring wear when comparing with upper Autogram and lower Autogram. Moreover, a novel method named as improved wavelet transform (IEWT) was proposed by Zheng et al. [13] in order to solve the segment over-decomposition obtained by the empirical wavelet transform (EWT). The proposed method was shown to be superior for eliminating the over-decomposition of the fault feature information.

3. Smart Flow Control in Valves

Valves play a significant role to change the flow rate, pressure and directions of fluids. Smart valves can turn out smart control of fluids.

Lei et al. [14] suggested a novel method depending on the Machine Learning Service (MLS) HUAWEI CLOUD to achieve accurate diagnosis of hydraulic valve faults. The method combines advantages of Principal Component Analysis (PCA) in dimensionality reduction and the eXtreme Gradient Boosting (XGBoost) algorithm and proves to be highly effective for identifying valve faults in the hydraulic directional valve.

Liu et al. [15] studied the throttling characteristics of the diaphragm valve. In order to identify the optimal design of the flow path profile, two-dimensional simulation of the Weir diaphragm valve flow field was conducted. The study shows that the flatting of the ridge side wall, widening of the ridge top and the gentle flatting of the internal protruding of the flow path prove to be three positive approaches for the improvement of the throttling characteristics.

Lu et al. [16] presented an investigation aiming at the oscillating flow field of the double-nozzle flapper servo valve pre-stage through Large Eddy Simulation (LES) turbulent modeling. Meanwhile, the User-Defined Function (UDF) was introduced to control the main stage movement. The results highlight the structure and flow parameter effect on the oscillating flow. In order to illustrate the damage caused by the increase of the injection pressure in the high pressure pump unloading valve ball, a theoretical calculation of the pressure relieve valve and the fatigue numerical simulation was carried out by Lu et al. [17]. Results indicate that the high pressure relief valve ball in the direct injection high pressure pump should not be a traditional structural damage under high pressure conditions and the surface damage of the valve ball is microscopic damage, such as fretting wear.

Qiu et al. [18] investigated the pressure drop and cavitation characteristics in the sleeve-regulating valve in different pressure differences and valve core displacements using the multiphase cavitation model. The results show that the decrease of the valve core displacement induces the enlargement of the vapor distribution region and the increase of the vapor density. The effects of the pressure difference on the cavitation intensity are more prominent with the decrease of the valve core displacement. The work provides valuable instructions for the cavitation control of the sleeve regulating valves.

Wu et al. [19] studied the flow and loss coefficients in a wedge-type double disk parallel gate valve. Effects of the Reynolds number, valve opening degree and groove depth were analyzed. The results suggest that a large groove depth should be selected to provide a large flow coefficient during the design process. However, during the machining process, the machining accuracy should be satisfied in order to avoid stress concentration of the bolt.

Besides, bileaflet mechanical heart valves (BMHVs) are widely used as the alternatives of diseased heart valves. Xu et al. [20] performed simulations of unsteady flow in a BMHV and pressure pulsation characteristics under different flow rates and leaflet fully opening angle conditions were investigated. The work provides a good reference for the alleviation of leaflet vibration phenomenon in BMHVs.

4. Smart Flow Control in Microfluidics

Droplet flow and microflow control in microfluidics are extensively studied.

Qian et al. [21] investigated the characteristics of droplets in a dynamic injection flow rate by the Volume of Fluid (VOF) method combined with UDF. The study presents a novel aspect of the droplet flow since the droplet generation is always at a constant flow rate of two phases in most researches.

Zhang et al. [22] studied the hydrodynamics of droplets passing through metal foam by the lattice Boltzmann method (LBM). The critical capillary number was identified. Results show that the droplet continues to be deformed until it breaks up when the capillary number is larger than 0.61. In order to avoid the calescence of the adjacent droplet, the distance between the droplets should be larger than three times the diameter of the droplet.

Li et al. [23] investigated the two-phase flow inside a grooved rotating-disk system both in experimental and numerical methods. Visualization tests indicated that the flow field of the system was an air–oil flow. The stable interface between the continuous oil phase and the two-phase area could be formed and observed.

Guan et al. [24] proposed a miniaturized, easily processed, and inexpensive xenon micro flow control device (XMFCD) in order to reduce the volume and weight of the traditional XMFCDs. The design of the proposed XMFCD is based on complex three-dimensional (3D) microfluidic channels while the fabrication process is based on low-temperature co-fired ceramic (LTCC) technology and it was illustrated in detail.

5. Smart Flow Control in Mechatronic Systems

In mechatronic systems, there are many flow control issues, and smart flow control can improve the efficiency of mechatronic systems significantly.

A rolling mill with a hydraulic system is widely used in the strip steel industry. The vertical vibration seriously affects the stability of the rolling mill system. Zhang et al. [25] analyzed the effects of the equivalent damping coefficient, leakage coefficient, and proportional coefficient of the controller on the hydraulic screw-down system of the rolling mill. Results suggest that in the closed-loop state, when Proportional–Integral–Derivative (PID) controller parameters are fixed, the system will have parameter uncertainty due to the change of the equivalent damping coefficient and internal leakage coefficient.

Yuan et al. [26] investigated the dynamics, flow responses and power consumption theoretically and experimentally in hydraulic systems using the switched inertance hydraulic converter (SIHC). Results highlight the superiority of the SIHC in operation involving high pressures and delivery-flow rates.

Qian et al. [27] proposed a static deformation-compensation method based on inclination sensor feedback for large-scale manipulators to reduce the deviation of the endpoint in manipulators with hydraulic actuation. Compared to the finite element method, the proposed method considers less boundary conditions, which are uncertain for flexible manipulators in most situations.

Zhu et al. [28] revealed the bifurcation characteristic of the load vertical vibration of the hydraulic automatic gauge control (HAGC) system through the investigation of the nonlinear factors such as excitation force, elastic force and damping force. Results point out that the resonance region can be effectively avoided by adjusting the nonlinear stiffness coefficient and the stability of the system will be promoted as well. In addition, Zhu et al. [29] described the function of the key position closed-loop system in HAGC. Results indicate that the absolute stability conditions of the position closed-loop system are derived whether the spool displacement is positive or negative.

Li et al. [30] introduced a new method for the evaluation of the blood cell damage and the observation of the real-time characteristics of blood flow patterns in vitro using rheometer and bionic microfluidic devices. The damaged erythrocytes were collected and injected into a bionic microfluidic device. Analysis of the captured images indicate that with the increase of shear stress suffered by the erythrocyte, the migration rate of damaged erythrocyte in bionic microchannel is significantly decreased.

Yuan et al. [31,32] investigated the natural frequency sensitivity and dynamic behaviors of the fire-fighting jet system. An adaptive gun-head design was proposed to achieve the fluid–structure interaction and discrete–continuous coupling characteristics and the sensitivity calculation formulas of the natural frequency was derived of the jet system to typical design parameters [31]. Focusing on the adaptive fire-fighting monitor, influence of the nonlinear fluid spring force on the dynamic characteristics was investigated. Results indicated that in the design of a fire-fighting system, the interval of the input shaft speed of the pump, and the pulsation frequency of the output fluid should be avoided [32].

Finally, in order to analyze the appropriate numerical simulation method for the investigation of the hydraulic performance, the mixing process and the flow law in the venturi injectors were considered by Li et al. [33]. Flow characteristics of the internal flow field obtained with and without the cavitation model were both compared with the experiments. Results indicate that the cavitation model has better agreement with experiments.

6. Conclusions

In this special issue, 33 papers are presented and they relate to smart flow control in pumps, valves, microfluidics and mechatronic systems. We believe that smart flow control, especially at microscales, will become more important and useful in the near future.

We would like to express our heartfelt gratitude to all the scientific contributors of the papers submitted to this Special Issue.

Author Contributions: Conceptualization, J.-y.Q. and J.Z.; Methodology, J.-y.Q. and Z.W.; Data Curation, J.-y.Q., J.Z. and Z.W.; Writing—Original Draft Preparation, J.-y.Q.; Writing—Review & Editing, J.-y.Q. and B.S. All authors have read and agreed to the published version of the manuscript.

Funding: This research was funded by the National Natural Science Foundation of China, through grant number 51922093 and 51805470; the National Key R&D Program of China, through grant number 2019YFB2005101; and the Yucai Project of Zhejiang Association for Science and Technology.

Conflicts of Interest: The authors declare no conflict of interest.

References

1. Bai, L.; Zhou, L.; Han, C.; Zhu, Y.; Shi, W. Numerical Study of Pressure Fluctuation and Unsteady Flow in a Centrifugal Pump. *Processes* **2019**, *7*, 354. [CrossRef]
2. Cao, W.; Jia, Z.; Zhang, Q. Near-Wall Flow Characteristics of a Centrifugal Impeller with Low Specific Speed. *Processes* **2019**, *7*, 514. [CrossRef]

3. Si, Q.; Wang, B.; Yuan, J.; Huang, K.; Lin, G.; Wang, C. Numerical and Experimental Investigation on Radiated Noise Characteristics of the Multistage Centrifugal Pump. *Processes* **2019**, *7*, 793. [CrossRef]

4. Si, Q.; Shen, C.; Ali, A.; Cao, R.; Yuan, J.; Wang, C. Experimental and Numerical Study on Gas-Liquid Two-Phase Flow Behavior and Flow Induced Noise Characteristics of Radial Blade Pumps. *Processes* **2019**, *7*, 920. [CrossRef]

5. Wang, C.; Hu, B.; Zhu, Y.; Wang, X.; Luo, C.; Cheng, L. Numerical Study on the Gas-Water Two-Phase Flow in the Self-Priming Process of Self-Priming Centrifugal Pump. *Processes* **2019**, *7*, 330. [CrossRef]

6. Jiao, W.; Cheng, L.; Xu, J.; Wang, C. Numerical Analysis of Two-Phase Flow in the Cavitation Process of a Waterjet Propulsion Pump System. *Processes* **2019**, *7*, 690. [CrossRef]

7. Luo, C.; Liu, H.; Cheng, L.; Wang, C.; Jiao, W.; Zhang, D. Unsteady Flow Process in Mixed Waterjet Propulsion Pumps with Nozzle Based on Computational Fluid Dynamics. *Processes* **2019**, *7*, 910. [CrossRef]

8. Cao, W.; Jia, Z.; Zhang, Q. The Simulation of Vortex Structures Induced by Different Local Vibrations at the Wall in a Flat-Plate Laminar Boundary Layer. *Processes* **2019**, *7*, 563. [CrossRef]

9. Jin, Y.; He, X.; Zhang, Y.; Zhou, S.; Chen, H.; Liu, C. Numerical and Experimental Investigation of External Characteristics and Pressure Fluctuation of a Submersible Tubular Pumping System. *Processes* **2019**, *7*, 949. [CrossRef]

10. Xue, L.; Wu, S.; Xu, Y.; Ma, D. A Simulation-Based Multi-Objective Optimization Design Method for Pump-Driven Electro-Hydrostatic Actuators. *Processes* **2019**, *7*, 274. [CrossRef]

11. Zhang, J.; Liu, B.; LÜ, R.; Yang, Q.; Dai, Q. Study on Oil Film Characteristics of Piston-Cylinder Pair of Ultra-High Pressure Axial Piston Pump. *Processes* **2020**, *8*, 68. [CrossRef]

12. Zheng, Z.; Li, X.; Zhu, Y. A Fault Feature Extraction Method for the Fluid Pressure Signal of Hydraulic Pumps Based on Autogram. *Processes* **2019**, *7*, 695. [CrossRef]

13. Zheng, Z.; Wang, Z.; Zhu, Y.; Tang, S.; Wang, B. Feature Extraction Method for Hydraulic Pump Fault Signal Based on Improved Empirical Wavelet Transform. *Processes* **2019**, *7*, 824. [CrossRef]

14. Lei, Y.; Jiang, W.; Jiang, A.; Zhu, Y.; Niu, H.; Zhang, S. Fault Diagnosis Method for Hydraulic Directional Valves Integrating PCA and XGBoost. *Processes* **2019**, *7*, 589. [CrossRef]

15. Liu, Y.; Lu, L.; Zhu, K. Numerical Analysis of the Diaphragm Valve Throttling Characteristics. *Processes* **2019**, *7*, 671. [CrossRef]

16. Lu, L.; Long, S.; Zhu, K. A Numerical Research on Vortex Street Flow Oscillation in the Double Flapper Nozzle Servo Valve. *Processes* **2019**, *7*, 721. [CrossRef]

17. Lu, L.; Xue, Q.; Zhang, M.; Liu, L.; Wu, Z. Non-Structural Damage Verification of the High Pressure Pump Assembly Ball Valve in the Gasoline Direct Injection Vehicle System. *Processes* **2019**, *7*, 857. [CrossRef]

18. Qiu, C.; Jiang, C.-H.; Zhang, H.; Wu, J.-Y.; Jin, Z.-J. Pressure Drop and Cavitation Analysis on Sleeve Regulating Valve. *Processes* **2019**, *7*, 829. [CrossRef]

19. Wu, H.; Li, J.-Y.; Gao, Z.-X. Flow Characteristics and Stress Analysis of a Parallel Gate Valve. *Processes* **2019**, *7*, 803. [CrossRef]

20. Xu, X.-G.; Liu, T.-Y.; Li, C.; Zhu, L.; Li, S.-X. A Numerical Analysis of Pressure Pulsation Characteristics Induced by Unsteady Blood Flow in a Bileaflet Mechanical Heart Valve. *Processes* **2019**, *7*, 232. [CrossRef]

21. Qian, J.-Y.; Chen, M.-R.; Wu, Z.; Jin, Z.-J.; Sunden, B. Effects of a Dynamic Injection Flow Rate on Slug Generation in a Cross-Junction Square Microchannel. *Processes* **2019**, *7*, 765. [CrossRef]

22. Zhang, J.; Yu, X.; Tu, S.-T. Lattice Boltzmann Simulation on Droplet Flow through 3D Metal Foam. *Processes* **2019**, *7*, 877. [CrossRef]

23. Li, C.; Wu, W.; Liu, Y.; Hu, C.; Zhou, J. Analysis of Air–Oil Flow and Heat Transfer inside a Grooved Rotating-Disk System. *Processes* **2019**, *7*, 632. [CrossRef]

24. Guan, C.-B.; Shen, Y.; Yao, Z.-P.; Wang, Z.-L.; Zhang, M.-J.; Nan, K.; Hui, H.-H. Design, Simulation, and Experiment of an LTCC-Based Xenon Micro Flow Control Device for an Electric Propulsion System. *Processes* **2019**, *7*, 862. [CrossRef]

25. Zhang, Y.; Jiang, W.; Zhu, Y.; Li, Z. Research on the Vertical Vibration Characteristics of Hydraulic Screw Down System of Rolling Mill under Nonlinear Friction. *Processes* **2019**, *7*, 792. [CrossRef]

26. Yuan, C.; Mao Lung, V.L.; Plummer, A.; Pan, M. Theoretical and Experimental Studies of a Digital Flow Booster Operating at High Pressures and Flow Rates. *Processes* **2020**, *8*, 211. [CrossRef]

27. Qian, J.; Su, Q.; Zhang, F.; Ma, Y.; Fang, Z.; Xu, B. Static Deformation-Compensation Method Based on Inclination-Sensor Feedback for Large-Scale Manipulators with Hydraulic Actuation. *Processes* **2020**, *8*, 81. [CrossRef]

28. Zhu, Y.; Tang, S.; Wang, C.; Jiang, W.; Yuan, X.; Lei, Y. Bifurcation Characteristic Research on the Load Vertical Vibration of a Hydraulic Automatic Gauge Control System. *Processes* **2019**, *7*, 718. [CrossRef]

29. Zhu, Y.; Tang, S.; Wang, C.; Jiang, W.; Zhao, J.; Li, G. Absolute Stability Condition Derivation for Position Closed-Loop System in Hydraulic Automatic Gauge Control. *Processes* **2019**, *7*, 766. [CrossRef]

30. Li, D.; Li, G.; Chen, Y.; Man, J.; Wu, Q.; Zhang, M.; Chen, H.; Zhang, Y. The Impact of Erythrocytes Injury on Blood Flow in Bionic Arteriole with Stenosis Segment. *Processes* **2019**, *7*, 372. [CrossRef]

31. Yuan, X.; Zhu, X.; Wang, C.; Zhang, L.; Zhu, Y. Natural Frequency Sensitivity Analysis of Fire-Fighting Jet System with Adaptive Gun Head. *Processes* **2019**, *7*, 808. [CrossRef]

32. Yuan, X.; Zhu, X.; Wang, C.; Zhang, L.; Zhu, Y. Research on the Dynamic Behaviors of the Jet System of Adaptive Fire-Fighting Monitors. *Processes* **2019**, *7*, 952. [CrossRef]

33. Li, H.; Li, H.; Huang, X.; Han, Q.; Yuan, Y.; Qi, B. Numerical and Experimental Study on the Internal Flow of the Venturi Injector. *Processes* **2020**, *8*, 64. [CrossRef]

Article

Theoretical and Experimental Studies of a Digital Flow Booster Operating at High Pressures and Flow Rates

Chenggang Yuan, Vinrea Lim Mao Lung, Andrew Plummer and Min Pan *

Centre for Power Transmission and Motion Control, Department of Mechanical Engineering, University of Bath, Claverton Down BA2 7AY, UK; cy466@bath.ac.uk (C.Y.); vlml20@bath.ac.uk (V.L.M.L.); arp23@bath.ac.uk (A.P.)
* Correspondence: m.pan@bath.ac.uk

Received: 18 December 2019; Accepted: 4 February 2020; Published: 10 February 2020

Abstract: The switched inertance hydraulic converter (SIHC) is a new technology providing an alternative to conventional proportional or servo-valve-controlled systems in the area of fluid power. SIHCs can adjust or control flow and pressure by means of using digital control signals that do not rely on throttling the flow and dissipation of power, and provide hydraulic systems with high-energy efficiency, flexible control, and insensitivity to contamination. In this article, the analytical models of an SIHC in a three-port flow-booster configuration were used and validated at high operating pressure, with the low- and high-pressure supplies of 30 and 90 bar and a high delivery flow rate of 21 L/min. The system dynamics, flow responses, and power consumption were investigated and theoretically and experimentally validated. Results were compared to previous results achieved using low operating pressures, where low- and high-pressure supplies were 20 and 30 bar, and the delivery flow rate was 7 L/min. We concluded that the analytical models could effectively predict SIHC performance, and higher operating pressures and flow rates could result in system uncertainties that need to be understood well. As high operating pressure or flow rate is a common requirement in hydraulic systems, this constitutes an important contribution to the development of newly switched inertance hydraulic converters and the improvement of fluid-power energy efficiency.

Keywords: digital hydraulics; switched inertance hydraulic systems; high-speed switching valves; pressure booster; flow booster; efficient fluid power

---

1. Introduction

Digital hydraulics is a new technology providing an alternative to conventional proportional or servo-valve-controlled systems in the area of fluid power. It promises hydraulic systems with high-energy efficiency, flexible control, and insensitivity to contamination [1–5]. The switched inertance hydraulic converter (SIHC) concept is a subdomain of digital hydraulics [5–7], which is analogous to the electrical buck converter. It makes use of the inherent reactive behaviour of hydraulic components, including high-speed switching valves (switch function), small diameter tubes (inductive effect), and accumulators (capacitive effect) acting as a switch, an inductor, and a capacitor in the electrical circuit. Figure 1 shows a schematic of a three-port flow booster, which is a typical configuration of SIHCs [8–11]. The three-/two-way high-speed switching valve alternatively switches between the high- and low-pressure supply port. When the high-speed valve connects to the supply pump, the high-velocity fluid passes from pump to load; when the valve switches from the pump to the low-pressure supply port, the momentum of the fluid in the inertance tube draws the continuous flow from the low-pressure supply port to the load despite the adverse pressure gradient. As long as the switching time of the valve is short, the reduction in delivery flow is very small, and the average delivery flow is boosted and could be significantly higher than the supply flow.

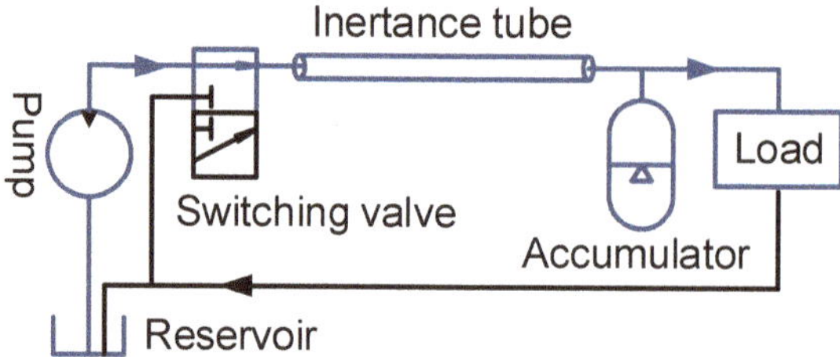

Figure 1. Schematic of three-port flow booster.

The concept of SIHCs was initially proposed by Brown et al. in 1987 [12]. The team proposed and investigated a series of SIHC configurations, including a step-down transformer (flow booster), step-up transformer (pressure booster), switching gyrator, and four-port SIHC analogously to electrically switch inductance transformers. They concluded that hydraulic transformers have clear potential to improve hydraulic-system bandwidth and energy efficiency on the basis of comprehensive theoretical and experimental studies [12–14]. However, due to the limitations of manufacturing high-speed switching valve in the 1980s, continuous work was limited. In the past decade, SIHC research came from Linz (Austria), the United States, Canada, the Nordic countries, Brazil, and Bath (UK).

This research can be categorised as SIHC characteristics, SIHC optimisation, and high-speed switching-valve design. Scheidl et al. designed a resonator, and found that a high-speed valve with a switching frequency of 1 kHz is required for a resonator with a length of 0.65 m [15]. They proved Brown's conclusion that the performance of a high-speed switching valve is a limiting factor in the development of SIHCs. To tackle this challenge, in 2006, Winkler and Scheidl designed a high-speed solenoid-controlled spool valve with a nominal flow rate of 45 L/min at a pressure drop of 5 bar. The valve had a fast response speed of 1 ms [16]. Winkler also designed an alternative poppet valve with a higher delivery flow rate of 90 L/min and a similar pressure drop of 5 bar [17]. However, this valve had a slower response speed of 2 ms. The first hydraulic switching converter was built by the team in Lehigh University [13]. Later, Kogler, and Scheidl reviewed two typical hydraulic switching converters (HBC) and concluded that the performance of hydraulic converters was affected by valve dynamics, parasitic effects, wave propagation along the pipe, system nonlinearities, and pressure pulsations [8]. The HBC was compactly designed [18], and then applied to control the leg of a quadruped robot [19,20] and caster-mould resonant drives [21].

In Bath in 2009, Johnston carried out theoretical and experimental work on the switched inertance device for the efficient control of pressure and flow [22]. He developed simulation models of a flow booster and a pressure booster in a three-port valve configuration. The experimental work was performed using a rotary valve that could only be used for very short periods due to extreme noise and vibration. Maximal experimental delivery pressure was 100 bar, and delivery flow rate was about 15 L/min.

In 2014, Pan et al. developed ideal analytical distributed models of a three-port SIHC that further enhanced the models, including switching-valve transition dynamics, nonlinearity, and leakage [11]. They validated the models in experiments using a commercial proportional directional valve from Parker Hannifin (DF*plus*), a high-speed linear valve, and a high-speed rotary valve. Experiment results proved the effectiveness of the analytical models. However, operating pressure and flow were low, with a maximal pressure of 60 bar and a delivery flow rate of 12 L/min. Pan et al. also developed analytical models of a four-port SIHC in which two inertance tubes were used.

Unlike the one-direction control of a three-port configuration, the four-port SIHC provides real four-quadrant operation and seamless directional changes [23]. The model was validated in experiments with an operating pressure of 32 bar and very small flow rates. Wu et al. developed a rotary hydraulic converter for variable load [24]. A new rotary module is designed as an inertia element for converter energy improvement. Simulated results showed good performance of the designed system. The simulated high and low pressures are 50 bar and 10 bar, and the experimental validation is needed [24].

Little work has been done to investigate the performance of SIHCs when they operate at high pressures and flow rates, both theoretically and experimentally. To bridge this gap, this article investigates the performance of a three-port flow booster with a high-speed rotary valve that operates at high pressure of 90 bar and flow rate of 21 L/min. It contributes to the digital hydraulics area by validating the high efficiency of SIHCs operating at high pressures and flow rates. The enhanced analytical model of the flow booster is briefly reviewed in Section 2, and used to analyse the experimental results in Section 3. The relationship between delivery pressure, switching ratio and frequency, and flow loss is discussed, followed by an investigation of system energy efficiency and power loss. The discussion and conclusions are presented in Section 4.

2. Previous Studies: Analytical Models and Experiment Investigations

The distributed analytical model of an ideal three-port SIHC was proposed in [10], and the enhanced model, including switching-valve transition dynamics, nonlinearity, and leakage was proposed in [11]. The analytical models were based on the mixed time-domain and frequency-domain approach in which linear parts (initial pressures, flow, and system impedance) are modeled in the frequency domain, while nonlinear and time-dependent parts (leakage, valve transition dynamics, and system nonlinearity) are modeled in the time domain; an iterative technique is used to link them. The models can effectively predict SIHC physical characteristics and provide a tool to aid in SIHC design.

A brief introduction is given here for the enhanced three-port SIHC analytical models that are used for further analysis in this paper. A detailed derivation can be found in [11]. A typical flow booster is shown Figure 2, where p_H is the high-pressure supply, p_L is the low-pressure supply, and p_d is the delivery pressure. Two pressure sources p_1 and p_2, were used to represent the open and closed ports of the three-port high-speed switching valve. P_3 is the inlet pressure of the inertance tube. R_tr is switching transition resistance, and R_u is the underlap/overlap resistance of the valve. R_non1 and R_non2 represent the nonlinear characteristics of the valve when it connects to the open and closed ports. Q_1 and q_2 are the flow rates from the open and closed ports when q_3 is the total outlet flow from valve to inertance tube. Tube resistance is R_tube and tube inertance is I_tube.

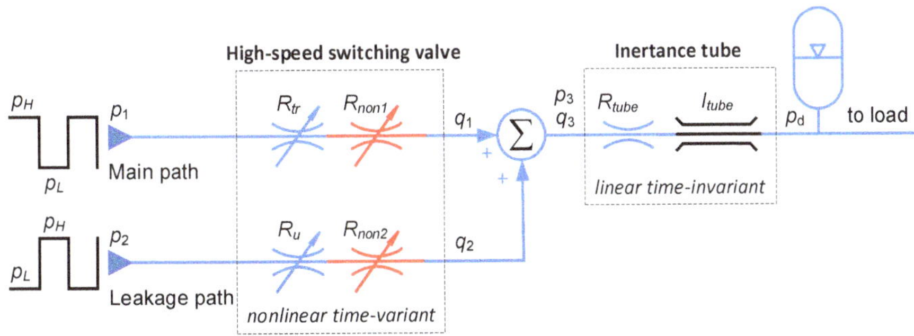

Figure 2. Schematic of enhanced analytical model of a three-port flow booster [11].

Assuming P_1 is the Fourier transform of p_1, the system flow of q_3 can be represented in the time domain as:

$$q_3 = \mathrm{IFFT}\left(\frac{P_1}{Z_E}\right), \tag{1}$$

where Z_E is entry impedance, which is the ratio of the pressure and flow of the entry to the circuit, and the entry impedance of Z_E in the frequency domain is

$$Z_E = jZ_0\xi \tan\left(\frac{\omega L\xi}{c}\right),$$ (2)

where j is the imaginary unit, Z_0 is the pipe characteristic impedance, c is the speed of sound and ξ is the viscous wave correction factor [25].

Therefore, flow from the leakage path and from the main path can be calculated by

$$q_2 = C_d A_{leak} \sqrt{\frac{2|p_2 - p_3|}{\rho}} \, \mathrm{sgn}(p_2 - p_3),$$ (3)

$$q_1 = q_3 - q_2$$ (4)

where c_d is discharge coefficient, A_{leak} is leakage area, and ρ is fluid density.

Iterative pressures are dependent on nonlinear time-variant parts

$$p'_{1(k+1)} = \lambda\left(p_1 - \frac{\rho q_{1(k)}|q_{1(k)}|}{2C_d^2 A^2}\right) + p'_{1(k)}(1 - \lambda)$$ (5)

$$p'_{2(k+1)} = p_2 - p'_{1(k+1)},$$ (6)

where A is the opening area of the valve, λ is the relaxation factor for model stability, and k is the index of iteration.

SIHC characteristics can be effectively predicted using Equations (1)–(6). The volume effects from the tube, valve chamber, and accumulators can also be included in the analytical model, as developed in [11]. Investigations of the nonlinearity of valve dynamics, switching-valve transition, leakage, and the accumulator-volume effect are fundamental in SIHC research.

Optimal switching frequencies and ratios are highly dependent on the wave-propagation effect. In our previous work [11], we concluded that the optimal switching frequency can be predicated using Equation (7):

$$f = \begin{cases} \frac{\alpha c}{2L} & 0 \leq \alpha \leq 0.5 \\ \frac{(1-\alpha)c}{2L} & 0.5 < \alpha \leq 1 \end{cases}$$ (7)

where f is the switching frequency, α is the switching ratio, L is the length of the inertance tube, and c is the speed of sound.

Recent experiment investigations based on a high-speed rotary valve and low operating pressure can be found in [1]. High pressure was set at 30 bar and low pressure at 20 bar, with a delivery flow of 7 and 20 L/min. The estimated overall resistance was about 0.36 bar/(L/min). The static delivery pressures agreed well with the analytical results, but a small right-hand-side shift was noticed in the experiment flow-loss curves. Theoretically, a symmetrical flow-loss curve is expected with a switching ratio from 0 to 1. High flow losses occurred with the predicated/calculated optimal switching frequencies and ratios when a large delivery-flow rate was applied, for example, 20 L/min, which were not expected according to theoretical analysis.

3. Flow Booster Operating at High Pressures and Flow Rates

The schematic of the test rig is shown in Figure 3, consisting of a high-speed rotary valve, an inertance tube, and a needle valve acting as load. The design and steady-state characteristics of the high-speed rotary valve can be found in [1]. A brushless servomotor (Baldor BSM50N-375AF) with a maximal speed of 5100 rpm was used to drive the high-speed switching valve with certain switching frequencies. The switching ratio of the valve was manually adjusted in experiments.

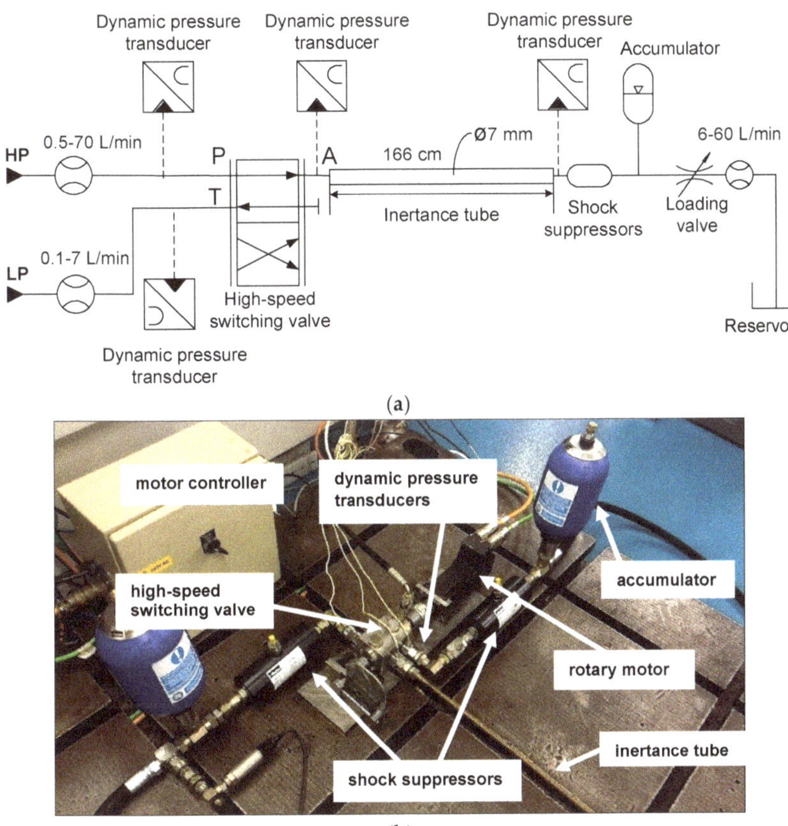

Figure 3. Test-rig (**a**) schematic and (**b**) photograph.

Four miniature piezoresistive pressure transducers (Measurement Specialties EPXseries) were used to measure high- and low-supply pressures, the inlet pressure of the inertance tube (A-port pressure), and the inlet pressure of the needle valve (loading pressure). Transducer ranges were 0–350, 0–35, 0–200, and 0–200 bar, respectively. A hydraulic power pack, including two gear pumps with a maximal supply pressure of 100 and 50 bar, was used as high and low-pressure supply, respectively. Three accumulators and three shock suppressors (Inline Pulse-Tone™ Shock Suppressors, Parker Hannifin, Ohio, United States) were used to eliminate pressure pulsations. The charging pressures of the high-pressure (HP), low-pressure (LP), and downstream accumulators were 30, 15, and 30 bar, and the charging pressures of the shock suppressors were 15, 7.5, and 15 bar, respectively. The length of the inertance tube was 1.66 m with a diameter of 7 mm. Two gear flow meters (0.5–70 and 0.1–7 L/min) were used to measure the dynamic high- and low-supply flow rates. Delivery flow rate was measured by using a turbine flow meter (6–60 L/min). The estimated speed of sound was 1300 m/s. More details about the measurement and effect of the speed of sound can be found in [26]. Parameters for the analytical model and experiments are listed in Table 1.

Table 1. Parameters for analytical model and experiments.

Parameters	Value (Unit)
Fluid viscosity	30 cSt
Fluid density	870 kg/m^3
High-supply pressure	90 bar
Low-supply pressure	30 bar
Delivery-flow rate	0, 14, 21 L/min
Switching frequency	200 Hz
Switching ratio	0–1
Inertance tube length	1.66 m (including fittings)
Inertance tube diameter	7 mm
Speed of sound	1300 m/s
Oil temperature	30 °C

3.1. Delivery Pressure and System-Flow Loss

To investigate the performance of the flow booster operating at high pressures, high- and low-supply pressures were set as 90 and 30 bar, with a delivery-flow rate of 0, 14, and 21 L/min. The pressure difference between the two supplies was 60 bar. Switching frequency was 200 Hz, and the switching ratio increased from 0 to 1. The relationship between delivery pressure, supply pressures, and delivery-flow rate was investigated through experiments, as shown in Figure 4. Theoretically, the delivery pressure should have a linear relationship with supply pressures p_H and p_L, and overall system resistance R, as shown in Equation (8):

$$p_d = p_H \alpha + (1 - \alpha) p_L - q_d R, \tag{8}$$

where α is switching ratio, q_d is delivery flow, and R is overall resistance.

Using Equation (8), it could be calculated that the overall resistance of testing system R was about 0.87 bar/(L/min), which was about 2.4 times the measurement in [1]. As a similar valve and inertance tube were used for the experiments, the high resistance may have been caused by the effects of high-pressure difference between the two pressure-supply lines and the switching transition of the valve. Linear experimental delivery pressure was achieved with a small deviation around the switching ratio of 0.1, which could be caused by insufficient inertia, leakage, or cavitation at such a low switching ratio for a high delivery-flow and supply-pressure difference. The largest deviation occurred at 21 L/min, and the smallest at 0 L/min.

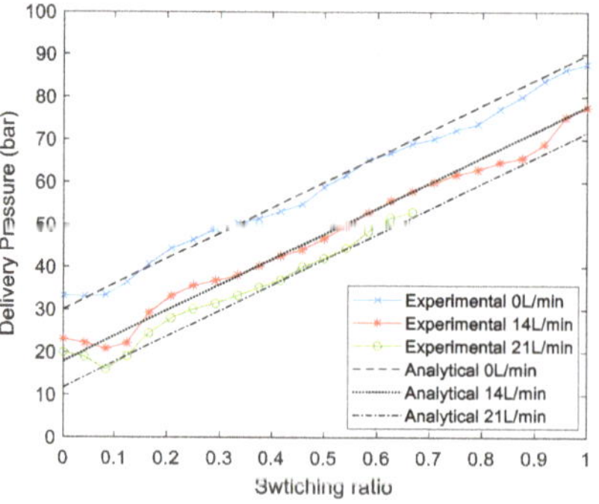

Figure 4. Experiment and analytical delivery pressure (delivery flow = 0, 14, and 21 L/min; switching frequency = 200 Hz; switching ratio from 0 to 1; overall resistance = 0.87 bar/(L/min)).

In practice, a reasonable operating range of the switching ratio could be between 0.2 and 0.8. Due to the capacity of the dual-pump supply, system pressure drops and unavoidable flow drops through the relief valves (accompanied by the supply pumps). The system was unable to achieve 21 L/min when the switching ratio was greater than 0.6. However, with a delivery flow rate of 21 L/min, linearly increasing delivery pressure could be predicted.

The theoretical relationship between flow loss and switching ratio varies with switching frequencies. Unlike the delivery-pressure curve, flow loss is dependent on switching frequency; details can be found in [11]. Flow loss is defined as the difference between actual and theoretical average flow rate from the high- and low-pressure supply ports [10]. Static leakage tests were performed at switching ratios of 0, 0.25, 0.5, and 1. High-pressure supply was 20 bar, and low-pressure supply was 3 bar. In the experiments, when the inertance tube was fully connected to the high-pressure port (switching ratio = 1), leakage flow of 0.57 L/min was measured from the high-pressure port to the low-pressure port and the leakage port of the valve. When the inertance tube was fully connected to the low-pressure supply port, a leakage flow of 0.50 L/min was measured from the low-pressure supply port to the leakage port. Figure 5 shows the experiment and analytical flow losses with delivery flow rates of 0, 14, and 21 L/min. Average flow losses were accurately predicted, but differences were seen between experiment and analytical flow losses, which was not predicted. The result was caused by the disturbance wave of the frequency of 134 Hz, which significantly influenced wave propagation along the pipeline and system performance. The harmonics of 134 Hz can be clearly seen in FFT (Fast Fourier transform) analysis. The 134 Hz wave, which was 2/3 of the switching frequency of the valve, could have been caused by rig vibrations. This experiment phenomenon has not been analytically explained. The reasons for the generation of the disturbance wave should be investigated. Figure 5b shows the analytical flow losses with a switching frequency of 134 and 200 Hz. Estimated leakage was 2.25 L/min with an operating pressure of 90 bar. The curve of 134 Hz agreed with the experiment flow losses. Experiment results also proved that flow loss is not directly affected by delivery-flow rate [10].

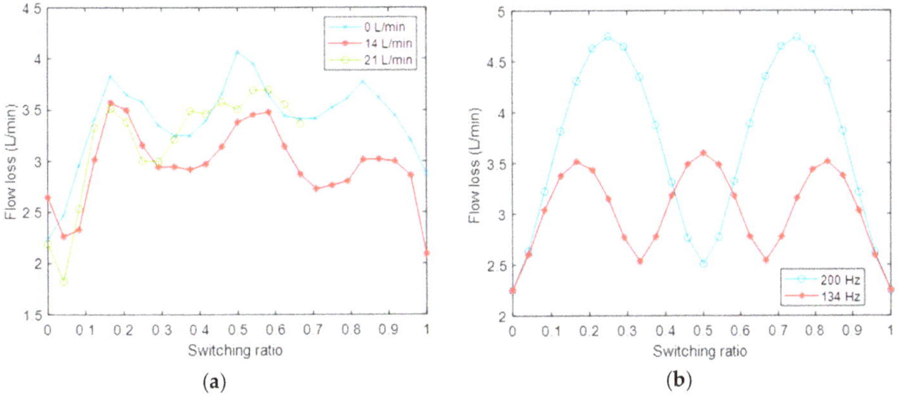

Figure 5. (**a**) Experiment flow losses with delivery-flow rate of 0, 14, and 21 L/min, and switching frequency of 200 Hz./ (**b**) Analytical estimations with switching frequency of 200 and 134 Hz.

3.2. System Efficiency and Power Loss

Analytical characteristics of the flow booster using the enhanced analytical model and experiment results [11] are shown in Figure 6 for different switching ratios (0.25, 0.5, and 0.67) and a fixed switching frequency of 200 Hz. The solid, dashed, and dotted lines represent the analytical results, while the squares, triangles, and circles represent the experiment results with a constant delivery-flow rate of 0, 14, and 21 L/min. Experimental delivery pressure followed the predicted trend well. The experiment high-pressure supply-flow rates were about 3 L/min higher than those pf the analytical results due to larger actual flow losses.

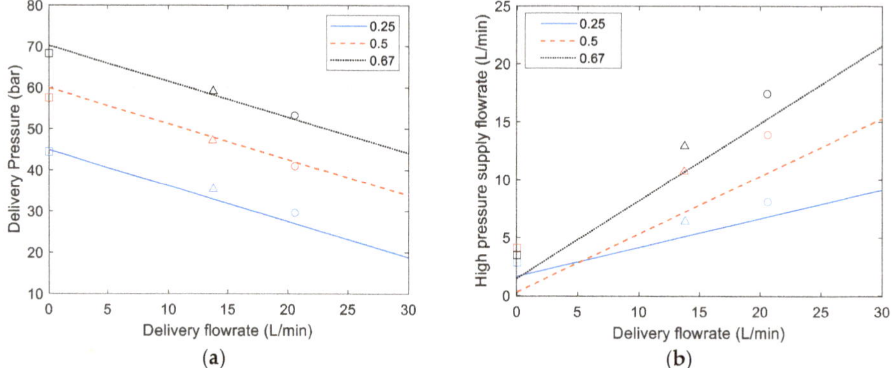

Figure 6. Experiment and analytical results of flow booster. (**a**) Delivery pressure versus delivery flow rate; (**b**) high-pressure supply flow rate versus delivery flow rate.

Figure 7 shows the experimental and analytical efficiency and power losses of a flow booster. The SIHC maintained high efficiency with an average above 65% with delivery flow rates of 14 and 21 L/min for a ratio of 0.25, 0.5, and 0.67, and a switching frequency of 200 Hz. The solid, dashed, and dotted lines represent the analytical results, while squares, triangles, and circles represent the experiment results with a constant delivery flow rate of 0, 14, and 21 L/min. Switching-valve transition was observed, and switching time was about 0.8 ms. Compared to a conventional valve-controlled hydraulic machine operating at a pressure of 90 bar and a delivery flow rate of 5, 14, and 21 L/min, a fixed-displacement pump with a constant flow of 30 L/min and a maximal pressure of 150 bar is assumed for use. Conventional system efficiency is 17%, 47%, and 70%, corresponding to power loss of 3750, 2400, and 1350 W. The SIHC provides much higher efficiency and lower energy losses with a delivery flow rate from 5 to 21 L/min. High energy efficiency can only be achieved for a conventional system when it operates around its maximal capacity. Actual efficiency was slightly lower than that in the analytical results, which could have been caused by the simplified prediction of the valve transition dynamics [11], and the neglecting of pressure pulsation (extra pressure harmonics occurred) and cavitation effects.

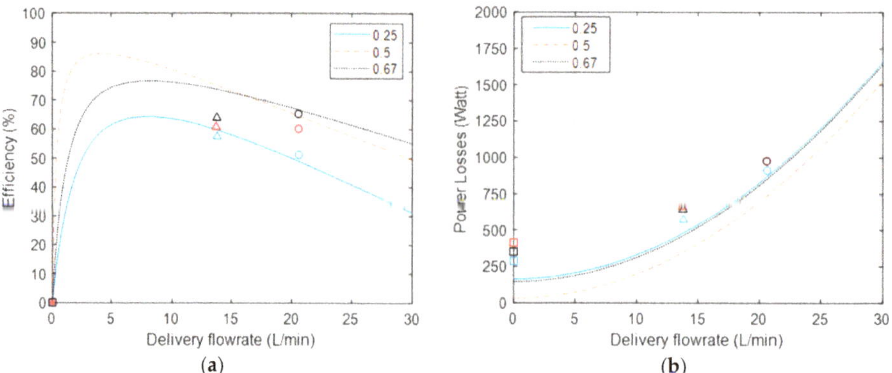

Figure 7. Experimental and analytical efficiency and power loss of a flow booster. (**a**) Efficiency versus delivery-flow rate; (**b**) power loss versus delivery flow rate.

4. Discussion and Conclusions

The main advantage of the SIHC is that the inherent reactive behaviour of a hydraulic tube is used to control flow and pressure, and improve system energy efficiency instead of relying on dissipation of power. High operating pressures with the low- and high-pressure supplies of 30 and 90 bar, and a high delivery flow rate of 21 L/min were applied to a flow booster in this work. Results were validated by using the three-port SIHC analytical models. Analytical and experiment results showed very good performance and efficiency for the SIHC when it operated at high pressures and delivery-flow rates, which are normally required for hydraulic applications such as in transportation, aerospace, and machinery for construction, industry, and agriculture. Flow loss and efficiency agreed with the prediction from the analytical model, which directly provided time-efficient solutions without running simulation models that were computationally intensive. Cavitation and vibrations were experienced when low-supply pressure was very low (0–10 bar). In practice, a pressurised tank could be useful for boosting low-supply pressure. Alternatively, a dual-pump system could be used to form two individual supply-pressure lines.

Experiment investigations were based on constant flow rates. We aim to apply a varying flow rate to the system in our continuing work to investigate SIHC adaptability and performance. The loading effect will be investigated and concluded for SIHC design and optimisation. Improving on the current system, a stepper motor will be used to automatically adjust the switching ratio, and a real-time control system will be used to adjust the switching frequency to maintain optimal operating parameters [11,26]. This can further improve system energy efficiency, which is important for a wide range of hydraulic sectors, including machinery for construction, industry, agriculture, transportation, oil and gas, and robotics.

Author Contributions: M.P. and C.Y. developed the analytical models, conceived and designed the experiments, analysed the data, and wrote the paper; V.L.M.L. carried out the experiments and instrumentation work; A.P. gave valuable technical advice and contributed to the manuscript editing. All authors have read and agreed to the published version of the manuscript.

Funding: This research was funded by the RAEng/The Leverhulme Trust Senior Research Fellowship, UK, grant number LTSRF1819\15\16, the RAEng Proof-of-Concept Award PoC1920/15, and the Open Foundation of the State Key Laboratory of Fluid Power and Mechatronic Systems, Zhejiang University, China, grant number GZKF-201801. Chenggang Yuan would like to thank China Scholarship Council for supporting his PhD (201706150102) while studying at the University of Bath, UK.

Conflicts of Interest: The authors declare no conflict of interest.

References

1. Scheidl, R.; Linjama, M.; Schmidt, S. Is the future of fluid power digital? *Proc. Inst. Mech. Eng. Part I J. Syst. Control Eng.* **2012**, *226*, 721–723. [CrossRef]
2. Yang, H.Y.; Pan, M. Engineering research in fluid power: A review. *J. Zhejiang Univ. Sci. A* **2015**, *16*, 427–442. [CrossRef]
3. Pan, M.; Plummer, A. Digital switched hydraulics. *Front. Mech. Eng.* **2018**, *13*, 225–231. [CrossRef]
4. Chao, Q.; Zhang, J.; Xu, B.; Huang, H.; Pan, M. A review of high-speed electro-hydrostatic actuator pumps in aerospace applications: Challenges and solutions. *J. Mech. Des.* **2019**, *141*. [CrossRef]
5. Yuan, C.; Pan, M.; Plummer, A. A review of switched inertance hydraulic converter technology. In Proceedings of the BATH/ASME 2018 Symposium on Fluid Power and Motion Control, Bath, UK, 12 September 2018.
6. Brandstetter, R.; Deubel, T.; Scheidl, R.; Winkler, B.; Zeman, K. Digital hydraulics and "Industries 4.0". *Proc. Inst. Mech. Eng. Part I J. Syst. Control Eng.* **2017**, *231*, 82–93. [CrossRef]
7. Yuan, C.; Pan, M.; Plummer, A. A review of switched inertance hydraulic converter technology. *ASME J. Dyn. Sys. Meas. Control* **2020**, 1–13. [CrossRef]
8. Kogler, H.; Scheidl, R. Two basic concepts of hydraulic switching converters. In Proceedings of the First Workshop on Digital Fluid Power, Tampere University of Technology, Tampere, Finland, 3 October 2008; Volume 3, pp. 7–30.

9. Winkler, B. Recent advances in digital hydraulic components and applications. In Proceedings of the Ninth Workshop on Digital Fluid Power, Aalborg, Denmark, 7–8 September 2017.

10. Pan, M.; Johnston, N.; Plummer, A.; Kudzma, S.; Hillis, A. Theoretical and experimental studies of a switched inertance hydraulic system. *Proc. Inst. Mech. Eng. Part I J. Syst. Control Eng.* **2014**, *228*, 12–25. [CrossRef]

11. Pan, M.; Johnston, N.; Plummer, A.; Kudzma, S.; Hillis, A. Theoretical and experimental studies of a switched inertance hydraulic system including switching transition dynamics, non-linearity and leakage. *Proc. Inst. Mech. Eng. Part I J. Syst. Control. Eng.* **2014**, *228*, 802–815. [CrossRef]

12. Brown, F.T. Switched reactance hydraulics: A new way to control fluid power. In Proceedings of the National Conference on Fluid Power, Chicago, IL, USA, 16–20 October 1987; pp. 25–34.

13. Brown, F.T.; Tentarelli, S.; Ramachandran, S. A hydraulic rotary switched-inertance servo-transformer. *J. Dyn. Syst. Meas. Control.* **1988**, *110*, 144–150. [CrossRef]

14. Liaw, C.J.; Brown, F.T. Nonlinear dynamics of an electrohydraulic flapper nozzle valve. *J. Dyn. Syst. Meas. Control.* **1990**, *112*, 298–304. [CrossRef]

15. Scheidl, R.; Schindler, D.; Riha, G.; Leitner, W. Basics for the energy-efficient control of hydraulic drives by switching techniques. In Proceedings of the 3rd Conference on Mechatronics and Robotics, Paderborn, Germany, 4–6 October 1995; pp. 118–131.

16. Winkler, B.; Scheidl, R. Optimization of a fast switching valve for big flow rates. In Proceedings of the Bath Workshop on Power Transmission and Motion Control, Bath, UK, 13–15 September 2006; pp. 387–399.

17. Winkler, B.; Scheidl, R. Development of a fast seat type switching valve for big flow rates. In Proceedings of the 10th Scandinavian International Conference on Fluid Power, Tampere, Finland, 21–23 May 2007.

18. Kogler, H.; Scheidl, R.; Ehrentraut, M.; Guglielmino, E.; Semini, C.; Caldwell, D.G. A compact hydraulic switching converter for robotic applications. In Proceedings of the Fluid Power and Motion Control, Bath, UK, 15 September 2010; pp. 55–66.

19. Guglielmino, E.; Semini, C.; Yang, Y.S.; Caldwell, D.; Kogler, H.; Scheidl, R. Energy efficient fluid power in autonomous legged robotics. In Proceedings of the ASME 2009 Dynamic Systems and Control Conference, Hollywood, CA, USA, 12–14 October 2009.

20. Peng, S.; Kogler, H.; Guglielmino, E.; Scheidl, R.; Branson, D.T.; Caldwell, D.G. The use of a hydraulic dc-dc converter in the actuation of a robotic leg. In Proceedings of the 2013 IEEE/RSJ International Conference on Intelligent Robots and Systems, Tokyo, Japan, 3 November 2013; pp. 5859–5864.

21. Kogler, H.; Scheidl, R. Hydraulic switching control of resonant drives. In Proceedings of the 12th Mechatronics Forum Biennial International Conference, Zurich, Switzerland, 28–30 June 2010; pp. 28–30.

22. Johnston, D.N. A switched inertance device for efficient control of pressure and flow. In Proceedings of the ASME 2009 Dynamic Systems and Control Conference, Hollywood, CA, USA, 1 January 2009; pp. 589–596.

23. Pan, M.; Plummer, A.; El Agha, A. Theoretical and experimental studies of a switched inertance hydraulic system in a four-port high-speed switching valve configuration. *Energies* **2017**, *10*, 780. [CrossRef]

24. Wu, G.; Yang, J.; Shang, J.; Fang, D. A rotary fluid power converter for improving energy efficiency of hydraulic system with variable load. *Energy* **2020**, *195*, 116957. [CrossRef]

25. Stecki, J.S.; Davis, D.C. Fluid transmission lines—Distributed parameter models part 1: A review of the state of the art. *Proc. Inst. Mech. Eng. Part A Power Process. Eng.* **1986**, *200*, 215–228. [CrossRef]

26. Pan, M.; Johnston, N.; Robertson, J.; Plummer, A.; Hillis, A.; Yang, H.Y. Experimental investigation of a switched inertance hydraulic system with a high-speed rotary valve. *J. Dyn. Syst. Meas. Control.* **2015**, *137*, 121003. [CrossRef]

Article

Static Deformation-Compensation Method Based on Inclination-Sensor Feedback for Large-Scale Manipulators with Hydraulic Actuation

Jianyong Qian, Qi Su *, Fu Zhang, Yun Ma, Zifan Fang and Bing Xu

State Key Laboratory of Fluid Power & Mechatronic Systems, Zhejiang University, Hangzhou 310027, China; 21825065@zju.edu.cn (J.Q.); zgzhangfu@outlook.com (F.Z.); 21825218@zju.edu.cn (Y.M.); zerofang@zju.edu.cn (Z.F.); bxu@zju.edu.cn (B.X.)
* Correspondence: suqi@zju.edu.cn

Received: 1 November 2019; Accepted: 19 December 2019; Published: 8 January 2020

Abstract: Modern large-scale manipulators with hydraulic actuation like mobile concrete pump manipulators are increasingly used in industrial, construction, and other fields. Due to the large span of these manipulators, the static deformation accumulation to the endpoint has seriously affected the precise control of the endpoint. In this paper, we propose a static deformation-compensation method based on inclination sensor feedback for large-scale manipulators to reduce the deviation of the endpoint. Compared with the finite element method, this method does not need to consider many boundary conditions that are uncertain for flexible manipulators in most situations. It has appropriate accuracy and is universal for large-scale manipulators of different sizes and working under different loads. Based on a 24m-3R mobile concrete pump manipulator, the parametric simulation is carried out. The reliability of the static deformation-compensation method is verified, and the error is analyzed. The validity of the static deformation-compensation method is verified by comparing the theoretical endpoint position with the actual endpoint position after static deformation compensation. The compensation error under different loads is obtained, and the universality of the compensation method for different loads is verified.

Keywords: static deformation; inclination sensor; large-scale manipulator; concrete pump; endpoint deviation

---

1. Introduction

Modern large-scale manipulators are increasingly used in industrial, construction, and other fields. With the increase in the operational speed requirements of large-scale manipulators and the demand for lightweight design, the flexibility of these manipulators have gradually increased. Compared to the conventional heavy and bulky manipulators, flexible link manipulators have the potential advantage of lower cost, larger working volume, faster operational speed, larger payload-to-manipulator-weight ratio, smaller actuators, lower energy consumption, better maneuverability, better transportability, and safer operation due to reduced inertia [1–4]. The concepts of modern light-weight construction enable the large-scale manipulators like mobile concrete pump manipulators with extended operating range and less static load. However, due to the reduced weight, the elasticity of the construction elements has a significant influence on the precise positioning of the endpoint [5,6]. The influence of large-scale manipulators' static deformation on the precise control of the endpoint cannot be ignored anymore [6,7].

The research on the motion control of the engineering manipulator has made good progress [8–17]. Many researchers have studied the control of the endpoint trajectory of the mobile concrete pump manipulator. Most of them considered the manipulator as rigid. Although the control of the endpoint trajectory was realized, there are still positioning errors (deviation of the endpoint) due to static deformation [18–20]. Despite the length of the mobile concrete pump manipulator ranging from 24 m to 53 m that are widely used now, the demand for the length of the mobile concrete pump manipulator will be higher with the development of the construction level and the influence of static deformation will be more serious.

In the literature, there are many types of research on the modeling of flexible multi-body systems. The existing methods are well developed and are presented in several textbooks, e.g., Bremer, Shabana, and many others [21]. For large-scale manipulators like mobile concrete pump manipulators, hydraulic actuators comprising hydraulic cylinders and valves are commonly used. Their dynamic behavior and the nonlinear characteristics have to be considered in the controller design. Henikl et al. and Lambeck et al. studied the combination of flexible multi-body systems and hydraulic actuators [22–24]. Zimmert et al. presented a control design considering the infinite-dimensional model of a flexible turntable ladder [25].

Many researchers have studied different schemes for modeling flexible link manipulators. Links are subjected to torsion, bending, and compression. The main concern is bending. For bending one may often use the Euler–Bernoulli equation, which ignores shearing and rotary inertia effects. These two effects may be incorporated using a Timoshenko beam element, which always is used if the beam is short relative to its diameter. However, since links may be considered as being rigid [26], in most models of flexible manipulators Euler–Bernoulli beams are used. In the literature [27,28], there are many well-established dynamic models in which three main modeling methods of the flexible link manipulators are the assumed mode method (AMM), the finite element method (FEM), and the lumped parameter model. AMM and FEM use either the Lagrangian formulation or the Newton–Euler recursive formulation.

The flexibility of the link is usually represented by a truncated finite modal series in terms of spatial mode eigenfunctions and time varying mode amplitudes in assumed mode model (AMM) formulation. This method's main disadvantage is the difficulty in finding modes for links with non-regular cross sections and multi-link manipulators [29]. Using the law of conservation of momentum, the Lagrangian principle was utilized to model the dynamic function of the space flexible manipulator incorporating the assumed modes method in Deng-Feng's research [30]. Subudhi and Morris [31] presented a dynamic modeling technique for a manipulator with multiple flexible links and flexible joints based on a combined Euler–Lagrange formulation and assumed modes method. Then, they controlled the system by formulating a singularly perturbed model and used it to design a reduced-order controller. In the finite element method (FEM), the elastic deformations are analyzed by assuming a known rigid body motion and later superposing the elastic deformation with the rigid body motion [32–37]. In order to solve a large set of differential equations derived by the finite element method, a lot of boundary conditions have to be considered, which are, in most situations, uncertain for flexible manipulators [38]. Using the assumed mode method to derive the equations of motion of the flexible manipulators, only the first several modes are usually retained by truncation and the higher modes are neglected. The lumped parameter model is the simplest one for analysis purposes; the manipulator is modeled as a spring and mass system, which does often not yield sufficiently accurate results [39–41]. Zhu et al. [42] employed a lumped model to simulate the tip position tracking control of a single-link flexible manipulator. Raboud et al. [43] showed the existence of multiple equilibrium solutions under a given load condition by studying the stability of very flexible cantilever beams.

Some researchers have paid attention to the static deformation of large-scale manipulators. Most of them are based on the application of finite element method (FEM) in the static deformation of flexible manipulators. Lee et al. reduced the endpoint deviation of the mobile concrete pump manipulator by 30% compared with steel structures by using carbon fiber material in the last link [44]. In order to better improve the trajectory tracking accuracy of the working platform at high altitude, Qing Hui Yuan considered the elastic deformation of the manipulator and the influence of vibration on the trajectory tracking control and introduces the deformation compensation strategy to eliminate the influence [45]. Xia Jijun et al., based on finite element deformation analysis under the actual load conditions of the mobile concrete pump manipulator system, established an expert database of full position and orientation deformation compensation of the manipulator and applied it to the trajectory control of the endpoint. The position deviation of the endpoint could be controlled within ±15 cm [46]. Zhao Xin et al. obtained the deformation compensation model of the manipulator and the vehicle body through the deformation analysis of the whole concrete pump truck under the full working condition, established the kinematics model of the concrete pump truck after the deformation compensation and used the control method of the cerebellar model neural network in the motion control, and well solved the dynamic detection and trajectory control of the manipulator position and orientation [47]. Wang Xiaoming et al. used finite element simulation to establish the data model of the deformation of the manipulator during the trajectory control process. Then, the BP neural network model was used to establish the deformation compensation algorithm, and the deformation law of the manipulator was obtained [48]. Pan Daoyuan et al. used FEM to analyze the variation of the acceleration of the manipulator and the force applied on the manipulator in different positions and orientations, and determined their influence on the deformation [49]. These studies have made some achievements and have made great progress in the deformation compensation of mobile concrete pump manipulators. However, these studies are based on the prototype studied by them, such as Xia Jijun et al. with Zoomlion 52m-6RZ concrete pump truck (RZ manipulator folding structure) as the research prototype [46]. Wang Xiaoming, Zhao Xin et al., and Pan Daoyuan et al. all built finite element simulation results databases based on their respective experimental prototypes [47–49]. All of the above studies needed to establish an accurate simulation model and needed to consider a lot of boundary conditions. The amount of calculation was very large, and the methods used to develop the prototype were not universal and thus are difficult to use in practical applications.

In this paper, we propose a static deformation-compensation method for large-scale manipulators based on inclination sensor feedback. Compared with the finite element method, this method does not need to consider many boundary conditions that are uncertain for flexible manipulators in most situations. It has appropriate accuracy and is universal for large-scale manipulators of different sizes and working under different loads.

2. The Structure and the Forward Kinematic Model of the Mobile Concrete Pump Manipulator

2.1. Structure of the Mobile Concrete Pump Manipulator

The mobile concrete pump manipulator is a multi-degree of freedom manipulator system. This paper takes the mobile concrete pump manipulator with three links as an example to explain the static deformation-compensation method. Its main components are rotation base, 1st joint, 1st link, 2nd joint, 2nd link, 3rd joint, 3rd link, and the concrete discharge hose. Given that the concrete discharge hose has little capacity to bear the load of concrete, it is only used as the concrete discharge guide, ignoring its degree of freedom. It is a manipulator composed of four joints, each link of which is a long-scale flexible link.

During the process of concrete placing during construction, the manipulator produces a large elastic deformation caused by complex stress due to the influence of gravity on the manipulator, the pumping concrete load, the concrete flow impact, and other factors. Due to the large span of the mobile concrete pump manipulator, the deformation accumulation to the end outlet has seriously affected the concrete placing accuracy. The elastic deformation varies in different positions and orientations. Figure 1 shows a structural diagram of the mobile concrete pump manipulator.

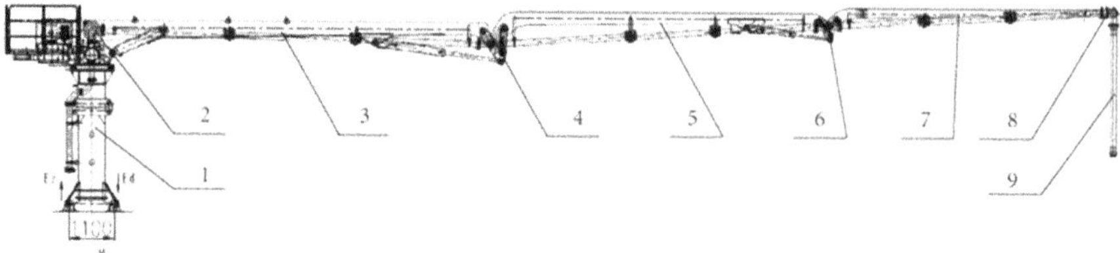

Figure 1. Structure diagram of the mobile concrete pump manipulator. 1. Rotation base; 2. 1st joint; 3. 1st link; 4. 2nd joint; 5. 2nd link; 6. 3rd joint; 7. 3rd link; 8. End outlet; 9. Concrete discharge hose.

2.2. Forward Kinematics Model of the Mobile Concrete Pump Manipulator

In order to study the influence of the joint compensation angle on the position of the endpoint, the relationship between the joint angle change of each joint and the position and orientation of the endpoint is established based on the D–H matrix (Denavit–Hartenberg Matrix) method. As shown in Figure 2, D–H coordinate system is established for the mobile concrete pump manipulator.

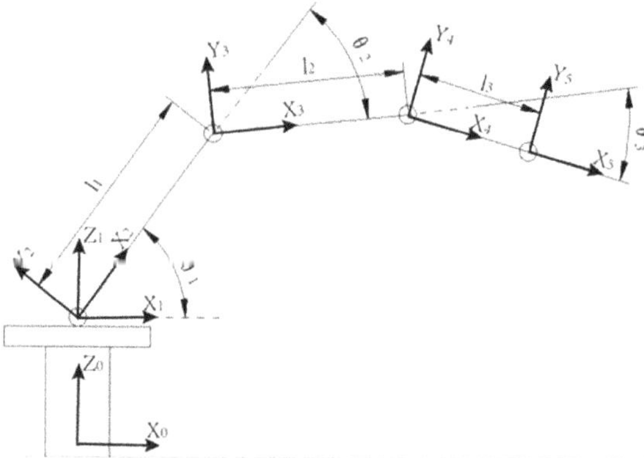

Figure 2. Denavit–Hartenberg (D–H) coordinate system of the mobile concrete pump manipulator.

The distance from the rotation axis of the rotation base to the rotation axis of the 1st link is negligible compared with the length of the whole manipulator (24 m), so let the origin of the rotation base coordinate system T_1 coincide with the origin of the 1st link coordinate system T_2. The base coordinate system T_0, the rotation base coordinate system T_1, the 1st link coordinate system T_2, the 2nd link coordinate system T_3, the 3rd link coordinate system T_4, and the end coordinate system T_5 is established. According to the D–H matrix method, the homogeneous coordinate transformation matrix is calculated. The joint angle $\theta_0\theta_1\theta_2\theta_3$ is positive clockwise and negative counterclockwise.

$$
{}^0_1T = \begin{bmatrix} c_0 & -s_0 & 0 & 0 \\ s_0 & c_0 & 0 & 0 \\ 0 & 0 & 1 & 0 \\ 0 & 0 & 0 & 1 \end{bmatrix} \quad
{}^1_2T = \begin{bmatrix} c_1 & -s_1 & 0 & 0 \\ 0 & 0 & -1 & 0 \\ s_1 & c_1 & 0 & 0 \\ 0 & 0 & 0 & 1 \end{bmatrix}
$$

$$
{}^2_3T = \begin{bmatrix} c_2 & -s_2 & 0 & l_1 \\ s_2 & c_2 & 0 & 0 \\ 0 & 0 & 1 & 0 \\ 0 & 0 & 0 & 1 \end{bmatrix} \quad
{}^3_4T = \begin{bmatrix} c_3 & -s_3 & 0 & l_2 \\ s_3 & c_3 & 0 & 0 \\ 0 & 0 & 1 & 0 \\ 0 & 0 & 0 & 1 \end{bmatrix} \quad
{}^4_5T = \begin{bmatrix} 1 & 0 & 0 & l_3 \\ 0 & 1 & 0 & 0 \\ 0 & 0 & 1 & 0 \\ 0 & 0 & 0 & 1 \end{bmatrix}
$$

$$_5^0T = \begin{bmatrix} c_3(c_0c_1c_2 - c_0s_1s_2) - s_3(c_0c_1s_2 + c_0c_2s_1) & -c_3(c_0c_1s_2 + c_0c_2s_1) - s_3(c_0c_1c_2 - c_0s_1s_2) & s_0 & l_2(c_0c_1c_2 - c_0s_1s_2) + l_3(c_3(c_0c_1c_2 - c_0s_1s_2) - s_3(c_0c_1s_2 + c_0c_2s_1)) + c_0c_1l_1 \\ c_3(c_1c_2s_0 - s_0s_1s_2) - s_3(c_1s_0s_2 + c_2s_0s_1) & -c_3(c_1s_0s_2 + c_2s_0s_1) - s_3(c_1c_2s_0 - s_0s_1s_2) & -c_0 & l_3(c_3(c_1c_2s_0 - s_0s_1s_2) - s_3(c_1s_0s_2 + c_2s_0s_1)) + l_2(c_1c_2s_0 - s_0s_1s_2) + c_1l_1s_0 \\ c_3(c_1s_2 + c_2s_1) + s_3(c_1c_2 - s_1s_2) & c_3(c_1c_2 - s_1s_2) - s_3(c_1s_2 + c_2s_1) & 0 & l_1s_1 + l_3(c_3(c_1s_2 + c_2s_1) + s_3(c_1c_2 - s_1s_2)) + l_2(c_1s_2 + c_2s_1) \\ 0 & 0 & 0 & 1 \end{bmatrix}$$

3. Joint Angle Independent Compensation Method Based on Inclination Sensor Feedback

3.1. Principle of Joint Angle Independent Compensation

The elastic deformation varies in different positions and orientations. Therefore, the elastic deformation of the entire mobile concrete pump manipulator is decomposed into the elastic deformation of each link. The position recovery of the endpoint of each link is achieved by compensation of the joint angle of each joint. Since the deformation angle is very small, the following simplification is made in the compensation angle solution: The deflection of the endpoint of each link with respect to the horizontal axis of the corresponding coordinate system is regarded as the compensation angle arc length. The basic principle of joint angle independent compensation can be simplified as the angle compensation of each link coordinate system around the z-axis rotation, as shown in Figure 3.

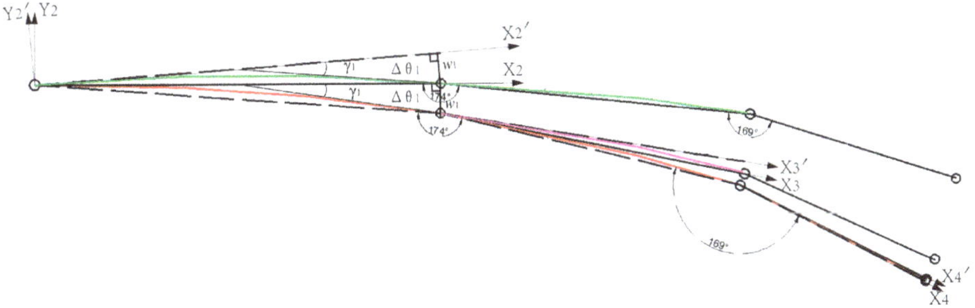

Figure 3. The basic principle of joint angle independent compensation.

The deviation of the 1st link's endpoint can be compensated by the 1st joint angle. The compensation angle of the 1st joint is as follows:

$$\Delta\theta_1 = \frac{w_1}{l_1} \tag{1}$$

where

w_1 is deflection of the endpoint of the 1st link;

l_1 is length of the 1st link;

γ_1 is tangential deformation angle of the 1st link.

The compensation angle of the 2nd joint and the 3rd joint can be calculated in the same way.

3.2. Joint Torque Solution Based on Jacobian Matrix

The deviation of the endpoint caused by gravity is often ignored in the existing static analysis of the manipulator, and most of the studies regard the manipulator as a rigid body. The relationship between the force on the endpoint of the manipulator and the torque of each joint has been established by the Jacobian matrix.

$$\tau = J^T \mathcal{F} \tag{2}$$

where

$\mathcal{F}$ is force vector and moment vector acting on the end actuator;

T is joint torque.

The gravity effect is simplified as follows: The Jacobian matrix of the three-link model and the two-link model is established separately without considering the rotation of the rotation base. When considering the joint torque, the gravity of the rear link of the joint is regarded as the load acting on the centroid of the latter link. The joint torque is calculated by the Jacobian matrix.

The manipulator's kinematic equation is as follows:

$$\begin{cases} x = l_1 cos\theta_1 + l_2 cos(\theta_1 + \theta_2) \\ y = l_1 sin\theta_1 + l_2 sin(\theta_1 + \theta_2) \end{cases}. \tag{3}$$

Then, the two sides are respectively derived from t as follows:

$$\begin{cases} \dot{x} = -l_1 sin\theta_1 \dot{\theta}_1 - l_2 sin(\theta_1 + \theta_2)(\dot{\theta}_1 + \dot{\theta}_2) \\ \dot{y} = l_1 cos\theta_1 \dot{\theta}_1 + l_2 cos(\theta_1 + \theta_2)(\dot{\theta}_1 + \dot{\theta}_2) \end{cases}. \tag{4}$$

The Jacobian matrix relative to the base coordinates is as follows:

$$J = \begin{bmatrix} -l_1 s_1 - l_2 s_{12} & -l_2 s_{12} \\ l_1 c_1 + l_2 c_{12} & l_2 c_{12} \end{bmatrix}. \tag{5}$$

Transposing the Jacobian matrix gives

$$J^T = \begin{bmatrix} -l_1 s_1 - l_2 s_{12} & l_1 c_1 + l_2 c_{12} \\ -l_2 s_{12} & l_2 c_{12} \end{bmatrix}. \tag{6}$$

For the two-link model, the distance between the centroid of the 2nd link and the 2nd joint is a_2 and the gravity is simplified to be applied to the centroid.

The torque of each joint produced by the gravity of the 2nd link is expressed as

$$\tau_{12} = (l_1 c_1 + a_2 c_{12}) m_2 \, g; \tag{7}$$

$$\tau_{22} = a_2 c_{12} m_2 \, g. \tag{8}$$

where

τ_{12} is torque on the 1st joint produced by the gravity of the 2nd link;

τ_{22} is torque on the 2nd joint produced by the gravity of the 2nd link.

Similarly, for the three-link model, the distance between the centroid of the 3rd link and the 3rd joint is a_3.

The torque of each joint produced by the gravity of the 3rd link is as follows:

$$\tau_{13} = (l_1 c_1 + l_2 c_{12} + a_3 c_{123}) m_3 \, g; \tag{9}$$

$$\tau_{23} = (l_2 c_{12} + a_3 c_{123}) m_3 \, g; \tag{10}$$

$$\tau_{33} = a_3 c_{123} m_3 \, g. \tag{11}$$

where

τ_{13} is torque on the 1st joint produced by the gravity of the 3rd link;

τ_{23} is torque on the 2nd joint produced by the gravity of the 3rd link;

τ_{33} is torque on the 3rd joint produced by the gravity of the 3rd link;

The gravity of the 1st link is simplified to the centroid. In summary, the total torque at each joint is finally obtained as follows:

$$\tau_1 = a_1 c_1 m_1\, g + (l_1 c_1 + a_2 c_{12}) m_2\, g + (l_1 c_1 + l_2 c_{12} + a_3 c_{123}) m_3\, g; \tag{12}$$

$$\tau_2 = a_2 c_{12} m_2\, g + (l_2 c_{12} + a_3 c_{123}) m_3\, g; \tag{13}$$

$$\tau_3 = a_3 c_{123} m_3\, g, \tag{14}$$

where

τ_1 is torque on the 1st joint;
τ_2 is torque on the 2nd joint;
τ_3 is torque on the 3rd joint,

where $c_{12} = \cos(\theta_1 + \theta_2)$; $c_{123} = \cos(\theta_1 + \theta_2 + \theta_3)$. The rest of the abbreviation has the same principle.

3.3. Compensation Angle Solution Based on Cantilever Model Combined Deformation of Compression and Bending

For the mobile concrete pump manipulator, its main feature is that the links are slender and have a large scale, especially the 1st link, which is the longest and has the largest deformation and compensation amount. The deformation caused by insufficient joint stiffness (piston–cylinder oil compression at the joint) is not considered at present, and each link is simplified to a cantilever beam model that is deformed by compression and bending.

The deformation analytical formulae of three links are as follows:

$$w_1 = \frac{\rho g A c_1}{24 EI} x^4 - \frac{\rho g A c_1 l_1}{6 EI} x^3 + \left(\frac{\tau_1}{2 EI} + \frac{\rho g A c_1 l_1^2}{4 EI} \right) x^2 \tag{15}$$

$$w_2 = \frac{\rho g A c_{12}}{24 EI} x^4 - \frac{\rho g A c_{12} l_2}{6 EI} x^3 + \left(\frac{\tau_2}{2 EI} + \frac{\rho g A c_{12} l_2^2}{4 EI} \right) x^2 \tag{16}$$

$$w_3 = \frac{\rho g A c_{123}}{24 EI} x^4 - \frac{\rho g A c_{123} l_3}{6 EI} x^3 + \left(\frac{\tau_3}{2 EI} + \frac{\rho g A c_{123} l_3^2}{4 EI} \right) x^2 \tag{17}$$

where

w is Deflection of the corresponding point of the corresponding link;
ρ is Equivalent density;
A is Equivalent cross-sectional area.

Deriving the deformation analytical formula to obtain the tangential deformation angle (the angle between the tangent and the x-axis) analytical formula yields

$$\gamma_1 = \arctan\left(\frac{\rho g A c_1}{6 EI} x^3 - \frac{\rho g A c_1 l_1}{2 EI} x^2 + \left(\frac{\tau_1}{EI} + \frac{\rho g A c_1 l_1^2}{2 EI} \right) x \right) \tag{18}$$

$$\gamma_2 = \arctan\left(\frac{\rho g A c_{12}}{6 EI} x^3 - \frac{\rho g A c_{12} l_2}{2 EI} x^2 + \left(\frac{\tau_2}{EI} + \frac{\rho g A c_{12} l_2^2}{2 EI} \right) x \right) \tag{19}$$

$$\gamma_3 = \arctan\left(\frac{\rho g A c_{123}}{6 EI} x^3 - \frac{\rho g A c_{123} l_3}{2 EI} x^2 + \left(\frac{\tau_3}{EI} + \frac{\rho g A c_{123} l_3^2}{2 EI} \right) x \right) \tag{20}$$

Take the 1st link as an example; the endpoint deflection is as follows:

$$w_1 = \frac{\rho g A c_1}{8EI} l_1{}^4 + \frac{\tau_1}{2EI} l_1{}^2.$$

(21)

The endpoint tangential deformation angle is

$$\gamma_1 = \arctan\left(\frac{\rho g A c_1}{6EI} l_1{}^3 + \frac{\tau_1}{EI} l_1\right).$$

(22)

The compensation angle is

$$\Delta\theta_1 = \frac{w_1}{l_1} = \frac{\rho g A c_1}{8EI} l_1{}^3 + \frac{\tau_1}{2EI} l_1.$$

(23)

According to the engineering practice, there are three methods to solve the compensation angle:

① Directly measure the endpoint deflection w of each link.

Take the 1st link as an example.

The compensation angle is calculated by Formula (1).

This is a joint angle compensation method based on the basic principle. In this method, direct measurement of the endpoint deflection of each link can be used to solve the joint compensation angle. Direct measurement of deflection reduces the error caused by deflection calculation using other methods. Nowadays, in the engineering practice, laser is adapted to measure the endpoint deflection, which has a high cost.

② Measure the tangential deformation angle γ at the end of each link.

The tangential deformation angle γ at the endpoint can be obtained by the difference of inclination sensor installed at both ends of the link.

Compensation angle:

$$\Delta\theta_1 = tan\gamma_1 - \frac{1}{24}\frac{\rho g A c_1}{EI} l_1{}^3 - \frac{\tau_1}{2EI} l_1.$$

(24)

This method has the error of theoretical modeling and calculation, with medium accuracy and low cost. However, considering the subtraction term and the deformation is small, when l is not large, γ can be used to replace $\Delta\theta$ directly. The simulation verification in the following text has adopted this method for the 2nd link and the 3rd link and achieved good results.

③ Reverse the tangential deformation angle analytical formula, then get the deformation analytical formula and calculate the endpoint deflection.

The analytical formula can be solved by the feedback value from four inclination sensors. Figure 4 shows the inclination sensors installed at three positions on the upper surface of the link and the inclination sensor installed at the head end of the link.

The three inclination sensor (x = a, b, c) measures the tangential deformation angle γ as γ_a, γ_b, γ_c, which are the differences between the feedback value from each inclination sensor and the feedback value from the inclination sensor at the head end of the link (reference inclination sensor) respectively.

According to Cramer's Rule, the coefficient matrix is as follows:

$$A = \begin{bmatrix} a^3 & a^2 & a \\ b^3 & b^2 & b \\ c^3 & c^2 & c \end{bmatrix},$$

(25)

and the constant term matrix is

$$\beta = \begin{bmatrix} tan\gamma_a \\ tan\gamma_b \\ tan\gamma_c \end{bmatrix}. \tag{26}$$

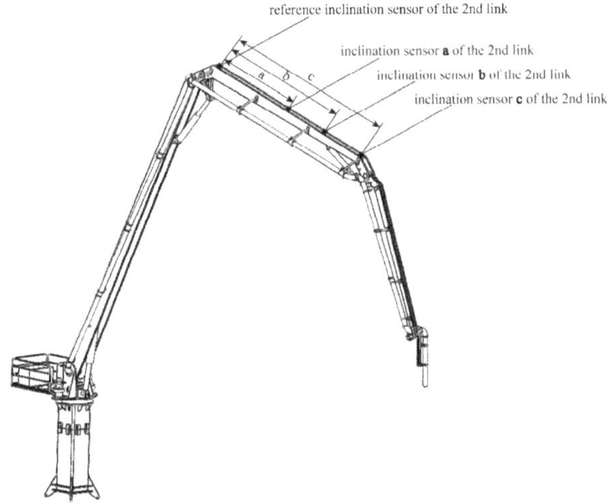

Figure 4. Schematic diagram of installation positions of inclination sensors.

As long as the coefficient matrix A is non-singular, the unique coefficient solution of the analytical formula of the tangential deformation angle can be obtained as follows:

$$A^{-1}\beta = \begin{bmatrix} -\dfrac{tan\gamma_a}{a^2b+a^2c-a^3-abc} - \dfrac{tan\gamma_b}{ac^2+bc^2-c^3-abc} - \dfrac{tan\gamma_c}{ab^2+b^2c-b^3-abc} \\ \dfrac{(b+c)tan\gamma_a}{a^2b+a^2c-a^3-abc} + \dfrac{(a+b)tan\gamma_b}{ac^2+bc^2-c^3-abc} + \dfrac{(a+c)tan\gamma_c}{ab^2+b^2c-b^3-abc} \\ -\dfrac{bctan\gamma_a}{a^2b+a^2c-a^3-abc} - \dfrac{abtan\gamma_b}{ac^2+bc^2-c^3-abc} - \dfrac{actan\gamma_c}{ab^2+b^2c-b^3-abc} \end{bmatrix}. \tag{27}$$

Then the compensation angle is obtained as follows:

$$\Delta\theta_1 = \frac{w_1}{l_1} = K_3 l_1{}^3 + K_2 l_1{}^2 + K_1 l_1 \tag{28}$$

where

$$K_3 = \frac{1}{4}\left(-\frac{tan\gamma_a}{a^2b+a^2c-a^3-abc} - \frac{tan\gamma_b}{ac^2+bc^2-c^3-abc} - \frac{tan\gamma_c}{ab^2+b^2c-b^3-abc}\right) \tag{29}$$

$$K_2 = \frac{1}{3}\left(\frac{(b+c)tan\gamma_a}{a^2b+a^2c-a^3-abc} + \frac{(a+b)tan\gamma_b}{ac^2+bc^2-c^3-abc} + \frac{(a+c)tan\gamma_c}{ab^2+b^2c-b^3-abc}\right) \tag{30}$$

$$K_1 = \frac{1}{2}\left(-\frac{bctan\gamma_a}{a^2b+a^2c-a^3-abc} - \frac{abtan\gamma_b}{ac^2+bc^2-c^3-abc} - \frac{actan\gamma_c}{ab^2+b^2c-b^3-abc}\right). \tag{31}$$

The functional relationship between the compensation angle and the inclination sensor installation positions a, b, c, and the tangential deformation angles γ_a, γ_b, γ_c is established.

This method has high accuracy and reasonable cost. The inclination sensor can be installed at any position on the upper surface of the link and only the accurate position data needs to be provided without the consideration of making the inclination sensor be close to the joint hinge. It is especially suitable for the link with a large scale and has large deformation.

4. Deformation Compensation Verification Based on ANSYS Workbench and MATLAB Co-Simulation

The finite element parametric simulation of a 24m-3R mobile concrete pump manipulator was carried out. The parameters of the mobile concrete pump manipulator are shown in Table 1.

Table 1. Parameters of 24m-3R mobile concrete pump manipulator.

Serial Number	1st Link	2nd Link	3rd Link
Link length (joint distance) (mm)	11,204	7800	4982
Joint angle range (°)	0°–90°	−180° to 0°	−180° to 40°

The mobile concrete pump manipulator has strong structural strength, light weight, and a no-load equivalent density of 7000 kg/m^3. Since there is no simple method for directly measuring the tangent slope of a point after deformation in ANSYS Workbench, in this paper, in order to simulate the inclination angle reading value measured by the inclination sensor in the actual experiment, the idea of limit was adopted.

Take two points that are very close to each other on the upper surface of one link (the distance between the two points is known), then the tangent angle of the midpoint of the two points is obtained by taking the deflection at these two points relative to the corresponding coordinate system. The approximate tangent obtained by this method has some errors. As shown in Figure 4, the measurement area in ANSYS is a rectangular area with a length of 200 mm. Compared with the length of each link of about 8000 mm, the distance between the two ends of the rectangular area is very small. The slope of the secant passing through the two ends of the small rectangle is equal to the tangent slope at the center of the small rectangle.

Use the maximum and minimum values of the deflection on the rectangle, as shown in Figure 5; then the tangent angle of the point in this coordinate system can be solved.

$$\beta = \arcsin\frac{MAX - MIN}{200} \tag{32}$$

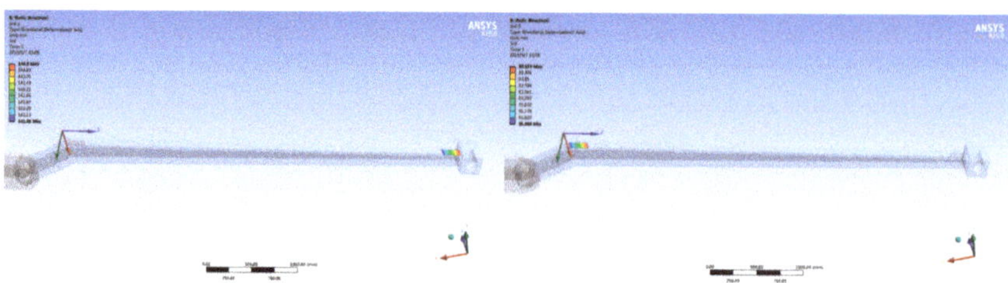

Figure 5. The tangent angle solution method.

The tangential deformation angle γ is obtained by making a difference with the tangent angle of the point at the head end in the coordinate system.

$$\gamma = \arcsin\frac{MAX_t - MIN_t}{200} - \arcsin\frac{MAX_0 - MIN_0}{200} \tag{33}$$

In the simulation, method ③ is applied to the 1st link, and the compensation angle values are calculated from the three inclination sensor readings at a, b, and c. For the 2nd link and 3rd link, method ② is directly adopted due to their small deformation, and γ is directly used to replace $\Delta\theta$.

4.1. ANSYS Workbench Parametric Simulation Verification

4.1.1. Parametric Simulation of Compensation Angle

Since θ_0 is the rotation angle of the rotation base, its influence on the deformation of the manipulator is little; take $\theta_0 = 0$ in the simulation.

The range of joint angles is $\theta_1 \in [0, 90]$; $\theta_2 \in [-180, 0]$; $\theta_3 \in [-180, 40]$, A total of 216 positions and orientations are simulated.

$$
\begin{bmatrix} \theta_1 & \theta_2 & \theta_3 \end{bmatrix}^{\mathrm{T}} =
\begin{bmatrix}
5 & 21 & 37 & 53 & 69 & 85 \\
-175 & -141 & -107 & -73 & -39 & -5 \\
-175 & -133 & -91 & -49 & -7 & 35
\end{bmatrix}
$$

The relationship between the endpoint's deviation of the mobile concrete pump manipulator and $\theta_1\ \theta_2\ \theta_3$ is obtained.

As shown in Figure 6, they are continuous three-dimensional relational surface diagrams between the endpoint's deviation and $\theta_1\ \theta_2$ when θ_3 takes $-175°, -133°, -91°, -49°, -7°, 35°$ respectively.

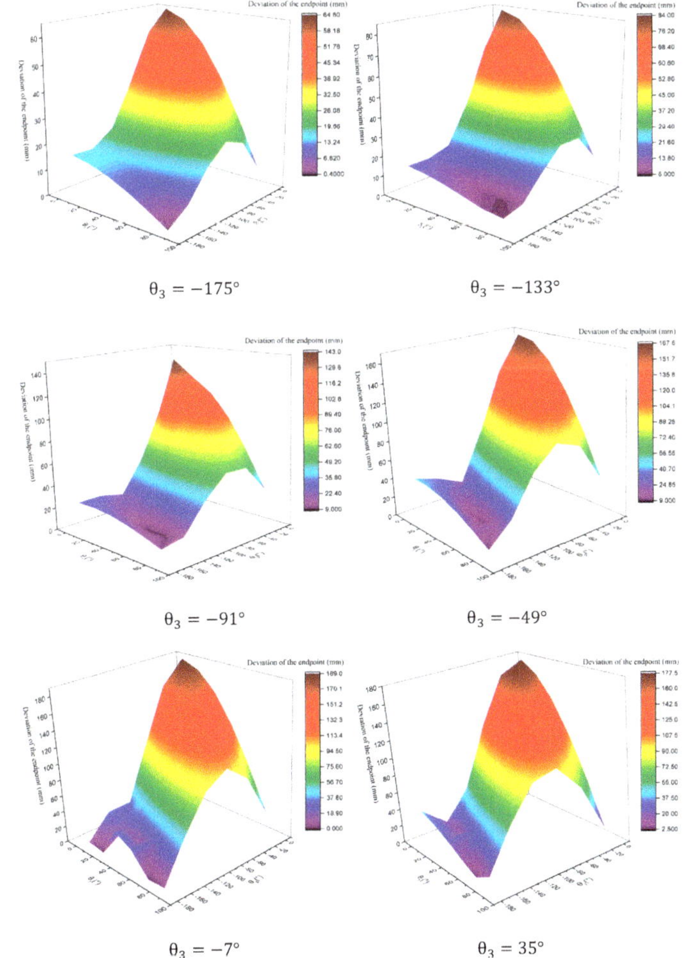

Figure 6. Relationship of endpoint deviation and each joint angle.

It can be seen that the larger the angle of the 2nd link unfolding, the larger the endpoint's deviation will become, and the sharper the endpoint's deviation will change when the 1st link's corresponding action is performed. The position and orientation of the 3rd link has a certain influence on the total deformation of the end. The effect on the deviation when the 3rd link is in the approximate symmetrical orientation (such as 35° and −49°) is approximately the same.

The relationship between the 1st joint compensation angle and the and θ_1 θ_2 θ_3 is obtained. As shown in Figure 7, they are continuous three-dimensional relational surface diagrams between the joint compensation angles $\Delta\theta_1$ and θ_1 θ_2 when θ_3 taking −175°, −133°, −91°, −49°, −7°, 35°.

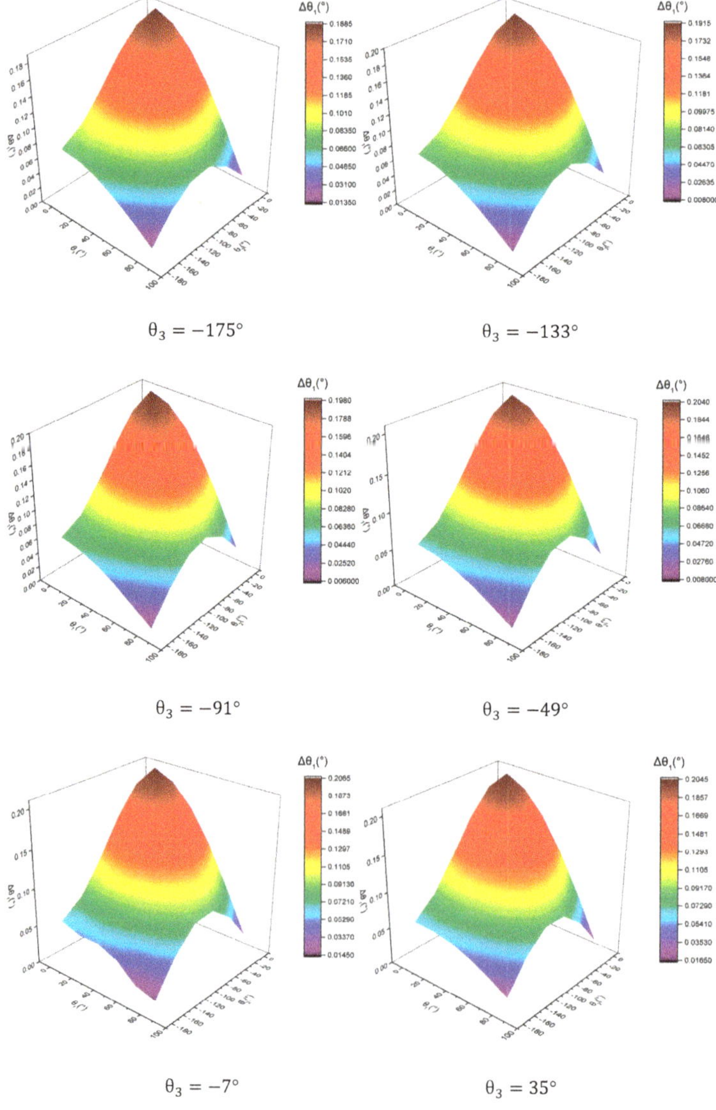

Figure 7. Relationship of compensation angle of the 1st joint angle and each joint angle.

It can be seen that the orientations of the 1st link and the 2nd link have a great influence on the compensation angle of the 1st joint and the orientation of the 3rd link has less influence. The compensation angle of the 1st joint reaches the maximum when the manipulator is fully extended.

4.1.2. Parametric Simulation to Verify the Reliability of the Compensation Method

Considering the ratio of deflection and link length as the actual value of angle compensation, the accuracy of the compensation method can be verified by comparing the actual value of angle compensation and the angle compensation value obtained by this static deformation-compensation method. In this paper, the accuracy of the compensation method is measured by the error rate (actual value − calculated value)/(actual value).

The relationship between the compensation error rate of the 1st link and the joint angles is obtained. As shown in Figure 8, they are continuous three-dimensional relational surface diagrams between error rate and θ_1 θ_2 when θ_3 takes $-133°$, $-7°$, $35°$.

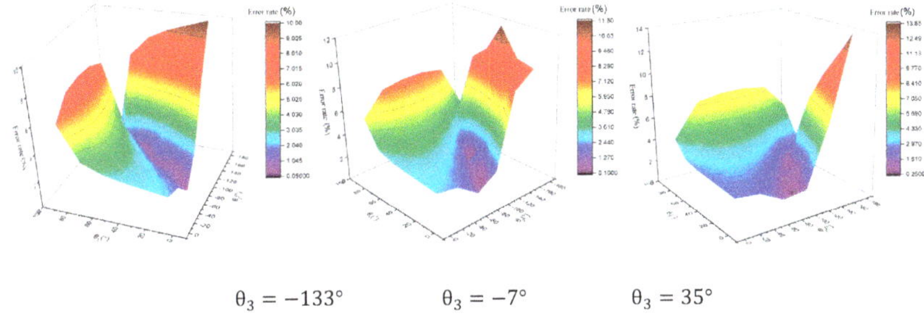

$\theta_3 = -133°$ $\theta_3 = -7°$ $\theta_3 = 35°$

Figure 8. Relationship of compensation error of the 1st joint angle and each joint angle.

It can be seen that when the manipulator is in various positions and orientations, the deformation compensation error rate is controlled within 14%, which verifies the reliability of the static deformation-compensation method based on the inclination sensor feedback. The orientations of the 1st and the 2nd links have a great influence on the deformation compensation error. It can be seen from the figure that the deformation compensation error is the smallest when the angles of the 1st and 2nd joints are in the middle of the value range, and there is a low error valley, which can be used as the research object in the future.

The main reason for the error is that using the secant slope approximates the tangent slope to obtain the inclination angle approximate value. In engineering practice, errors will be of two types: γi experimental determination and accuracy of the coefficient matrix (Equation (26)). In order to compensate the endpoint deviation of the concrete truck manipulator with a length of 24 m up to 200 mm, it is recommended that the accuracy of the inclination sensor is 0.01° to achieve better results.

4.2. Verification of Deformation Compensation Effect Based on ANSYS Workbench and MATLAB Co-Simulation

The six common orientations and six different loads of the 24m-3R mobile concrete pump manipulator were simulated by ANSYS Workbench and MATLAB co-simulation. The validity of the static deformation-compensation method was verified by comparing the theoretical endpoint position of the manipulator with the actual endpoint position (simulation result) after deformation compensation. The compensation error under different loads was obtained, and the universality of the compensation method to different loads was verified.

4.2.1. Verification of the Static Deformation-Compensation Method's Validity

The six common orientations of the link were simulated by ANSYS Workbench and MATLAB co-simulation. ANSYS Workbench parametric simulation obtained the deformation value before and after deformation compensation and obtained the simulation values of inclination sensors; then the results were input to MATLAB, MATLAB according to the set joint angle and ANSYS Workbench input data real-time calculate the manipulator's endpoint position. As shown in Table 2, the deformation compensation effects of six common orientations were obtained.

The deformation compensation error of the mobile concrete pump manipulator in different orientations under no-load operation can be controlled within 50 mm. Compared with the endpoint deviation of 177 mm before compensation, the effect of deformation compensation is obvious. The validity of this method was verified.

4.2.2. Verification of the Static Deformation-Compensation Method's Universality for Different Loads

Six different loads of the 24m-3R mobile concrete pump manipulator in a certain orientation were simulated by ANSYS Workbench and MATLAB co-simulation. In engineering practice, in order to make the concrete pumpable, the concrete in the pipeline is a fluid liquid. In the simulation analysis of concrete pumping, there is a consensus on the treatment of the liquid concrete in the pipeline. In the analysis and calculation, the rigidity of the liquid concrete is not considered, and it is directly applied to the pipeline arch rib as an external load. This method is supported by many engineering application backgrounds, and the calculated results are consistent with the corresponding measured results. Considering the concrete load has the same uniformity as gravity when verifying the static deformation-compensation method is universal for different loads, the equivalent density of the manipulator is changed in the parametric modeling. The simulation results are shown in Table 3, and the selected orientation was (5, −5, 35).

By changing the equivalent density, the deformation compensation effect of the manipulator under different materials and loads was obtained.

The compensation error increased with the increase of the equivalent density, but the endpoint deviations were all reduced to about 20% of the original endpoint deviations after the compensation. The static deformation-compensation method's universality for different loads was verified.

Table 2. Deformation compensation effect of six common orientations.

Orientation (°) $(\theta_1,\theta_2,\theta_3)$	Endpoint Deviation (mm)	Compensation Angle (°) $(\Delta\theta_1,\Delta\theta_2,\Delta\theta_3)$	Endpoint Target Position (mm) (x,z)	Endpoint Position before Compensation (mm) (x,z)	Endpoint Position after Compensation (mm) (x,z)	Compensation Error (Residual Deviation) (mm) (x,z)
$(5, -5, 35)$	177.0635	(0.204, 0.181, 0.042)	(23042, 3834)	(23079.22, 3660.73)	(23055.25, 3784.12)	(13.25, 49.88)
$(37, -5, 35)$	140.1612	(0.163, 0.138, 0.022)	(17509, 15462)	(17606.04, 15360.49)	(17539.17, 15432.31)	(30.17, 29.69)
$(53, -5, 35)$	104.5001	(0.123, 0.100, 0.010)	(12569, 19689)	(12659.12, 19635.76)	(12596.90, 19673.30)	(27.90, 15.70)
$(53, -73, -49)$	105.5360	(0.147, 0.146, 0.015)	(15858, 1629)	(15814.44, 1533.23)	(15829.77, 1596.88)	(28.23, 32.12)
$(69, -73, -49)$	101.2360	(0.120, 0.171, 0.027)	(14794, 5937)	(14773.62, 5837.99)	(14777.04, 5901.81)	(16.96, 35.19)
$(85, -73, -49)$	89.38721	(0.083, 0.183, 0.037)	(12585, 9785)	(12585.90, 9695.61)	(12578.28, 9753.63)	(6.72, 31.37)

Table 3. Deformation compensation effect of different equivalent density.

Equivalent Density (kg/m^3)	Endpoint Deviation (mm)	Compensation Angle (°) $(\Delta\theta_1,\Delta\theta_2,\Delta\theta_3)$	Endpoint Target Position (mm) (x,z)	Endpoint Position before Compensation (mm) (x,z)	Endpoint Position after Compensation (mm) (x,z)	Compensation Error (Residual Deviation) (mm) (x,z)
7000	177.0635	(0.204, 0.181, 0.042)	(23042, 3834)	(23079.22, 3660.73)	(23055.25, 3783.82)	(13.25, 50.18)
8272	209.2331	(0.241, 0.214, 0.049)	(23042, 3834)	(23085.98, 3629.25)	(23058.42, 3774.71)	(16.42, 59.29)
11116	281.1570	(0.324, 0.288, 0.066)	(23042, 3834)	(23101.10, 3558.87)	(23063.69, 3754.57)	(21.69, 79.43)
13959	353.0808	(0.407, 0.361, 0.083)	(23042, 3834)	(23116.21, 3488.48)	(23070.30, 3734.47)	(28.30, 99.53)
16803	425.0046	(0.490, 0.435, 0.100)	(23042, 3834)	(23131.33, 3418.10)	(23076.25, 3714.50)	(34.25, 119.50)
19646	496.9285	(0.573, 0.509, 0.117)	(23042, 3834)	(23146.45, 3347.72)	(23082.54, 3694.66)	(40.54, 139.34)

5. Conclusions and Future Work

In this paper, we propose a static deformation-compensation method based on inclination sensor feedback for large-scale manipulators with hydraulic actuation like mobile concrete pump manipulators, maritime crane systems, and so on to reduce the deviation of the endpoint. Compared with the finite element method, this method does not need to consider many boundary conditions that are uncertain for flexible manipulators in most situations. It has appropriate accuracy and is universal for large-scale manipulators of different sizes and working under different loads.

Based on a 24m-3R mobile concrete pump manipulator, the parametric simulation based on ANSYS is carried out. The relationship between the endpoint's deviation of the mobile concrete pump manipulator and θ_1 θ_2 θ_3 and the relationship between the compensation error rate of the 1st link and the joint angles are obtained. When the manipulator is in various positions and orientations, the deformation compensation error rate is controlled within 14%. The reliability of the static deformation-compensation method is verified, and the error of this method is analyzed.

Based on the ANSYS and MATLAB co-simulation, we compared the theoretical endpoint position with the actual endpoint position after deformation compensation. The deformation compensation error of the mobile concrete pump manipulator in different orientations under no-load operation can be controlled within 50 mm. Compared with the endpoint deviation of 177 mm before compensation, the effect of deformation compensation is obvious. The validity of this method is verified.

Besides, the compensation error under different loads is obtained. The compensation error increases with the increase of the equivalent density, but the endpoint deviations are all reduced to about 20% of the original endpoint deviations after the compensation. The universality of the compensation method for different loads is verified.

In the future, experiments based on this method will be performed to verify the feasibility of the method and to evaluate the deformation-compensation effectiveness of this method in practical application.

Author Contributions: J.Q., Q.S., and B.X. conceived and designed the study. J.Q., F.Z., Y.M. and Z.F. performed the simulations. J.Q. wrote the paper. Q.S., and F.Z. reviewed and edited the manuscript. All authors have read and agreed to the published version of the manuscript.

Funding: This research was funded by [Fundamental Research on Hydraulic Stepping Drive Technology for Multi-joint Hydraulic Manipulators] grant number [51905473] and [the National Natural Science Foundation of China] grant number [91748210]. And the APC was funded by [the National Natural Science Foundation of China].

Acknowledgments: This work was supported by Fundamental Research on Hydraulic Stepping Drive Technology for Multi-joint Hydraulic Manipulators (grant number 51905473) and the National Natural Science Foundation of China (grant number 91748210).

Conflicts of Interest: The authors declare no conflict of interest.

References

1. Dwivedy, S.K.; Eberhard, P. Dynamic analysis of flexible manipulators, a literature review. *Mech. Mach. Theory* **2007**, *41*, 749–777. [CrossRef]
2. Book, W.J.; Majette, M. Controller design for flexible distributed parameter mechanical arm via combined state-space and frequency domain techniques. *Trans. ASME: J. Dyn. Syst. Meas. Contr.* **1983**, *105*, 245–254. [CrossRef]
3. Azad, A.K.M. Analysis and Design of Control Mechanisms for Flexible Manipulator Systems. Ph.D. Thesis, Department of Automatic Control and Systems Engineering, The University of Sheffield, Sheffield, UK, 1994.
4. Poerwanto, H. Dynamic Simulation and Control of Flexible Manipulator Systems. Ph.D. Thesis, Department of Automatic Control and Systems Engineering, The University of Sheffield, Sheffield, UK, 1998.
5. Hartmut, B. Mechydronic for Boom Control on Truck Mounted Concrete Pumps. *Tech. Symp. Constr. Equip. Technol.* **2003**, *1*, 12.

6. Henikl, J.; Kemmetmüller, W.; Meurer, T.; Elsevier, A.K. Infinite-dimensional decentralized damping control of large-scale manipulators with hydraulic actuation. *Automatica* **2016**, *63*, 101–115. [CrossRef]

7. Liu, J.; Dai, L.; Zhao, L.; Cai, J.; Zhang, W. Flexible multi-body dynamics modeling and simulation of concrete pump truck boom. *J. Mech. Eng.* **2007**, *1*, 131–135. [CrossRef]

8. Keating, S.J.; Leland, J.C.; Cai, L. Toward site-specific and self-sufficient robotic fabrication on architectural scales. *Sci. Robot.* **2017**, *2*, eaam8986. [CrossRef]

9. Cheng, M.; Zhang, J.; Xu, B.; Ding, R. An Electrohydraulic Load Sensing System based on flow/pressure switched control for mobile machinery. *ISA Trans.* **2019**. [CrossRef]

10. Bradley, D.A.; Seward, D.W. Developing real-time autonomous excavation-the LUCIE story. In Proceedings of the 34th IEEE Conference on Decision and Control, New Orleans, LA, USA, 13–15 December 1995; pp. 3028–3033.

11. Bradley, D.; Seward, D. The development, control and operation of an autonomous robotic excavator. *J. Intell. Robot. Syst.* **1998**, *21*, 73–97. [CrossRef]

12. Nguyen, Q.H.; Ha, Q.P.; Rye, D.C.; Durrant-Whyte, H.F. Force/position tracking for electro-hydraulic systems of a robotic excavator. In Proceedings of the 39th IEEE Conference on Decision and control, Sydney, NSW, Australia, 12–15 December 2000; pp. 5224–5229.

13. Ha, Q.P.; Nguyen, Q.H.; Rye, D.C.; Durrant-Whyte, H.F. Fuzzy sliding-mode controllers with applications. *Ieee Trans. Ind. Electron.* **2001**, *48*, 38–46. [CrossRef]

14. Lee, S.U.; Chang, P.H. Control of a heavy-duty robotic excavator using time delay control with integral sliding surface. *Control Eng. Pract.* **2002**, *10*, 697–711. [CrossRef]

15. Chang, P.H.; Lee, S.J. A straight-line motion tracking control of hydraulic excavator system. *Mechatronics* **2002**, *12*, 119–138. [CrossRef]

16. Budny, E.; Chlosta, M.; Gutkowski, W. Load-independent control of a hydraulic excavator. *Autom. Constr.* **2003**, *12*, 245–254. [CrossRef]

17. Araya, H.; Makoto, K.; Kinugawa, H.; Tatsuo, A. Level luffing control system for crawler cranes. *Autom. Constr.* **2004**, *13*, 689–697. [CrossRef]

18. Guo, L.; Zhang, G.; Li, J. Kinematics analysis and trajectory planning control modeling and simulation of concrete pump truck. *Constr. Mach.* **2000**, *4*, 29–32.

19. Guo, L.; Zhao, M.; Zhang, G.; Huang, Y. Automatic pumping and process simulation of concrete pump truck cloth mechanism. *J. Northeast. Univ. (Nat. Sci. Ed.)* **2000**, *21*, 617–619.

20. Zhipeng, Y. *Research on the end Track Control of the Concrete Pump Truck Boom*; Wuhan University of Science and Technology: Wuhan, China, 2018.

21. Bremer, H. *Elastic Multibody Dynamics: A Direct Ritz Approach*; Springer: Dordrecht, The Netherlands, 2008.

22. Henikl, J.; Kemmetmüller, W.; Bader, M.; Kugi, A. Modeling, simulation and identification of a mobile concrete pump. *Math. Comput. Model. Dyn. Syst.* **2015**, *21*, 180–201. [CrossRef]

23. Henikl, J.; Kemmetmüller, W.; Kugi, A. Modeling and control of a mobile concrete pump. In Proceedings of the 6th IFAC Symposium of Mechatronic Systems, Hangzhou, China, 10–12 April 2013; pp. 91–98.

24. Lambeck, S.; Sawodny, O.; Arnold, E. Trajectory tracking control for a new generation of fire rescue turntable ladders. In Proceedings of the 2nd IEEE Conference on Robotics, Automation and Mechatronics, Orlando, FL, USA, 15–19 May 2006; pp. 847–852.

25. Zimmert, N.; Pertsch, A.; Sawodny, O. 2-DOF control of a fire-rescue turntable ladder. *IEEE Trans. Control Syst. Technol.* **2012**, *20*, 438–452. [CrossRef]

26. Book, W.J. Modeling, design, and control of flexible manipulator arms: A tutorial review. In Proceedings of the IEEE Conference on Decision and Control, Honolulu, HI, USA, 5–7 December 1990; pp. 500–506.

27. Rahimi, H.N.; Nazemizadeh, M. Dynamic analysis and intelligent control techniques for flexible manipulators: A review. *Adv. Robot.* **2013**, *28*, 63–76. [CrossRef]

28. Kiang, T.; Spowage, A.; Yoong, C.K. Review of control and sensor system of flexible manipulator. *J. Intell. Robot. Syst.* **2015**, *77*, 187–213. [CrossRef]

29. Theodore, R.J.; Ghosal, A. Comparison of the assumed modes and finite element models for flexible multi-link manipulators. *Int. J. Robot. Res.* **1995**, *14*, 91–111. [CrossRef]

30. Huang, D.; Chen, L. Partitioned neural network control and fuzzy vibration control for free-floating space flexible manipulator. *Eng. Mech.* **2012**, *37*. [CrossRef]

31. Subudhi, B.; Morris, A.S. Dynamic modelling, simulation and control of a manipulator with flexible links and joints. *Rob. Auton. Syst.* **2002**, *41*, 257–270. [CrossRef]

32. Usoro, P.B.; Nadira, R.; Mahil, S.S. A finite element/lagrangian approach to modeling light weight flexible manipulators. *ASME J. Dyn. Syst. Meas. Contr.* **1986**, *108*, 198–205. [CrossRef]

33. Zienkiewicz, O.C.; Taylor, R.L.; Zhu, J.Z. *The Finite Element Method, Its Basis and Fundamentals*; Butterworth-Heinemann: Oxford, UK, 2005.

34. Geradin, M.; Cardona, A. Kinematics and dynamics of rigid and flexible mechanisms using finite elements and quaternion algebra. *Comput. Mech.* **1988**, *4*, 115–135. [CrossRef]

35. Bayo, E. A finite-element approach to control the end point motion of a single link flexible robot. *Rob. Syst.* **1987**, *4*, 63–75. [CrossRef]

36. Jonker, B. A finite element dynamic analysis of flexible spatial mechanisms with flexible links. *Comput. Methods Appl. Mech. Eng.* **1989**, *76*, 17–40. [CrossRef]

37. Jonker, B. A finite element dynamic analysis of flexible manipulators. *Int. J. Rob. Res.* **1990**, *9*, 59–74. [CrossRef]

38. Hastings, G.G.; Book, J.W. Verification of a linear dynamic model for flexible robotic manipulators. In Proceedings of the IEEE International Conference on Robotics and Automation, Francisco, CA, USA, 7–10 April 1986; pp. 1024–1029.

39. Kim, J.S.; Uchiyama, M. Vibration mechanism of constrained spatial flexible manipulators. *JSME Ser. C* **2003**, *46*, 123–128. [CrossRef]

40. Feliu, V.; Rattan, K.S.; Brown, H.B. Modeling and control of single-link flexible arms with lumped masses. *ASME J.Dyn. Syst. Meas. Contr.* **1992**, *114*, 59–69. [CrossRef]

41. Khalil, W.; Gautier, M. Modeling of mechanical systems with lumped elasticity. *Proc. IEEE Int. Conf. Rob. Autom.* **2000**, *4*, 3964–3969.

42. Zhu, G.; Ge, S.S.; Lee, T.H. Simulation studies of tip tracking control of a single-link flexible robot based on a lumped model. *Robotica* **1999**, *17*, 71–78. [CrossRef]

43. Raboud, D.W.; Lipsett, A.W.; Faulkner, M.G.; Diep, J. Stability evaluation of very flexible cantilever beams. *Int. J.Nonlinear Mech.* **2001**, *36*, 1109–1122. [CrossRef]

44. Lee, S.J.; Chung, I.S.; Bae, S.Y. Structural design and analysis of CFRP boom for concrete pump truck. *Mod. Phys. Lett. B* **2019**, *33*, 787–792. [CrossRef]

45. Yuan, Q.H.; Lew, J.; Piyabongkarn, D. Motion control of an aerial work platform. In Proceedings of the 2009 American control conference, AACC 2009, Hyatt Regency Riverfront, St. Louis, MO, USA, 10–12 June 2009; pp. 2873–2878.

46. Xia, J.; Gang, G.; Xin, Z. Research on deformation compensation and trajectory control technology of concrete boom. *Constr. Mach.* **2016**, *5*, 70–76.

47. Xin, Z.; Liang, W.; Jiaqian, W.; Shiyang, H. Research on trajectory control technology of concrete pump truck based on vehicle deformation compensation. *J. Mech. Eng.* **2015**, *13*, 492–496.

48. Wang, X. Research on Trajectory Control Technology of Concerte Pump Arm Frame based on Arm Frame Denaturalization Compensation. Master's Thesis, Qingdao University of Science and Technology, Qingdao, China, 2018.

49. Pan, D.; Xing, J.; Wang, B. Static Analysis and Dynamic Simulation of Concrete Pump Truck Boom. *J. Chongqing Univ. Technol. (Nat. Sci.)* **2018**, *32*, 86–91, 123.

 *processes*

Article

Study on Oil Film Characteristics of Piston-Cylinder Pair of Ultra-High Pressure Axial Piston Pump

Jin Zhang [1,2,3,*], Baolei Liu [1], Ruiqi LÜ [1], Qifan Yang [1] and Qimei Dai [1]

[1] College of Mechanical Engineering, Yanshan University, Qinhuangdao 066004, China; liubaoleino1@163.com (B.L.); LYURUIQI@126.com (R.L.); yqffzr@163.com (Q.Y.); zhangjinno1@163.com (Q.D.)

[2] China Key Laboratory of Advanced Forging & Stamping Technology and Science, Yanshan University, Qinhuangdao 066004, China

[3] School of Mechanical Engineering, Yanshan University, Qinhuangdao 066004, China

[*] Correspondence: zhangjin@ysu.edu.cn; Tel.: +86-0335-8077-4618

Received: 13 December 2019; Accepted: 23 December 2019; Published: 3 January 2020

Abstract: The piston-cylinder pair is the key friction pairs in the piston pump. Its performance determines the volume efficiency of piston pump. With the increase of load pressure, the leakage at the clearance of piston-cylinder pair will also increase. In order to reduce leakage, the clearance of the piston-cylinder pair of the ultra-high pressure piston pump is smaller than that of the medium-high pressure piston pump. In order to explore whether the piston will stuck in the narrow gap, it is necessary to study the oil film characteristics of the piston-cylinder pair under the condition of ultra-high pressure, so as to provide a theoretical basis for the optimal design of the piston-cylinder pair of ultra-high pressure axial piston pump. In this paper, an ultra-high pressure axial piston pump is taken as the research object, and its structural characteristics are analyzed. The mathematical model of the oil film thickness of the piston-cylinder pair is established by using the cosine theorem in the cross section of the piston. The finite volume method is used to discretize the Reynolds equation of the oil film of the piston-cylinder pair, and the over relaxation iteration method is used to solve the discrete equations, and the mathematical model of the oil film pressure of the piston-cylinder pair is obtained. The mathematical model of oil film thickness and pressure field of piston-cylinder pair is solved by programming. The dynamic change process of oil film thickness and pressure field of the plunger pair of the ultra-high pressure axial piston pump under the load of 20 MPa and 70 MPa is obtained. Under the two conditions, the thinnest area of the oil film reaches 3 μm and 2 μm dangerous area respectively; the oil film pressure reaches 20 MPa and 70 MPa respectively when the swashplate rotates 10° and continues to increase with the increase of swashplate rotation angle. When the rotation angle reaches 90°, the oil film pressure also reaches the maximum value, but there is no pressure spike phenomenon. The oil film pressure characteristics of ultra-high pressure axial piston pump under conventional and ultra-high pressure conditions were obtained by modification and experimentation.

Keywords: ultra-high pressure; axial piston pump; piston-cylinder pair; oil film characteristics; experimental study on pump

---

1. Introduction

Compared with other transmission modes, hydraulic transmission system has been widely accepted because of its advantages of large power-mass ratio, convenient control, and smooth transmission [1–3]. Ultra-high pressure is one of the development directions in hydraulic field at present. Hydraulic pump as the power component of the system, the ultra-high pressure will inevitably have a certain impact on its own structure [4]. During the operation of piston pump, the friction pairs bears such functions as sealing and lubrication, so the performance of the friction pairs has a key impact on the reliability and service life of the pump [5–7]. As the core component of energy conversion of piston pump, the design of piston-cylinder pair is more stringent with the ultra-high pressure of working conditions [8,9]. In order to ensure the sealing performance, the distance between piston chambers is only 3–5 mm. Therefore, under ultra-high pressure condition, the failure of oil film of piston-cylinder pair causes many problems. For example, the wear degree between the piston-cylinder pair is aggravated, the volume efficiency is reduced, and the piston is stuck. In order to prevent the above situation of the piston-cylinder pair under ultra-high pressure condition, it is necessary to analyze the oil film characteristics of the piston-cylinder pair. In 1975, Professors Yamaguchi and Takaoka of Japan carried out theoretical and experimental analysis to explore the oil film characteristics of piston-cylinder pair. The motion of the piston is analyzed by perturbation method, and the distribution of the oil film pressure of the piston-cylinder pair is solved. A test-bed is built and the experimental results are compared with the theoretical analysis. The results are basically consistent [10]. In 1998 Tanaka et al. found that the innovation was the application of displacement sensor and force sensor. By measuring the oil film characteristics and friction characteristics of the piston-cylinder pair, the correctness of the conjecture that the piston rotates around its axis was confirmed. It was also noted that the angular velocity of the piston rotation is similar to that of the spindle [11–13]. Xiaofeng He of Huazhong University of Science and Technology established an experimental device for piston-cylinder pair of piston pump in 2001. However, because of the complexity of the motion mechanism of piston-cylinder pair, the experimental device can only be used to evaluate the wear stage of piston-cylinder pair [14]. Professor Monica I. of Purdue University, USA, has written the CASPAR program. CASPAR program is a tool for calculating the oil film characteristics of the friction pairs clearance of an anisotropic axial piston pump. The hydrodynamic, dynamic, and temperature characteristics of the oil film at the piston-cylinder pair, slipper pairs, and distributor pairs are studied. In 2009, Zhang Bin of Zhejiang University discussed the experimental method based on the oil film characteristics of the piston-cylinder pair under actual working conditions, and established the virtual prototype simulation model of the axial piston pump. The signal is sent to the data acquisition system by the sensor, and the pressure distribution of the piston oil film is tested by the piston oil film characteristic test rig. The experimental results are in good agreement with the simulation results, and the pressure field distribution of the oil film of the piston-cylinder pair can be well displayed [15]. Professor Bergada of Catalonia University of Technology in Spain has set up a test rig for dynamic oil film pressure of piston-cylinder pair. The test rig is mainly used to analyze the pressure fluctuation of the piston-cylinder pair oil film caused by the change of rotational speed, outlet pressure and inclined plate angle [16]. Professor Monica I. has built a single piston model pump experiment. The device uses a single piston pump experimental platform with swashplate rotation and fixed cylinder block to measure the oil film lubrication characteristics. Pressure distribution is measured by the pressure sensor, temperature distribution is measured by the temperature sensor, and oil film thickness is measured by the displacement sensor [17]. Xu Bing and Zhang Junhui of Zhejiang University used virtual prototyping technology to simulate the piston-cylinder pair of axial piston pump. The data transmission of sub-module was carried out by a software, and the fluid-structure coupling and rigid-flexible coupling of the piston-cylinder pair simulation model were also carried out. The validity of the model is verified by the test results, which proves that the virtual prototype simulation platform of the axial piston pump has a strong guiding role in the design of the axial piston pump [18]. Xu Bing and Zhang Junhui of Zhejiang University describe the piston state of oblique axial piston pump of electro-hydrostatic actuator accurately by solving the discrete oil film Reynolds equation and force balance equation iteratively, and obtain the leakage of piston-cylinder pair under high speed and pressure. It provides appropriate theoretical guidance for the design of EHA pump [19].

Nowadays, scholars from all over the world have made a detailed study of the piston-cylinder pair. However, because of the limitation of experimental conditions and the consideration of safety when the piston pump speed is too high, the pressure load setting value generally does not exceed 10 MPa. There are few studies on the oil film characteristics of piston-cylinder pair under ultra-high pressure conditions. In this paper, the oil film thickness and pressure characteristics of the piston-cylinder pair of ultra-high pressure piston pump under different working conditions are obtained through mathematical analysis and simulation analysis, and the oil film pressure characteristics of the piston-cylinder pair are obtained through experiments.

2. Analysis of Mechanical Characteristics of Ultra-High Pressure Piston Pump

Figure 1 shows the structure of the ultra-high pressure axial piston pump analyzed in this paper. Its rated working pressure is 70 MPa. The swashplate of the ultra-high pressure piston pump is integrated with the main shaft. It is composed of transmission shaft, pump cover, cycloid pump (refueling), pressure valve screw (valve distribution), and so on. Ultra-high pressure axial piston pump is driven by a motor to rotate the drive shaft, which causes the synchronous rotation of the swashplate and the thrust ball bearing on its surface. The return mechanism tightly compresses the piston ball head with the thrust ball bearing surface, and then drives the piston to move in the axis direction.

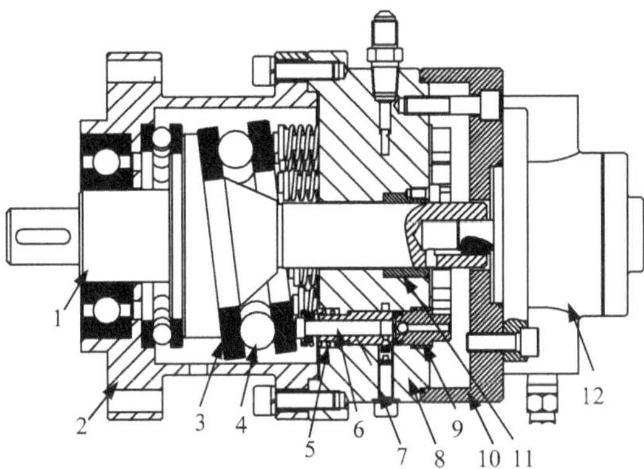

Figure 1. Structural sketch of piston pump. 1—drive shaft; 2—pump cover; 3—swash plate; 4—thrust ball bearing; 5—return mechanism; 6—piston; 7—hydraulic valve screw; 8—hydraulic cylinder; 9—marble seat; 10—lower end cover; 11—copper sleeve; 12—cycloid pump.

The piston is subjected to hydraulic pressure of oil in the cavity, inertia force of linear motion, viscous friction force, support force of thrust ball bearing, friction force, and self-gravity. When the piston moves under their combined action, the center line of the piston and the center line of the piston cavity will not be in the same line, which will produce a certain deviation as shown in Figure 2.

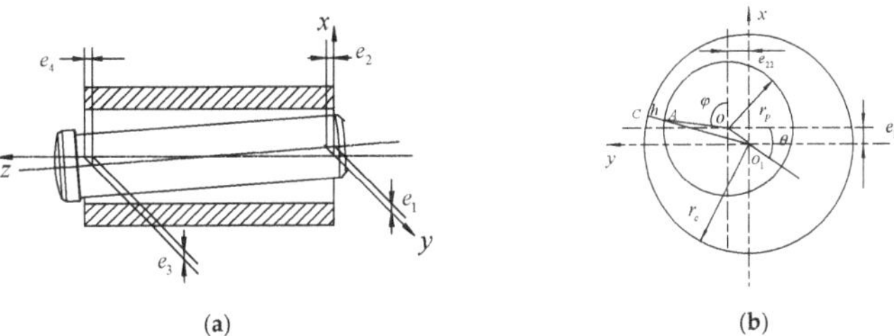

(a)   (b)

Figure 2. (a) Attitude diagram of piston in piston cavity; (b) variable schematic diagram at cross section of piston end face.

In order to describe the inclined state of piston, the attitude inclination degree of piston in piston cavity is represented by introducing (e_1, e_2, e_3, e_4) and named (e_1, e_2, e_3, e_4) as offset coordinates. (e_1, e_2) are near the end of the piston cavity and (e_3, e_4) are near the piston ball.

The thickness of the piston-cylinder pair at any position in the axial direction of the piston satisfies:

$$h = r_c - o_1 A \tag{1}$$

The unknown quantity $o_1 A$ in the above formula can be obtained by using cosine theorem for triangle $oo_1 A$. The expression of cosine theorem is as follows:

$$o_1 A = \sqrt{r_p^2 + oo_1^2 + 2r_p oo_1 \sin(\varphi + \rho)} \tag{2}$$

In the formula, $o_1 A$ is the distance between the center of the piston cavity section o_1 and any point A on the piston section; oo_1 is the distance between the center of the piston section o_1 and the center of the piston cavity section o_1; φ is the angle between the piston center and any point on the piston and the positive direction of the x axis; and oo_1 is the angle between the center of the piston and the center of the piston cavity and the negative direction of the y axis.

Formula (2) is brought into Formula (1) to obtain:

$$h = r_c - \sqrt{r_p^2 + oo_1^2 + 2r_p oo_1 \sin(\varphi + \rho)} \tag{3}$$

After simplification, we can get:

$$h = r_c - r_p - oo_1 + 2r_p \sin(\varphi + \rho) \tag{4}$$

In Figure 2b, we can see that:

$$\tan \sigma = \frac{e_{11}}{e_{22}} \tag{5}$$

$$oo_1^2 = e_{11}^2 + e_{22}^2 \tag{6}$$

e_{11} is projected in the x-axis direction from the center of the piston section to the center of the piston cavity, e_{22} is projected in the y-axis direction from the center of the piston section to the center of the piston cavity.

If Formulas (5) and (6) are introduced into the simplified formula of oil film thickness, the expression h of oil film thickness at any point in the axial direction of piston is only a function of e_{11} and e_{22}. Therefore, according to the values of e_{11} and e_{22}, the oil film thickness at any position in the axial direction under the graphical attitude of piston can be obtained.

The e_{11} and e_{22} in the above formula are related to the migration coordinates (e_1, e_2, e_3, e_4). The specific relations between them are as follows in Figure 3.

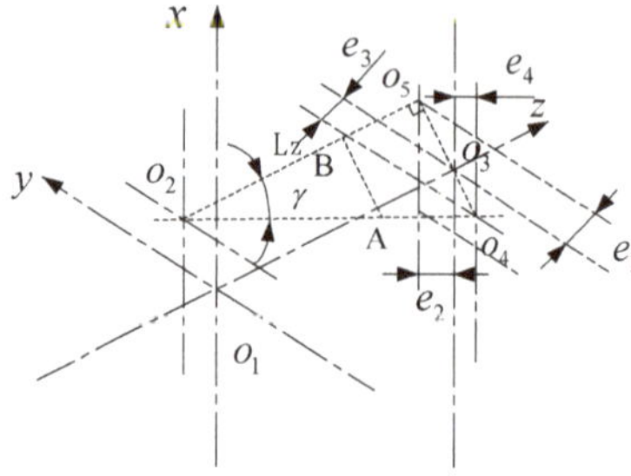

Figure 3. Diagram of relationship between displacement and displacement coordinates at piston section.

The central deviation AB of any point in the axis direction of the piston:

$$\frac{AB}{o_4 o_5} = \frac{L_Z}{L_{CZ}} \tag{7}$$

AB projection in horizontal and vertical directions:

$$e_{11} = e_1 - \frac{L_Z(e_1 - e_2)}{L_{CZ}} \tag{8}$$

$$e_{22} = e_2 - \frac{L_Z(e_2 - e_4)}{L_{CZ}} \tag{9}$$

The mathematical model of oil film thickness of piston pairs is as follows:

$$h = r_c - r_p - \left[e_2 - \frac{L_Z(e_2 - e_4)}{L_{CZ}}\right]\sin\varphi - \left[e_1 - \frac{L_Z(e_1 - e_3)}{L_{CZ}}\right]\cos\varphi \tag{10}$$

Under the condition of ultra-high pressure, the flow of oil film in piston-cylinder pair belongs to crevice flow. Based on the theory of crevice flow and laminar flow characteristics, the following assumptions are put forward:

(1) Ignoring the proportion of mass force; (2) ignoring the inertial force of fluid; (3) ignoring the curvature of oil film, replacing rotational velocity with translation velocity; (4) ignoring the change of oil film pressure in thickness direction; (5) oil flow velocity and viscosity do not change with the change of oil film height; (6) it is assumed that the curvature radius of the interface contacted with the oil is much larger than the thickness of the oil film.

For incompressible viscous fluids, Navier–Stokes equation (N–S equation for short) characterizes the motion characteristics of fluids and is its differential equation of motion. N–S equation reflects the relationship between mass force, viscous force, and motion parameters of viscous fluid at any point in the flow process. In space rectangular coordinates, the expression of N-S equation is as follows:

$$\begin{cases} \rho f_x - \dfrac{\partial p}{\partial x} + \mu\left(\dfrac{\partial^2 \mu_x}{\partial x^2} + \dfrac{\partial^2 \mu_x}{\partial y^2} + \dfrac{\partial^2 \mu_x}{\partial z^2}\right) \\ \rho f_y - \dfrac{\partial p}{\partial y} + \mu\left(\dfrac{\partial^2 \mu_y}{\partial x^2} + \dfrac{\partial^2 \mu_y}{\partial y^2} + \dfrac{\partial^2 \mu_y}{\partial z^2}\right) \\ \rho f_z - \dfrac{\partial p}{\partial z} + \mu\left(\dfrac{\partial^2 \mu_z}{\partial x^2} + \dfrac{\partial^2 \mu_z}{\partial y^2} + \dfrac{\partial^2 \mu_z}{\partial z^2}\right) \end{cases} \tag{11}$$

In the formula, f_x is the mass force of a fluid with unit mass in the x-axis direction, f_y is the mass force of a fluid with unit mass in the y-axis direction, f_z is the mass force of a fluid with unit mass in the z-axis direction, p is the fluid pressure, ρ is the density of fluid, and μ is the motion viscosity of the fluid.

Under the above assumptions, the simplified N-S equation is:

$$\begin{cases} \dfrac{\partial p}{\partial x} = \mu\dfrac{\partial^2 \mu_x}{\partial z^2} \\ \dfrac{\partial p}{\partial y} = \mu\dfrac{\partial^2 \mu_y}{\partial z^2} \end{cases} \tag{12}$$

According to hypothesis (6), the annular oil film of piston pairs is expanded as follows in Figure 4.

The oil film of expanded piston-cylinder pair is analyzed in Cartesian coordinate system. Because the piston of the ultra-high pressure pump rotates in the direction of the non-winding transmission shaft, and because the motion between the swashplate and the cylinder block is reciprocal. Therefore, in order to express the velocity boundary condition of the oil film of the piston-cylinder pair, the actual situation of cylinder block fixing and swashplate rotating are transformed into that of cylinder block rotating and swashplate fixing.

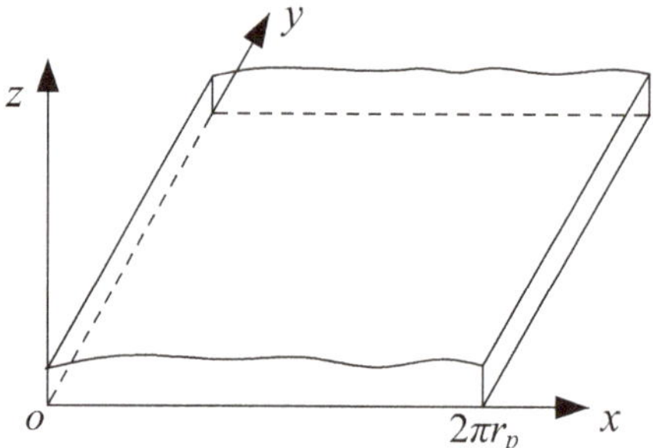

Figure 4. Piston pairs oil film unfolding diagram.

Mathematical model expression of oil film pressure of piston pairs:

$$\frac{\partial}{\partial x}\left(\frac{\partial P}{\partial x}\frac{h^3}{\mu}\right) + \frac{\partial}{\partial y}\left(\frac{\partial P}{\partial y}\frac{h^3}{\mu}\right) = 6r_p\omega_p\frac{\partial h}{\partial x} + 6v_p\frac{\partial h}{\partial y} + 12\frac{\partial h}{\partial t} \tag{13}$$

In the formula: h is the thickness of the oil film of the piston-cylinder pair; μ is the viscosity of the oil film of the piston-cylinder pair; P is the pressure of the oil film of the piston-cylinder pair.

In view of the very small film thickness of the ultra-high pressure piston pump, in order to simplify the solution process, the expanded oil film is equivalent to a plane form. In this paper, the Reynolds equation of oil film is solved by the finite volume method. The expanded oil film is meshed as follows (Figure 5a). By integrating the Reynolds equation of oil film in each control volume, a set of discrete equations about pressure can be solved. Therefore, by solving the discrete equations, the pressure distribution at each point on the oil film of the piston-cylinder pair under ultra-high pressure can be obtained. Establish a two-dimensional control volume structure as shown in Figure 5b.

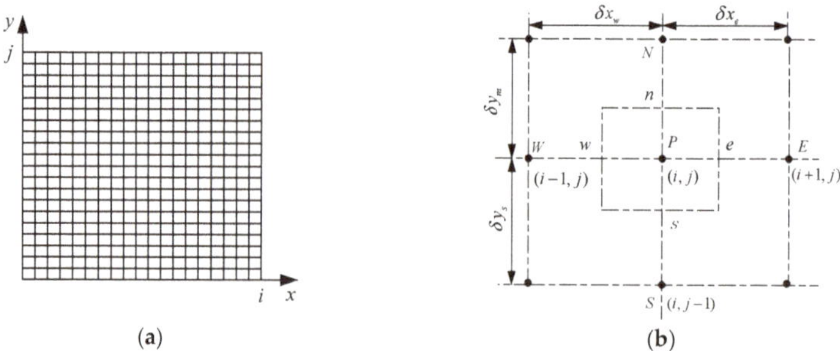

(a)   (b)

Figure 5. (a) Schematic diagram of oil film plane mesh generation; (b) control volume structure diagram.

According to the idea of finite volume method, the Reynolds equation of oil film is integrated into the control volume n-w-s-e in Figure 5b, which is expressed as follows:

$$\int_V\left(\frac{\partial}{\partial x}\left(\frac{\partial P}{\partial x}\frac{h^3}{\mu}\right) + \frac{\partial}{\partial y}\left(\frac{\partial P}{\partial y}\frac{h^3}{\mu}\right)\right)dV = \int_V\left(6r_p\omega_p\frac{\partial h}{\partial x} + 6v_p\frac{\partial h}{\partial y} + 12\frac{\partial h}{\partial t}\right)dV \tag{14}$$

The volume integral of formula is transformed into area integral by using the Gauss theorem. The expression of the Gauss theorem is as follows:

$$\int_s^n\int_w^e\left(\frac{\partial p}{\partial x} + \frac{h^3}{\mu}\right)dxdy + \int_s^n\int_w^e\left(\frac{\partial p}{\partial y}\frac{h^3}{\mu}\right)dxdy - \oiint\left(6r_p w_p\frac{\partial h}{\partial x} + 6v_p\frac{\partial h}{\partial y} + 12\frac{\partial h}{\partial t}\right)dxdy = 0 \tag{15}$$

In the control volume n-w-s-e, the horizontal e and w points are:

$$\left(\frac{\partial p}{\partial x}\frac{h^3}{\mu}\right)_e = \frac{h_e^3}{\mu}\frac{P_N - P_P}{\delta x_{PE}}, \left(\frac{\partial p}{\partial x}\frac{h^3}{\mu}\right)_w = \frac{h_w^3}{\mu}\frac{P_P - P_W}{\delta x_{WP}} \tag{16}$$

In the control volume n-w-s-e, the vertical n and s points are:

$$\left(\frac{\partial p}{\partial x}\frac{h^3}{\mu}\right)_n = \frac{h_n^3}{\mu}\frac{P_N - P_P}{\delta x_{NP}}, \left(\frac{\partial p}{\partial x}\frac{h^3}{\mu}\right)_s = \frac{h_s^3}{\mu}\frac{P_P - P_N}{\delta x_{PS}} \tag{17}$$

Bring the formula into the formula and simplify it into:

$$a_P P_P = a_N P_N + a_S P_S + a_E P_E + a_W P_W + S \tag{18}$$

$$a_P = a_N + a_S + a_E + a_W \tag{19}$$

$$a_N = \frac{h_n^3}{\mu}\frac{\delta x}{\delta y}, a_S = \frac{h_s^3}{\mu}\frac{\delta x}{\delta y}, a_E = \frac{h_e^3}{\mu}\frac{\delta x}{\delta y}, a_W = \frac{h_w^3}{\mu}\frac{\delta x}{\delta y} \tag{20}$$

$$S = -6\left(\omega r_p(h_e - h_w)\delta y - v_p(h_n - h_s)\delta x\right) - 12\int_w^e \int_s^n h' dx dy \tag{21}$$

In the formula, h_n is the oil film thickness at n point in n-w-s-e, h_s is the oil film thickness at s point in n-w-s-e, h_e is the oil film thickness at e point in n-w-s-e, and h_w is the oil film thickness at n point in n-w-s-e, δx is the x-direction length of each micro-grid in the oil film division area of the piston-cylinder pair; δy is the y-direction length of each micro-grid in the oil film division area of the piston-cylinder pair; h' is the change rate of the oil film thickness of the piston-cylinder pair with time; and μ is the motion viscosity of the fluid.

In this paper, the successive over-relaxation iteration method (SOR iteration method) is selected to solve the discrete equation. Over-relaxation method is a modification of the Gauss–Seidel algorithm (GS algorithm). In the iteration process, the results of each iteration and the changes of each iteration are weighted and then brought into the next calculation. Therefore, successive over-relaxation iteration method greatly optimizes the convergence rate. The oil film discrete equation of the piston pairs is substituted into the iteration equation to obtain:

$$P_p^{k+1} = \omega\left(\frac{a_N P_N^k + a_S P_S^k + a_W P_W^k + a_E P_E^k + S}{a_P} - P_P^k\right) + P_P^k \tag{22}$$

In this paper, the convergence criteria for discrete computation of equations are as follows:

$$\frac{\sum_{i=2}^{m-1}\sum_{j=1}^{n}\left|P_{i,j}^k - P_{i,j}^{k-1}\right|}{\sum_{i=2}^{m-1}\sum_{j=1}^{n}\left|P_{i,j}^k\right|} \leq error \tag{23}$$

In order to ensure better convergence and calculation accuracy, the convergence accuracy $error = 10^{-3}$ and the value range of relaxation factor ω are chosen between 1.6 and 1.8.

3. Simulation and Analysis of Oil Film Characteristics of Ultra-High Pressure Axial Piston Pump

In this paper, the oil film characteristics of ultra-high pressure axial piston pump are simulated and analyzed by using the simulation software of MATLAB. In order to compare the oil film characteristics of piston-cylinder pair under ultra-high pressure with those of ordinary medium and high pressure piston pumps, the oil film characteristics of piston-cylinder pair were simulated under 20 MPa and 70 MPa respectively. Parameter settings are shown in Table 1.

Table 1. Simulation parameter setting.

Parameter	Numerical Value
Piston cavity radius mm	4.005
Piston radius mm	4
Number of pistons	13
Bushing length mm	41
Piston distribution circle radius mm	100
Swash plate inclination angle °	7
Piston quality kg	195.8×10^{-3}
Longest contact length of piston pairs mm	34.3
Distance from piston center to end mm	48
Speed of piston pump r/min	1500
Pressure of oil chamber MPa	70
Suction chamber pressure MPa	2.5
Pressure inside the shell MPa	0.2
Oil viscosity Pa·s	0.046
Axial meshing number	40
Number of circumferential meshes	40
Pressure of oil chamber MPa	70
Suction chamber pressure Mpa	2.5
Pressure in shell MPa	0.2
Oil viscosity Pa·s	0.046
Axial grid fraction	40
Circumferential grid fraction	40

Under constant pressure (20 MPa), the oil film thickness expression of the piston–cylinder pair obtained above is brought into the simulation software of MATLAB, and the results are shown in Figure 6 (the point of coordinate (0,0) in the figure below corresponds to point A in Figure 2b).

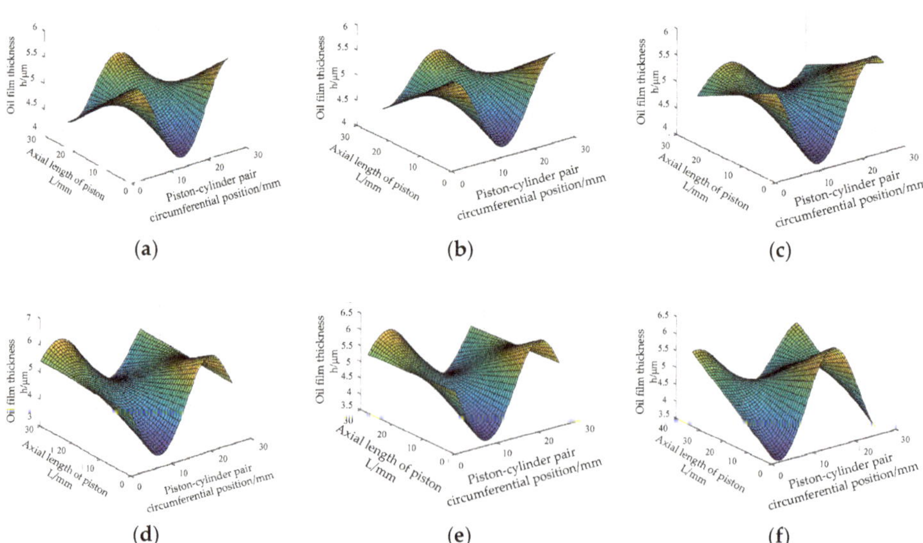

Figure 6. The figure shows the result of oil film thickness at different angle of inclined plate rotation: (**a**) Oil film thickness distribution with swashplate rotation of 1°; (**b**) oil film thickness distribution with swashplate rotation of 10°; (**c**) oil film thickness distribution with swashplate rotation of 45°; (**d**) oil film thickness distribution with swashplate rotation of 90°; (**e**) oil film thickness distribution with swashplate rotation of 110°; (**f**) oil film thickness distribution with swashplate rotation of 210°.

When the swashplate rotates 1°, the plunger is still near the starting point, the inclination of the plunger is very small, and the oil film thickness is about 5 μm. When the swashplate rotates 10°, the plunger cavity enters the oil pressure area, and the pressure in the plunger cavity increases, resulting in the increase of the eccentric load on the plunger and the increase of the inclination of the plunger. When the swashplate rotates 45°, the eccentric load on the plunger continues to increase, resulting in the increase of the inclination of the plunger. When the swashplate rotates 90°, the movement speed of the plunger reaches the maximum. At this time, the eccentric load of the plunger is the maximum, the inclination degree of the plunger reaches the maximum, the oil film thickness of the cylinder-piston pair reaches the minimum, and the oil film thickness is 3–3.5 μm. When the swashplate rotates 110°, the eccentric load on the plunger is relieved and the inclination of the plunger is reduced. When the swashplate rotates 210°, the plunger enters into the low-pressure oil absorption area, and the degree of eccentric load of the plunger is very small, and the oil film distribution of the cylinder-piston pair is relatively average.

Under constant pressure (20 MPa), the expression of the oil film pressure gauge of the piston-cylinder pair obtained above is brought into the simulation software of MATLAB, and the results are shown in Figure 7.

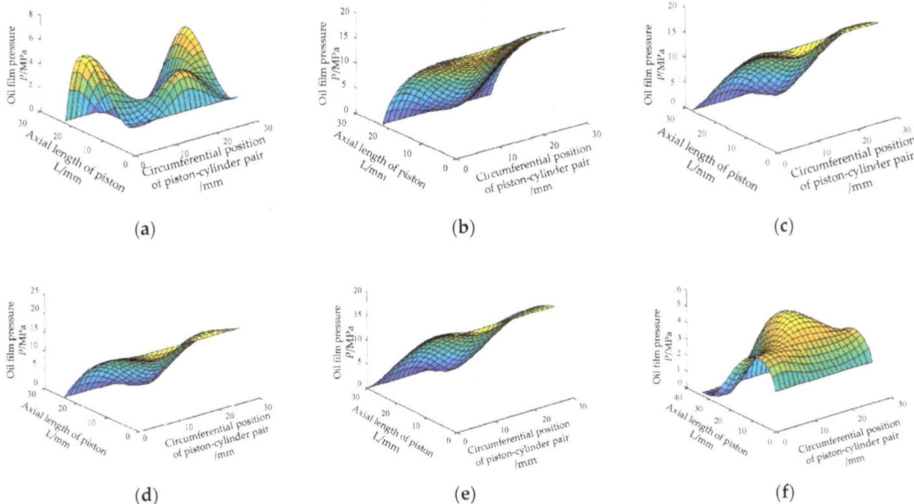

Figure 7. The figure shows the distribution of oil film pressure at different inclined plate rotation angles under constant pressure conditions: (**a**) Oil film pressure distribution with swashplate rotation of 1°; (**b**) oil film pressure distribution with swashplate rotation of 10°; (**c**) oil film pressure distribution with swashplate rotation of 45°; (**d**) oil film pressure distribution with swashplate rotation of 90°; (**e**) oil film pressure distribution with swashplate rotation of 110°; (**f**) oil film pressure distribution with swashplate rotation of 210°.

When the swashplate rotates 1°, because the plunger is still near the starting point, the pressure in the plunger cavity is not established and the pressure is small. When the swashplate rotates for 10°, the plunger cavity enters the oil pressure area, and the inclination degree of the plunger increases, so the oil film pressure reaches 20 MPa. When the swashplate rotates 45°, with the increase of the pressure in the plunger cavity, the inclination degree of the plunger continues to increase. Therefore, in order to balance the increased eccentric load of the plunger, the oil film pressure continues to increase. When the swashplate rotates to 90°, the movement speed of the plunger is the maximum, the eccentric load degree of the plunger reaches the maximum, and the oil film appears the thinnest area in the whole movement cycle. Therefore, in order to balance the eccentric load, the oil film pressure field of the cylinder-piston pair reaches the maximum, and the oil film pressure reaches 25 MPa. When the swashplate rotates to 110°, the eccentric load on the plunger decreases and the pressure field of the oil film decreases. When the swashplate rotates 210°, the plunger cavity has left the oil pressure area and stepped into the low-pressure oil absorption area. The plunger tends to be in a stable state. Therefore, the oil film pressure distribution is small. The oil film characteristics of the ultrahigh pressure plunger pump studied in this paper are basically the same as those in reference [15]. When pressing oil, when the inclination of plunger increases, the oil film pressure increases, and when the inclination of plunger decreases, the oil film pressure also decreases. In the process of oil absorption, the pressure distribution of the plunger is also small because of its small inclination. However, the oil film gap of the cylinder-piston pair in this paper is only 5 μm, which is quite different from the oil film gap (17 μm) in reference [15]. Therefore, although the inclination degree of the plunger will increase, the space (5 μm) for the inclination of the plunger is not as large as that in reference [15] (17 μm). Therefore, the oil film of the cylinder-piston pair in this paper is squeezed, but the oil film is squeezed membrane pressure does not have a sharp edge. The pressure spike shown in Figure 8 does not occur.

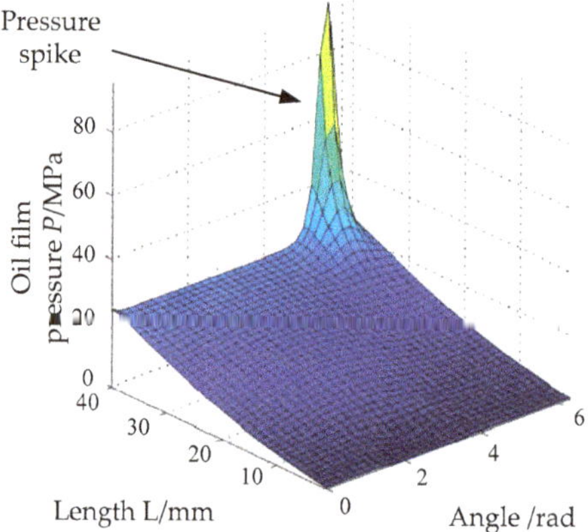

Figure 8. The figure shows pressure spike of plunger auxiliary oil film.

Under the condition of ultra-high pressure (70 MPa), the oil film thickness expression of the piston-cylinder pair obtained above is brought into the simulation software of MATLAB, and the results are shown in Figure 9.

Compared with the normal pressure condition, the change trend and distribution of the oil film thickness under the ultra-high pressure condition are the same. Before the swashplate rotates 45°, the oil film thickness is almost the same. But after the swashplate rotates 45°, the oil film thickness under the ultra-high pressure condition is obviously smaller than that under the normal pressure condition. The minimum thickness of oil film is about 2 μm under ultra-high pressure condition. At this time, the bearing limit area appears in the auxiliary oil film of the plunger. The inclination degree of the plunger is greater than that under normal pressure condition, and the failure probability of the auxiliary oil film of the plunger is greater. When the swashplate rotates more than 90°, the oil film thickness under the two conditions is basically the same.

Under the condition of ultra-high pressure (70 MPa), the expression of the oil film pressure gauge of the piston-cylinder pair obtained above is brought into the simulation software of MATLAB, and the results are shown in Figure 10.

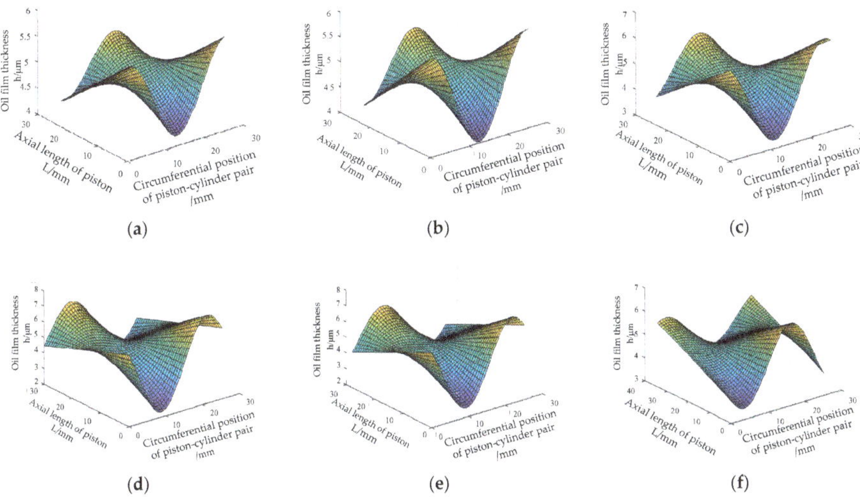

Figure 9. The figure shows the distribution of oil film thickness under different inclined plate rotation angles under ultra-high pressure conditions: (**a**) Oil film thickness distribution with inclined disk rotation of 1°; (**b**) oil film thickness distribution with inclined disk rotation of 10°; (**c**) oil film thickness distribution with inclined disk rotation of 45°; (**d**) oil film thickness distribution with inclined disk rotation of 90°; (**e**) oil film thickness distribution with inclined disk rotation of 110°; (**f**) oil film thickness distribution with inclined disk rotation of 210°.

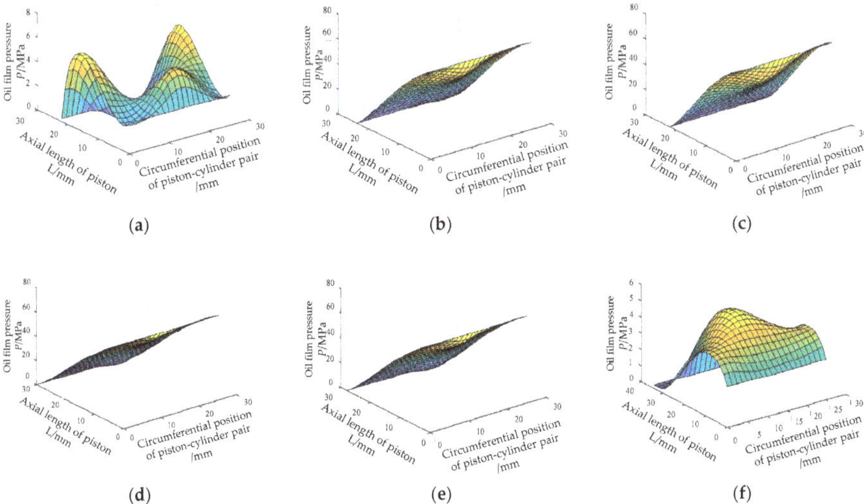

Figure 10. The figure shows the distribution of oil film pressure under different inclined plate rotation angles under ultra-high pressure conditions: (**a**) Oil film pressure distribution with swashplate rotation of 1°; (**b**) oil film pressure distribution with swashplate rotation of 10°; (**c**) oil film pressure distribution with swashplate rotation of 45°; (**d**) oil film pressure distribution with swashplate rotation of 90°; (**e**) oil film pressure distribution with swashplate rotation of 110°; (**f**) oil film pressure distribution with swashplate rotation of 210°.

The change trend of oil film pressure under the condition of ultra-high pressure is the same as that under the condition of normal pressure, but there is a big difference in the value. The oil film pressure under the condition of ultra-high pressure is far greater than that under the condition of normal pressure, In this paper, the oil film gap of the cylinder-piston pair of the ultra-high pressure pump is only 5 μm, so whether the oil film thickness is squeezed to 3 μm under the normal pressure condition or the oil film gap is squeezed to 2 μm under the ultra-high pressure condition, the piston tilt space of the two is not large, so there is no pressure spike phenomenon that the normal pressure piston pump will appear when the swashplate rotates 90°.

The oil film thickness and pressure distribution trend of the piston-cylinder pair under 70 MPa condition are basically the same as that under 20 MPa condition. All of them are as follows: when oil is pressed, the axial force of piston increases with the increase of the angle of inclined plate rotation, which leads to the increase of eccentric load on the piston, the increase of piston tilt, and the increase of oil film pressure; when the swashplate rotates at 90°, the axial velocity and force of the piston are the largest, and the eccentric load of the piston is the most serious. The dangerous area of the oil film thickness is less than 3 μm, and the oil film pressure reaches the maximum at this time; when the swashplate rotates at 125°, the axial force acting on the piston is relieved, the inclination degree of the piston is relatively reduced, and the oil film pressure is also reduced. When the piston chamber is sucking oil, the piston tends to be stable and the oil film pressure distribution is small.

4. Experimental Study on Film Pressure of Piston Pairs in Ultra-High Pressure Axial Piston Pump

Because the oil film gap of piston-cylinder pair is very small, the existence of processing errors and the difficulty of installing pressure sensors lead to the measurement of oil film thickness is difficult and the reliability is low, so this paper only tests the oil hydraulic pressure. In the experiment, we choose the way of reforming the solid pump to analyze the oil film of the cylinder–piston pair, and directly measure the pressure of the oil film pair on the solid pump. When selecting the experimental pressure measurement points, the more intensive the experimental measurement points are selected, the more representative the comparison is with the simulation results. Because of the structure of the pump itself, limited by the installation position, the volume of the sensor itself, and the existence of the oil outlet check valve, only one pressure sensor can be installed on the cylinder body for each plunger. In this paper, three pressure measuring points are actually selected on the cylinder block to measure the axial pressure of three different plungers in the same position. Finally, the pressure measurement results of the three measuring points on the cylinder are compared with the simulation results. Figure 11 is the schematic diagram of the measurement of the oil film pressure of the cylinder-piston pair.

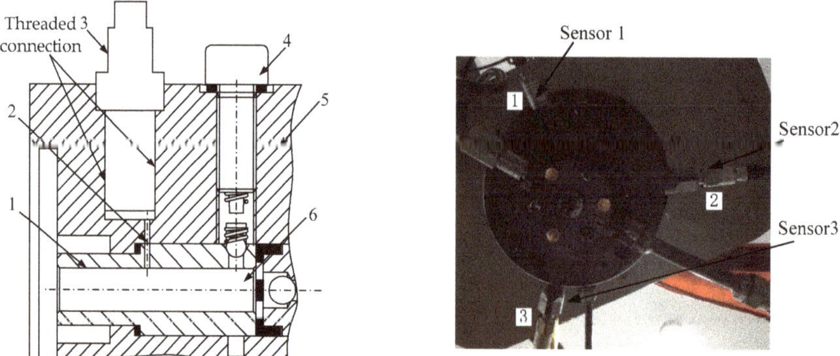

Figure 11. Schematic diagram of plunger auxiliary oil film pressure measurement. 1—plunger sleeve; 2—oil drain hole; 3—pressure sensor; 4—pressure valve screw; 5—cylinder block; 6—plunger cavity.

In this section, an experimental platform is designed for the pump, which can provide different load pressures. The experimental platform can complete the measurement of the oil film pressure of the piston-cylinder pair under different load pressures. The principle of the experimental platform is shown in the Figure 12.

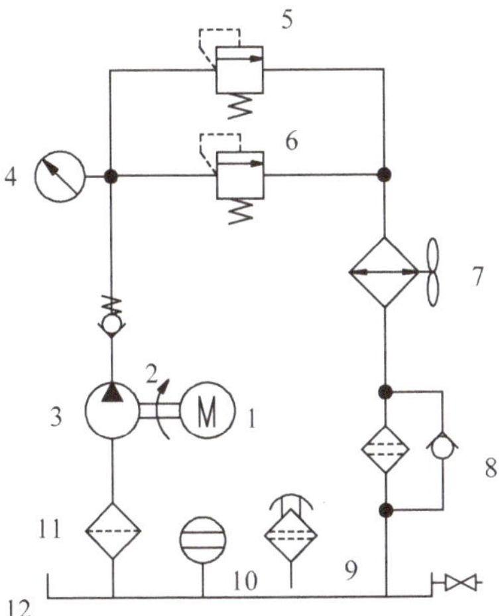

Figure 12. The figure is the schematic diagram of the test bed. 1—motor; 2—coupling; 3—pump; 4—pressure gauge; 5—high pressure relief valve; 6—relief valve (relief valve); 7—air cooler; 8—oil return filter; 9—air filter; 10—level gauge; 11—filter; 12—fuel tank.

The assembled test device is shown in Figure 13.

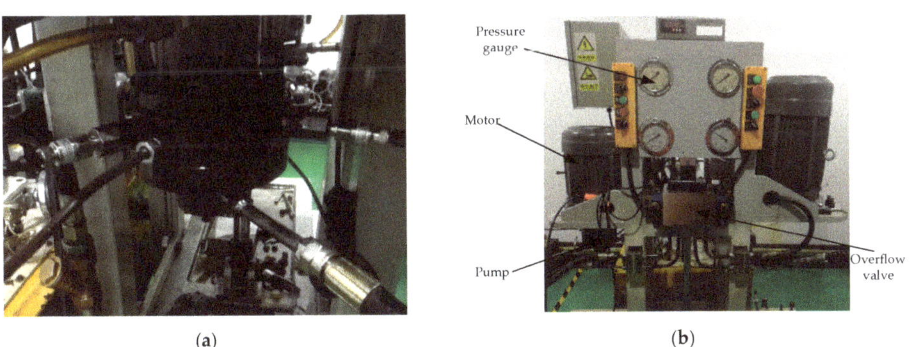

(a) (b)

Figure 13. The figure above shows the experimental object: (**a**) Physical drawing of pump; (**b**) physical drawing of pump test bed.

This paper uses NI industrial computer and multi-functional data acquisition system, pressure sensor, measurement and acquisition software LabVIEW, and data processing and analysis in data acquisition software LabVIEW.

The experimental results are shown in Figure 14.

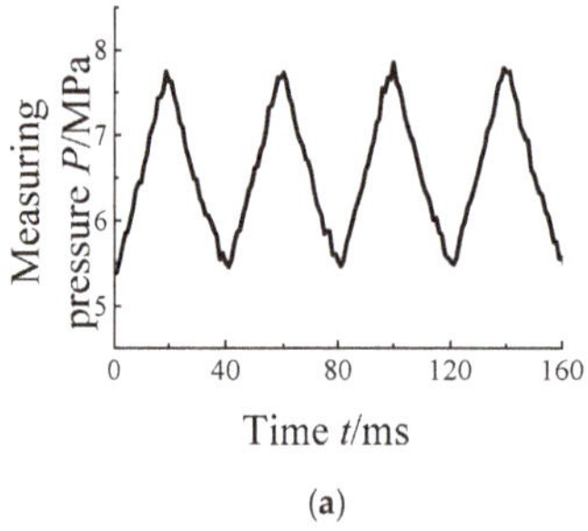
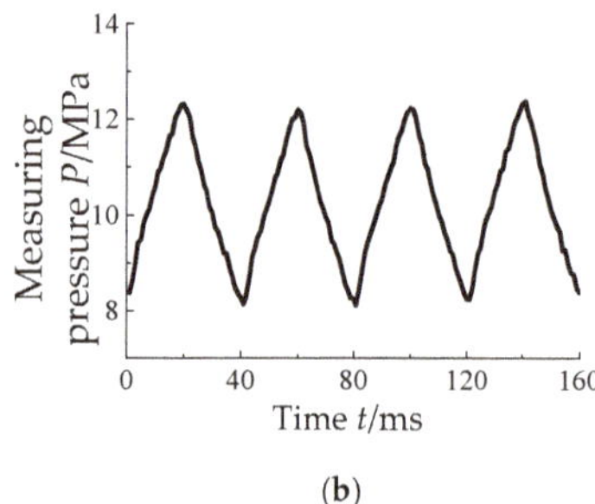

Figure 14. This figure is the experimental result: (**a**) Change of oil film pressure under 20 MPa load; (**b**) change of oil film pressure under 70 MPa load.

In order to better compare the test results with the simulation, this paper also simulates the oil film pressure of the plunger pair under the condition of the rated speed of 1500 r/min and the load pressure of 20 MPa and 70 MPa. According to the position of the measurement point in the test, the pressure at the simulation point is connected into a curve, and the simulation pressure curve is shown in the Figure 15.

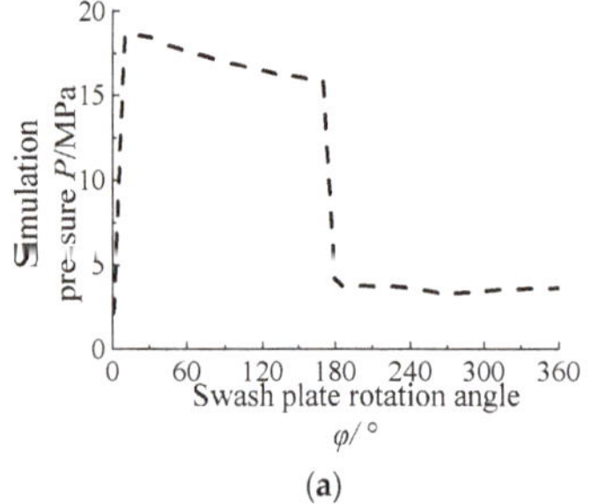
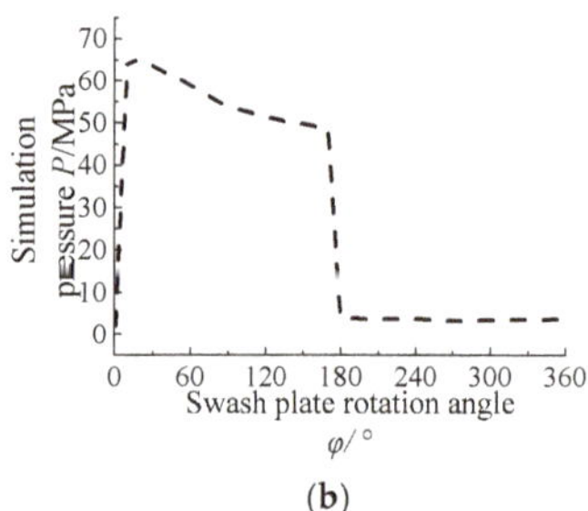

Figure 15. This figure is the simulation result: (**a**) Simulation curve of oil film pressure with load pressure of 20 MPa; (**b**) simulation curve of oil film pressure with load pressure of 70 MPa.

According to test Figure 14 and simulation Figure 15, the following conclusions can be obtained: (1) The change trend of the test pressure curve is basically the same as that of the simulation pressure curve, showing the trend of first increasing and then decreasing; (2) the test pressure curve shows a triangular trend, and the simulation pressure curve shows a trapezoid like trend, that is to say, when the test pressure curve is pressurized, the pressure rises slowly, and at the same time, it is unable to hold the high pressure; (3) with the increase of the load pressure, the pressure values measured in the test are less than the simulation pressure. The larger the load pressure is, the greater the difference between the pressure values obtained in the test and the simulation results is.

The main reason for the above (2) and (3) phenomena is that the oil drawing hole is processed too much. In the measurement process, when the oil reaches the bottom of the threaded hole where the pressure sensor is installed through the oil drain hole, the excessive diameter of the oil drain hole will cause the oil to leak along the wall of the threaded hole, causing a large pressure loss and slow pressure growth. In the process of oil pressure, the larger the load pressure is, the larger the oil leakage through the threaded hole wall of the pressure sensor is, and the larger the pressure measurement error is. Therefore, in the test process, with the increase of load pressure, the measured pressure values are smaller.

5. Conclusions

(1) The attitude of piston in piston chamber under ultra-high pressure is described, and the mathematical model of oil film thickness of piston-cylinder pair is established. The fluid continuity equation and N-S equation are combined, and Reynolds equation is discretized by finite volume method. The mathematical model of oil film pressure of piston-cylinder pair under ultra-high pressure (70 MPa) is obtained.

(2) The dynamic characteristics of oil film of piston-cylinder pair are simulated and analyzed under 20 MPa and 70 MPa respectively. The distribution of oil film pressure and thickness at different rotating angles is obtained. The oil film pressure reaches the maximum when the inclined plate rotates to 90° and the oil film thickness reaches the minimum at this time. Under 70 MPa, the oil film thickness reaches 2 mm when the oil film thickness is the minimum, and at this time the oil film thickness reaches 2 mm. Oil film failure occurs easily when the pressure is higher than 70 MPa.

(3) The dynamic value of the oil film pressure of the piston-cylinder pair was collected and compared with the simulation results. It was found that the change trend of the oil film test pressure curve was basically the same as that of the simulation pressure curve, which increased first and then decreased. However, there is a gap between the measured pressure and the simulation because of the processing error of the oil intake hole in the test process.

Author Contributions: Conceptualization, J.Z.; methodology, J.Z., B.L.; experiments, Q.D., R.L.; data analysis, B.L., Q.Y.; supervision, Q.Y., R.L.; writing—review and editing, J.Z., B.L., R.L., and Q.Y. All authors have read and agreed to the published version of the manuscript.

Funding: This research was supported by the National Natural Science Foundation of China project (51805467).

Conflicts of Interest: The authors declare no conflicts of interest.

References

1. Wang, Z.J. Development Status and Trend of Hydraulic Technology. *J. Appl. Technol.* **2016**, *20*, 85–89.
2. Li, G.B.; Han, L.W.; Han, B. Status and developing trend of hydraulic technology. *J. Jiangsu Sci. Technol. Inf.* **2016**, *24*, 74.
3. Manring, D.; Mehta, S.; Nelson, B.E.; Graf Kevin, J.; Kuehn Jeff, L. Increasing the Power Density for Axial-Piston Swash-Plate. *J. Mech. Des.* **2013**, *135*, 071002. [CrossRef]
4. Xu, S.W. Development Trend of Axial Piston Pump and Motor Hyd. *J. Pneum. Seals* **2003**, *34*, 10–15.
5. Shen, L.L.; Lu, P.F. Research on Friction Performance Study of Piston Pairs of Swash Plate Axial Piston Pump and Port Plate Pairs. *J. Mach. Des. Manuf.* **2014**, *63*, 167–173.
6. Ma, L.Y.; Tian, M.T.; Song, Y.B.; Li, R.W. The Force Analysis and Reliability Study on Friction Pairs for Ram-type Pump. *J. Mod. Manuf. Technol. Equip.* **2016**, *115*, 8–10.
7. Ma, J.M.; Huang, Y.H.; Guo, J.; Shi, Y.Y.; Song, Y.H. Review of Wear Analyses Research for Main Pairss in Hydraulic Axial Piston Pump. *J. Chin. Hydraul. Pneum.* **2017**, *134*, 42–47.
8. Tan, G. Discussion on Ultra-high Pressure Hydraulic Technology in Construction Machinery. *J. Sci. Technol. Inf.* **2014**, *38*, 100–103.
9. Qiao, P.P. Discussed and Application of Super High Pressure Hydraulic Seal Method. *J. Hydraul. Pneum. Seals* **2014**, *34*, 31–35.
10. Yamaguchi, A.; Tanioka, Y. Motion of Pistons in Piston-Type Hydraulic Machines: 1st Report, Theoretical Analysis. *J. Trans. Jpn. Soc. Mech. Eng.* **1975**, *41*, 2399–2406. [CrossRef]
11. Tanaka, K.; Nakahara, T.; Kyogoku, K.; Momozono, S. Oil Whirl of Piston in Axial Piston Pump and Motor. Numerical Calculation under Mixed Lubrication. *J. Trans. Jpn. Soc. Mech. Eng.* **1998**, *64*, 653–661. [CrossRef]
12. Tanaka, K.; Nakahara, T.; Kyogoku, K. Experimental Verification of Oil Whirl of Piston in Axial Piston Pump and Motor. *JSME Int. J. Ser. C Mech. Syst. Mach. Elem. Manuf.* **2002**, *44*, 230–236. [CrossRef]
13. Tanaka, K.; Kyogoku, K.; Nakahara, T. Lubrication Characteristics on Sliding Surfaces between Piston and Cylinder in a Piston Pump and Motor. Effects of Running-In, Profile of Piston Top and Stiffness. *J. Trans. Jpn. Soc. Mech. Eng. C* **1998**, *64*, 3959–3967. [CrossRef]

14. He, X.F.; Zhang, T.H.; Yang, S.D. An Experimental Research Method for Piston Pairss in Water Hydraulic Piston Pump. *J. Hydraul. Pneum. Seals* **2001**, *2*, 7–8.

15. Zhang, B. Study on Virtual Prototype and Pressure Characteristics of Oil Film for Axial Piston Pump. Ph.D. Thesis, Zhejiang University, Hangzhou, China, 2009.

16. Bergada, J.M.; Kumar, S.; Davies, D.L.; Watton, J. A complete analysis of axial piston pump leakage and output flow ripples. *J. Appl. Math. Model.* **2012**, *36*, 1731–1751. [CrossRef]

17. Ivantysynova, M.; Lasaar, R. An Investigation into Micro- and Macrogeometric Design of Piston/Cylinder Assembly of Swash Plate Machines. *J. Int. Fluid Power* **2004**, *5*, 23–36. [CrossRef]

18. Xu, B.; Zhang, J.H.; Yang, H.Y. Simulative analysis of piston-cylinder pairs of axial piston pump based on virtual prototype. *J. Lanzhou Univ. Technol.* **2010**, *36*, 31–37.

19. Lü, F.; Xu, B.; Zhang, J.H. Simulative Analysis of Piston Posture and Piston/Cylinder Interface Leakage of EHA Pumps by the Influence of Rotating Speed. *J. Mech. Eng.* **2018**, *54*, 123–130.

Article

Numerical and Experimental Study on the Internal Flow of the Venturi Injector

Hao Li [1,2], Hong Li [1,*], Xiuqiao Huang [2], Qibiao Han [2], Ye Yuan [1] and Bin Qi [3,*]

[1]   Research Center of Fluid Machinery Engineering and Technology, Jiangsu University,
       Zhenjiang 212013, China; leehao.caas@gmail.com (H.L.); student_yy@126.com (Y.Y.)
[2]   Farmland Irrigation Research Institute, Chinese Academy of Agricultural Sciences, Xinxiang 453002, China;
       huangxq626@126.com (X.H.); hanbiaoedu@126.com (Q.H.)
[3]   School of Agricultural Engineering and Food Science, Shandong University of Technology,
       Zibo 255049, China
*   Correspondence: hli@ujs.edu.cn (H.L.); qibin@sdut.edu.cn (B.Q.)

Received: 30 October 2019; Accepted: 27 December 2019; Published: 2 January 2020

Abstract: To study the appropriate numerical simulation methods for venturi injectors, including the investigation of the hydraulic performance, mixing process, and the flowing law of the two internal fluids, simulations and experiments were conducted in this study. In the simulations part, the cavitation model based on the standard k–ε turbulence and mixture models was added, after convergence of the calculations. The results revealed that the cavitation model has good agreement with the experiment. However, huge deviations occurred between the experimental results and the ones from the calculation when not considering the cavitation model after cavitation. Thus, it is inferred that the cavitation model can exactly predict the hydraulic performance of a venturi injector. In addition, the cavitation is a crucial factor affecting the hydraulic performance of a venturi injector. The cavitation can ensure the stability of the fertilizer absorption of the venturi injector and can realize the precise control of fertilization by the venturi injector, although it affects the flow stability and causes energy loss. Moreover, this study found that the mixing chamber and throat are the main areas of energy loss. Furthermore, we observed that the internal flow of the venturi injector results in the majority of mixing taking place at the diffusion and outlet sections.

Keywords: venturi injector; cavitation; numerical investigation; mixing process; internal flow

---

1. Introduction

Fertigation is becoming increasingly common, and fertilizer devices are becoming increasingly important [1]. A venturi injector, a commonly used device for fertilizer application, uses the turbulent diffusion of the jet to transfer energy and mass. This injector is broadly applied in fertigation systems because of its advantages such as simple structure, convenient operation, low cost, and no need for external power [2–4]. However, the internal flow involves the mixing of two flows with different pressures, although the internal structure of the venturi injector is simple with no moving parts. Thus, the internal flow is complex, energy loss is large, and the mass transfer energy efficiency is low [5], making it necessary to assess the flow characteristics of the venturi injector.

To date, venturi injectors have been studied widely, especially focusing on fertilizer absorption performance [6–8]. Neto and Porto [9] observed that the area ratio of a venturi injector exerts a major impact on the fertilizer suction efficiency; the authors also presented a simple methodology for the design and construction of low-cost ejectors from PVC, to reduce costs and enhance the fertilizer suction performance of fertigation systems. Ozkan et al. [10] investigated venturi injectors' structure parameters, including the impact of the inlet diameter, the diameter of the suction pipe, and the ratio of the throat diameter to the inlet diameter. In a field experiment, Parish et al. [11] investigated the injection methods, injection rate, and solution volume on the fertigation uniformity and reported that venturi injectors have a better fertilizer distribution and that the injection rate exerts a significant impact on the fertilizer distribution uniformity. Li et al. [12] investigated the performance of three different fertilization devices (venturi injector, proportional pump, and differential pressure tank) in laboratory and field experiments of a micro-irrigation system; they reported that the type of fertilization device and the manufacturing variability of emitters exert a considerable impact on the fertilizer distribution uniformity.

Using dynamics theory [13,14] for model analysis and modern testing and signal processing technology [15,16] has become a common research tool these days. With the technological advancement of computers and computational fluid dynamics (CFD), complex flows that could previously only be acquired by experimental methods can be simulated precisely [17–19], especially the internal flow fields in fluid machinery and microfluidics [20,21], such as two-phase flow [22–24], pressure fluctuation [25], energy loss [26–28], and so on. Other studies have mainly focused on the flow inside venturi injectors. Huang et al. [29,30] numerically analyzed the influence of the structural parameters on the absorption capacity. Yan et al. [31] used a high-speed video camera to investigate the development of the cavitation inside a venturi injector. Zwart et al. [32] presented a new multiphase flow algorithm to predict cavitation and validate the transient cavitation in a venturi. Simpson and Ranade [33] developed CFD models to simulate the cavitating flow in various venturi injectors. Shi et al. [34] established a semi-empirical model to predict cavitation in different venturi injectors. Furthermore, Dastane et al. [35] developed a CFD modeling scheme to successfully simulate flows in a cavitating venturi. Various study reports on venturi injectors revealed that CFD methods have been used extensively to investigate the impact of key structure and working parameters on the performance, including the diffusion, shape of the nozzle, ratio of the throat length to diameter, and contraction ratio [36–39]. However, few studies have focused on the effect of the mixing process between water and fertilizer liquid. Of note, the mixing process seriously affects the uniformity of the water and fertilizer distribution in the irrigation system, necessitating further investigation. Hence, this study aims to extend the solution method and test the reliability of calculated models. Moreover, this study investigates the mixing of two flows with different pressures.

2. Experimental Setup

We studied the working process and the fertilizer suction/hydraulic performance in a venturi injector using water as a working fluid. In this study, a closed-loop system was considered to assess the venturi injector. The system contained water circulation and measuring subsystems. Figure 1 presents the configuration of the closed-loop system.

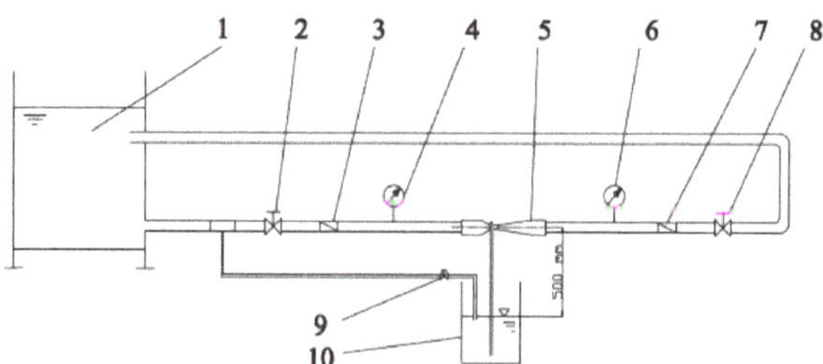

Figure 1. Schematic of the experimental system. **1.** Variable-frequency, constant-pressure water supply device; **2.** Valve 1; **3.** Turbine flowmeter; **4.** Pressure gauge; **5.** Venturi injector; **6.** Pressure gauge; **7.** Turbine flowmeter; **8.** Valve 2; **9.** Valve 3; **10.** Tank.

In the experiment, water was driven by a variable-frequency, constant-pressure water supply device. We mounted two pressure gauges (precision: 0.4%) on the inlet and outlet lines of the venturi injector to measure the local pressures accurately. Likewise, the flow rates at the inlet and outlet lines of the venturi injector were measured by two turbine flowmeters (precision: 0.2%). In addition, valves of the main pipeline were used to regulate the flow rate of the experimental system and control the import and export pressures of the venturi injector. The tank's water level was maintained constant to isolate the suction flow rate from the water level impact. Accordingly, a water pipe was set from the main pipeline to the water tank. Notably, the vertical distance between the water level and the venturi injector axis was 500 mm.

3. Analysis Model

Figure 2 shows the configuration and dimensions of the venturi injector; it contains the following seven parts: entrance, exit, contraction section, throat, mixing chamber, diffuser, and suction chamber. Table 1 presents the geometric parameters of the venturi injector.

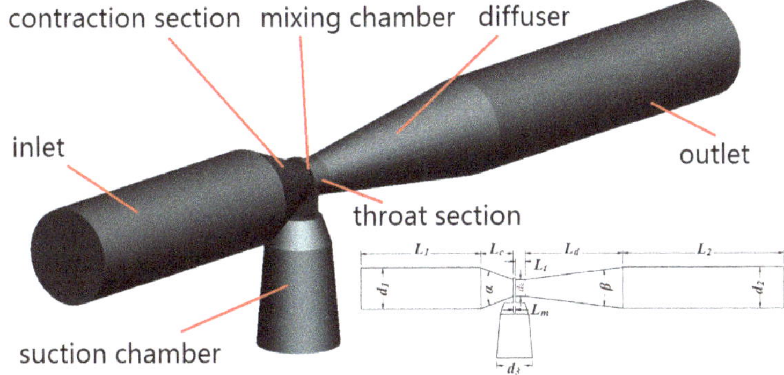

Figure 2. Configuration of the venturi injector.

Table 1. Basic geometric parameters of the venturi injector.

Geometric Parameter	Value	Geometric Parameter	Value
Inlet diameter d_1 (mm)	20	Inlet length L_1 (mm)	59.5
Outlet diameter d_2 (mm)	20	Outlet length L_2 (mm)	80
Throat diameter dt (mm)	8	Contraction section length L_c (mm)	16
Suction diameter d_3 (mm)	17.5	Mixing chamber length L_m (mm)	1
Contraction angle α (°)	41	Throat length L_t (mm)	5
Diffusion angle β (°)	14	Diffuser length L_d (mm)	48.5

4. Mesh and Boundary Conditions

4.1. Mesh Independence Study

To ensure a mesh-independent solution, we used four mesh sizes for the simulation. Table 2 shows the results under the $P_1 = 350$ kPa and $P_2 = 100$ kPa flow conditions. We found that the solution was mesh-independent beyond 400K elements; hence, this mesh was used for all further simulations.

Table 2. Grid independence test of the venturi injector under the flow conditions of $P_1 = 350$ kPa and $P_2 = 100$ kPa.

Grid No.	1	2	3	4
Number of grid cells/10^4	20	30	40	60
$Q_1/(\text{m}^3/\text{h})$	4.170418	4.200837	4.306605	4.320276
$Q_s/(\text{m}^3/\text{h})$	0.166950	0.186685	0.239519	0.243264

4.2. Comparison of Turbulence Models

We constructed four turbulence models—standard k–ε model, RNG k–ε model, realizable k–ε model, and standard k–ω model—to select the optimum turbulence model for the interflow simulations and compared their results with the experimental results to arrive at the most suitable turbulence model. Table 3 presents numerical and experimental results of the venturi with different turbulence models under the flow conditions of P_1 = 350 kPa and P_2 = 100 kPa. We found that the standard k–ε model offered better accuracy among these models; hence, this model was selected in this study.

Table 3. Numerical and experimental results with different turbulence models under P_1 = 350 kPa and P_2 = 100 kPa flow conditions.

Grid No.	Standard k–ε	RNG k–ε	Realizable k–ε	Standard k–ω	Test Value
$Q_1/(m^3/h)$	4.306605	4.322436	4.305457	4.352721	4.32
$Q_s/(m^3/h)$	0.239519	0.186685	0.18506	0.165261	0.21

4.3. Boundary Conditions

Structured meshing with hexahedral elements was generated using ANSYS ICEM CFD software (ANSYS, Inc., Commonwealth of Pennsylvania, USA) to calculate the flow of the computation domain more effectively. In addition, numerical simulations for the venturi injector were executed in ANSYS FLUENT 17.2 software (ANSYS, Inc., Commonwealth of Pennsylvania, USA). The separation solver and the absolute velocity formula were used to deal with the steady-state problem. The conventional fluid dynamics were computed by solving the Reynolds-averaged Navier–Stokes equations. We applied the SIMPLE algorithm to solve the coupling velocity and pressure problems. Furthermore, the finite volume method was used to discretize the governing equations.

In addition, we used the standard k–ε turbulence model in the simulations. The mixture model was selected as a multiphase model to assess the flow law of the fertilizer solution in the working process of the venturi injector. Besides this, the Zwart–Gerber–Belamri model, based on the Rayleigh–Plesset equation, was added after the numerical calculation was convergent to attain a more accurate and appropriate simulation method. Moreover, non-slip wall boundary conditions were applied. The pressure inlet and pressure outlet boundary conditions were applied at the two inlets (Inlet 1 and Inlet 2 denote the inlet and suction chamber, respectively) and outlet boundaries. In the mixture model, the calculation medium is Newtonian fluid, and the water liquid model was applied for the primary phase, while the urea liquid model was applied for the secondary phase. Residuals of $<10^{-5}$ were set as the convergence criterion. Based on the experiment, different pressures for Inlet 1 were set in the simulations. Different outlet pressures corresponding to different pressure points were set at the outlet to assess the hydraulic performance under different working conditions of the venturi injector. Figure 3 shows grid details and boundary conditions for the whole domain.

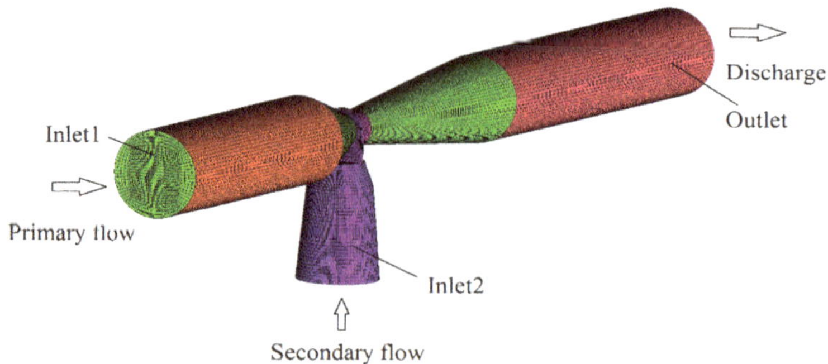

Figure 3. Computational grids and boundary conditions.

5. Results and Discussion

5.1. Comparison of the Numerical and Experimental Results

In this study, the experiments were conducted at the Center of Water-Saving Irrigation Equipment Quality Inspection, Ministry of Water Resources. We considered four inlet pressures in the experiment (range: 50–350 kPa). Meanwhile, numerical calculations for the venturi injectors were implemented with the same working conditions. Figure 4 presents the results obtained from the experiment and numerical simulation. In this study, we conducted two sets of numerical simulations. The cavitation model was applied in the first simulation, while the other simulation was performed without the cavitation model.

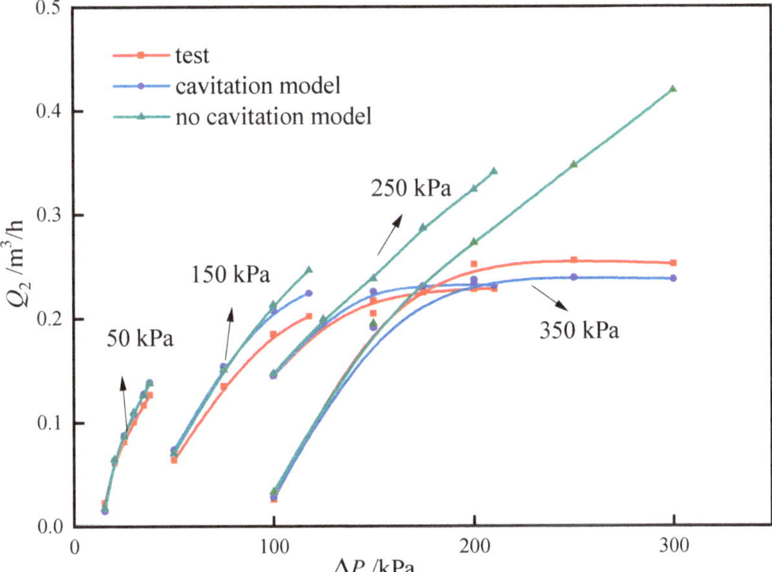

Figure 4. Comparison of experimental and numerical simulation results.

Typically, applying the cavitation model results in predictions which are in better agreement with the results from the experiment. It is feasible to use CFD simulation based on the mixture model to estimate the performance of a venturi injector, whereas the cavitation model could make the simulation results more reliable.

The numerical and experimental results corroborate each other well for cases with low inlet pressure (i.e., $P_1 = 50$ kPa) or low pressure difference. However, in other conditions, the deviations between experimental values and calculations are large due to ignoring cavitation. Moreover, each performance curve of the simulation that considered the cavitation model had a bending point; they all tended to flatten when the amount of absorbed fluid reached approximately 0.2 m^3/h after the bending point; this is because when the venturi injector works with a large inlet pressure and pressure difference, the cavitation is more intense. Affected by cavitation, the mixing of the working fluid and the absorbed liquid becomes increasingly complicated. Of note, cavitation occurs when the internal local pressure is lower than the saturated vapor pressure of the working fluid at the corresponding temperature. The occurrence of cavitation leads to stability in the amount of fertilizer absorbed by the venturi injector applicator. Thus, the amount of fertilizer absorbed can be controlled in cavitation conditions to provide precise fertilization in terms of quantity.

5.2. Analysis of the Mixing Process

We used two parameters to indicate the performance and thereby study the mixing process of two fluids in a venturi injector. The parameters were the local pressure ratio (p_x) and the flow ratio (q), which are defined as follows:

$$p_x = \frac{P(x) - P_s}{P_1 - P_s}$$

$$q = \frac{Q_s}{Q_1}$$

where x is the position along the flow and $P(x)$, P_1, P_s, Q_s, and Q_1 are the local pressure at x section, inlet pressure, suction inlet pressure, suction flow, and workflow, respectively.

The local pressure ratio (p_x) is the ratio between energy surplus with the absorbed fluid and the energy loss with the working fluid at the x cross section. In addition, the flow ratio (q) presents the ratio of the suction flow to the working flow.

We used five different working conditions at P_1 = 350 kPa in the calculations to study the characteristics of the internal flow field of the venturi injector upstream and downstream of the cavitation zone. Table 4 shows the performance obtained from the numerical simulation with the cavitation model in different flow conditions (i.e., P_2 = 50, 100, 150, 200, and 250).

Table 4. Simulation results under conditions of P_1 = 350 kPa.

P_1 (kPa)	P_2 (kPa)	ΔP (kPa)	Q_1 (m^3/h)	Q_s (m^3/h)	q
	50	300	4.3066	0.2376	0.0552
	100	250	4.3066	0.2396	0.0556
350	150	200	4.3057	0.2372	0.0551
	200	150	4.0651	0.1912	0.0470
	250	100	3.7182	0.0284	0.0076

5.3. Development of Cavitation Distributions

Notably, the occurrence of cavitation depends on the local pressure. We assessed the cavitation creation for fixed inlet pressure P_1 = 350 kPa, as shown in Figure 5. No cavitation occurred under the pressure conditions of P_2 = 250 and 200 kPa. The cavitation primarily occurred in the throat and diffuser sections. In addition, the cavitation intensity increased as the outlet pressure decreased. Figure 5 shows that the most intense cavitation takes place for P_2 = 50 kPa.

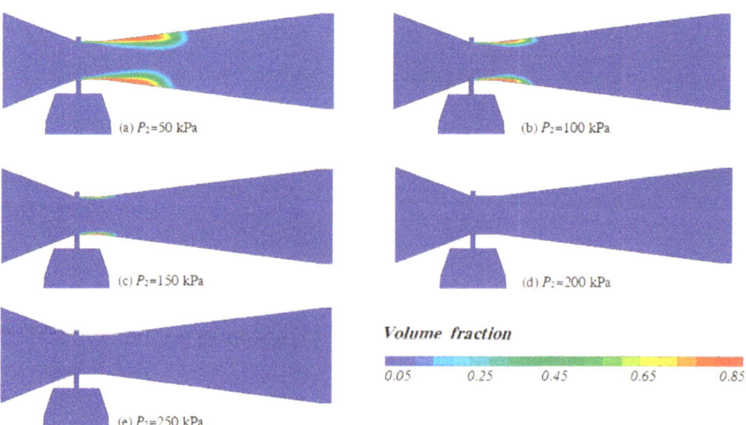

Figure 5. Cavitation contours in the z = 0 section.

Table 2 shows that the flow ratio exhibited a small change when cavitation occurred for the pressure points of $P_2 = 50$, 100, and 150 kPa. These changes were similar for $P_2 = 50$–150 kPa, and there was a peak in the operating point under 100 kPa. Thus, the suction flow and flow ratio exhibited a minor fluctuation when the cavitation was created. Notably, the cavitation was critical to the suction flow and flow ratio, where both increased as the cavitation degree increased. Conversely, both parameters decreased when the cavitation reached a certain extent.

5.4. Development of Pressure and Velocity Distributions

We simulated the pressure and velocity distributions to analyze the flow structure inside the venturi injector. Figures 6 and 7 show the pressure and velocity distributions of the venturi injector at the $z = 0$ section for $P_1 = 350$ kPa. Figure 6 shows that the pressure decreased at the beginning but then increased along the flow direction. Negative pressure was formed at the throat, where the pressure reached its minimum value. Figure 6a–c indicates that for cases with a high pressure difference, low pressure areas were present in the throat, diffuser, and mixing chamber, where the suction liquid enters the venturi injector; in other words, there was pressure loss at these places, proving the distribution of cavitation (Figure 5).

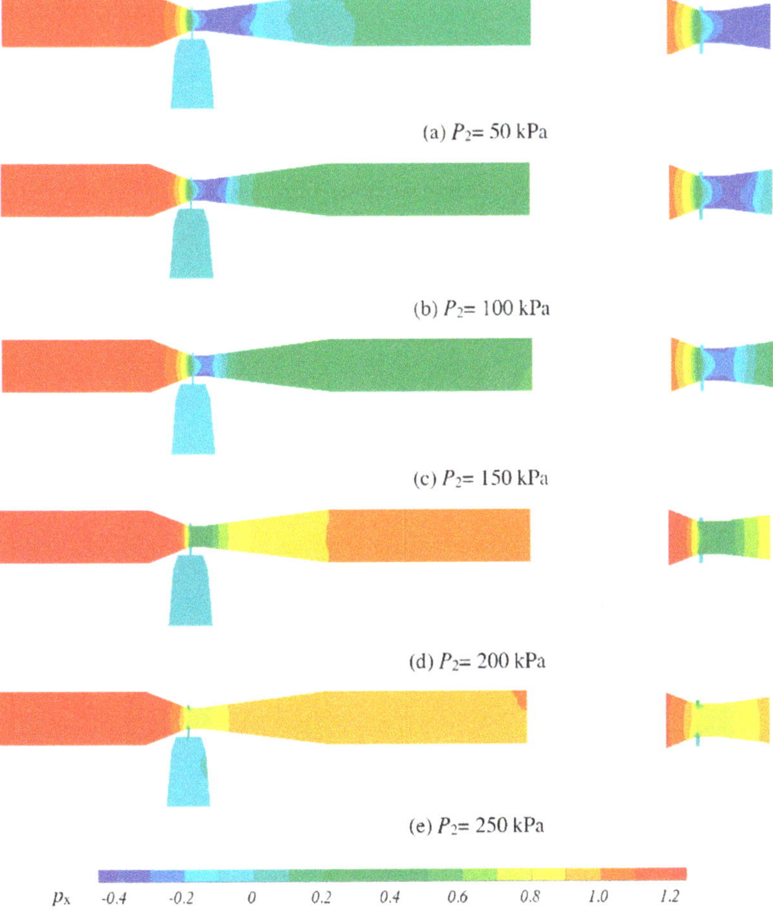

Figure 6. Local pressure ratio contours in the $z = 0$ section.

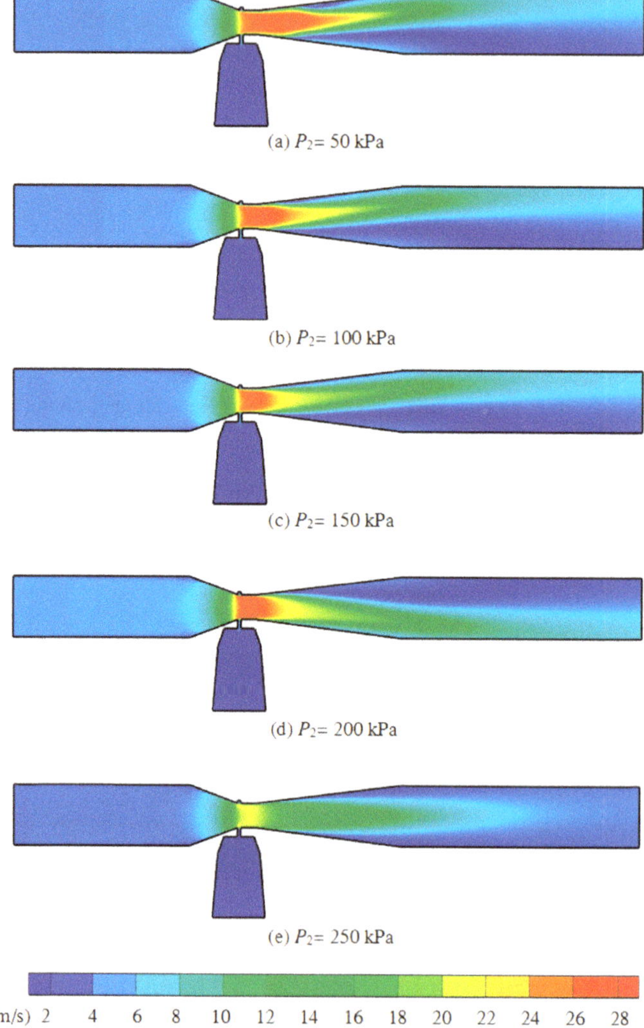

Figure 7. Velocity contours in the $z = 0$ section.

Figure 7 illustrates that the velocity changed abruptly from the contraction to diffuser sections; it increased steadily in the contraction section and reached its peak at the throat. During flow in the diffuser and outlet, a reduction trend was observed owing to the impact of the suction flow and the variability of the flow cross-sectional area. In addition, the velocity presented different variation trends under different outlet pressures.

Of note, the velocity distribution is symmetric only for the outlet pressure $P_2 = 250$ kPa. For lower outlet pressures ($P_2 \leq 150$ kPa), the velocity near the upper wall was higher than that in the other areas of the diffuser and outlet; however, this was reversed for $P_2 = 200$ kPa.

Figure 8 presents the streamlines in the $z = 0$ section for different outlet pressures; it shows that the flow was steady only for the outlet pressure $P_2 = 250$ kPa, and there were vortices at other outlet pressures. Among these conditions, the difference is the position of the vortex; the presence of vortices near the side of the suction tubes for low outlet pressures is contrary to their absence at higher outlet pressure. This could be attributed to the nonuniformity of the flow and the velocity distribution after the throat section. Conversely, when the second fluid was absorbed into the working fluid, an energy conversion occurred between them, which led to the different velocity field near the suction side of the diffuser compared with those in other regions. Thus, the vortex was formed in a different area, suggesting that the formation of a vortex inevitably leads to energy loss. This phenomenon correlates with the structure of the suction chamber. Owing to the lateral inlet, the suction fluid flows around the nozzle. In addition, the flow at the throat entrance is absorbed asymmetrically into the throat, which then becomes different in different regions. Thus, a symmetric suction chamber is necessary and preferred.

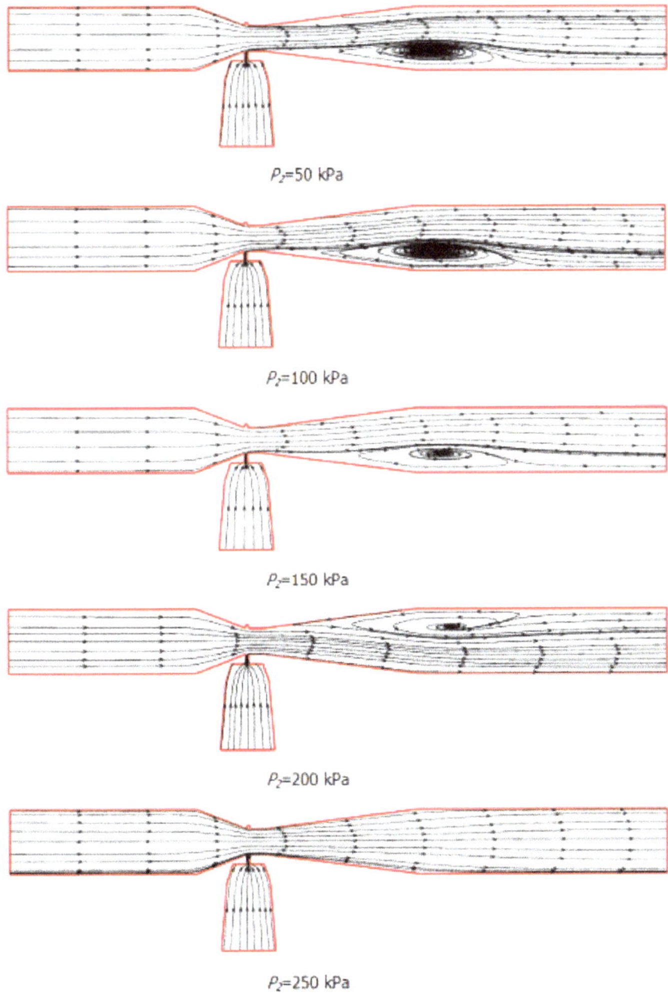

Figure 8. Streamlines in the $z = 0$ section.

5.5. Analysis of the Local Pressure Ratio

Figure 9 illustrates the local pressure ratio along the length of the venturi injector under different outlet pressure conditions, suggesting that the local pressure ratio increased gradually as the outlet pressure increased. In addition, the curves revealed that the local pressure ratio began to decrease in the contraction section. We observed that the local pressure ratio markedly decreased in the throat, as the outlet pressure decreased. Furthermore, the curves were almost coincident in the throat, and the outlet pressure was <150 kPa. With the mixing of two flows, the energy of the suction fluid increased gradually. Finally, in the diffuser, the energy obtained by the suction flow reached a maximum in the unit volume; in other words, the pressure ratio of the mixture reached a peak value and then tended to be stable.

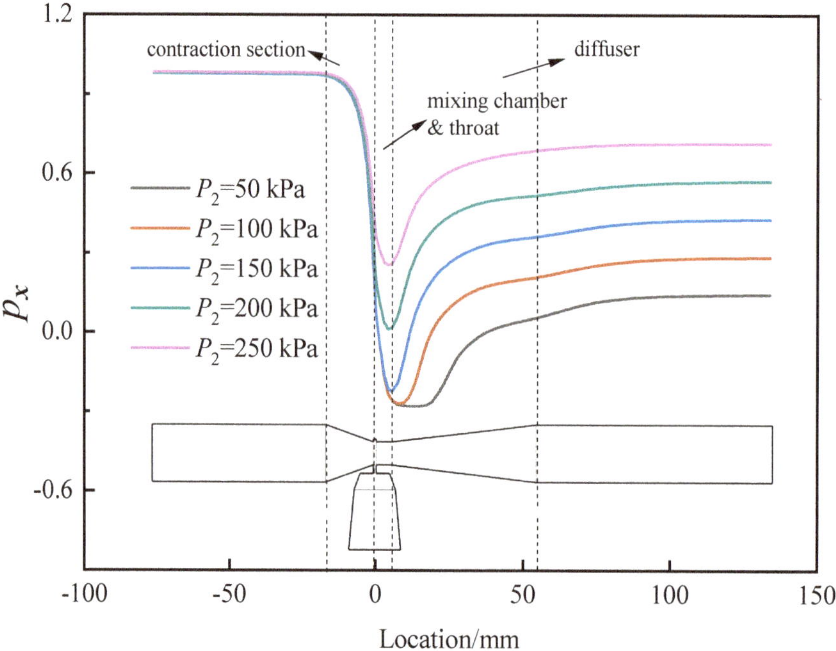

Figure 9. Local pressure ratios at different flow ratios.

After contrasting with Figure 6, we found that cavitation occurred for small outlet pressures. The cavitation region occupied a wide range of the diffuser, and the phenomenon of cavitation was higher when the outlet pressure was 50 kPa. Hence, we infer a certain impact of cavitation on fertilizer absorption performance.

5.6. Development of the Turbulent Kinetic Energy

Reportedly, the turbulent kinetic energy is a key parameter to illustrate cavitation [40]. When cavitation occurs, there is strong disturbance in the flow, and the turbulent kinetic energy increases rapidly. In addition, the variation of the turbulent kinetic energy depicts the cavitation intensity [41].

Figure 10 describes the trend of the turbulent kinetic energy at different outlet pressures along the venturi injector, also illustrating that the turbulent kinetic energy remained constant from the contraction to the throat sections, while it increased in the diffuser section, where the position of the turbulent kinetic energy was approximately the same as that of the cavitation. We observed that the turbulent kinetic energy reached its maximum value after a sharp increase in the diffuser, but then gradually decreased to its ultimate value (about 2 m^2 s^{-2}) in the outlet. Furthermore, the effect of the turbulent kinetic energy was enhanced as the pressure difference increased, which could be attributed to the presence of cavitation in this region.

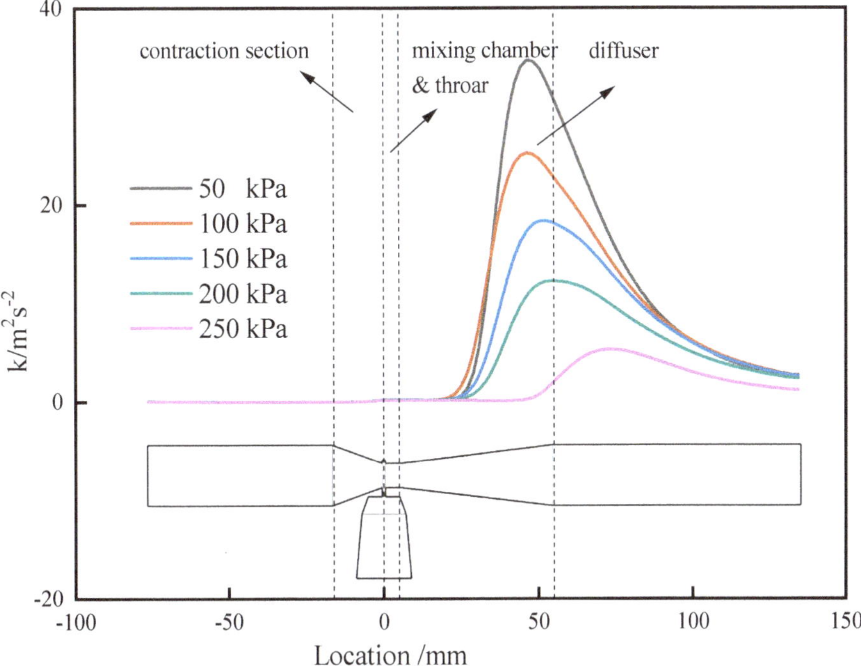

Figure 10. Turbulent kinetic energy at different outlet pressures.

5.7. Development of the Volume Fraction of the Suction Flow

Figure 11 illustrates the development of the volume fraction of the suction flow in different conditions. Accordingly, the curves present three distinct trends in different conditions along the venturi injector. Initially, there were plenty of similarities in the curves for low outlet pressure. The volume fractions showed a considerable increase in the diffuser, and their rate of increase decreased in the outlet but reached a high point at 3.5%. Moreover, the curve exhibited the same trend in the diffuser for outlet pressure equal to 200 kPa; however, the volume fraction tended to rise first but then decrease in the outlet. Finally, Figure 11 shows that the volume fraction only grew after the diffuser for $P_2 = 250$ kPa.

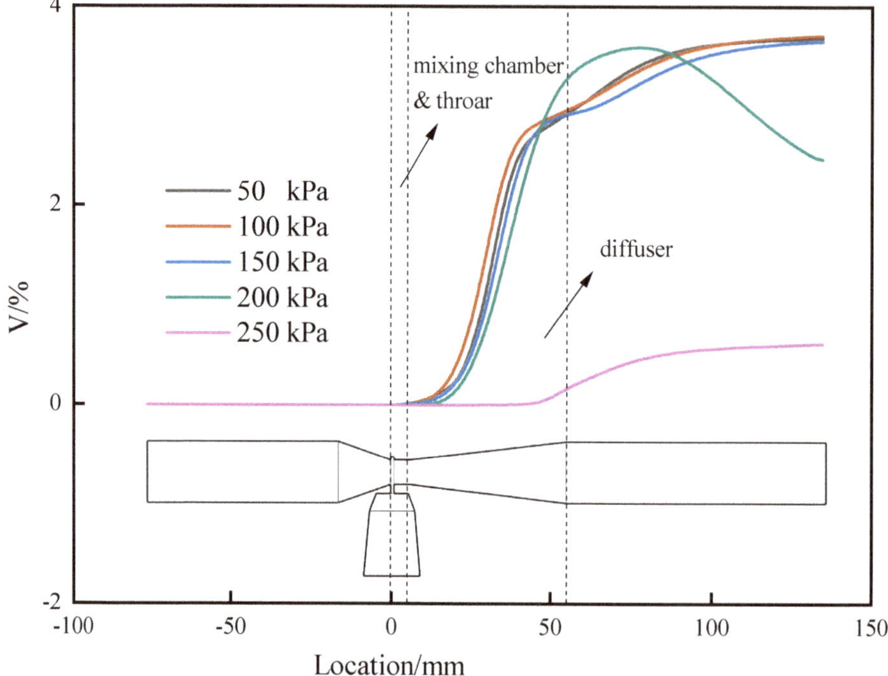

Figure 11. The volume fraction of suction flow at different outlet pressures.

Accordingly, the flow inside the venturi injector is a non-negligible factor. The flow did not completely mix in the throat because of its large velocity. Furthermore, the variation in the velocity gradient in the diffusion section made the suction flow spread rapidly in the working flow. Thus, the velocity and flow ratio can be considered to be the major factors that affect the mixing speed.

6. Conclusions

This study obtained the following conclusions. Numerical simulation can precisely depict the hydraulic performance of a venturi injector. To ensure the reliability of numerical results, it is key to select a suitable calculation model based on the flow characteristics of the internal flow field. For cases with cavitation, the results calculated without considering the cavitation model presented a significant error when compared with the experimental results, where the maximum error exceeded 50%. However, the largest deviation between the calculated results obtained from the cavitation model and the experiment results was just 3%. Thus, CFD methods with an appropriate computational model can be used to reliably estimate the hydraulic performance and the flow field of a venturi injector. This study offers information that could be used to perfect the numerical simulation method for venturi injectors.

The internal flow field analysis revealed that cavitation is a key factor affecting the hydraulic performance of the venturi injector. The occurrence of cavitation makes the flow of the absorbed liquid remain steady at around 0.23 m³/h for different working pressures. In conclusion, cavitation might enhance the performance of a venturi injector in applications such as agricultural irrigation and engineering. In addition, the flow field inside the venturi injector could be adjusted according to the inlet condition. For fixed entrance conditions, the suction flow and the flow ratio increase as the pressure difference increases. However, the two flows do not mix well inside the throat because the flow ratio value is relatively small. When liquid enters the diffuser, the flow speed decreases.

Meanwhile, the complex flow in the mixing chamber and the throat, as well as the flow area of the abrupt change, results in a drastic change in the velocity gradient, causing the reduction of the local pressure ratio. Hence, the main energy loss correlates with the region.

Author Contributions: H.L. (Hao Li) and H.L. (Hong Li) conceived and designed the experiments; Y.Y. and Q.H. performed the experiments and simulation; B.Q. analyzed the data; H.L. (Hao Li) wrote the paper; X.H. funding acquisition. All authors have read and agreed to the published version of the manuscript.

Funding: This research was funded by National Key R&D Program of China (2016YFC0400202, 2016YFC0400104) and Central Public-Interest Scientific Institution Basal Research Fund (Y2019PT72, FIRI2016-19, FIRI2016-22, FIRI2017-22).

Conflicts of Interest: The authors declare no conflict of interest.

References

1. Ryan, J.; Saleh, M. Use of phosphorus fertilizers in pressurized irrigation systems: Problems and possible solutions. In *Natural Resource Management Program*; Alexandria University: Alexandria, Egypt, 1998.
2. Kong, L.; Fan, X. Analysis on the influencing factors of throat negative pressure for Venturi injector. *Agric. Res. Arid Areas* **2013**, *31*, 78–82.
3. Li, J.; Hong, T.; Feng, R.; Yue, X.; Luo, Y. Design and experiment of Venturi variable fertilizer apparatus based on pulse width modulation. *Trans. Chin. Soc. Agric. Eng.* **2012**, *28*, 105–110.
4. Jin, Y.; Xia, C.; Fang, B. Research and Development of Venturi Fertilizer Applicator Series. *China Rural Water Hydropower* **2006**, *5*, 14–16.
5. Fan, X.; Kong, L. Relationship of energy conversion for Venturi injector. *J. Drain. Irrig. Mach. Eng.* **2013**, *31*, 528–533.
6. Han, Q.; Huang, X.; Liu, H.; Wu, W.; Fan, Y. Comparative Analysis on Fertilization Performance of Six Venturi Injectors. *Trans. Chin. Soc. Agric. Mach.* **2013**, *44*, 113–117.
7. Yan, H.; Chu, X.; Wang, M.; Ma, Z. Injection performance of Venturi injector in micro-irrigation system. *J. Drain. Irrig. Mach. Eng.* **2010**, *28*, 251–255, 264.
8. Feitosa Filho, J.C.; Botrel, T.A.; Pinto, J.M. Performance of the venturi type injectors joined to pipe under free and pressurized flow. *Eng. Agric.* **1998**, *17*, 20–35.
9. Lima Neto, I.E.; de Melo Porto, R. Performance of Low-Cost Ejectors. *J. Irrig. Drain. Eng.* **2004**, *130*, 122–128. [CrossRef]
10. Ozkan, F.; Ozturk, M.; Baylar, A. Experimental investigations of air and liquid injection by venturi tubes. *Water Environ. J.* **2006**, *20*, 114–122. [CrossRef]
11. Parish, R.L.; Rosendale, R.M.; Bracy, R.P. Fertigation Uniformity Affected by Injector Type. *Horttechnology* **2003**, *13*, 103–105.
12. Li, J.; Meng, Y.; Li, B. Field evaluation of fertigation uniformity as affected by injector type and manufacturing variability of emitters. *Irrig. Sci.* **2007**, *25*, 117–125. [CrossRef]
13. Ye, S.; Zhang, J.; Xu, B.; Zhu, S. Theoretical investigation of the contributions of the excitation forces to the vibration of an axial piston pump. *Mech. Syst. Signal Process* **2019**, *129*, 201–217. [CrossRef]
14. Zhu, Y.; Qian, P.; Tang, S.; Jiang, W.; Li, W.; Zhao, J. Amplitude-frequency characteristics analysis for vertical vibration of hydraulic AGC system under nonlinear action. *AIP Adv.* **2019**, *9*, 035019. [CrossRef]
15. Zhu, Y.; Tang, S.; Wang, C.; Jiang, W.; Yuan, X.; Lei, Y. Bifurcation Characteristic research on the load vertical vibration of a hydraulic automatic gauge control system. *Processes* **2019**, *7*, 718. [CrossRef]
16. Zhu, Y.; Tang, S.; Quan, L.; Jiang, W.; Zhou, L. Extraction method for signal effective component based on extreme-point symmetric mode decomposition and Kullback-Leibler divergence. *J. Braz. Soc. Mech. Sci. Eng.* **2019**, *41*, 100. [CrossRef]
17. Yan, H.; Chen, Y.; Chu, X.; Xu, Y.; Wang, Z. Effect of structural optimization on performance of Venturi injector. In *Conference Series: Earth and Environmental Science*; IOP Publishing: Bristol, UK, 2012; Volume 15, p. 072014.
18. Sun, Y.; Niu, W. Simulating the Effects of Structural Parameters on the Hydraulic Performances of Venturi Tube. *Model. Simul. Eng.* **2012**, *2012*, 1–7. [CrossRef]
19. Baylar, A.; Aydin, M.C.; Unsal, M.; Ozkan, F. Numerical modeling of venturi flows for determining air injection rates using FLUENT V6. 2. *Math. Comput. Appl.* **2009**, *14*, 97–108.

20. Qian, J.Y.; Chen, M.R.; Liu, X.L.; Jin, Z.J. A numerical investigation of the flow of nanofluids through a micro Tesla valve. *J. Zhejiang Univ. Sci. A* **2019**, *20*, 50–60. [CrossRef]
21. Nguyen, T.; Devaraj, V.D.M.; Albert, V.D.B.; Eijkel, J.C.T. Investigation of the effects of time periodic pressure and potential gradients on viscoelastic fluid flow in circular narrow confinements. *Microfluid. Nanofluid.* **2017**, *21*, 37. [CrossRef]
22. Wang, C.; Hu, B.; Zhu, Y.; Wang, X.; Luo, C.; Cheng, L. Numerical study on the gas-water two-phase flow in the self-priming process of self-priming centrifugal pump. *Processes* **2019**, *7*, 330. [CrossRef]
23. Wang, C.; He, X.; Zhang, D.; Hu, B.; Shi, W. Numerical and experimental study of the self-priming process of a multistage self-priming centrifugal pump. *Int. J. Energy Res.* **2019**, *43*, 4074–4092. [CrossRef]
24. Lu, Y.; Zhu, R.; Wang, X.; Wang, Y.; Fu, Q.; Ye, D. Study on the complete rotational characteristic of coolant pump in the gas-liquid two-phase operating condition. *Ann. Nucl. Energy* **2019**, *123*, 180–189.
25. Wang, C.; Chen, X.; Qiu, N.; Zhu, Y.; Shi, W. Numerical and experimental study on the pressure fluctuation, vibration, and noise of multistage pump with radial diffuser. *J. Braz. Soc. Mech. Sci. Eng.* **2018**, *40*, 481. [CrossRef]
26. Wang, C.; He, X.; Cheng, L.; Luo, C.; Xu, J.; Chen, K.; Jiao, W. Numerical simulation on hydraulic characteristics of nozzle in waterjet propulsion system. *Processes* **2019**, *7*, 915. [CrossRef]
27. He, X.; Jiao, W.; Wang, C.; Cao, W. Influence of surface roughness on the pump performance based on Computational Fluid Dynamics. *IEEE Access* **2019**, *7*, 105331–105341. [CrossRef]
28. Nguyen, T.; Tran, T.; De Boer, H.; Albert, V.D.B.; Eijkel, J.C. Rotary-atomizer electric power generator. *Phys. Rev. Appl.* **2015**, *3*, 034005. [CrossRef]
29. Huang, X.; Li, G.; Wang, M. CFD simulation to the flow field of Venturi injector. In Proceedings of the International Conference on Computer and Computing Technologies in Agriculture II, Volume 2, Beijing, China, 18–20 October 2009; pp. 805–815.
30. Wang, M.; Huang, X.; Li, G. Numerical simulation of characteristics of Venturi Injector. *Trans. Chin. Soc. Agric. Eng.* **2006**, *22*, 27–31.
31. Yan, H.; Wang, Z.; Chen, Y. High speed photography analysis on cavitation of Venturi Injector. *J. Drain. Irrig. Mach. Eng.* **2014**, *32*, 901–905, 920.
32. Zwart, P.J.; Andrew, G.G.; Thabet, B. A two-phase flow model for predicting cavitation dynamics. In Proceedings of the Fifth international conference on multiphase flow, Yokohama, Japan, 30 May–4 June 2004; p. 152.
33. Simpson, A.; Ranade, V.V. Modeling hydrodynamic cavitation in venturi: Influence of venturi configuration on inception and extent of cavitation. *AIChE J.* **2019**, *65*, 421–433. [CrossRef]
34. Shi, H.; Li, M.; Nikrityuk, P.; Liu, Q. Experimental and numerical study of cavitation flows in venturi tubes: From CFD to an empirical model. *Chem. Eng. Sci.* **2019**, *207*, 672–687. [CrossRef]
35. Dastane, G.G.; Thakkar, H.; Shah, R.; Perala, S.; Raut, J.; Pandit, A.B. Single and multiphase CFD simulations for designing cavitating venturi. *Chem. Eng. Res. Des.* **2019**, *149*, 1–12. [CrossRef]
36. Brinkhorst, S.; Lavante, E.V.; Wendt, G. Experimental and numerical investigation of the cavitation-induced choked flow in a herschel venturi-tube. *Flow Meas. Instrum.* **2017**, *54*, 56–67. [CrossRef]
37. Bashir, T.A.; Soni, A.G.; Mahulkar, A.V.; Pandit, A.B. The CFD driven optimisation of a modified venturi for cavitational activity. *Can. J. Chem. Eng.* **2011**, *89*, 1366–1375. [CrossRef]
38. Ashrafizadeh, S.M.; Ghassemi, H. Experimental and numerical investigation on the performance of small-sized cavitating venturis. *Flow Meas. Instrum.* **2015**, *42*, 6–15. [CrossRef]
39. Saharan, V.K. Computational study of different venturi and orifice type hydrodynamic cavitating devices. *J. Hydrodyn. Ser. B* **2016**, *28*, 293–305.
40. Wang, C.; Wang, M.; Yu, Y.; Wang, L.; Wei, W. CFD Simulation of Venturi Tube Hydraulic Cavitation. *Pipeline Tech. Equip.* **2013**, *1*, 10–12.
41. Wang, Y.; Zhao, L.; Deng, C.; Zhu, M.; Chen, P.; Su, H.; You, X. Numerical simulation of cavitation effect of the composite cavitation generator based on the Orifice plate and Venturi tube. *Environ. Eng.* **2012**, *S2*, 458–460.

Article

Research on the Dynamic Behaviors of the Jet System of Adaptive Fire-Fighting Monitors

Xiaoming Yuan [1,2], **Xuan Zhu** [1], **Chu Wang** [1,*], **Lijie Zhang** [1] and **Yong Zhu** [3,*]

[1] Hebei Key Laboratory of Heavy Machinery Fluid Power Transmission and Control, Yanshan University, Qinhuangdao 066004, China; yuanxiaoming@ysu.edu.cn (X.Y.); zx00712@126.com (X.Z.); zhangljys@126.com (L.Z.)

[2] State Key Laboratory of Fluid Power & Mechatronic Systems, Zhejiang University, Hangzhou 310027, China

[3] National Research Center of Pumps, Jiangsu University, Zhenjiang 212013, China

* Correspondence: wangchu@stumail.ysu.edu.cn (C.W.); zhuyong@ujs.edu.cn (Y.Z.)

Received: 30 October 2019; Accepted: 9 December 2019; Published: 12 December 2019

Abstract: Based on the principles of nonlinear dynamics, a dynamic model of the jet system for adaptive fire-fighting monitors was established. The influence of nonlinear fluid spring force on the dynamic model was described by the Duffing equation. Results of numerical calculation indicate that the nonlinear action of the fluid spring force leads to the nonlinear dynamic behavior of the jet system and fluid gas content, fluid pressure, excitation frequency, and excitation amplitude are the key factors affecting the dynamics of the jet system. When the excitation frequency is close to the natural frequency of the corresponding linear dynamic system, a sudden change in vibration amplitude occurs. The designed adaptive fire-fighting monitor had no multi-cycle, bifurcation, or chaos in the range of design parameters, which was consistent with the stroboscopic sampling results in the dynamic experiment of jet system. This research can provide a basis for the dynamic design and optimization of the adaptive fire-fighting monitor, and similar equipment.

Keywords: adaptive control; fire-fighting monitor; jet system; dynamics; duffing equation; flow control

---

1. Introduction

A fire-fighting monitor is a piece of fire-fighting equipment with a large flow and long-range, and is mainly composed of a barrel and a gun head [1,2]. Through the electromechanical control system of the barrel, horizontal and pitching rotation of the fire-fighting monitor can be realized, so that fires can be extinguished rapidly due to the burning object being sprayed directly with fluid. The gun head is the end effector of the fire-fighting jet system, which converts the pressure energy of the fluid into kinetic energy. The two jet states, spray jet and straight jet, can be switched by adjusting the electromechanical control system of the gun head [2–4]. The nozzle opening of the traditional fire-fighting monitor remains unchanged during the jet flow, and the flow, pressure, and nozzle opening of the jet system are not well matched. Based on the principle of valve components in a hydraulic system [5,6], an adaptive fire-fighting monitor with an elastic adaptive adjustment mechanism at the front end of the nozzle was designed. The nozzle opening of the monitor could be automatically adjusted according to the changes in the flow and pressure of the jet system to improve the performance of the jet system, and its fire extinguishing efficiency, in a wider flow range.

The spray medium of the fire-fighting monitor is generally water, or a mixture of water and foam. During the working process of the fire-fighting jet system, the centrifugal pump, which works as the power source, tends to exhibit pressure pulsation [7–9]. Under the joint action of the pump and the load pressure, the dynamic fluid spring is easily formed due to the compressibility of the fluid [10,11]. The fluid spring and load mass constitute a dynamic system of the fluid spring and mass The nonlinearity of spring stiffness will cause the natural frequency of the jet system to be non-constant, which will nonlinear vibration of the load mass of the jet system, leading to a loud noise, and even the destruction of the entire structure [12–14]. Similar to the jet system, the valve components in the hydraulic system will also have nonlinear dynamic behaviors such as multi-cycle and chaos under the action of the pressure pulsation, which will cause leakage of the hydraulic valve and other failures [15–17]. Additionally, unstable motion occurs when the railway vehicle reaches high speeds. In the literature [18], the effects of changing parameters with different lateral stiffnesses on nonlinear hunting behavior were analyzed. It was found that the system with bogie and wheelset had less critical speeds than the wheelset system alone, and the increase of the wheelset mass made the hunting behavior even worse. Therefore, it is necessary to explore the effect of a nonlinear fluid spring on the dynamic characteristics of the jet system of adaptive fire-fighting monitors.

At present, the research on the dynamic characteristics of fluid spring systems is generally carried out by methods of analysis, simulation, and experimentation [19–22]. Generally, when using the analytical method, the nonlinear factors of the system are linearized first, and then the system is analyzed by linear control theory, such as the root locus and Bode's chart methods [23–25]. The linearized analysis of nonlinear systems can reveal essential relationships between system parameters and performance evaluation indexes, such as stability, accuracy, and rapidity of the system. However, to some extent, the conclusions are often different from real world observations. It is difficult to explain some of the abnormal phenomena existing in real world dynamic tests, such as complex time-domain waveforms and numerous frequency domain peaks [26]. With the development of nonlinear differential equation solutions, such as the Runge-Kutta algorithm, scholars often use simulation methods to directly calculate the nonlinear system, describing the dynamics of the system more accurately [27–29]. The experimental method is mainly used to verify the results of systematic analysis and simulations, to verify the accuracy of nonlinear system analysis and simulation models [30,31].

In this paper, the adaptive fire-fighting monitor was taken as the research object, and the influence of the nonlinear fluid spring force on the dynamic characteristics of the jet system was the focus of the study. Based on the principles of nonlinear dynamics, a nonlinear dynamic model of the jet system was established. The numerical calculation was used to determine whether the jet system will have dynamic behaviors of deteriorating systems such as multi-cycle, bifurcation, and chaos within the range of operating parameters. Then the measured experimental data was analyzed with nonlinear dynamic research methods, to verify the rationality of the mechanical performance of the designed adaptive fire-fighting monitor.

2. Dynamic Modeling of the Jet System

The gun head of the fire-fighting monitor can automatically adjust the nozzle opening with a change of the flow or pressure of the incident fluid, as it can be divided into a traditional diversion gun head and an adaptive gun head. The structure of a traditional diversion gun head is shown in Figure 1. During the working process, except for the outer nozzle which is used for adjusting the state of the jet, the other components are in a relatively static state. Therefore, the nozzle opening cannot be changed with the change of the jet flow and pressure, and an aggravation of turbulent flow and sudden increase of pressure is prone to occur, resulting in a reduction of fire extinguishing efficiency. The structure of an adaptive gun head is shown in Figure 2. Inside the adaptive gun head, an adaptive mechanism is added, consisting of a spray core, an end cap, a core rod, and a spring. When the flow of the incident fluid in the gun head increases, the fluid force on the left side of the spray core increases. When the fluid force is greater than the spring force on the right side of the spray core, the spray core moves to the right, and the nozzle opening increases. In contrast, When the flow of the incident fluid decreases, the spray core moves to the left, and the nozzle opening decreases. Therefore, an adaptive fire-fighting monitor with an adaptive mechanism can automatically adjust the nozzle opening according to changes in inlet flow and pressure, thereby achieving good jet performance in a wider flow range.

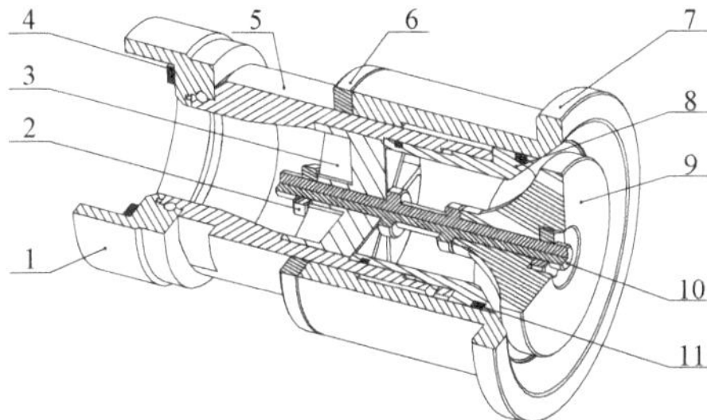

Figure 1. Structure of traditional diversion gun head. The labels are as follows: 1. Joint, 2. Nut, 3. Regulator, 4. Gasket, 5. Enclosure, 6. Ring, 7. Outer nozzle, 8. Inner nozzle, 9. Spray core, 10. Endcap, and 11. Core rod.

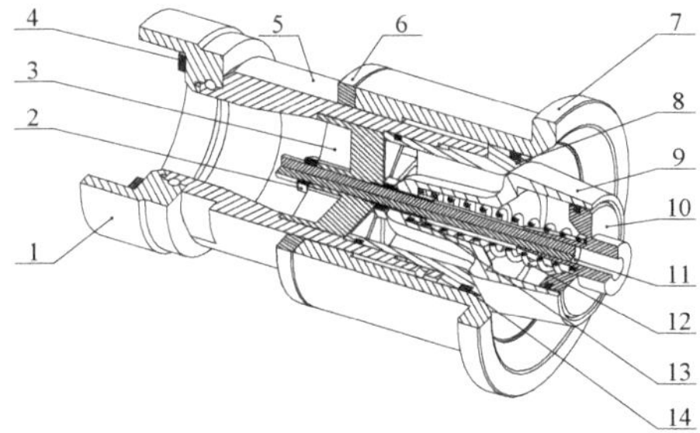

Figure 2. Structure of adaptive gun head. The labels are as follows: 1. Joint, 2. Nut, 3. Regulator, 4. Gasket, 5. Enclosure, 6. Ring, 7. Outer nozzle, 8. Inner nozzle, 9. Spray core, 10. Endcap, 11. Core rod, 12. Spring, 13. Core sleeve, and 14. Seal ring.

The connection between the adaptive gun head, the barrel, and the pipeline is shown in Figure 3. The electromechanical system installed on the barrel can realize the horizontal and pitching rotation of the fire-fighting monitor.

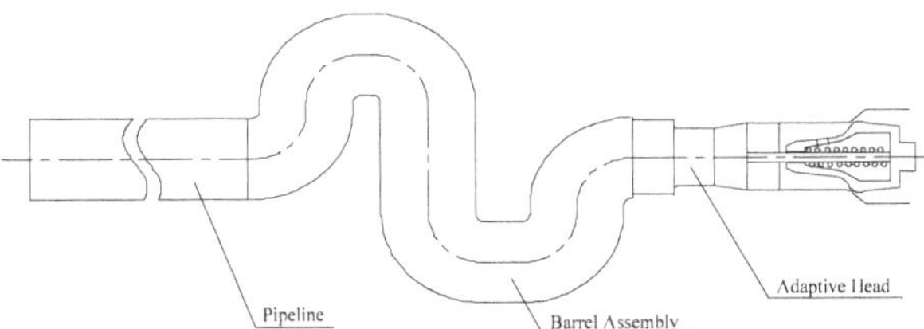

Figure 3. The jet system of an adaptive fire-fighting monitor.

The jet system of the adaptive fire-fighting monitor has better jet performance due to the addition of the adaptive mechanism in the gun head. However, the mechanical spring introduced in the mechanism reduces the stiffness of the jet system, and the compressibility of the fluid changes the stiffness of the jet system dynamically, increasing the complexity of the dynamic behaviors of the jet system. When the excitation frequency is close to the natural frequency of the jet system, the vibration of the spray core is intensified, which will significantly reduce the fire extinguishing efficiency. The internal structure of the adaptive gun head is very similar to that of the valve element in a hydraulic system. Therefore, referring to the dynamic analysis of the relief valve, the working principle of the jet system of an adaptive fire-fighting monitor is shown in Figure 4.

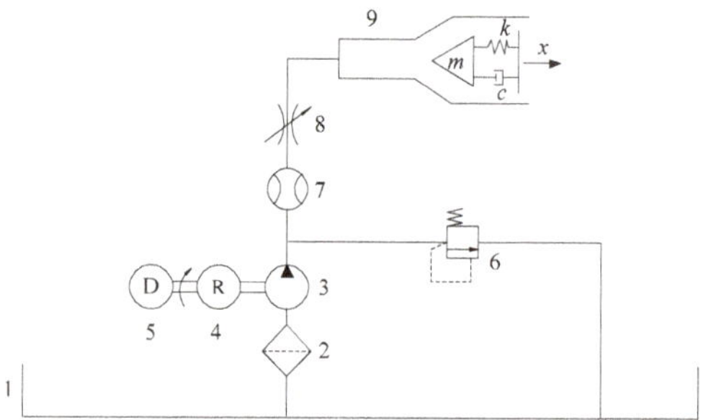

Figure 4. Working principle of the jet system of an adaptive fire-fighting monitor. The labels are as follows: 1. Water tank, 2. Filter, 3. Pump, 4. Reducer, 5. Diesel, 6. Relief valve, 7. Flowmeter, 8. Throttle valve, and 9. Adaptive fire-fighting monitor.

It can be seen in Figure 4 that the diesel and the reducer installed on the fire engine drive the main spindle of the pump to rotate, and the water in the water tank is filtered by the filter, and then sucked into the interior by the pump. The water discharged from the pump enters the adaptive fire-fighting monitor through the flowmeter and the throttle valve, and finally shoots at the fire point.

In Figure 4, m is the mass of the spray core. Analyzing the forces of the spray core, the dynamic equation of the spray core is

$$m\ddot{x} + F_c + F_s = F \tag{1}$$

where, x is the displacement of the spray core under the action of jet fluid, F_c is the viscous resistance, F_s is the spring force, and F is the fluid force on the spray core.

The total equivalent stiffness of the jet system, as shown in Figure 4, is made up of the stiffness of the mechanical spring inside the spray core and the stiffness of the fluid unit on the left side of the spray core. During the operation of the jet system, the movement of the spray core causes a change in the length of the fluid unit on the left side, which in turn causes a change in the stiffness of the fluid unit, ultimately resulting in a change in the total spring stiffness of the jet system. Therefore, the variation law of total stiffness is

$$k(x) = \frac{B_f S}{L + x} + k_s \tag{2}$$

where B_f is the bulk elastic modulus of the fluid. Considering the compressibility of the gas-containing fluid, B_f is calculated by the bulk elastic modulus formula [32]. S is the equivalent cross-sectional area of the fluid unit, L is the equivalent length of the fluid unit, and k_s is the stiffness of the mechanical spring inside the spray core.

Let y be the vibration displacement near the working point of the spray core, x, that is $y = \Delta x$. According to the Taylor series, the total stiffness of the jet system near the operating point can be expressed as:

$$k(x + y) = k(x) + \dot{k}(x)y + \ddot{k}(x)y^2/2 + o\left(y^2\right) \tag{3}$$

Assuming $k(x) = k_1$, $\dot{k}(x) = k_2$, and $\ddot{k}(x) = k_3$, then substituting them into Equation (3):

$$k(x + y) = k_1 + k_2 y + k_3 y^2 + o\left(y^2\right) \tag{4}$$

Omitting the infinitesimal of higher order $o(y^2)$ in Equation (4), the total stiffness of the jet system can be expressed as:

$$F_s = k(x + y)y = k_1 y + k_2 y^2 + k_3 y^3 \tag{5}$$

Since the elastic potential energy U of the spring has symmetry, the total elastic potential energy of the jet system can be expressed as:

$$U = k_1 y^2 / 2 + k_3 y^4 / 4 \tag{6}$$

The nonlinear spring force of the jet system can be further expressed by Equation (6):

$$F_s = dU/dy = k_1 y + k_3 y^3 \tag{7}$$

This paper mainly studies the influence of nonlinear spring force on the dynamic characteristics of the jet system, so the nonlinear factors such as friction and system damping are not considered. Then the dynamic equation of the jet system near the working point x is

$$m\ddot{y} + c\dot{y} + k_1 y + k_3 y^3 = F_0 \cos(\omega t + \varphi_0) \tag{8}$$

where, c is the linear damping coefficient of the system, which is the sum of the system structural damping coefficient c_0, and the fluid damping coefficient c_1. $F_0 \cos(\omega t + \varphi_0)$ is the external periodic excitation caused by the pressure pulsation of the incident fluid, F_0 is the amplitude of the external excitation, ω is the angular frequency of external excitation, and φ_0 is the initial phase angle of external excitation.

In order to analyze the jet system more conveniently and intuitively, the mass unit of Equation (8) can be normalized as

$$\ddot{y} + 2\xi\omega_0\dot{y} + \omega_0^2 y + \beta y^3 = F_1 \cos(\omega t + \varphi_0) \tag{9}$$

where $\xi = \frac{c}{2\sqrt{k_1 m}}$, $\omega_0 = \sqrt{\frac{k_1}{m}}$, $\beta = \sqrt{\frac{k_3}{m}}$, $F_1 = \frac{F_0}{m}$. ξ is the linear damping ratio, ω_0 is the natural frequency of the linear harmonic oscillator when the nonlinear term coefficient β is equal to 0, and F_1 is the amplitude of external excitation received by the unit mass.

3. Dynamic Analysis of the Jet System

It can be seen from Equation (9) that the dynamic model of the jet system of the adaptive fire-fighting monitor can be described by a Duffing equation with damping. Therefore, the basic laws of the jet system can be revealed by the characteristics of the Duffing equation. The design parameters of the jet system are shown in Table 1:

Table 1. Design parameters of the jet system.

Parameter Name	Parameter Name	Unit	Value
Mass of spray core	m	kg	0.3163
Structural damping coefficient	c_0	N/(m/s)	0.01
Fluid damping coefficient	c_1	N/(m/s)	0.1
Equivalent section of fluid unit	S	m^2	0.0056
Equivalent length of fluid unit	L	m	1.6679
Displacement of spray core working point	x	mm	4.284
Stiffness of spring in the gun head	k_s	kN/m	18
Temperature	T	K	293

According to the theory of bulk elastic modulus of gas-liquid fluid, the gas content of the fluid (α_0), and the fluid pressure (p), are the main factors affecting the bulk elastic modulus of the fluid. It can be seen from Equation (2) that these two factors also have direct effects on the stiffness of the jet system. Combining the amplitude (F_0) and frequency (ω) of the external excitation in Equation (9), the dynamic characteristics of the jet system can be researched with variables α_0, p, ω, and F_0.

3.1. Resonance Analysis of the Jet System

When the incident fluid gas content (α_0) is 2%, the fluid pressure (p) is 0.6 MPa, and the external excitation amplitude (F_0) values of 1 N, 3 N, and 5 N, respectively. The bifurcation, with a frequency of external excitation (ω), is shown in Figure 5.

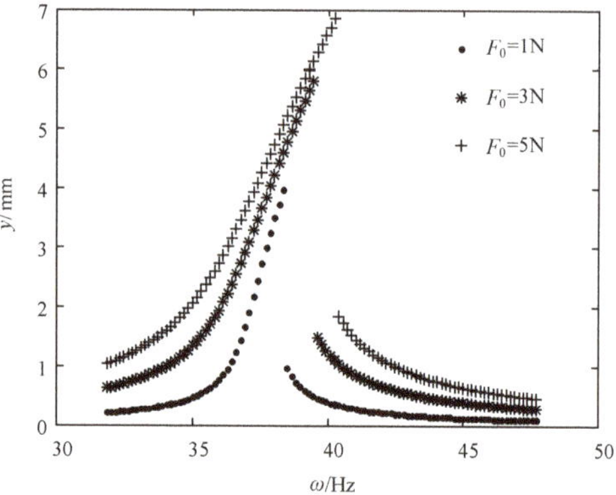

Figure 5. Bifurcation of nonlinear dynamic equations of a jet system with a frequency of external excitation (ω).

It can be seen from Figure 5, that when ω and F_0 have different values, the motion of the jet system is a single-cycle vibration. Under the action of different external excitation amplitudes, the amplitude of the jet system first increases and then decreases with the increase of the external excitation frequency. Moreover, the larger the external excitation amplitude, the larger the amplitude of the jet system. There are amplitude mutations in all three curves, and the larger the external excitation amplitude, the higher the external excitation frequency when the amplitude mutation occurs.

Taking $\alpha_0 = 2\%$, $p = 0.6$ MPa, $\omega = 46.8$ Hz, and $F_0 = 3$ N as an example, the Rouge-Kutta method was used to calculate in order to reflect the operating state of the jet system vividly. The time course, the phase diagram, the power spectrum density, and the Poincare map are shown in Figure 6.

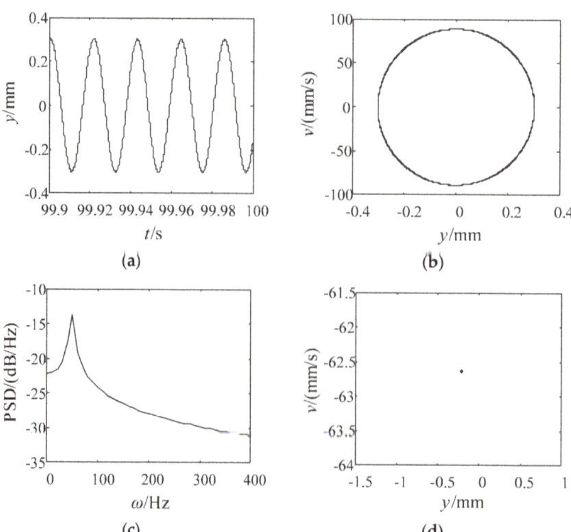

Figure 6. Results of dynamics calculation of the jet system. (**a**) Time course, (**b**) Phase diagram, (**c**) Power spectrum density, (**d**) Poincare map.

It can be seen from Figure 6 that the time course is periodically repeated, i.e., the phase diagram is repeated in a finite region, which is a closed curve, that is, there exists a limit cycle; the power spectrum density has a peak at an external excitation frequency of 46.8 Hz; Poincare map has only one point in a certain area. The above are obvious single-cycle vibration characteristics, indicating that when α_0 is 2%, p is 0.6 MPa, ω is 46.8 Hz, and F_0 is 3 N, the jet system is in a single-cycle vibration state. The results of the bifurcation of a large number of samples show that when α_0 varies from 0 to 5%, p ranges from 0 to 5 MPa, ω ranges from 0 to 100 Hz, and F_0 ranges from 0 to 5 N, the jet system operates steadily in a single-cycle.

3.2. Analysis of the Amplitude Mutation of the Jet System

In order to explore the cause of the amplitude mutation of the jet system shown in Figure 5, the amplitude curve of the jet system with the external excitation frequency shown in Figure 7 was plotted with $p = 0.6$ MPa, $\alpha_0 = 2\%$, $F_0 = 3$ N, and the nonlinear coefficient β in Equation (9) had values of 0 and 2849.5 N/(mm^3·t), respectively. In Figure 7, y_1 is the amplitude when $\beta = 2849.5$ N/(mm^3·t), and y_2 is the amplitude when $\beta = 0$.

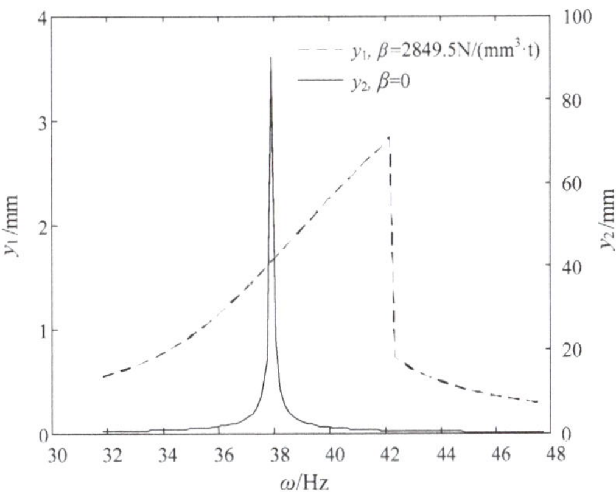

Figure 7. Amplitude curve of the jet system with different nonlinear coefficients.

When $\beta = 0$, the dynamic equation of the jet system shown in Equation (9) can be regarded as a linear one. As can be seen from Figure 7, at the natural frequency of about 37.35 Hz, the amplitude reaches its maximum, which is 90.25 mm. Meanwhile, there is no amplitude mutation. In contrast, when $\beta = 2849.5$ N/(mm^3·t), the dynamic equation is nonlinear, and the amplitude mutates from 2.833 mm to 0.7359 mm at the external excitation frequency of 42.18 Hz, which is greater than the natural frequency of the linear equation. From the above comparison, it is the nonlinear term that causes the amplitude mutation when the external excitation frequency changes. When the amplitude mutation occurs, the external excitation frequency is greater than the natural frequency of the corresponding linear system, and the maximum amplitude of the jet system is significantly reduced.

3.3. Analysis of the Influence of Incident Fluid Pressure and External Excitation Amplitude

When p is 0.6 MPa and F_0 is 3 N, and changing α_0 and ω, the curve of amplitude of the jet system with α_0 and ω under steady conditions is shown in Figure 8.

It can be seen from Figure 8 that the amplitude mutation occurs in the jet system with an increase of ω. The higher the α_0, the lower the ω and the greater the amplitude, when an amplitude mutation occurs. When α_0 is low, the mutation excitation frequency decreases rapidly with the increase of α_0. While the rate of decrease of the mutation excitation frequency gradually slows down with the increase of α_0 when α_0 is high. When ω is 38.2 Hz and α_0 is 5%, the amplitude reaches the maximum, which is 7.739 mm.

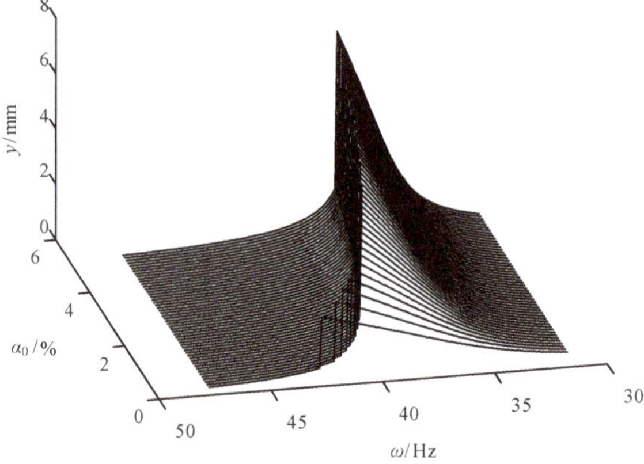

Figure 8. Curve of amplitude of the jet system with fluid gas content (α_0) and ω.

It can also be seen from Figure 8 that when ω is less than 38.2 Hz or greater than 42.7 Hz, the amplitude of the jet system changes smoothly with the change of α_0. When ω is between 38.2 Hz and 42.7 Hz, the amplitude of the system has a mutation, showing that the amplitude at this stage is more sensitive to the frequency change.

3.4. Analysis of the Influence of Incident Fluid Gas Content and External Excitation Amplitude

When α_0 is 2% and F_0 is 3 N, changing p and ω, the curve of amplitude of the jet system with p and ω under steady conditions is shown in Figure 9.

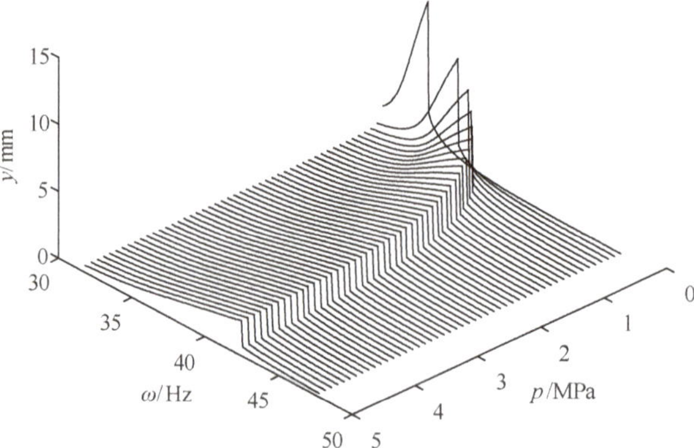

Figure 9. Curve of amplitude of the jet system with fluid pressure (p) and ω.

It can be seen from Figure 9 that an amplitude mutation occurs in the jet system with an increase of ω. Additionally, the higher the p, the higher the ω, and the smaller the amplitude when the amplitude mutation occurs. When p is low, the mutation excitation frequency increases rapidly with the increase in p. While the rate of increase of the mutation excitation frequency gradually slows down, with an increase of p when p is high. When p is 0.2 MPa and ω is 34.78 Hz, the amplitude reaches the maximum, which is 11.48 mm.

It can also be seen from Figure 9 that when ω is greater than 42.5 Hz, the amplitude of the jet system changes smoothly with the change of p. When ω is between 31.8 Hz and 42.5 Hz, there is a mutation in the amplitude of the system. At this stage, the amplitude was more sensitive to the frequency change, and the sensitive frequencies differed when the pressure is different.

3.5. Analysis of the Influence of Incident Fluid Pressure and the Incident Fluid Gas Content

When ω is 46.8 Hz and F_0 is 3 N, changing p and α_0, the curve of amplitude of the jet system with p and α_0 under steady conditions is shown in Figure 10.

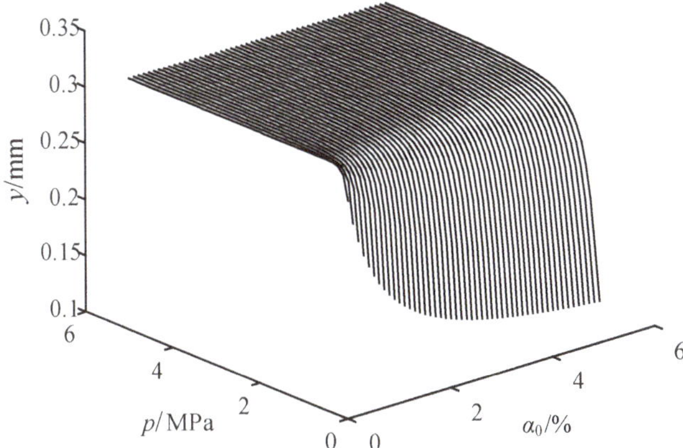

Figure 10. Curve of amplitude of the jet system with p and α_0.

It can be seen from Figure 10 that when p and α_0 change simultaneously, there is no amplitude mutation in the jet system, which is because ω remains unchanged at 46.8 Hz and is not within the frequency mutation interval, and the interval should also be avoided in the actual jet system to keep jet performance unaffected. When p is low, the higher the α_0, the lower the system amplitude at the same pressure. When p is high, the amplitude of the jet system is almost unchanged regardless of the changing gas content.

It can also be seen from Figure 10 that when p is greater than 1.2 MPa, the amplitude of the jet system changes smoothly. When p is less than 1.2 MPa, the amplitude of the system decreases in the form of step, and the amplitude at this stage is more sensitive to the change of p.

In summary, when the four parameters α_0, p, ω, and F_0, are within the design range, the motion of the jet system is a typical single-cycle vibration with no multi-cycle, bifurcation, and chaos, indicating that the vibration of the jet system of the designed adaptive fire-fighting monitor is regular and predictable in the current parameter range. When ω is in a certain interval, the jet system has amplitude mutation, and the amplitude of the jet system near the mutation is large. Therefore, in the design of the fire-fighting jet system, the input shaft speed of the pump and the pulsation frequency of the output fluid should avoid the interval. Compared with the wheelset system with obvious bifurcation dynamics in [18], the motion of the adaptive fire-fighting jet system in this paper is single-cycle, when within the range of the design parameters, indicating that the design of the jet system was reasonable.

4. Sensitivity Analysis of the Jet System

The steady-state amplitude of the jet system is mainly related to α_0, ω, and p. In order to analyze the influence of the three variables on the steady-state amplitude, the sensitivity of the steady-state amplitude to the three variables is calculated by the difference method. Due to the interactions between the three variables, the sensitivity of the jet system is analyzed according to the conditions in Section 3.3, Section 3.4, and Section 3.5.

When p is 0.6 MPa and F_0 is 3 N, the sensitivity variation curves of the steady-state amplitude y with respect to ω and α_0 are shown in Figures 11 and 12, respectively.

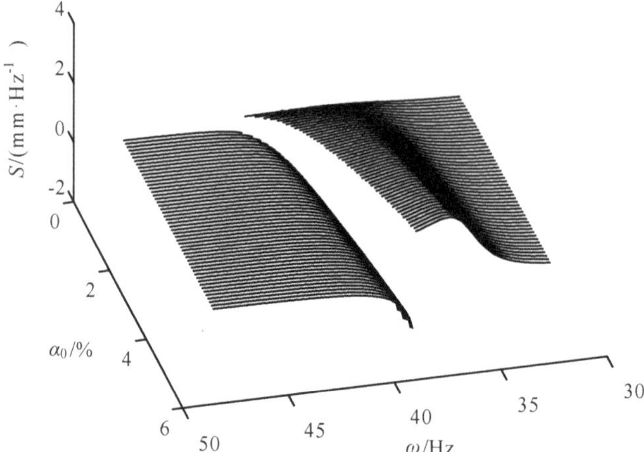

Figure 11. Sensitivity variation curve of the steady-state amplitude with respect to ω.

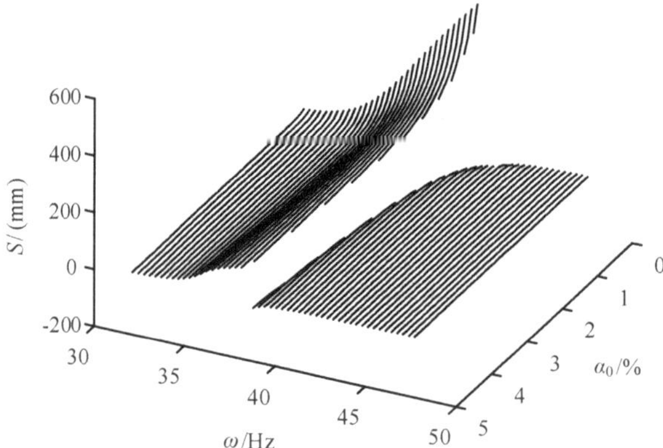

Figure 12. Sensitivity variation curve of the steady-state amplitude with respect to α_0.

It can be seen from Figure 11 that in the low-frequency range before the amplitude mutation, the sensitivity of the amplitude to ω is positive. When α_0 is constant, the sensitivity increases first and then decreases gradually as ω increases. When α_0 increases, the sensitivity under the same frequency also increases gradually. Meanwhile, the larger the α_0, the shorter the frequency range corresponding to the sensitivity variation curve of the low frequency and the greater the maximum sensitivity. In the high-frequency range after the amplitude mutation, the sensitivity of the amplitude to ω is negative. When α_0 is constant, the sensitivity gradually increases and approaches zero as the frequency increases. When α_0 increases gradually, the sensitivity slightly increases. Besides, the larger the α_0, the longer the frequency range, corresponding to the sensitivity variation curve of the high frequency and the smaller the minimum sensitivity.

It can be seen from Figure 12, that in the low-frequency range before the amplitude mutation, the sensitivity of the amplitude to α_0 is positive. When ω is constant, the sensitivity decreases first and then increases as α_0 increases. When ω increases, the sensitivity under the same gas content also increases gradually. Meanwhile, the higher the ω, the shorter the gas-content range corresponding to the sensitivity variation curve of the low frequency and the greater the maximum sensitivity. In the high-frequency range after the amplitude mutation, the sensitivity of the amplitude to a_0 is negative. When ω is constant, the sensitivity gradually increases and approaches zero as a_0 increases. When ω gradually increases, the sensitivity under the same gas content also increases. Besides, when the frequency is relatively low, the lower the ω, the shorter the gas-content range, corresponding to the sensitivity variation curve of the high frequency and the smaller the maximum sensitivity.

When a_0 is 2% and F_0 is 3 N, the sensitivity variation curves of the steady-state amplitude y with respect to p and ω are shown in Figures 13 and 14, respectively.

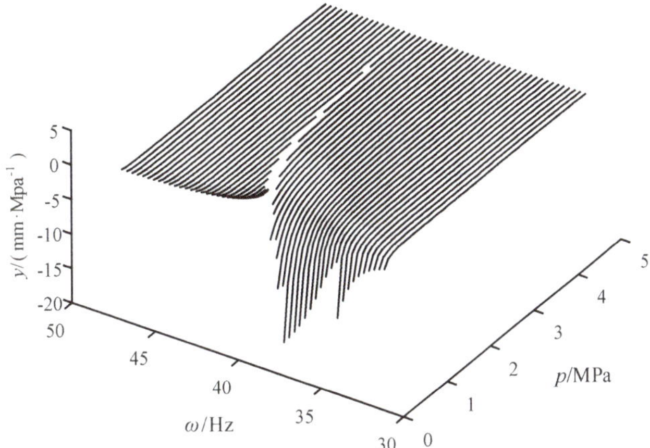

Figure 13. Sensitivity variation curve of the steady-state amplitude with respect to p.

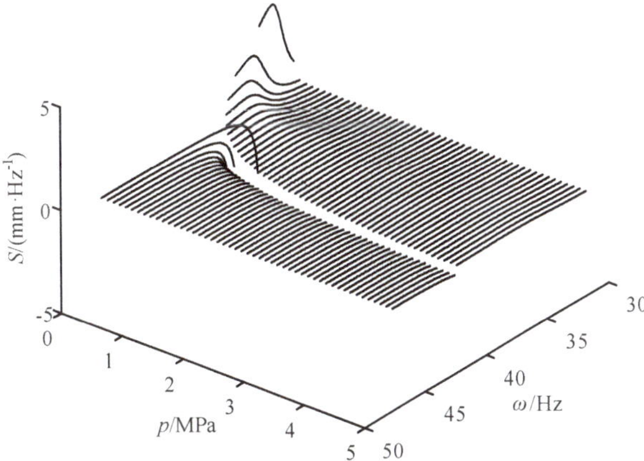

Figure 14. Sensitivity variation curve of the steady-state amplitude with respect to ω.

It can be seen from Figure 13 that in the low-frequency range before the amplitude mutation, the sensitivity of the amplitude to p is negative. When ω is constant, the sensitivity gradually increases and approaches zero, as p increases. When ω gradually increases, the sensitivity shows a stepwise distribution in the low-pressure region, which corresponds to the amplitude-variation tendency under low frequency and pressure, as shown in Figure 9. In the high-pressure region, the sensitivity gradually increases and approaches zero. In the high-frequency range after the amplitude mutation, the sensitivity of the amplitude to p is positive. When ω is constant, the sensitivity gradually decreases and approaches zero as p increases. When ω gradually increases, the sensitivity under the same pressure gradually decreases. Besides, when the frequency is relatively low, the shorter the pressure variation range, corresponding to the sensitivity variation curve of the high frequency and the larger the maximum sensitivity.

It can be seen from Figure 14 that in the low-frequency range before the amplitude mutation, the sensitivity of the amplitude to ω is positive. When p is constant, the sensitivity increases first and then decreases as ω increases. When p gradually increases, the sensitivity under the same frequency gradually decreases. Meanwhile, the larger the p, the longer the frequency range, corresponding to the sensitivity variation curve of the low frequency and the smaller the maximum sensitivity. In the high-frequency range after the amplitude mutation, the sensitivity of the amplitude to ω is negative. When p is constant, the sensitivity gradually increases and approaches zero as ω increases. When p gradually increases, the sensitivity under the same frequency gradually decreases. Besides, the larger the p, the shorter the frequency range, corresponding to the sensitivity variation curve of the high frequency and the smaller the maximum sensitivity.

When ω is 46.8 Hz and F_0 is 3 N, the sensitivity variation curves of the steady-state amplitude y with respect to p and α_0 are shown in Figures 15 and 16, respectively.

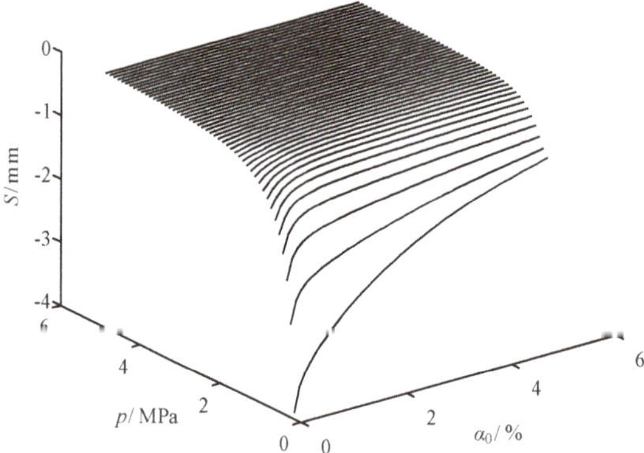

Figure 15. Sensitivity variation curve of the steady-state amplitude with respect to α_0.

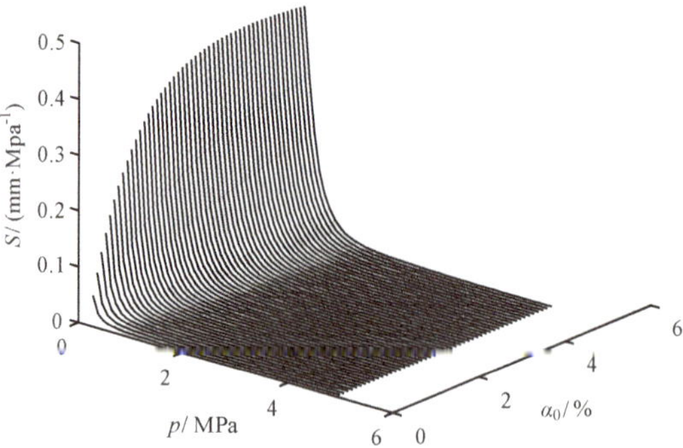

Figure 16. Sensitivity variation curve of the steady-state amplitude with respect to p.

It can be seen from Figure 15 that the sensitivity of the amplitude to α_0 is negative. When p is constant, the sensitivity gradually increases with the increase of α_0. Meanwhile, the lower the p, the greater the change of the sensitivity variation curve. When p gradually increases, the sensitivity under the same gas content gradually increases.

It can be seen from Figure 16 that the sensitivity of the amplitude to p is positive. When α_0 is constant, the sensitivity gradually decreases as p increases. Besides, the larger the α_0, the greater the change of the sensitivity variation curve. When α_0 gradually increases, the sensitivity under the same pressure gradually increases.

From the above sensitivity analysis, it is known that when the external excitation frequency variation range includes the mutation frequency, the sensitivity curve will mutate, i.e., a positive sensitivity value changes to be a negative one or a negative value becomes a positive one, which corresponds to the mutation in amplitude variation curve.

It can be seen from Figures 11–16 that within the normal working range, the sensitivity of the jet system amplitude to α_0 ranges from −43.62 mm to 519 mm, the sensitivity to ω ranges from −3.34 mm/Hz to 4.39 mm/Hz, and the sensitivity to p ranges from −18.03 mm/MPa to 3.543 mm/MPa. Since the dimensions of the three variables are different, we cannot compare the sensitivity of the amplitude with respect to different factors. When the design parameters of the jet system vary within a given range, the degree of influence of the parameters on the amplitude can be analyzed according to the sensitivity variation under different parameters.

5. Dynamic Experiment of the Jet System

Dynamic research methods are used in this section to analyze the dynamic data of the jet system of the fire-fighting monitor, and to verify the rationality of the dynamic performance of the designed adaptive fire-fighting monitor.

5.1. Composition of the Experimental System

According to Figure 17, a platform for the dynamic experiment of the jet system of the fire-fighting monitor was built, which could collect data of the flow, pressure and acceleration of the jet system under different working conditions. During the experiment, the throttle valve was fully opened, and the diesel was used to adjust the system flow. Signals including the pressure signal at the entrance of the gun head were collected by the pressure sensor, and the acceleration signal of the enclosure the gun head collected by the acceleration sensor were transmitted to the DAQ card, and then the computer for data processing and display. The pressure sensor 10 was a MIK-P300 acceleration sensor from MEACON, the acceleration sensor 11 was a 603C01 single-axis acceleration sensor from PCB, and the data acquisition card 12 was a spider20E four-channel dynamic signal analyzer from Crystal Instruments. The fire-fighting monitor prototype and sensors are shown in Figure 18. The acceleration sensor 2 in Figure 18 could collect acceleration data in three directions, which could be used to evaluate the axial vibration state of the jet system. The acceleration sensor 2 was the 356a24 three-axis acceleration sensor of PCB Piezotronics, Inc.

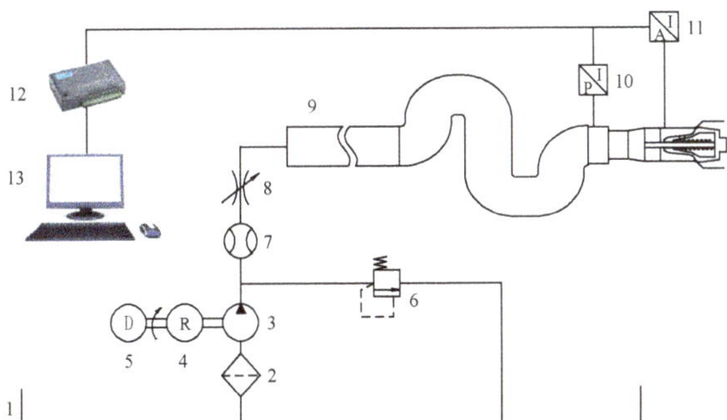

Figure 17. Experimental system for the dynamic test of the jet system. The labels are as follows: 1. Water tank, 2. Filter, 3. Pump, 4. Reducer, 5. Diesel, 6. Relief valve, 7. Flowmeter, 8. Throttle valve, 9. Adaptive fire-fighting monitor, 10. Pressure sensor, 11. Acceleration sensor, 12. Data Acquisition card, and 13. Computer.

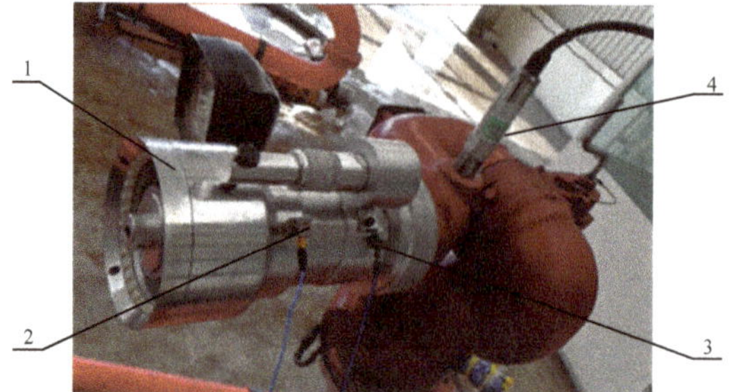

Figure 18. The fire-fighting monitor prototype and sensors. The labels are as follows: 1. Fire-fighting monitor 2. Acceleration sensor (three directions), 3. Acceleration sensor (one direction), and 4. Pressure sensor.

5.2. Acquisition and Analysis of Signals

5.2.1. Acquisition and Analysis of the Pressure Signal

Under the normal working condition of the jet system, the dynamic test of the fire-fighting monitor was carried out under different flow shown in Table 2. With the recorded data of the pressure at the entrance of the gun head under the corresponding flow, the average pressure P, the fluctuation value ΔP, and the load fluctuation value ΔF of the entrance of the gun head were obtained by calculation.

Table 2. State parameters of the entrance of the gun head under different flow.

Flow/(L/s)	P/(MPa)	ΔP/(KPa)	ΔF/(N)
40	0.59	0.68	4.23
50	0.63	0.62	3.86
60	0.71	0.6	3.73

It can be seen from Table 2, that as the flow of the jet system increase, the pressure at the entrance of the gun head gradually increases, but the fluctuation range tends to gradually decrease, which is because the damping of the jet system becomes larger as the load pressure increases.

5.2.2. Acquisition and Analysis of the Acceleration Signal

The acceleration sensor was used to collect the axial vibration signal during the operation of the adaptive fire-fighting monitor. The data processing methods such as zero-mean processing, wavelet denoising and frequency domain integration were used to preprocess the acquired signal. Then data analysis was carried out by time course, stroboscopic sampling, and power spectrum methods, which are commonly used in nonlinear dynamics.

The relationship between the axial displacement signal and time when the flow of the jet system was 40 L/s, 50 L/s, and 60 L/s is shown in Figure 19. It was found that with the increase of the flow, the amplitude of the axial vibration of the measuring position decreased slightly, but the change was not obvious. The change of the amplitude of the vibration was mainly caused by the change of the pressure.

The power spectrum of the jet system at flows of 40 L/s, 50 L/s, and 60 L/s is shown in Figure 20. It can be seen in the figure that the power spectrum under different working conditions fluctuated gently, indicating that the fluid spring had no obvious alternating transformation between the soft one and the hard one, and there was no cavitation inside the fire-fighting monitor, or the possibility of cavitation was low.

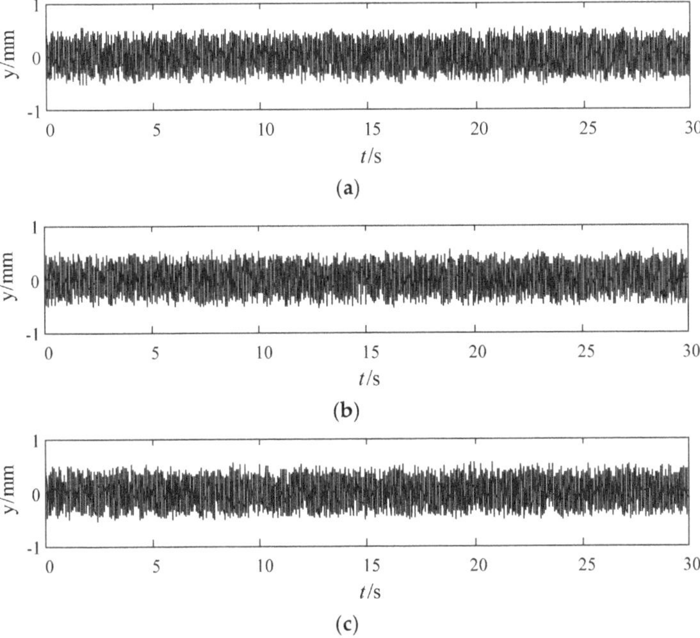

Figure 19. Time-domain waveform of vibration displacement signal under different flow. (**a**) Flow of 40 L/s, (**b**) Flow of 50 L/s, (**c**) Flow of 60 L/s.

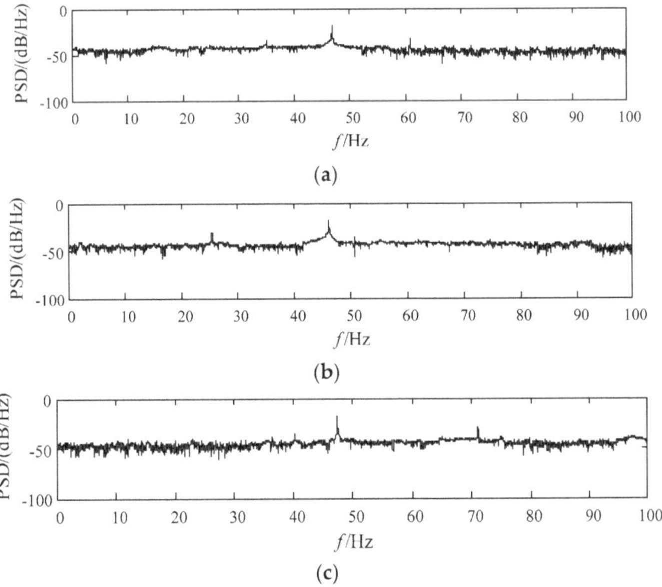

Figure 20. Power spectrum of the jet system under different flows. (**a**) Flow of 40 L/s, (**b**) Flow of 50 L/s, (**c**) Flow of 60 L/s.

The stroboscopic sampling of the jet system at flows of 40 L/s, 50 L/s, and 60 L/s are shown in Figure 21. It can be seen that the stroboscopic sampling of the jet system under different flows is gathered in the fixed area rather than dispersedly distributed in the whole plane, and all the points on the stroboscopic sampling pattern form an oblique elliptical shape with limit-cycle oscillation, indicating that the designed adaptive fire-fighting monitor had no multi-cycle, bifurcation, or chaos under the corresponding design parameters and external excitation. Besides, the single-cycle characteristic result achieved with the experiment is consistent with the simulation results.

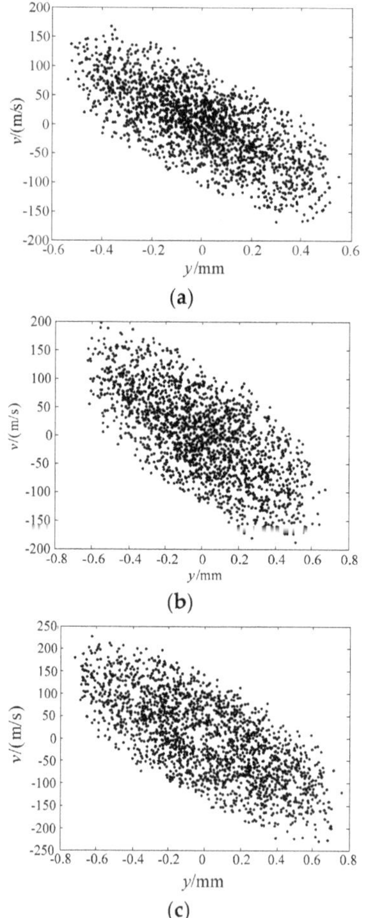

Figure 21. Stroboscopic sampling of the jet system under different flows. (a) Flow of 40 L/s, (b) Flow of 50 L/s, (c) Flow of 60 L/s.

6. Conclusions

(1) During the operation of the adaptive fire-fighting monitor, the fluid spring stiffness changes dynamically with the gas content and pressure of the jet system, and the nonlinear effect of the fluid spring stiffness can be described by the Duffing equation with damping.

(2) Compared with the dynamic system composed of linear spring and linear damping, the soft spring characteristic of the fluid reduces the vibration amplitude of the jet system of the adaptive fire-fighting monitor at the equilibrium position, which to some extent, can weaken the vibration tendency of the spray core. When the external excitation frequency continuously changes, amplitude mutation occurs near the natural frequency of the corresponding linear system, and the amplitude of

the jet system is large near the mutation frequency. Therefore, in the design of a fire-fighting jet system, the input shaft speed of the pump and the pulsation frequency of the output fluid should avoid the interval where the mutation happens.

(3) Under the corresponding design parameters and external excitation, the designed adaptive fire-fighting monitor always maintained single-cycle motion without multi-cycle, bifurcation, or chaos, which was consistent with the stroboscopic sampling results of the dynamics experiment, verifying the rationality of the design of the adaptive fire-fighting monitor.

Author Contributions: Conceptualization, X.Y. and C.W.; Methodology, X.Z.; Investigation, X.Z. and C.W.; Writing-Original Draft Preparation, X.Y.; Writing-Review & Editing, C.W. and X.Z.; Supervision, L.Z. and Y.Z.

Funding: This work was funded by the National Natural Science Foundation of China (No. 51805468, 51805214), the Natural Science Foundation of Hebei Province (No. E2017203129), the Open Foundation of the State Key Laboratory of Fluid Power and Mechatronic Systems (No. GZKF-201820), the Basic Research Special Funding Project of Yanshan University (No. 16LGB001), and the China Postdoctoral Science Foundation (No. 2019M651722). The authors would also like to thank the reviewers for their valuable suggestions and comments.

Conflicts of Interest: The authors declare no conflict of interest.

References

1. Zhu, J.S.; Li, W.; Lin, D.; Zhao, G. Study on water jet trajectory model of fire monitor based on simulation and experiment. *Fire Technol.* **2018**, *55*, 773–787. [CrossRef]

2. Hu, G.L.; Chen, W.G.; Gao, Z.G. Flow analysis of spray jet and direct jet nozzle for fire water monitor. *Adv. Mater. Res.* **2010**, *139–141*, 913–916. [CrossRef]

3. Wanning, S.; Süverkrüp, R.; Lamprecht, A. Jet-vortex spray freeze drying for the production of inhalable lyophilisate powders. *Eur. J. Pharm. Sci.* **2017**, *96*, 1–7. [CrossRef]

4. Ebrahimzadeh, S.; Maragkos, G.; Beji, T.; Merci, B. Large eddy simulations of a set of experiments with water spray-hot air jet plume interactions. *Flow Turbul. Combust.* **2019**, *103*, 203–223. [CrossRef]

5. Qian, J.Y.; Chen, M.R.; Liu, X.L.; Jin, Z.J. A numerical investigation of the flow of nanofluids through a micro tesla valve. *J. Zhejiang Univ. Sci. A* **2019**, *20*, 50–60. [CrossRef]

6. Qian, J.Y.; Gao, Z.X.; Liu, B.Z.; Jin, Z.J. Parametric study on fluid dynamics of pilot-control angle globe valve. *ASME J. Fluids Eng.* **2018**, *140*, 111103. [CrossRef]

7. Wang, C.; He, X.K.; Shi, W.D.; Wang, X.K.; Wang, X.L.; Qiu, N. Numerical study on pressure fluctuation of a multistage centrifugal pump based on whole flow field. *AIP Adv.* **2019**, *9*, 035118. [CrossRef]

8. He, X.K.; Jiao, W.X.; Wang, C.; Cao, W.D. Influence of surface roughness on the pump performance based on computational fluid dynamics. *IEEE Access* **2019**, *7*, 105331–105341. [CrossRef]

9. Wang, C.; Chen, X.X.; Qiu, N.; Zhu, Y.; Shi, W.D. Numerical and experimental study on the pressure fluctuation, vibration, and noise of multistage pump with radial diffuser. *J. Braz. Soc. Mech. Sci. Eng.* **2018**, *40*, 481. [CrossRef]

10. Zhu, Y.; Qian, P.; Tang, S.; Jiang, W.; Li, W.; Zhao, J. Amplitude-frequency characteristics analysis for vertical vibration of hydraulic AGC system under nonlinear action. *AIP Adv.* **2019**, *9*, 035019. [CrossRef]

11. Zhu, Y.; Tang, S.; Quan, L.; Jiang, W.; Zhou, L. Extraction method for signal effective component based on extreme-point symmetric mode decomposition and Kullback-Leibler divergence. *J. Braz. Soc. Mech. Sci. Eng.* **2019**, *41*, 100. [CrossRef]

12. Zhang, J.; Xia, S.; Ye, S.; Xu, B.; Song, W.; Zhu, S.; Xiang, J. Experimental investigation on the noise reduction of an axial piston pump using free-layer damping material treatment. *Appl. Acoust.* **2018**, *139*, 1–7. [CrossRef]

13. Ye, S.; Zhang, J.; Xu, B.; Zhu, S. Theoretical investigation of the contributions of the excitation forces to the vibration of an axial piston pump. *Mech. Syst. Signal Process.* **2019**, *129*, 201–217. [CrossRef]

14. Zhang, T.; Zhang, Y.O.; Ouyang, H. Structural vibration and fluid-borne noise induced by turbulent flow through a 90° piping elbow with/without a guide vane. *Int. J. Press. Vessel. Pip.* **2015**, *125*, 66–77. [CrossRef]

15. Licskó, G.; Champneys, A.; Hős, C. Nonlinear analysis of a single stage pressure relief valve. *IAENG Int. J. Appl. Math.* **2009**, *39*, 4–12.

16. Lyu, L.; Chen, Z.; Yao, B. Energy saving motion control of independent metering valves and pump combined hydraulic system. *IEEE/ASME Trans. Mechatron.* **2019**, *24*, 1909–1920. [CrossRef]

17. Lyu, L.; Chen, Z.; Yao, B. Development of pump and valves combined hydraulic system for both high tracking precision and high energy efficiency. *IEEE Trans. Ind. Electron.* **2019**, *66*, 7189–7198. [CrossRef]
18. Serajian, R. Parameters' changing influence with different lateral stiffnesses on nonlinear analysis of hunting behavior of a bogie. *J. Meas. Eng.* **2013**, *1*, 195–206.
19. Dasgupta, K.; Murrenhoff, H. Modelling and dynamics of a servo-valve controlled hydraulic motor by bondgraph. *Mech. Mach. Theory* **2011**, *46*, 1016–1035. [CrossRef]
20. Lan, Z.K.; Su, J.; Xu, G.; Cao, X.N. Study on dynamical simulation of railway vehicle bogie parameters test-bench electro-hydraulic servo system. *Appl. Mech. Mater.* **2012**, *644–650*, 863–866. [CrossRef]
21. Zhu, Y.; Tang, S.; Wang, C.; Jiang, W.; Yuan, X.; Lei, Y. Bifurcation characteristic research on the load vertical vibration of a hydraulic automatic gauge control system. *Processes* **2019**, *7*, 718. [CrossRef]
22. Zhu, Y.; Tang, S.; Wang, C.; Jiang, W.; Zhao, J.; Li, G. Absolute stability condition derivation for position closed-loop system in hydraulic automatic gauge control. *Processes* **2019**, *7*, 766. [CrossRef]
23. Milić, V.; Šitum, Ž.; Essert, M. Robust H∞ position control synthesis of an electro-hydraulic servo system. *ISA Trans.* **2010**, *49*, 535–542. [CrossRef]
24. Guo, Q.; Zhang, Y.; Celler, B.G.; Su, S.W. Backstepping control of electro-hydraulic system based on extended-state-observer with plant dynamics largely unknown. *IEEE Trans. Ind. Electron.* **2016**, *63*, 6909–6920. [CrossRef]
25. Fallahi, M.; Zareinejad, M.; Baghestan, K.; Tivay, A.; Rezaei, S.M.; Abdullah, A. Precise position control of an electro-hydraulic servo system via robust linear approximation. *ISA Trans.* **2018**, *80*, 503–512. [CrossRef]
26. Toufighi, M.H.; Sadati, S.H.; Najafi, F.; Jafari, A.A. Simulation and experimentation of a precise nonlinear tracking control algorithm for a rotary servo-hydraulic system with minimum sensors. *J. Dyn. Syst. Meas. Control* **2013**, *135*, 061004. [CrossRef]
27. Chen, C.T. Hybrid approach for dynamic model identification of an electro-hydraulic parallel platform. *Nonlinear Dyn.* **2012**, *67*, 695–711. [CrossRef]
28. Guo, Q.; Li, X.; Jiang, D. Full-State error constraints based dynamic surface control of electro-hydraulic system. *IEEE Access* **2018**, *6*, 53092–53101. [CrossRef]
29. Saeedzadeh, A.; Tivay, A.; Zareinejad, M.; Rezaei, S.M.; Rahimi, A.; Baghestan, K. Energy-efficient hydraulic actuator position tracking using hydraulic system operation modes. *Proc. Inst. Mech. Eng. E J. Process Mech. Eng.* **2018**, *232*, 49–64. [CrossRef]
30. Jarosik, M.W.; Szczęśniak, R.; Durajski, A.P.; Kalaga, J.K.; Leonski, W. Influence of external extrusion on stability of hydrogen molecule and its chaotic behavior. *Chaos Interdiscip. J. Nonlinear Sci.* **2018**, *28*, 013126. [CrossRef]
31. Cao, J.; Chugh, R. Chaotic behavior of logistic map in superior orbit and an improved chaos-based traffic control model. *Nonlinear Dyn.* **2018**, *94*, 959–975.
32. Imagine, S.A. *HYD Advanced Fluid Properties*; Technical Bulletin No. 117 Rev. 7; SIEMENTS: Munich, Germany, 2007.

Article

Numerical and Experimental Investigation of External Characteristics and Pressure Fluctuation of a Submersible Tubular Pumping System

Yan Jin [1],*, Xiaoke He [2], Ye Zhang [1], Shanshan Zhou [1], Hongcheng Chen [3] and Chao Liu [1]

[1] College of Hydraulic Science and Engineering, Yangzhou University, Yangzhou 225009, China; ydyzhang@163.com (Y.Z.); ZhouD_D@163.com (S.Z.); liuchao@yzu.edu.cn (C.L.)

[2] School of Electric Power, North China University of Water Resources and Electric Power, Zhengzhou 450045, China; hexiaoke@ncwu.edu.cn

[3] Jiangsu Surveying and Design Institute of Water Resources Co., Ltd., Yangzhou 225009, China; chenhc_yz@163.com

* Correspondence: jinyan_yz@163.com or jinyan@yzu.edu.cn; Tel.: +86-13813144081

Received: 31 October 2019; Accepted: 9 December 2019; Published: 12 December 2019

Abstract: This paper presents an investigation of external flow characteristics and pressure fluctuation of a submersible tubular pumping system by using a combination of numerical simulation and experimental methods. The steady numerical simulation is used to predicted the hydraulic performance of the pumping system, and the unsteady calculation is adopted to simulate the pressure fluctuation in different components of a submersible tubular pumping system. A test bench for a model test and pressure pulsation measurement is built to validate the numerical simulation. The results show that the performance curves of the calculation and experiment are in agreement with each other, especially in the high efficiency area, and the deviation is minor under small discharge and large discharge conditions. The pressure pulsation distributions of different flow components, such as the impeller outlet, middle of the guide vane, and guide vane outlet and bulb unit, are basically the same as the measurement data. For the monitoring points on the impeller and the wall of the guide vane especially, the main frequency and its amplitude matching degree are higher, while the pressure pulsation values on the wall of the bulb unit are quite different. The blade passing frequency and its multiples are important parameters for analysis of pressure pulsation; the strongest pressure fluctuation intensity appears in the impeller outlet, which is mainly caused by the rotor–stator interaction. The farther the measuring point from the impeller, the less the pressure pulsation is affected by the blade frequency. The frequency amplitudes decrease from the impeller exit to the bulb unit.

Keywords: submersible tubular pumping system; external characteristics; pressure fluctuation; numerical simulation; measurement

1. Introduction

A submersible tubular pump is a kind of horizontal pump with a low head and large discharge, which uses a postpositive tubular-type structure with a motor and pump coaxial. It has the advantages of low cost, high efficiency, and easy-to-realize automatic or semi-automatic control [1,2]. In recent years, this type of pump has been widely used in small and medium-sized pumping stations, such as in agricultural irrigation and urban flood control, especially in the plain areas [3–5]. All pumps experience pressure pulsation due to changes, discontinuities, and variations that occur in their pumping or pressure generating action, and submersible tubular pumps are no exception. The pressure pulsation and unstable flow in the vane pump is mainly caused by the rotor–stator interaction between the impeller and the guide vane. These pulsations can sometimes be very severe and cause damage to the piping or other components in a pumping system, which may give rise to vibration [6–9], generate hydraulic noise [10–12], and affect the performance of the pumping system, thus affecting the stable operation of the pumping system. Therefore, the distribution of pressure pulsations in the pump needs to be studied to ensure the safe, efficient, and stable operation of the pumping station.

With the development of computational fluid dynamics (CFD) technology, more scholars are using CFD to study complex flow fields in pumps [13–15], but the results of numerical simulation need to be verified by experimental data. Therefore, combining numerical simulation and experiments is more reliable. To ensure the safe and stable operation of pumping stations, many researchers are paying attention to the pressure pulsation and the unsteady flow inside the centrifugal pumps [16–23] and axial-flow pumps [24–29]. Studies on tubular pumps are relatively rare. Yang et al. [30] studied the pressure fluctuation of an S-shaped shaft extension tubular pumping system by CFD and experimentation, where the pressure fluctuations at 21 measurement locations in inlet and outlet passages were obtained and analyzed in time and frequency domains for three typical working conditions of different flow rates. Zhang et al. [31] investigated the three-dimensional turbulent flow and the pressure fluctuation in a submersible axial-flow pump by adopting the RNG (Renormalization Group) k-ε turbulence model and the SIMPLEC (Semi-Implicit Method for Pressure-Linked Equation) algorithm, with which the pressure pulsation distribution of the impeller inlet and outlet was obtained.

In this paper, an experimental system for model and pressure pulsation tests is built to validate the numerical simulation results using six transient pressure sensors in different sections of the pump. Unsteady numerical simulations are used to reveal the complex flow fluctuations, and the fast Fourier transform (FFT) method is used to obtain the amplitudes of pressure fluctuations. The results can provide references for further analysis of the pressure fluctuation of submersible tubular pumps, and ensure the safe and stable operation of submersible tubular pump stations.

2. Numerical Simulation

2.1. Pump Geometry

The simulated object is a submersible turbine pump device. Figure 1 shows a single-line diagram of the pump used in the numerical simulation and experiment, including the inlet passage, impeller, guide vane, bulb unit, and outlet passage. The dimensions given in the figure are values relative to the diameter D of the impeller. The main geometric parameters of the pump device are shown in Table 1.

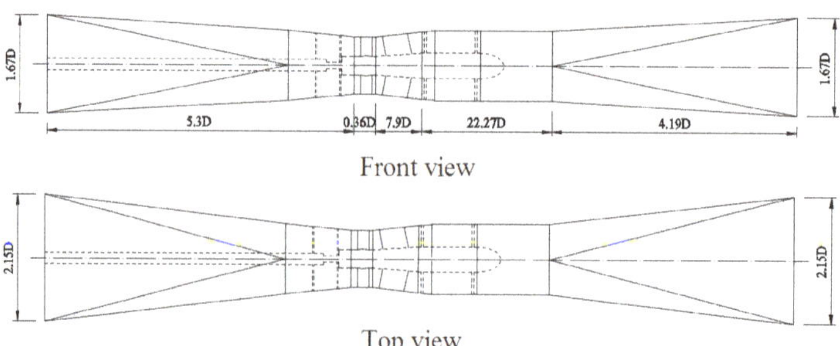

Figure 1. Single-line diagram of the pumping system.

Table 1. Parameters of the model pump.

Impeller diameter (mm)	120
Rotational speed (r/min)	1450
Hub to tip ratio	0.4
Discharge range (L/s)	14~24
Number of blades	3
Number of guide vane blades	5
Blade angle (°)	0
Tip clearance (mm)	0.2

2.2. Pump Modeling

The numerical simulation study in this paper is for the entire submersible tubular pump device, in which the inlet passage, the outlet passage, and the bulb unit are modeled by Unigraphics NX (11.0, Siemens PLM Software, Shanghai, China, 2016) for 3-D solid modeling, as shown in Figure 2a. The impeller and guide vane components are generated automatically in the TurboGrid software (14.5, ANSYS Inc., Pittsburgh, PA, USA, 2013); as shown in Figure 2b, the distance between the blade tip and the impeller chamber is set to 0.2 mm.

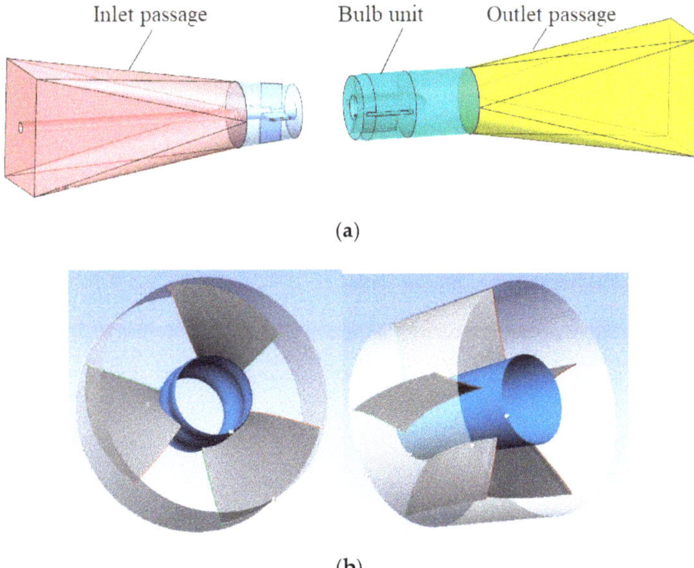

Figure 2. A 3-D model of the pump device: (**a**) passages and bulb unit; (**b**) impeller and guide vane.

2.3. Numerical Model and Grid Generation

The numerical simulation of this paper adopts the current common commercial CFD software ANSYS CFX-14.5 (14.5, ANSYS Inc., Pittsburgh, PA, USA, 2013), which performs the steady and unsteady calculations for the submerged tubular pump devices under different working conditions. The three-dimensional Reynolds-averaged Navier–Stokes equations were solved by CFX code. The turbulence effects were modeled by the standard k-ε turbulence model. The pressure–velocity coupling was performed using the SIMPLEC algorithm. The criterion for convergence was considered to be 10^{-4}, allowing an optimal number of iterations for each time step.

In this calculation, structured hexahedral cells were used to define the computational domain. The grids of the inlet passage, bulb unit, and outlet passage were generated by ICEM-CFD (14.5, ANSYS

Inc., Pittsburgh, PA, USA, 2013), while the grids of the impeller and guide vane were generated by TurboGrid (14.5, ANSYS Inc., Pittsburgh, PA, USA, 2013). In order to ensure the grid quality, the grid independence calculation was carried out, and the total grid number was about 4.19×10^6. Figure 3 shows the grid details for each component of the pumping system.

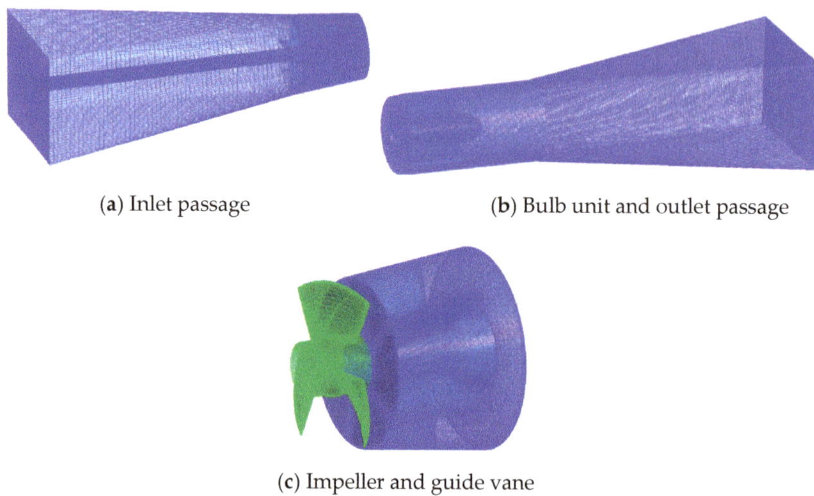

(**a**) Inlet passage   (**b**) Bulb unit and outlet passage

(**c**) Impeller and guide vane

Figure 3. Grid generation.

The boundary conditions were set as follows: the inlet pressure was specified at the entrance of the inlet passage. The inlet adopts the total pressure and the pressure is set to 1 for the atmosphere. The mass outflow condition was defined at the exit of the outlet passage. The transient rotor–stator model was used for the unsteady calculation. The shroud of the impeller was set as absolutely stationary, and the blade and hub of the impeller were relatively stationary. No slip boundary conditions or wall functions were used for the solid walls.

In this paper, the result of the steady calculation was taken as the initial flow field of the unsteady calculation, and then the unsteady numerical simulation was carried out using the sliding mesh technique. The time step of the unsteady calculations was $\Delta t = 3.4483 \times 10^{-4}$ s. The impeller rotated 3° at each time step, so it took 120 steps to complete the rotation. The chosen time step was small enough to get the necessary time resolution.

3. Experiment System

3.1. Test Bench

In order to test the external and internal characteristics of the pump, a small submersible tubular pump test bench for model tests and pressure pulsation tests was established. The test cycle piping system is shown in Figure 4, which had a length of 4.2 m and a height of 1.2 m (excluding the pressure tank and the suspended height). The main dimensions of the test system are shown in Figure 4 (unit is mm). The pipeline included a thick pipe section with an inner diameter of 200 mm, a tapered section with an inner diameter of 200 mm to 120 mm, and a pipe section with an inner diameter of 120 mm. The test bench layout was divided into two layers. The upper layer included the submersible pump, the pressure tank, the torquemeter, and the motor. From the inlet passage to the outlet passage, the entire submersible tubular pump unit was made of plexiglass for flow visualization and internal flow field measurements based on laser testing technology. The lower part contained an electromagnetic flowmeter, auxiliary pump, butterfly valve, and other pipe accessories. The electromagnetic flowmeter satisfies the installation requirements as the water inlet pipe was greater than 10 D and the water outlet pipe was longer than 5 D.

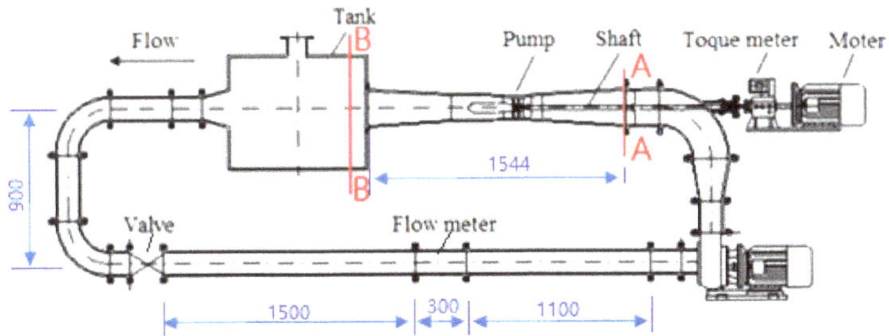

Figure 4. Single-line diagram of pump test stand.

The external characteristic parameters of the pump device include discharge, lift head, power, and efficiency. The instruments generally used for testing are mainly electromagnetic flowmeters, torque meters, and differential pressure transmitters.

The discharge was measured by an electromagnetic flowmeter. The average discharge obtained during a period of time was used as the discharge value under this operating condition.

Pressure measurement sections A-A and B-B in the system can be seen in Figure 4. The equation for lift head is written as [32]:

$$H = \left(\frac{p_1}{\rho g} + \frac{V_1^2}{2g} + Z_1 \right) - \left(\frac{p_2}{\rho g} + \frac{V_2^2}{2g} + Z_2 \right) \tag{1}$$

The torque moment and the rotational speed values are read from the tacho-torquemeter directly, and converted to power through Equation (2):

$$P = M\omega, \ \omega = \frac{2\pi n}{60} \tag{2}$$

When the above data are obtained, the efficiency is computed through Equation (3):

$$\eta = \frac{\rho g Q H}{p} \tag{3}$$

3.2. Pressure Pulsation Measurement

The pump device pressure pulsation test mainly uses multiple dynamic pressure sensors to collect pulsation data. The micropressure sensor used in this test was a CYG1505GSLF made by Kunshan Shuangqiao Sensor Measurement Controlling Company (Kunshan, China, 2016). The basic parameters are given in Table 2. The SQCJ-USB-36 data acquisition instrument was also produced by Kunshan Shuangqiao Sensor Measurement Controlling Company (Kunshan, China, 2016). The number of analog channels was 36 channels, and the sampling frequency was 100 kHz.

Table 2. Parameters of the pressure sensor.

Sensor Model	CYG1505GSLF
Range (kPa)	50
Output (V)	0~5
Accuracy (%)	0.25

In order to compare the different pressure pulsations in the various flow components of the submersible pump device, dynamic pressure sensors were arranged at the impeller outlet (P1), middle of the guide vane (P2), the guide vane outlet (P3), and the bulb unit (P4, P5 and P6) to monitor the pressure pulsation, as shown in Figure 5. All the measuring points were arranged on the line where the horizontal longitudinal section intersected the wall surface.

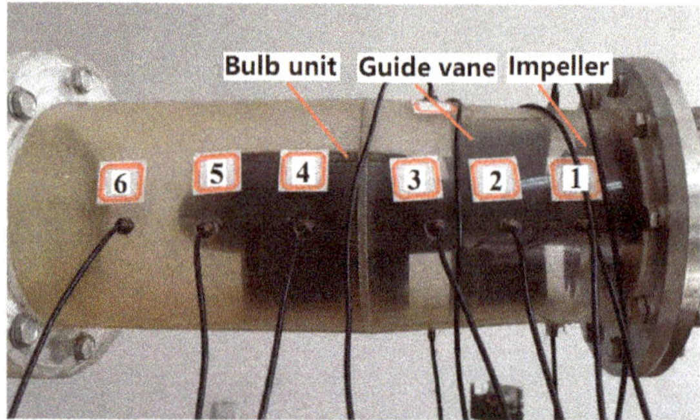

Figure 5. Location of pressure measurements.

4. Results and Analysis

4.1. External Characteristics of Pumping System

The external characteristic data of the submersible pumping system obtained by the model test was compared with the performance of the pumping system predicted by CFD calculation. The comparison results of the two are shown in Figure 6 (the solid square points in the figure are the data points obtained by the model test, and the hollow square points are the data points obtained by the numerical simulation. These points are respectively fitted with a quadratic curve to obtain the final performance curve). It can be seen that the numerical simulation results agree well with the experimental results, especially in the high efficiency area, where the high efficiency point appears in the same flow rate at about 19 L/s, and the corresponding maximum efficiency is 74.4%. Under the small flow condition, the numerical calculation results were higher than the model test values, while under large flow conditions, the model test value was slightly higher than the calculated value. From the comparison of calculated and measured results, the calculated data can be considered accurate and reliable.

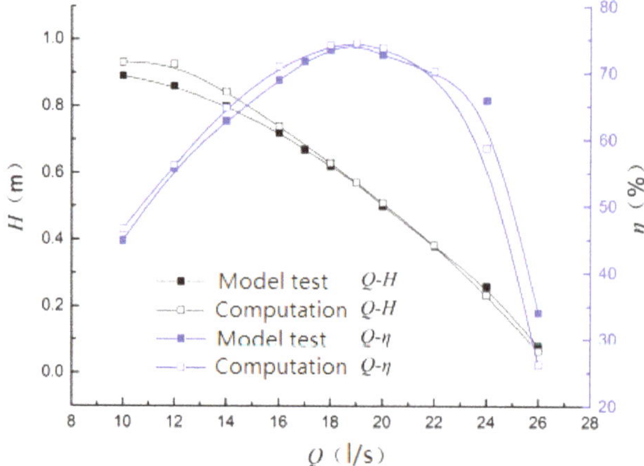

Figure 6. Hydraulic performance of the pumping system.

4.2. Pressure Pulsation Analysis

In order to compare the results of the pressure pulsation measurements, the monitoring points set in the numerical calculation were the same as those in the experiment (P1, P2, P3, P4, P5, and P6), and additionally the monitoring points inside the pump, which cannot be measured by testing, were added to obtain more pulsation information for the pump. The detailed location of the monitoring points is shown in Figure 7. Since the actual measurement points are distributed from the plane parallel to the ground plane, this arrangement makes each point on the line appears to coincide, and only one point can be seen in the front view. In order to show the location of the measuring points, Figure 7 is the actual calculation domain rotated 90° clockwise along the axis.

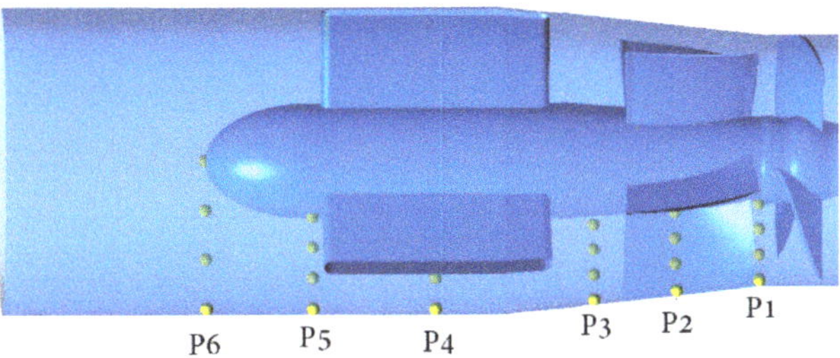

Figure 7. Pressure monitoring points.

The pressure pulsation data of each point was subjected to fast Fourier transform analysis to obtain the frequency domain diagram of each point. The results were compared with the experimental data in different sections, as shown in Figure 8. (The red points in the figure are the pulsation data points measured experimentally, while the black points are the pulsation data points obtained by numerical calculation. Each point is connected with a straight line to obtain the final pressure pulsation curve.) Only the pressure pulsation values for the six points P1–P6 were taken: P1 is the impeller outlet, P2 is the middle guide vane, P3 is the guide vane outlet, P4 is the middle of the bulb unit, P5 is the bulb unit tapered section, and P6 is the bulb unit tail in the optimum condition (Q_{opt} = 19 L/s), where the abscissa is the rotational frequency multiple, N_F, and the ordinate is the amplitude, A. Here, N_F is defined as follows:

$$N_F = 60 \, zF/n = F/F_n \tag{4}$$

Figure 8 shows that the pressure pulsation measurement at the monitoring point P1 has a significant amplitude at the full-fold frequency, and the numerical calculation of the frequency domain map is only due to small adverse effects, such as noise interference. The blade passing frequency (BPF) and its integer multiples have a large amplitude. However, both of them obtain the maximum amplitude at BPF. The amplitude of the pressure pulsation test at the impeller exit monitoring point is 1.14 times the unsteady calculated amplitude. The reasons for the error may be: (1) The influence of dynamic and static interference. (2) The submerged cross-flow pump has a small bulb unit, and the motor is placed outside the pump device and connected to the impeller through the rotating shaft. During the rotation of the impeller, the rotating shaft also drives the water in the inflow passage to rotate, so under test conditions, the pressure pulsation value is too large. (3) There are many interference factors in the test case. (4) There is a certain relationship between the form of the Fourier transform and the selection and length of the window function. However, the overall pressure pulsation test and numerical simulation have a high degree of agreement for the main frequency. At monitoring point P2, due to the restriction effect of the inlet vane on the water flow, the amplitude of the blade frequency is significantly reduced with respect to the impeller outlet, and the amplitudes of the 1x rotation frequency and the 2x rotation frequency do not change much. The amplitude of the model test at the blade frequency is 1.07 times the unsteady calculated amplitude. At the integer frequency of the blade frequency, the amplitude of the test and the numerical calculation is in good agreement, and the model test is caused by the influence of bubbles. The frequency domain map has more components at high frequencies. At monitoring point P3 (the vane outlet), the amplitude of the model test and the unsteady calculation result of the blade frequency are somewhat reduced, and the amplitude of the model test at this point is 1.23 times the unsteady calculated amplitude. The amplitudes of the measurement and the unsteady calculation result are obviously increased with the 12-fold frequency shift, which may cause undesirable flow states. such as backflows and vortexes at the exit of the guide vane.

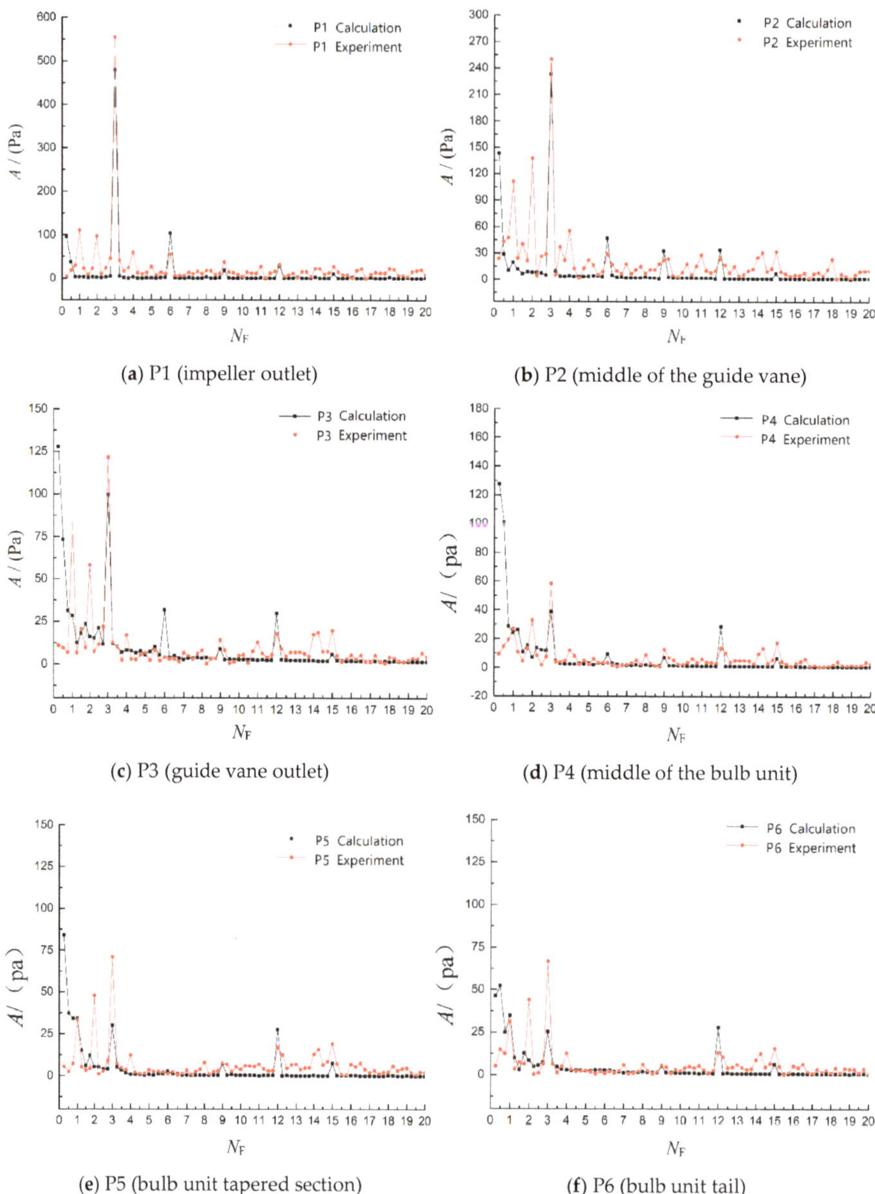

(**a**) P1 (impeller outlet)

(**b**) P2 (middle of the guide vane)

(**c**) P3 (guide vane outlet)

(**d**) P4 (middle of the bulb unit)

(**e**) P5 (bulb unit tapered section)

(**f**) P6 (bulb unit tail)

Figure 8. Comparison of numerical calculation results of pressure pulsation for each part with experimental data.

The bulb unit also has a certain inhibitory effect on the pressure pulsation due to the rectifying action of the supports. At monitoring point P4 (in the middle of the bulb unit), the pressure pulsation amplitude is significantly reduced relative to the vane section. At the blade frequency, the amplitude of the pressure pulsation test is 1.5 times the amplitude of the unsteady calculation. The amplitude of the pressure pulsation test is significantly larger than the unsteady calculation result, but the difference between the amplitudes of these two is smaller at 1x frequency. Probably because the monitoring point is far from the impeller, the influence of the impeller on the pressure pulsation is reduced, which leads to a large numerical error in the calculation of the amplitude at the 3x frequency shift. The frequency domain diagram of monitoring point P5 is shown at the blade frequency, and the amplitude of the pressure pulsation test is 2.4 times the amplitude of the unsteady calculation. The amplitude difference between the test and the digital mode gradually increases, and the distance from the pulsating source impeller also increases. The amplitude becomes smaller, and the influence of interference on the test results is more obvious. At the same time, the flow path of the gradual section is widened, and certain bad flow patterns occur when the water flow is fast. The model test has a large amplitude at high frequencies. At the end of bulb section monitoring point P6, the test and numerical calculations have large amplitudes at low frequencies. At the blade frequency, the amplitude of the model test is 2.8 times the unsteady calculated amplitude. Due to the poor flow pattern in the tail of the bulb body, the frequency domain diagram of monitoring point P6 obtained as measured by the pressure pulsation test also has a large amplitude at high frequencies.

In summary, under the optimal working conditions, the pressure pulsation values of the monitoring points obtained by the pressure pulsation test are basically the same as the general trend for the unsteady calculation results. In the impeller and vane parts especially, the monitoring points are at the main frequency and amplitude. The upper abundance is higher, and the amplitude of the blade frequency increases up to 11%. It can be seen that the pressure pulsation test and the pressure pulsation data obtained by the nonfixed constant value simulation are more accurate, which further studies the internal characteristics of the pump device through a nonfixed constant value simulation.

5. Conclusions

The study presents a numerical simulation and experiment of external characteristics and pressure pulsation of a submersible tubular pumping system. The following results are obtained:

(1) Comparing the numerical simulation results and experimental data of pump performance under different working conditions, the results of the calculation and the experiment are in good agreement, especially in the high efficiency area, and some deviations under high flow and small flow conditions.

(2) The pressure pulsation values of the six monitoring points in the axial direction of the wall of the pumping system calculated by the unsteady simulation are basically the same as the overall trend of the pressure pulsation test results. The pressure fluctuation intensity is strongest in the impeller outlet, and then it gradually weakens. The main frequency and the amplitude matching degree are higher, and the pressure pulsation values on the wall of the bulb unit are quite different.

(3) The blade passing frequency and its multiples play a leading role in pressure pulsation. Due to the rotor–stator interaction at the exit of the impeller, the influence of the blade frequency is obvious. The farther the measuring point is from the impeller, the less affected the pressure pulsation is by the blade frequency. The frequency amplitudes decrease from the impeller exit to the bulb unit.

(4) Based on this research, the experimental pressure pulsation monitoring points can be increased in the future to further reveal the internal pulsation distribution law of submersible tubular pumping systems.

Author Contributions: Y.J. conceived and designed the experiments. X.H. participated in numerical simulation and proposed the modification suggestions to the original manuscript. H.C., Y.Z., and S.Z. performed the experiments and simulation. H.C. and Y.Z. analyzed the data. Y.J. wrote the paper. C.L. performed funding acquisition.

Funding: This work was supported by a project funded by the Priority Academic Program Development of Jiangsu Higher Education Institutions (PAPD), Nature Science Foundation of China (Grant No. 51609210), and Nature Science Foundation of Jiangsu Province (Grant No. SBK2019041842).

Conflicts of Interest: The authors declare no conflicts of interest.

Nomenclature

A	Amplitude, Pa
F	Actual frequency after Fourier transform, Hz
F_n	Blade passing frequency, Hz
g	Gravity, m/s^2
H	Lift head, m
M	Torque, Nm
n	Rotational speed, r/min
N_F	Rotational frequency multiple
p_1, p_2	Static pressure of section A-A and B-B, Pa
P	Shaft power, kW
Q	Discharge, L/s
Q_{opt}	Discharge of optimal condition, L/s
T	Period, s
V_1, V_2	Flow velocity of section A-A and B-B, m/s
z	Blade number of impeller, 3
η	Efficiency
ρ	Density of water, kg/m^3
ω	Angular velocity of rotation, rad/s
Δt	Time step, s

References

1. Wu, C.; Jin, Y.; Wang, D.; Yang, H.; Yang, F.; Liu, C. Study and application of large bidirectional submersible tubular pump system. *J. Hydroelectr. Eng.* **2012**, *31*, 265–270.
2. Liu, C.; Zhou, Q.; Qian, J.; Jin, Y. Flow characteristics of two way passage vertical submersible pump system. *Trans. Chin. Soc. Agric. Mach.* **2016**, *47*, 59–65.
3. Tang, F.; Liu, C.; Xie, W. Experimental studies on hydraulic Models for a reversible, Tubular, and Submersible axial-flow P ump Installation. *Trans. Chin. Soc. Agric. Mach.* **2004**, *35*, 74–77.
4. Yang, F.; Jin, Y.; Liu, C. Performance test and numerical analysis of bidirectional submersible tubular pump. *J. Agric. Eng.* **2012**, *28*, 60–67.
5. Xia, C. Research on the flow pattern of low-lift diving tubular pumping system by 3D CFD. *Earth Environ. Sci.* **2016**, *49*, 032010. [CrossRef]
6. Zhu, Y.; Tang, S.; Wang, C.; Jiang, W.; Yuan, X.; Lei, Y. Bifurcation Characteristic research on the load vertical vibration of a hydraulic automatic gauge control system. *Processes* **2019**, *7*, 718. [CrossRef]
7. Zhu, Y.; Qian, P.; Tang, S.; Jiang, W.; Li, W.; Zhao, J. Amplitude-frequency characteristics analysis for vertical vibration of hydraulic AGC system under nonlinear action. *Aip Adv.* **2019**, *9*, 035019. [CrossRef]
8. Ye, S.; Zhang, J.; Xu, B.; Zhu, S. Theoretical investigation of the contributions of the excitation forces to the vibration of an axial piston pump. *Mech. Syst. Signal Process.* **2019**, *129*, 201–217. [CrossRef]
9. Khalifa, A.E.; Al-Qutub, A.M.; Ben-Mansour, R. Study of pressure fluctuations and induced vibration at blade-passing frequencies of a double volute pump. *Arab. J. Sci. Eng.* **2011**, *36*, 1333–1345. [CrossRef]
10. Zhang, J.; Xia, S.; Ye, S.; Xu, B.; Song, W.; Zhu, S.; Xiang, J. Experimental investigation on the noise reduction of an axial piston pump using free-layer damping material treatment. *Appl. Acoust.* **2018**, *139*, 1–7. [CrossRef]
11. Chu, S.; Dong, R.; Katz, J. Relationship between unsteady flow, pressure fluctuations, and noise in a centrifugal pump—part A: Use of PDV data to compute the pressure field. *J. Fluids Eng.* **1995**, *117*, 24–29. [CrossRef]
12. Chu, S.; Dong, R.; Katz, J. Relationship between unsteady flow, pressure fluctuations, and noise in a centrifugal pump—Part B: Effects of blade-tongue interactions. *J. Fluids Eng.* **1995**, *117*, 30–35. [CrossRef]
13. Parrondo, J.; Pérez, J.; Barrio, R.; Gonzales, J. A simple acoustic model to characterize the internal low frequency sound field in centrifugal pumps. *Appl. Acoust.* **2011**, *72*, 59–64. [CrossRef]

14. He, X.; Jiao, W.; Wang, C.; Cao, W. Influence of surface roughness on the pump performance based on Computational Fluid Dynamics. *IEEE Access* **2019**, *7*, 105331–105341. [CrossRef]
15. Wang, C.; Hu, B.; Zhu, Y.; Wang, X.; Luo, C.; Cheng, L. Numerical study on the gas-water two-phase flow in the self-priming process of self-priming centrifugal pump. *Processes* **2019**, *7*, 330. [CrossRef]
16. Si, Q.; Yuan, J.; Yuan, S.; Wang, W.; Zhu, L.; Bois, G. Numerical investigation of pressure fluctuation in centrifugal pump volute vased on SAS model and experimental validation. *Adv. Mech. Eng.* **2014**, *6*, 972081. [CrossRef]
17. Li, Y.; Li, X.; Zhu, Z.; Li, F. Investigation of unsteady flow in a centrifugal pump at low flow rate. *Adv. Mech. Eng.* **2016**, *8*, 1–8. [CrossRef]
18. Dai, C.; Kong, F.; Dong, L. Pressure fluctuation and its influencing factors in circulating water pump. *J. Cent. South Univ.* **2013**, *20*, 149–155. [CrossRef]
19. Wang, C.; He, X.; Zhang, D.; Hu, B.; Shi, W. Numerical and experimental study of the self-priming process of a multistage self-priming centrifugal pump. *Int. J. Energy Res.* **2019**, *43*, 4074–4092. [CrossRef]
20. Wang, C.; He, X.; Shi, W.; Wang, X.; Wang, X.; Qiu, N. Numerical study on pressure fluctuation of a multistage centrifugal pump based on whole flow field. *Aip Adv.* **2019**, *9*, 035118. [CrossRef]
21. Feng, J.; Benra, F.K.; Dohmen, H.J. Unsteady flow investigation in rotor-stator interface of a radial diffuser pump. *Forsch. Ingen.* **2010**, *74*, 233–242. [CrossRef]
22. Wang, C.; Shi, W.; Wang, X.; Jiang, X.; Yang, Y.; Li, W.; Zhou, L. Optimal design of multistage centrifugal pump based on the combined energy loss model and computational fluid dynamics. *Appl. Energy* **2017**, *187*, 10–26. [CrossRef]
23. Pavesi, G.; Cavazzini, G.; Ardizzon, G. Time-frequency characterization of the unsteady phenomena in a centrifugal pump. *Int. J. Heat Fluid Flow* **2008**, *29*, 1527–1540. [CrossRef]
24. Xie, C.; Tang, F.; Zhang, R.; Zhou, W.; Zhang, W.; Yang, F. Numerical calculation of anial-flow pump's pressure fluctuation and model test analysis. *Adv. Mech. Eng.* **2018**, *10*, 1–13. [CrossRef]
25. Feng, J.; Luo, X.; Guo, P.; Wu, G. Influence of tip clearance on pressure fluctuations in an axial flow pump. *J. Mech. Sci. Technol.* **2016**, *30*, 1603–1610. [CrossRef]
26. Wang, F.; Zhang, L.; Zhang, Z. Analysis on pressure fluctuation of unsteady flow in axial-flow pump. *J. Hydraul. Eng.* **2007**, *38*, 1003–1009.
27. Zhang, D.; Wang, H.; Shi, W.; Pan, D.; Shao, P. Experimental investigation of pressure fluctuation with multiple flow rates in scaled axial flow pump. *Trans. Chin. Soc. Agric. Mach.* **2014**, *45*, 139–145.
28. Zhang, D.; Geng, L.; Shi, W.; Pan, D.; Wang, H. Experimental investigation on pressure fluctuation and vibration in axial-flow pump model. *Trans. Chin. Soc. Agric. Mach.* **2015**, *46*, 66–72.
29. Gao, Q. *Study of Pressure Fluctuation in a Mixed-Flow Pump with Vaned Diffuser*; Lanzhou University of Technology: Lanzhou, China, 2014.
30. Yang, F.; Liu, C.; Tang, F.; Zhou, J.; Cheng, L. Numerical analysis on pressure fluctuations of time-varying turbulent flow in tubular pumping system with S-shaped shaft extension. *J. Hydroelectr. Eng.* **2015**, *34*, 175–180.
31. Zhang, Y.; Wang, X.; Ding, P.; Tang, X. Numerical analysis of pressure fluctuation of internal flow in submersible axial-flow pump. *J. Drain. Irrig. Mach. Eng.* **2014**, *32*, 302–307.
32. Jin, Y.; Liu, C.; Qiu, Z.; Zhou, J.; Cheng, L. Numerical simulation and experimental study on postpositive bulb tubular pump. In Proceedings of the 6th International Symposium on Fluid Machinery and Fluid Engineering, Wuhan, China, 22–25 October 2014.

Article

Experimental and Numerical Study on Gas-Liquid Two-Phase Flow Behavior and Flow Induced Noise Characteristics of Radial Blade Pumps

Qiaorui Si [1], Chunhao Shen [1], Asad Ali [1], Rui Cao [1], Jianping Yuan [1] and Chuan Wang [2],*

[1] National Research Center of Pumps, Jiangsu University, Zhenjiang 212013, China; siqiaorui@ujs.edu.cn (Q.S.); shencunhaoujs@163.com (C.S.); asadam969@gmail.com (A.A.); caoruiujs@163.com (R.C.); yh@ujs.edu.cn (J.Y.)

[2] College of Hydraulic Science and Engineering, Yangzhou University, Yangzhou 225009, China

* Correspondence: wangchuan@ujs.edu.cn; Tel.: +86-1505-085-4169

Received: 31 October 2019; Accepted: 25 November 2019; Published: 4 December 2019

Abstract: Miniature drainage pumps with a radial blade are widely used in situations with critical constant head and low noise requests, but the stable operation state is often broken up by the entraining gas. In order to explore the internal flow characteristics under gas–liquid two phase flow, pump performance and emitted noise measurements were processed under different working conditions. Three-dimensional numerical calculations based on the Euler inhomogeneous model and obtained experimental boundaries were carried out under different inlet air void fractions (IAVFs). A hybrid numerical method was proposed to obtain the flow-induced emitted noise characteristics. The results show there is little influence on pump characteristics when the IAVF is less than 1%. The pump head slope degradation was found to increase with air content. The bubbles adhere to the impeller hub on the blade's suction side and spread to the periphery with a big IAVF, leading to unstable operation. It is obvious that vortices appear inside the impeller flow passage as IAVF reaches 6.5%. The two-phase flow pattern has a small effect on the characteristic frequency distribution of pressure fluctuation and emitted noise, but the corresponding pulsation intensity and noise level will increase. The study could provide some reference for low noise design of the drainage pump.

Keywords: miniature drainage pump; gas-liquid two-phase flow; radiate noise; pressure

1. Introduction

Pumps are classified as general machinery with varied applications [1–3]. In addition to energy saving and emission reduction, pumps now attract attention in terms of vibration and noise reduction [4,5]. Mini pumps with radial vane impellers are widely used in many fields such as electronic industries, blood pumps for medical industries, aerospace industries, and in some home appliances [6,7]. These pumps are known as mini-pumps because their sizes are smaller than 50 mm. Drainage pumps for home appliances such as drum type washing machines are mini pumps that consist of a synchronous motor with a splined shaft and a radial impeller connected with the splined shaft and the pressure chamber [8,9]. The radial impellers are easy to manufacture and present a stable head [10,11]. At present, the most important qualities of the washing machine are the water saving, power saving, and low noise ability, similar to other home appliances. The stable operation state of the centrifugal pumps is always broken up by the entraining gas, resulting in performance degradation and a higher level of the noise [12]. Therefore, it is significant to study gas–liquid two-phase flow behavior and the relationship to pump performance and flow induced noise characteristics of radial blade pumps.

The drainage pump is the prime drainage power component in a drum type washing machine. Initial research from Jang and Lim [13] points out that the drainage pump is one of the main noise sources in the operation of a washing machine, which affects the quality of the equipment seriously. They invented an apparatus that can control the vibration and noise during the startup period. However, the pump falls in the state of air water mixed transportation, namely empty discharge with the decrease of water level, which has a higher impact on the noise and vibration levels. An unsteady flow pattern of the air water mixture will induce gas pockets and cause performance degeneration when the pump is working in this period. Murakami and Minemura [14,15] studied experimentally the influence of air–water two-phase flow on centrifugal pump performance related to the internal flow-field by visualization and proposed a one-dimensional modeling method. This was the first study which focused on the flow pattern inside the impeller in relation with pump performance. Subsequently, Pessoa and Prado [16] observed cyclical head variation of the centrifugal pump under gas–liquid two-phase flow and pointed out that this may be caused by the periodic variation of the air pocket in the pump. Flow visualization by a high-speed camera makes it easy to identify the flow patterns and bubble size distribution when analyzing the internal flow-field under gas–liquid two phase flow. Serena and Bakken [17] performed flow instability analysis on a multiphase pump model working under higher gas void fractions and described the complex flow behavior in terms of length, location, and time scale. Further, their studies provide detailed information about the unsteady two-phase flow behavior and surging phenomena. Schäfer et al. [18] performed investigations using gamma-ray tomography techniques to find local discontinuities in pump characteristics. The handling ability for gas entraining of a centrifugal pump is still not clear to the pump designer, which is related to the two-phase flow behaviors inside the flow passage. Verde et al. [19] investigated how both flow and head modification can be related to flow pattern modifications inside the impeller. The above research, mainly experimental, focused on a traditional bend centrifugal impeller, rarely on radial blade pumps. Computational fluid dynamics (CFD) has been used widely in many engineering fields with the rapid development of computer technology [20–22] which could comprehensively and physically understand complicated rotating flow phenomena combined with advance signal processing technology [23–25] Three-dimensional Reynolds-averaged Navier–Stokes equations (3D-URANS) CFD approaches based on Eulerian-Eulerian two-phase momentum equations and drag force models are widely adopted to process the simulation. The size and coalescence of the bubble has been found to play an important role in modeling the flow characteristic of pump under two-phase conditions [26,27]. The review paper by Si et al. [28] introduces a lot of related research works on flow-induced noise of pumps, most of which believe that the intensive pressure pulsations are an important source of hydrodynamic excitation force that in turn produces fluid and structure-borne noise. Further, fluid-borne noise is a major contributor to radiated noise as well as increased fatigue in the system components. However, related research introduced in [28] mainly focused on the internal flow and induced noise under pure water conditions. More attention should be paid to the influence of air–water mixed flow behavior to fluid borne noise on such radial blade pumps under two-phase flow.

In the present work, pump performance and noise characteristics were experimentally tested on a mini drainage pump used in washing machines. Pump performance and emitted noise measurements were both processed to obtain a relation between basement and boundary conditions for the next simulation works. 3D-URANS numerical calculation based on Euler inhomogeneous model and the obtained experimental boundary conditions are applied for four flow rates and different inlet air void fractions. The obtained air phase distribution and velocity field are presented for exploring the influence of the IAVF on the head degeneration and pressure pulsation variation of the model pump. The results could provide some theoretical guidance in optimizing the mini pump with low noise.

2. Experimental Setup and Calculation of Boundary Conditions for Pump Model

A single-stage, single-suction centrifugal pump with radial blade impeller and annular chamber volute was used for the study. This type of drainage pump is widely used on the drum washing machine because it could produce the same head when reversed rotate, thereby effectively avoiding the entanglement of debris. As shown in Figure 1, it provides kinetic energy and potential energy for the liquid after washing, which is necessary for the drainage process of washing machine. The required pump head H_r for draining is 0.8m. Design parameters of the pump are shown in Table 1. With the decrease in water level, the pump undergoes in the state of empty discharge expressed as pump head falling and high level of noise generation with intuitive adverse feeling of humans.

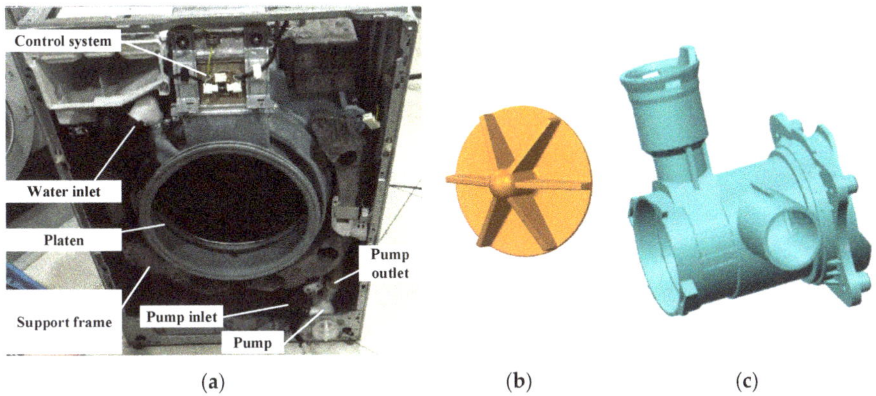

Figure 1. Selected pump model. (**a**) Working scenario of the pump; (**b**) Impeller; (**c**) Casing.

Table 1. Design parameters of the model pump.

Parameters	Values	Parameters	Values
Design Flow rate Q_d	1.08 m^3/h	Blade thickness δ	1.5 mm
Head H_d	1.2 m	Blades number Z	6
Rotating speed n	3000 r/min	Specific speed n_s	68
Impeller outlet diameter D_2	37 mm	Volute base circle diameter D_3	50 mm
Blade outlet width b_2	6.25 mm	Diameter of outlet pipe D_4	30 mm

2.1. Test Equipment and Loop

In order to simulate the working condition of the model drainage pump used in the washing machine, an open type test rig was built, as shown in Figure 2. The hydraulic performance and noise experiments of drainage pump are performed in the semi-anechoic chamber with size of 4.2 × 3.2 × 3 m. The background noise of the semi-anechoic chamber is 20 dB and the cut-off frequency is 50 Hz.

In this open loop, the valve between the upper tank and lower tank is kept open when processing pump performance measurements under different IAVF. The air injection system is driven at the constant 0.5 bar pressure by a compressor combined with a regulator tank. The air mass flow rate is measured and well controlled by a Bürkert 8107 fluid control system which contains micro-electromechanical system flow sensors that could supply volume air flowrate value on standard conditions (25°, 101,325 Pa). The initial air volume flow rate is calculated by ideal gas equation of state after measuring the pump inlet pressure. Then, further calculation of the IAVF could be processed by the obtained air and pure water volume flowrate. The air–water mixed fluid is sucked into the pump, goes through the regulating valve, and finally arrives into the upper tank. Air bubbles inside the mixed fluid exhaust to the external space stay in the upper tank and the left pure water runs to the lower tank. The flow rate of pure water was measured by an electromagnetic flowmeter set between the lower tank and the mixer. Pump performance curves at pure water fluid are obtained when the compressor stopped.

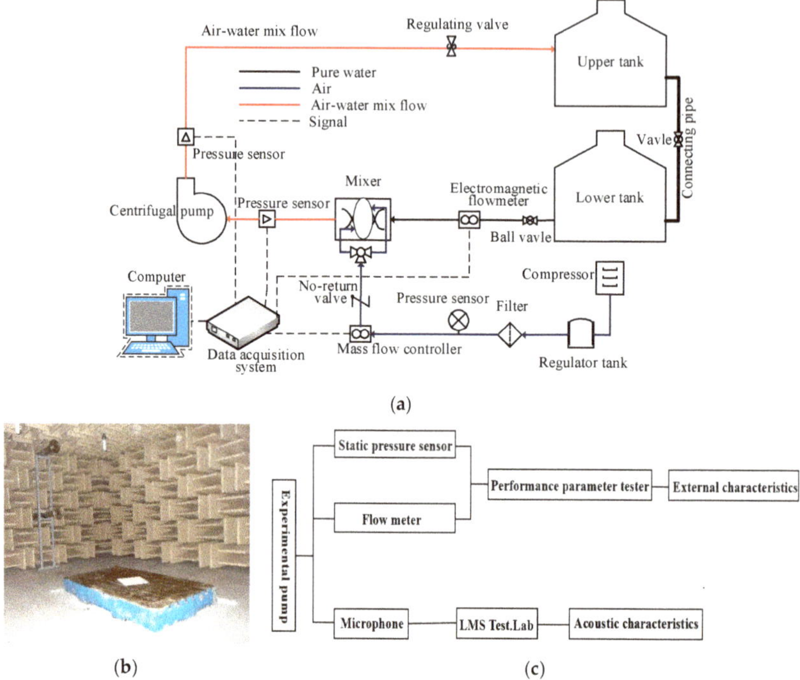

Figure 2. Experiment system list are (**a**) Test rig; (**b**) semi anechoic chamber; (**c**) Data acquisition system.

Pump noise characteristics are also measured as they are used in washing machines, which means working for a drainage period. The valve between the upper tank and lower tank keeps close to simulate the drainage of the water inside the lower tank. Three microphones typed as PCB 14,043 (3.15 Hz~20 kHz frequency response, 146 dB maximum sound pressure level) located around the pump as a circle with 1 m to collect noise signal together with equipment and software of LMS test Lab platform (Leuven Measurement & System international, Leuven, The Kingdom of Belgium).

Two static pressure sensors with an accuracy of 0.25% were set at the pump inlet and outlet to calculate the pump head. Digital power meter with accuracy 0.4% were used to measure the hydraulic shaft power after dividing it by a constant motor efficiency coefficient. The pump head and efficiency also could be obtained following the procedure in ISO 9906: 2012 [29]. The biggest uncertainties of the measurement are ±1.8% error of pump head, ±5% error of pump efficiency, ±0.5% error of water flow rate, and ±0.5% of air void fraction calculated by instrument precision.

2.2. Experimental Results Analysis

2.2.1. Pump Performance

According to the Bernoulli equation, the parameters such as head, efficiency, IAVF, and density of the air–water mixer fluid are defined as follows.

$$H = \frac{p_2 - p_1}{\rho g} + \frac{v_2^2 - v_1^2}{\rho g} + (z_2 - z_1) \tag{1}$$

$$\eta = \frac{\rho g H Q}{P_e} \tag{2}$$

$$IAVF = \frac{Q_g}{Q_g + Q_l} \tag{3}$$

$$\rho = (1 - \alpha)\rho_l + \rho_g \tag{4}$$

where the subscript g means air and l means water. Shaft power P_e is calculated by multiplication of the obtained motor current and voltage values and its constant efficiency value (namely 0.65).

According to the Euler formula, the radial type impeller could provide a constant head whatever the flow rate. Figure 3a shows the measured hydraulic performance curve of the washing machine drain pump under pure water. Inconsistent with theory, the head curve of the pump shows a linear degradation trend with a small slope when the flow rate increases. The head shows a maximum value of 2.3 times H_r when the pump is running at zero flow rate. The maximum flow rate that could supply enough head for draining is no more than 1.4 Q_d. A maximum efficiency value of 22.4% concerns mostly the design flow rate vicinity. The above phenomenon can be explained in a way that the hydraulic loss in the inlet and outlet part of pumps increases with the increasing flow rate in actual flow conditions, which leads to the drop in the pump head. Figure 3b presents the pump head degradation results under initial flow rate as Q_d and constant regulating valve position when the IAVF increases. The result shows that the pump can still work when IAVF maximum reaches 70%. However, the required head H_r cannot be supplied by the pump when the gas content is greater than 25%, which might result in reflux flow from the outlet pipe of the pump. As shown in Figure 3c are the pump head degradation results under three constant water flow rates wherein the pump head drops in all three conditions, but a small flow rate could reach a bigger IAVF because of the higher initial head. When the IAVF is greater than 15%, the slope of the head degradation curve under 0.8 Q_d deteriorates rapidly, by 10% under 0.6 Q_d and 7% under 0.4 Q_d. This might refer to the gas entrained handing ability difference of the impeller under different initial flow conditions.

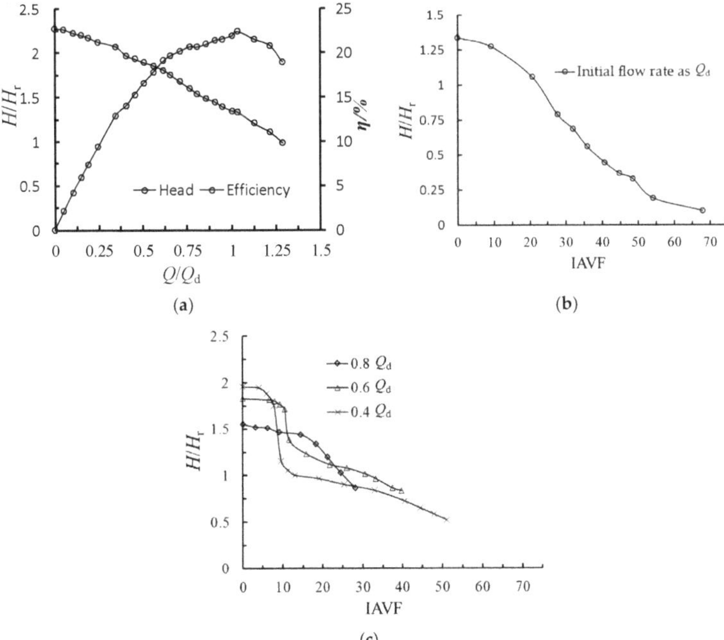

Figure 3. Pump performance characteristics curve under list condition: (**a**) IAVF = 0; (**b**) Initial flow rate as Q_d and Valve position kept constant; (**c**) Water flow rate kept constant.

2.2.2. Radiated Noise of the Drain Pump at Different Measuring Points

Figure 4 shows the averaged radiated noise characteristic of the model pump during one cycle of operation. The horizontal axis corresponds to the running time of the drainage pump and the ordinate corresponds to the total sound pressure level of the model pump. The sound pressure level (SPL) can reflect the acoustic field characteristics in terms of sound pressure distribution, which is defined as:

$$SPL = 20 \lg \frac{p}{p_{ref}} \tag{5}$$

where p means sound pressure and p_{ref} equal to 10^{-5} Pa. During the noise test process, the valve between the upstream tank and lower tank is kept close in order to simulate the actual draining process of the drainage pump. The LMS noise acquisition module starts at the same time as the drainage pump is working. The results show that the radiated noise curve of model pump can be divided into three parts corresponding to the different running states of the washing machine, namely the starting period, normal draining period, and gas–liquid mixture transportation period (empty period). The motor starts, and the rotor produces significant noise at the beginning of start-up period. Then, the rotational speed of the pump gradually reaches the designed one, leading to normal flow rate working conditions and the SPL gradually reduces to a low level, about 38.5 dB(A). In this period, namely the normal drainage period, the pump transports pure water. As the water level inside the lower tank drops, some of the air draw into the drainage pump to form a gas-liquid two-phase flow state. The SPL of the pump increases sharply and reaches about 50 dB (A). Due to the performance deterioration of the pump during the empty discharge period, the head of the pump does not reach the required 0.8 m and the fluid in the outlet pipe is continuously returning to the impeller and generating fluid borne noise, thereby keeping the radiated noise of the pump at a higher level. The difference in the above sound pressure level is related to the pressure pulsation phenomenon corresponding to the specific flow state of the two-phase flow inside the pump passage.

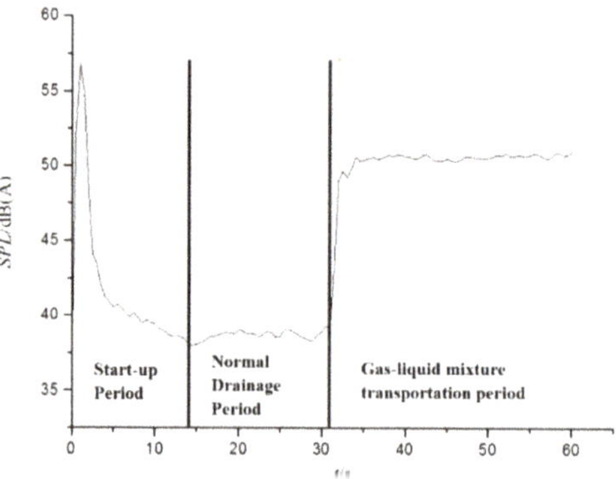

Figure 4. The SPL of radiated noise of the pump.

In addition, Figure 5 shows the spectrum comparison in frequency between the normal drainage period and the empty discharge period. The ordinate represents the SPL amplitude and the abscissa is processed as a multiple of the shaft frequency (f_n). For the normal drainage period and the empty discharge period, an expected discrete noise characteristic appears and the axial frequency and its multiple is the main frequency, which means the discrete noise is related to the motor rotational speed. For a two-phase flow state, the unsteady flow in the pump gives rise to the instability of the pump rotor, which in turn makes the discrete noise of the empty discharge period higher. Thus, we could conclude that SPL is bigger at the end of the drainage period because of gas-liquid two phase flow. This happens due to the unsteady flow inside the rotating part of the pump that causes the flow induced structure vibration and turns to radiated noise.

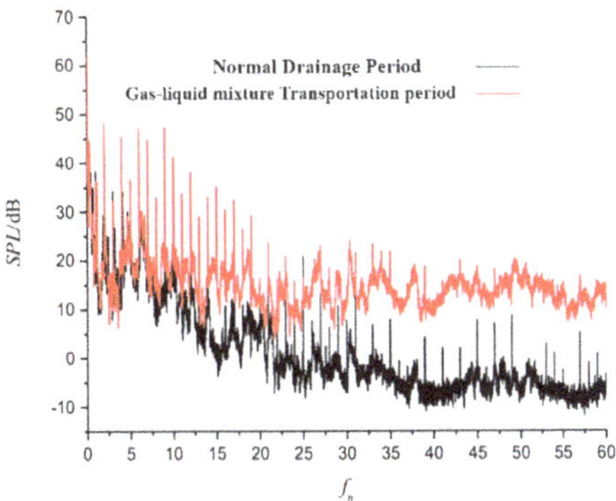

Figure 5. Frequency responses of the two working periods.

3. Numerical Simulation Study on the Inner Flow Behavior

Presented here is a very helpful tool to solve basic equations that model the flow movement by CFD, which give us more movement characteristics of the flow than experimental investigations [30,31]. In recent years, in order to provide detailed two-phase flow results in terms of flow structure, phase distribution and slippage, CFD has become increasingly important with the rapid development of computer technology, which can be used to obtain physical understanding of the noise generated by gas–liquid two phase flow.

3.1. The Eulerian–Eulerian Inhomogeneous Two-Phase Flow Modeling Method

The numerical calculation of the model pump flow field concerns pumps working under pure and air–water mixed flow conditions at different flow rates. Normally, multiphase flow models are subdivided into homogeneous and inhomogeneous kind. In the inhomogeneous model, both the velocity slip and the interphase mass and momentum transfer terms are solved. Each phase has its own fluid field and passes through the phase transfer unit. Because the homogeneous model does not consider any velocity slip between the two phases, this research adopts the inhomogeneous model regardless of the temperature field, for which the liquid phase is continuous and the gas phase is discrete. The particle model assumes that the gas–liquid two-phase flow pattern corresponds to a bubbly flow, meeting the principle of the mass and momentum conservation. Gas–liquid here always refer to air–water in the following.

$$\frac{\partial}{\partial t}(\alpha_k \rho_k) + \nabla \cdot (\alpha_k \rho_k \boldsymbol{w}_k) = 0 \tag{6}$$

$$\frac{\partial}{\partial t}(\alpha_k \rho_k \boldsymbol{w}_k) + \nabla \cdot (\alpha_k \rho_k \boldsymbol{w}_k \otimes \boldsymbol{w}_k) = -\alpha_k \nabla p_k + \nabla \cdot (\alpha_k \mu_k (\nabla \boldsymbol{w}_k + (\nabla \boldsymbol{w}_k)^{\mathrm{T}})) + \boldsymbol{M}_k + \boldsymbol{f}_k \tag{7}$$

where k is phase (l-liquid, g-gas); ρ_k is density of the k phase, kg/m^3; p_k is pressure of k phase, Pa; α_k is the void fraction of k phase; μ_k is the dynamic viscosity of k phase, Pa·s; $\boldsymbol{w}_k$ is the relative velocity of the k phase fluid, m/s; $\boldsymbol{M}_k$ is due to the interphase drag force; and $\boldsymbol{f}_k$ refers to the added mass force related to the contribution of the impeller rotation.

For this two-phase flow approach, the liquid phase is considered the continuous phase using the renormalization group (RNG) k-ε turbulence model. Meanwhile, the gas phase is considered as the discrete phase using the zero-equation theoretical model, which means that the action between the two phases only considers the so-called interfacial drag coefficient through the following relations [32]:

$$M_l = -M_g = \frac{3}{4}c_D\frac{\rho_l}{d_B}\alpha_g\left(w_g - w_l\right)\left|w_g - w_l\right|$$ (8)

with:

$$c_D = \begin{cases} \frac{24}{R_e}(1 + 0.15Re^{0.687}) & (R_e \leq 1000) \\ 0.44 & (R_e > 1000) \end{cases}$$ (9)

and

$$Re = \rho_l\frac{\left|w_g - w_l\right|}{\mu_l}d_B$$ (10)

where d_B is the diameter of the bubble and c_D refers to the resistance coefficient.

3.2. Computational Domain, Meshing and Boundary Setting

The flow field computational domain includes the inlet chamber, impeller, pressure chamber, and filter chamber. ANSYS ICEM 14.0 was used to carry out this work. As shown in Figure 6, the impeller and the filter chamber are meshed with a hexahedron structure, which can better control the density of the grid boundary layer and ensure the orthogonality of the grid. Due to the complexity of the geometric models of the inlet chamber and the pressure chamber, unstructured meshes with better geometric adaptation are adopted. The resulting pump model that consisted of 2,377,605 elements in total (about 2.4×10^6) was chosen for rotating and stationary domains after grid independence analysis. It can be seen from Figure 7 that when the number of meshes is greater than this value, the change in head coefficient of the model pump working under Q_d is less than 1%.

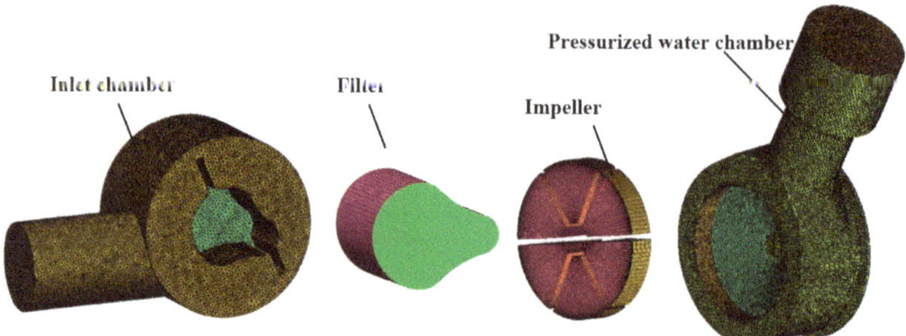

Figure 6. Meshing results of the computational domain.

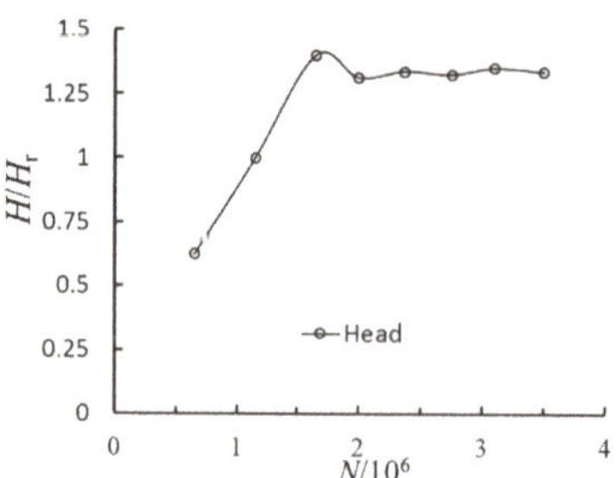

Figure 7. Grid independence analysis with pure water under Q_d.

Considering the requirements of the test, the inlet boundary conditions are set according to the pressure measured in the experiment, and the outlet boundary conditions are set as the mass flow rate. A certain amount of air phase is defined as initial bubble with a diameter of 0.2 mm which is equivalent to the experimental injector device. The liquid boundary adopts the non-slip solid wall condition, and the gas boundary adopts the free-slip solid wall condition. The smooth wall condition is used for the near-wall function. The time step was set as 5.56×10^{-5} s, corresponding to $1°$ impeller rotation for each step. The accuracy of convergence is set to 10^{-4}. A total simulation time was set as 0.2 s in order to further process a data reduction of the unsteady flow field, corresponding to 10 impeller revolutions. The direction of impeller rotation is set as counterclockwise.

The previous experimental results show that the system cannot maintain a very constant flow rate of pure water when inject air at high flow rates. Pump performance deteriorates quickly, leading the drain pump to go into an empty discharge operation period when IAVF is bigger than 7%. Big radiated noise during the empty discharge period is caused by the mixing and impacting process from the outlet backflow to the impeller blade. Moreover, pre-research of the modeling method used in this calculation by Si et al. [33] suggests that it is more accurate for low IAVF. Therefore, simulation processes at selected flow rates smaller than 1.2 Q_d and several conditions IAVF lower than 7% were carried out.

3.3. Simulation Results Analysis from CFD

3.3.1. Pump Performance Degradation

A comparison between the experimental and the numerical results of the pump head ratio and efficiency under the pure water conditions are shown in Figure 8. Values of each point are obtained by the average of 360 timesteps from the unsteady calculation. It can be seen that the experimental results show the same trend as the simulation results in the range from 0.1 Q_d to 1.2 Q_d. Simulated head ratio of the model pump is almost identical to the test one when flow rate below 0.4 Q_d. There is a certain difference with bigger values when the flow rate is above 0.4 Q_d, which might be due to the smooth wall boundary setting. Deviation between them increases gradually with the increase of the flow rate. The relative error between the two works is below 5%, indicating that the results of simulation are credible. Simulated head ratios under eight flow rates and three different IAVFs are also presented in Figure 8. Here, the pump head ratio decreases faster as the flow rate increases at big IAVF. The pump head ratio has little change at small flow rates when the gas content is lower than 7%. This might occur because the radial type blades are conducive to transporting air bubbles into the cavity without causing blockage under a low flow rate.

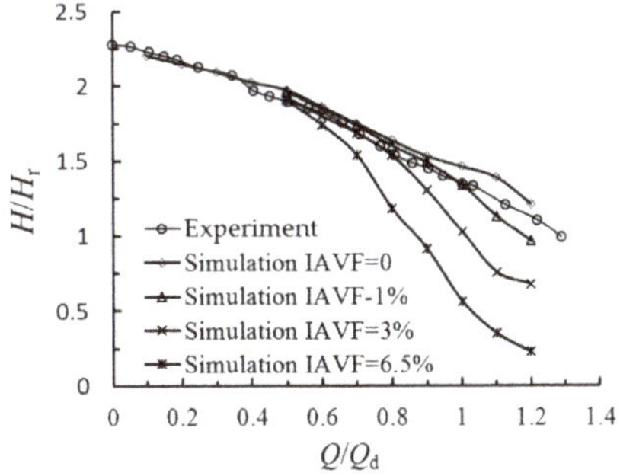

Figure 8. Numerical pump performance.

3.3.2. Two Phase Flow Behavior Analysis

Figure 9 shows the volume distribution of the air phase inside the pump flow passage under Q_d at three different IAVF conditions. When the inlet gas content is low, the bubble volume is relatively small and mainly distributed near the suction surface of the impeller. With the increase of inlet gas content, the bubble near the suction surface of impeller will develop along the hub direction and radial direction of the impeller, and then block some part of the flow channel. When the IAVF reaches 6.5%, the impeller channel is partly blocked, which reduces the working capability of the model pump.

The air pocket formed in the filter chamber occupies the upper half of the flow passage, which further deteriorate the impeller flow condition.

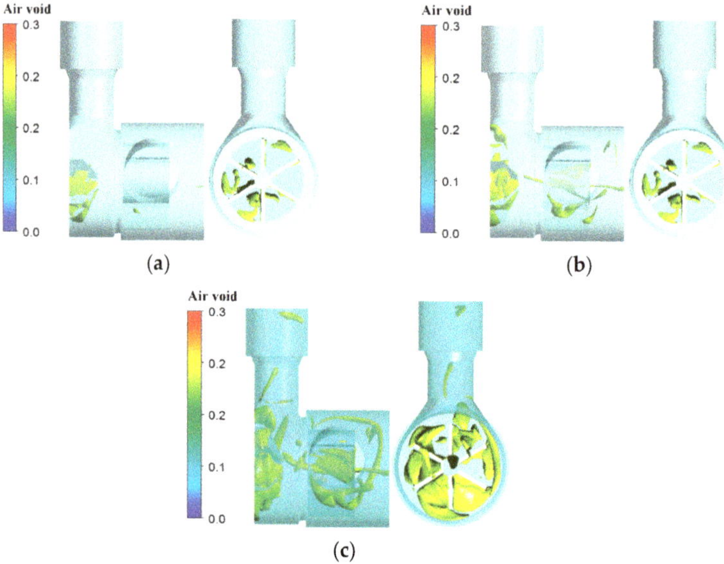

Figure 9. Air phase distribution at different IAVF under Q_d: (**a**) 1%; (**b**) 3%; (**c**) 6.5%.

Figure 10 shows the air velocity distribution inside the impeller middle section under four different flow rates when IAVF equal to 6.5%. Seen from it, the flow rate has obvious effect on the air phase velocity, the larger the flow rate, the bigger the velocity. The air phase with high velocity mainly appears near the pressure side in the middle of the flow passage under small flow rate. And also, partly bubbles with high velocity separate from the tip of the impeller, which is related to the centrifugal force generated by the rotating impeller. As the flow rate increases, the area of air phase with high velocity becomes larger and moves to the middle of the flow passage and trailing edge position. Under 0.8 Q_d, the distribution of air phase velocity streamline is quite disordered, and there are obvious swirls in some blade passage. With the increase of flow rate, the number of swirls increase, and the position of the vortex also has a corresponding deviation. The corresponding gas velocities are higher in the vortex region above, which indicates that the swirl motion in the flow has a great influence on gas accumulation as well as on its position, and then effects the working capability of the model pump.

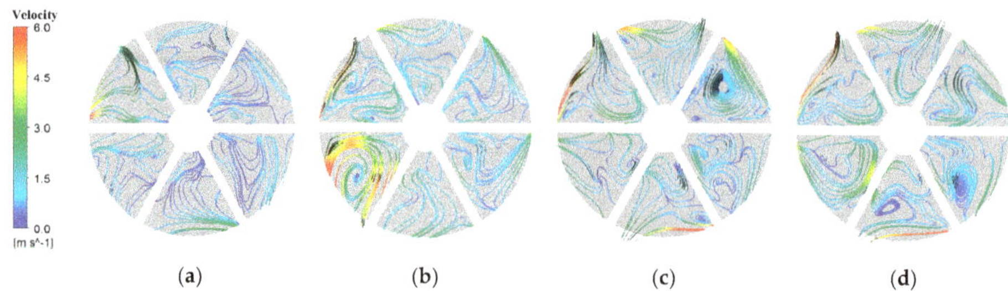

Figure 10. Air phase velocity distribution under different flow rates when IAVF = 6.5%: (**a**) 0.6 Q_d; (**b**) 0.8 Q_d; (**c**) Q_d; (**d**) 1.2 Q_d.

Turbulent kinetic energy is used to describe the degree of turbulence pulsation, and it is often used to describe the viscous dissipation of fluid and the range of pulsation and diffusion. Figure 11 shows the distribution of the turbulent kinetic energy of the impeller in the model pump under the conditions of an inlet gas content of 6.5% and different flow rates. With the increase of flow rate, the turbulent kinetic energy in the impeller passage increases obviously, and the high turbulent kinetic energy region mainly appears near the impeller outlet. Under the design conditions, the turbulent kinetic energy of the flow is relatively uniform. On the other hand, under off-design conditions, the high turbulent kinetic energy is mainly concentrated in the gas accumulation area due to the decrease of fluid flow stability in the channel, which is caused by the gas accumulation, and then it leads to high energy loss, which reduces the working capability of the model pump.

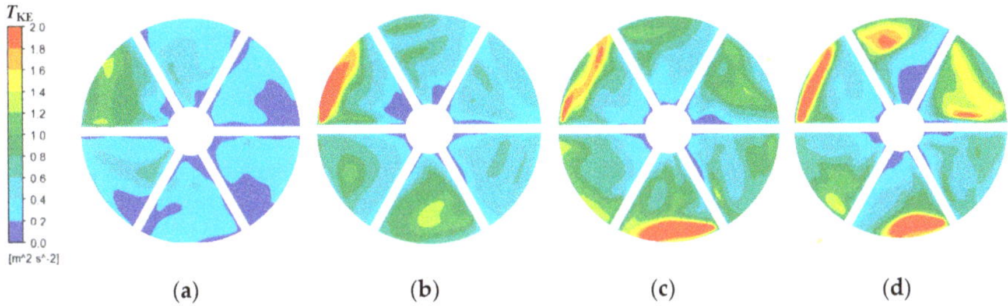

(a)   (b)   (c)   (d)

Figure 11. Turbulent kinetic energy distribution under different flow rates when IAVF = 6.5%: (**a**) 0.6 Q_d; (**b**) 0.8 Q_d; (**c**) Q_d; (**d**) 1.2 Q_d.

3.3.3. Analysis of Pressure Fluctuations

From the above turbulent kinetic energy (T_{KE}) analysis, the void fraction of bubbles is closely related to the intensity of T_{KE}. Future analysis on frequency domain are needed to express the flow behavior. The inner flow characteristics, including rotor–stator interaction, eddy current, and backflow, which is closely related to pump noise, induce a dynamic response of the pump in terms of pressure pulsation. Therefore, several monitoring points are set as shown in Figure 12 during the simulation. Among them, P1, P2, P3, and P4 are evenly distributed on the outlet of the impeller, whereas P5 is locate at the middle of outlet chamber, P6 and P7 are located near the volute tongue of the drainage pump. The pressure pulsation of each monitoring point is analyzed by fast Fourier transform (FFT) under Q_d at IAVF equal to 0% and 6.5%.

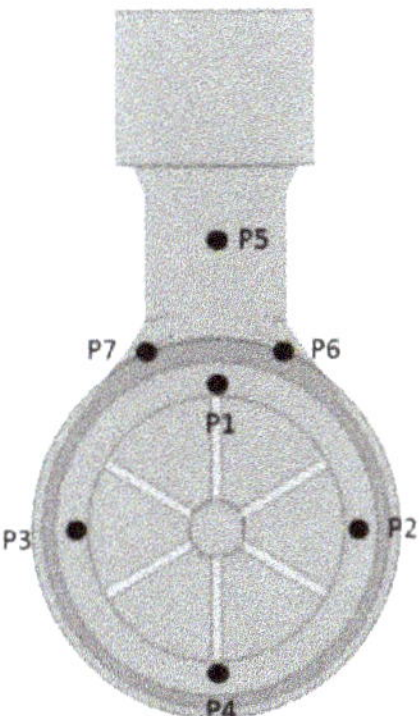

Figure 12. Location of monitoring points.

In order to present this in a normalized form, a nondimensional pressure coefficient $C_P{}^*$ expressed as follows is introduced.

$$C_P{}^* = \frac{\left(P - \overline{P}\right)}{0.5\rho u_2{}^2} \tag{11}$$

where P is the static pressure of the monitoring points, $\overline{P}$ is the average of the static pressure, ρ is the fluid density, and u_2 is the circumferential velocity component at the impeller outlet.

Figure 13 is the frequency representation of the pressure pulsation spectrum. The results shown in Figure 13a,b indicate that the amplitude of pressure pulsation is larger in the low frequency region (below 300 Hz) and it is mainly caused by the flow separation, backflow, and turbulence inside the pump flow passage under pure water conditions. No matter whether the inlet is pure water or gas–liquid two-phase flow, there are obvious discrete frequencies in the impeller outlet region, which are 300 Hz and frequency multiplier. From Figure 13c,d, the peak value of the frequency multiplier decreases faster than that of blade frequency, which indicates that the gas–liquid two-phase flow does not affect the characteristic frequency of pressure pulsation. Compared with pure water, the amplitude of pressure pulsation increases obviously, which indicates that the existence of bubble makes the flow instability of impeller outlet increase. Close to the tongue area of the volute, the pressure pulsation intensity in the low frequency region of the two-phase flow is obviously higher than that in the pure water conditions, inducing greater noise.

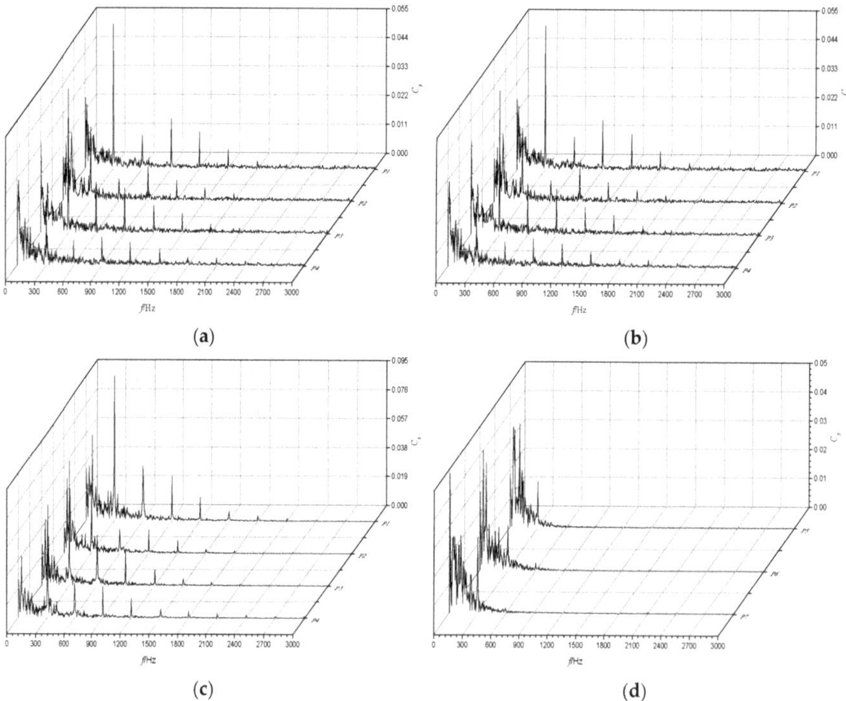

Figure 13. Spectrum of pressure fluctuation in the model pump: (**a**) P1-P4, IAVF = 0; (**b**) P5-P7, IAVF = 6.5%; (**c**) P1-P4, IAVF = 0; (**d**) P5-P7, IAVF = 6.5%.

4. Radiated Noise Simulation

In an air–water two phase flow situation, the complex flow behavior is expressed as pressure pulsation with big intense generation of strong hydraulic excitation which in turn causes radiated noise problems. Hence, the relationship between hydraulics and radiated noise of the pump running at the above conditions must be well understood. A numerical simulation of the model pump working under Q_d and air–water two phase flow are presented to provide detailed information.

4.1. Computational Acoustic Theory and Method

A hybrid numerical method mentioned by Gao et al. [34] is proposed to describe the flow-induced noise generation and dissemination based on Lighthill acoustic analogy theory, which subdivides the computational process into computational fluid dynamics (CFD) and computational acoustics (CA). The CFD process calculates the characteristics of the sound source and puts out the flow field behavior such as velocity, density, and pressure for acoustic simulation. Acoustic analogies are derived from the Navier–Stokes equations, which govern both the flow field and corresponding acoustic filed. It can be transformed into a Lighthill function, which is expressed as:

$$\int_\Omega \left(\frac{\partial^2}{\partial t^2}(\rho - \rho_0)\delta\rho + c_0^2 \frac{\partial}{\partial x_i}(\rho - \rho_0)\frac{\partial(\delta\rho)}{\partial x_i} \right) dx = -\int_\Omega \frac{\partial T_{ij}}{\partial x_j}\frac{\partial(\delta\rho)}{\partial x_i} dx + \int_\Gamma \frac{\partial \Sigma_{ij}}{\partial x_j} n_i \delta\rho d\Gamma(x) \qquad (12)$$

The first part of the right term is the volume source, and the second part is the surface source. In the acoustic simulation, the solution of the Navier-Stokes equation firstly assumes the water as incompressible to calculate the flow-induced acoustic source. Then, the compressibility of the water is to be considered to solve the acoustic wave propagation. Combined with acoustic wave equation, the final frequency response of the radiated noise could be calculated.

During the noise simulation procedure, the unsteady simulation of the pump flow field was re-obtained using the detached eddy simulation (DES) method, a modification of a RANS (Reynolds Average Navier-Stokes) model in which switches to a sub-grid scale formulation in regions fine enough for large eddy simulation [35]. The detailed flow information, such as flow velocity, pressure, density, etc., under two phase flow conditions, are extracted and transformed as the sound source in the acoustic simulation. Meanwhile, the acoustic calculation domains were built and the acoustic meshes were generated containing an interface setting. Finally, the radiated noise calculation was completed by the acoustic finite element method using Actran12.0 software. The flow chart of the acoustic simulation work is shown in Figure 14.

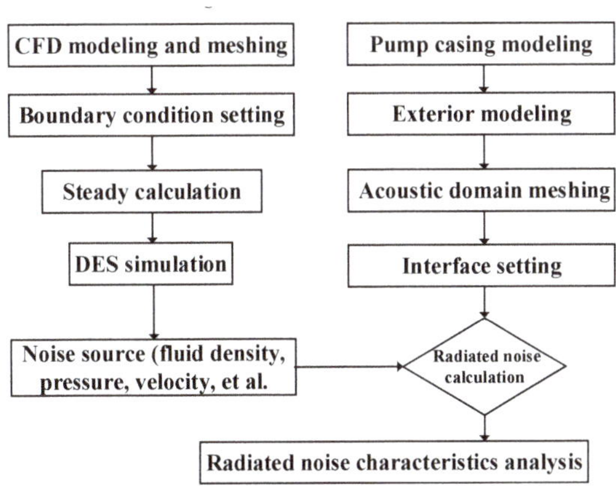

Figure 14. The flow chart of the radiated noise calculation of the multi-stage centrifugal pump.

4.2. Computational Domain, Mesh Generation and Boundary Condition of the Acoustic

The computational domain for the acoustic simulation includes the structural domain and air-borne domain, as shown in Figure 15. Unstructured mesh, which has better adaptability to the geometry, is applied in the acoustic simulation. To guarantee the precision of the acoustic computation, the maximum mesh size should meet the Equation (13). The data transmission of the interface between the structural domain and the air-borne domain is accomplished with the integral interpolation method. Also shown in Figure 15, the maximum mesh size of the structural domain is 0.0002 m and air-borne domain is 0.0006 m in this study. The inner surface between the structural domain and flow field domain is loaded with information of the unsteady flow to yield the sound source in the subsequent simulation, and the definition of the radiated air-borne sound domain is set to get the distribution of the radiated noise in the subsequent simulation.

$$L_{\max} < \frac{c}{6 f_{\max}}$$ (13)

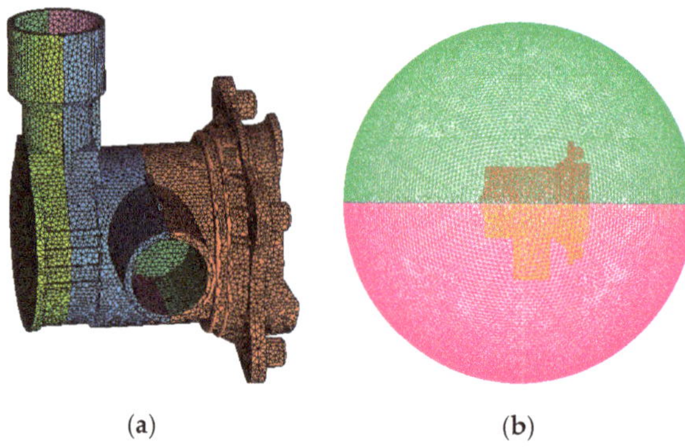

(a) (b)

Figure 15. The mesh for sound field calculation: (**a**) Structural domain; (**b**) Air-borne domain.

The material properties of the structure domain are shown in Table 2. Considering the total time of the unsteady CFD simulation and the time step, the frequency range of the acoustic simulation is set from 0–3000 Hz and the resolution is set as 1 Hz.

Table 2. Material properties of the structure domain.

Material	Density/(kg/m^3)	Young's Modulus/GPa	Poisson's Ratio
PP	910	0.896	0.4103

4.3. Acoustic Field Results

In order to analyze the properties of the radiated noise, 60 monitor points are mounted equally in the mid-span surface of the pump casing surface with a radius equal to 1 m. The SPL of the monitoring points set in the sound field is obtained by using the ACTRAN PLTViewer module, and the frequency response spectrum results are shown in Figure 16.

The radiation noise under pure water, as can be seen in the above figure, contains certain discrete frequency amplitude. Confirmed with the pressure pulsation analysis above, it is known that this discrete frequency is caused by the rotor stator interaction between the impeller and the water chamber. Compared with the discrete noise under the pure water condition, the two-phase flow condition with IAVF of 6.5% present a higher level in the entire frequency domain, which is consistent with the experimental results. In addition, in the two-phase flow condition, the discrete frequency of the radiated noise is no longer obvious, which is consistent with the pressure pulsation analysis above.

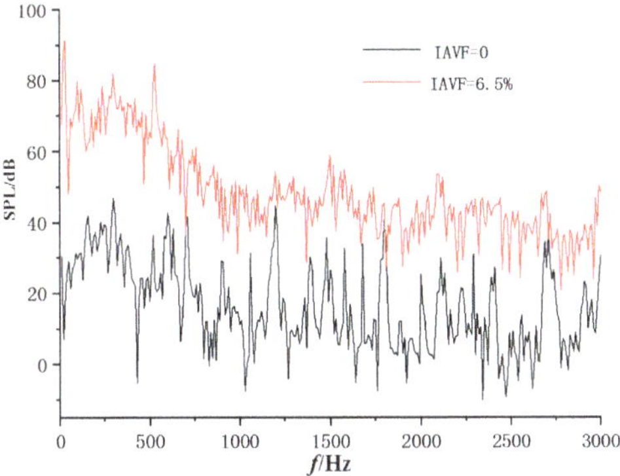

Figure 16. Frequency response of the radiated noise at monitoring point of the sound field.

Figure 17 shows the corresponding cloud image of the radiation noise at blade passing frequency (300 Hz) in pure water and IAVF equal to 6.5% condition. The left column with "map pressure" means SPL distribution. The scale represents the magnitude of the SPL values. In Figure 17a, the SPL near the wall of the shell is the largest, and the noise gradually decreases with the increase of the radius. The distribution of noise sound pressure levels across both the section of the impeller and the axial section exhibit dipole characteristics, which are related to the dynamic and static interference phenomena in the model pump. Among them, the light blue band region can be regarded as a dipole axis, and the dipole axis normal region has the highest sound pressure level. The larger radiated sound pressure is mainly in the vicinity of the model pump outlet and the model pump chamber. Compared with Figure 17a, the SPL of the pump noise working under an air–water two-phase flow condition increases significantly as shown in Figure 17b, and the dipole characteristics are no longer obvious. This phenomenon indicates that the interaction between the air–water two-phase flow and the model pump casing is intensified, the excitation force of the pump casing is obviously increased, and the dipole source is no longer the main source of radiated noise. Under two-phase flow conditions, the bubble volume inside the pump passage changes with time, which occurs owing to the influence of the monopole.

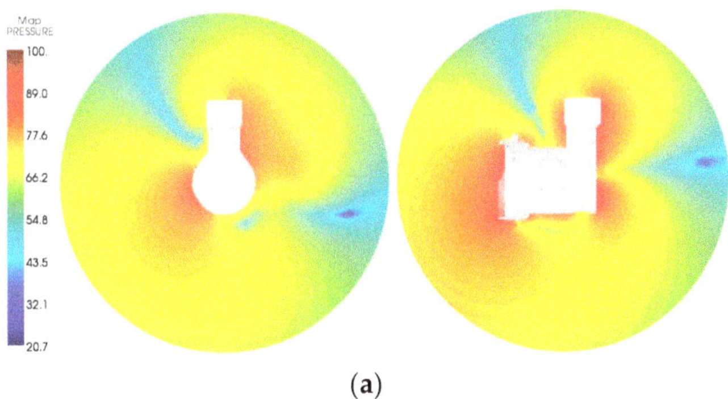

(a)

Figure 17. *Cont.*

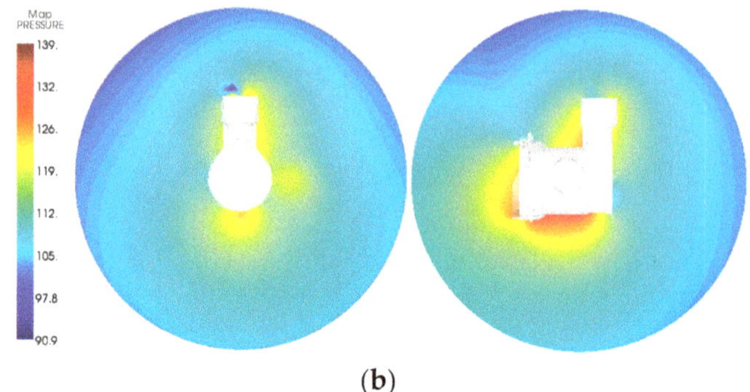

(b)

Figure 17. The SPL contour at blade passing frequency: (**a**) IAVF = 0; (**b**) IAVF = 6.5%.

5. Conclusions

The study presents an experimental and numerical study on gas–liquid two-phase flow behavior and flow induced noise characteristics of radial blade pumps. Three-dimensional numerical calculations based on the Euler–Euler inhomogeneous model and obtained experimental boundaries are carried out under different inlet air void fraction (IAVF). A hybrid numerical method is proposed to obtain the flow-induced emitted noise characteristics. The following results have been obtained:

(1) Under small flow rate conditions, the change of inlet gas content has a little effect on the head characteristics of drainage pump. When the IAVF is 1%, the influence of gas phase on the head characteristics of the drainage pump is small. With the increase of IAVF, the slope of the flow-head descending curve of the pump becomes larger. The influence of gas void on pump performance gradually increases with an increase of flow rate. When IAVF reaches 6.5%, the head curve decreases obviously at all flow rates.

(2) The sound pressure level is bigger at the end of drainage period (empty discharge period) because of air-water two phase flow. The flow induced structure vibration and noise is due to the unsteady flow inside the rotating part of the pump. Further, under two-phase flow conditions, the unsteady flow in the pump gives rise to the instability of the pump rotor, which in turn makes the discrete noise of the empty discharge period higher.

(3) As the IAVF increases at different flow rate conditions, the bubble adheres to the impeller hub and the backside of the blade, and develops around it. In addition, the formation of vapor pockets is easy in the inlet chamber, which could reduce the working capability of the drainage pump. When the IAVF is 6.5%, the air velocity increases with an increase of the flow rate. At small flow rates, flow separation will occur, and when the flow rate is 80% of the nominal value, an obvious vortex appears in the impeller flow channel.

(4) The air–water two-phase flow has a little effect on the characteristic frequency distribution of both the impeller outlet pressure pulsation and radiated noise SPL. However, the corresponding pulsation intensity and SPL values will increase with an increase in IAVF, which indicates that the existence of bubbles increases the instability of the impeller outlet flow. Thus, the influence of flow state should be taken into account in the hydraulic optimization of drainage pump impellers.

Author Contributions: Q.S. and C.W. conceived and designed the experiments; C.S. and R.C. performed the experiments and simulation; A.A. and Q.S. analyzed the data; Q.S. wrote the paper; J.Y. provided funding acquisition.

Funding: This research was funded by National Key Research and Development Program of China (2018YFB0606103), National Natural Foundation of China (51976079, 51779107), Training Project for Young Core Teacher of Jiangsu University and Senior Talent Foundation of Jiangsu University (15JDG048), China Postdoctoral Science Foundation (2019M661745) and Open Foundation of National Research Center of Pumps, Jiangsu University (NRCP201604).

Conflicts of Interest: The authors declare no conflict of interest.

Nomenclature

IAVF	inlet air void fraction
b	impeller blade width
d	diameter of bubble
D	diameter
H	pump head
M	shaft torque
n	rotational speed
p	local static pressure
P	shaft power
Q	volume water flow rate
R	radius
SPL	sound pressure level
v	velocity
z	height level
Z	impeller blade number

Greek Symbols

α	local gas void fraction
η	global efficiency of the pump
ρ	density of fluid mixture
ω	angular velocity
δ	blade thickness

Subscripts

B	bubble
d	design condition
g	gas
l	liquid
r	required
1	inlet of the impeller
2	outlet of the impeller
3	inlet of the volute
4	outlet of the pump

References

1. Wang, C.; Hu, B.; Zhu, Y.; Wang, X.; Luo, C.; Cheng, L. Numerical Study on the Gas-Water Two-Phase Flow in the Self-Priming Process of Self-Priming Centrifugal Pump. *Processes* **2019**, *7*, 330. [CrossRef]
2. He, X.K.; Jiao, W.X.; Wang, C.; Cao, W.D. Influence of Surface Roughness on the Pump Performance Based on Computational Fluid Dynamics. *IEEE Access* **2019**, *7*, 105331–105341. [CrossRef]
3. Wang, C.; He, X.K.; Shi, W.D.; Wang, X.K.; Wang, X.L.; Qiu, N. Numerical study on pressure fluctuation of a multistage centrifugal pump based on whole flow field. *AIP Adv.* **2019**, *9*, 035118. [CrossRef]
4. Zhang, J.; Xia, S.; Ye, S.; Xu, B.; Song, W.; Zhu, S.; Xiang, J. Experimental investigation on the noise reduction of an axial piston pump using free-layer damping material treatment. *Appl. Acoust.* **2018**, *139*, 1–7. [CrossRef]
5. Ye, S.G.; Zhang, J.H.; Xu, B.; Zhu, S.Q.; Xiang, J.; Tang, H. Theoretical investigation of the contributions of the excitation forces to the vibration of an axial piston pump. *Mech. Syst. Signal Process.* **2019**, *129*, 201–217. [CrossRef]
6. Kaigala, G.V.; Hoang, V.N.; Stickel, A.; Lauzon, J.; Manage, D.; Pilarski, L.M.; Backhouse, C.J. An inexpensive and portable microchip-based platform for integrated RT–PCR and capillary electrophoresis. *Analyst* **2008**, *133*, 331–338. [CrossRef] [PubMed]
7. Yu, S.C.; Ng, B.T.; Chan, W.K.; Chua, L.P. The flow patterns within the impeller passages of a centrifugal blood pump model. *Med. Eng. Phys.* **2000**, *22*, 381–393. [CrossRef]
8. Orue, R. Drain Pump for Home Appliances. U.S. Patent 2007/0071617 A1, 29 March 2007.

9.   Marioni, E.; Cavalcante, V. Centrifugal Pump for Electric Household Appliances such as Washing Machines, Dishwashers and the Like. U.S. Patent 4861240 A, 29 August 1989.

10. Luo, X.W.; Nishi, M.; Yoshida, K.; Dohzono, H.; Miura, K. Cavitation Performance of a Centrifugal Impeller Suitable for a Mini Turbo-Pump. In Proceedings of the Fifth International Symposium on Cavitation, Osaka, Japan, 1–4 November 2003.

11. Zhuang, B.; Luo, X.; Zhang, Y.; Wang, X.; Xu, H.; Nishi, M. Design optimization for a shaft-less double suction mini turbo pump. *IOP Conf. Ser. Earth Environ. Sci.* **2010**, *12*, 012049. [CrossRef]

12. Jiang, Q.F.; Heng, Y.G.; Liu, X.B.; Zhang, W.B.; Bois, G.; Si, Q.R. A review on design considerations of centrifugal pump capability for handling inlet gas-liquid two-phase flows. *Energies* **2019**, *12*, 1078. [CrossRef]

13. Jang, J.H.; Lim, D.B. Washing Machine Having a Drain Pump to Reduce Vibration. U.S. Patent 7966848 B2, 28 June 2011.

14. Murakami, M.; Minemura, K. Effects of Entrained Air on the Performance of a Centrifugal Pump: 1st Report, Performance and Flow Conditions. *Bull. JSME* **1974**, *17*, 1047–1055. [CrossRef]

15. Murakami, M.; Minemura, K. Effects of Entrained Air on the Performance of Centrifugal Pumps: 2nd Report, Effects of Number of Blades. *Bull. JSME* **1974**, *17*, 1286–1295. [CrossRef]

16. Pessoa, R.; Prado, M. Two-Phase Flow Performance for Electrical Submersible Pump Stages. *SPE Prod. Facil.* **2003**, *18*, 13–27. [CrossRef]

17. Serena, A.; Bakken, L.E. Design of a Multiphase Pump Test Laboratory Allowing to Perform Flow Visualization and Instability Analysis. In Proceedings of the ASME 2015 Power Conference, San Diego, CA, USA, 28 June–2 July 2015.

18. Schäfer, T.; Bieberle, A.; Neumann, M.; Hampel, U. Application of gamma-ray computed tomography for the analysis of gas holdup distributions in centrifugal pumps. *Flow Meas. Instrum.* **2015**, *46*, 262–267. [CrossRef]

19. Verde, W.M.; Biazussi, J.L.; Sassim, N.A.; Bannwart, A.C. Experimental study of gas-liquid two-phase flow patterns within centrifugal pumps impellers. *Exp. Therm. Fluid Sci.* **2017**, *85*, 37–51. [CrossRef]

20. Wang, C.; Shi, W.D.; Wang, X.K.; Jiang, X.P.; Yang, Y.; Li, W.; Zhou, L. Optimal design of multistage centrifugal pump based on the combined energy loss model and computational fluid dynamics. *Appl. Energy* **2017**, *187*, 10–26. [CrossRef]

21. Wang, C.; He, X.K.; Zhang, D.S.; Hu, B.; Shi, W.D. Numerical and experimental study of the self-priming process of a multistage self-priming centrifugal pump. *Int. J. Energy Res.* **2019**, *43*, 4074–4092. [CrossRef]

22. Qian, J.Y.; Chen, M.R.; Liu, X.L.; Jin, Z.J. A numerical investigation of the flow of nanofluids through a micro Tesla valve. *J. Zhejiang Univ. Sci. A* **2019**, *20*, 50–60. [CrossRef]

23. Qian, J.-Y.; Gao, Z.-X.; Liu, B.-Z.; Jin, Z.-J. Parametric Study on Fluid Dynamics of Pilot-Control Angle Globe Valve. *J. Fluids Eng.* **2018**, *140*, 111103. [CrossRef]

24. Zhu, Y.; Tang, S.; Wang, C.; Jiang, W.; Yuan, X.; Lei, Y. Bifurcation Characteristic Research on the Load Vertical Vibration of a Hydraulic Automatic Gauge Control System. *Processes* **2019**, *7*, 718. [CrossRef]

25. Zhu, Y.; Tang, S.; Quan, L.; Jiang, W.; Zhou, L. Extraction method for signal effective component based on extreme-point symmetric mode decomposition and Kullback-Leibler divergence. *J. Braz. Soc. Mech. Sci. Eng.* **2019**, *41*, 100. [CrossRef]

26. Zhu, J.; Zhang, H.-Q. Numerical Study on Electrical Submersible Pump Two-Phase Performance and Bubble-Size Modeling. *SPE Prod. Oper.* **2017**, *32*, 267–278. [CrossRef]

27. Si, Q.R.; Bois, G.; Jiang, Q.F.; He, W.T.; Ali, A.; Yuan, S.Q. Investigation on the Handling Ability of Centrifugal Pump Under Air-Water Two-Phase Inflow: Model and Experimental Validation. *Energies* **2018**, *11*, 3048. [CrossRef]

28. Si, Q.R.; Asad, A.; Yuan, J.P.; Ibra, F.; Yasin, M.F. Flow-Induced Noises in a Centrifugal Pump: A Review. *Sci. Adv. Mater.* **2019**, *11*, 909–924. [CrossRef]

29. International Organization for Standardization (ISO). *ISO 9906: 2012 Rotodynamic Pumps-Hydraulic Performance Acceptance Tests—Grades 1, 2 and 3*; ISO: Geneva, Switzerland, 2012.

30. Wang, C.; Chen, X.X.; Qiu, N.; Zhu, Y.; Shi, W.D. Numerical and experimental study on the pressure fluctuation, vibration, and noise of multistage pump with radial diffuser. *J. Braz. Soc. Mech. Sci. Eng.* **2018**, *40*, 481. [CrossRef]

31. Jiao, W.X.; Cheng, L.; Zhang, D.; Zhang, B.W.; Su, Y.P.; Wang, C. Optimal Design of Inlet Passage for Waterjet Propulsion System Based on Flow and Geometric Parameters. *Adv. Mater. Sci. Eng.* **2019**, *2019*, 2320981. [CrossRef]

32. Clift, R.; Grace, J.R.; Weber, M.E. *Bubbles, Drops and Particles*; Academic Press: New York, NY, USA, 1978.
33. Si, Q.R.; He, W.T.; Cui, Q.L.; Bois, G.; Yuan, S.Q. Experimental and numerical studies on inner flow characteristics of centrifugal pump under air-water inflow. *Int. J. Fluid Mach. Syst.* **2019**, *12*, 31–38.
34. Gao, M.; Dong, P.X.; Lei, S.H.; Turan, A. Computational Study of the Noise Radiation in a Centrifugal Pump When Flow Rate Changes. *Energies* **2017**, *10*, 221. [CrossRef]
35. Si, Q.; Wang, B.; Yuan, J.; Huang, K.; Lin, G.; Wang, C. Numerical and Experimental Investigation on Radiated Noise Characteristics of the Multistage Centrifugal Pump. *Processes* **2019**, *7*, 793. [CrossRef]

Article

Unsteady Flow Process in Mixed Waterjet Propulsion Pumps with Nozzle Based on Computational Fluid Dynamics

Can Luo [1,2,*], Hao Liu [1], Li Cheng [1,*], Chuan Wang [1], Weixuan Jiao [1] and Di Zhang [1]

[1] College of Hydraulic Science and Engineering, Yangzhou University, Yangzhou 225009, China; yzuliuhao@163.com (H.L.); wangchuan198710@126.com (C.W.); DX120170049@yzu.edu.cn (W.J.); DX120180054@yzu.edu.cn (D.Z.)

[2] Key Laboratory of Fluid and Power Machinery of Ministry of Education, Xihua University, Chengdu 610039, China

* Correspondence: luocan@yzu.edu.cn (C.L.); chengli@yzu.edu.cn (L.C.)

Received: 20 October 2019; Accepted: 27 November 2019; Published: 3 December 2019

Abstract: The unsteady flow process of waterjet pumps is related to the comprehensive performance and phenomenon of rotating stall and cavitation. To analyze the unsteady flow process on the unsteady condition, a computational domain containing nozzle, impeller, outlet guide vane (OGV), and shaft is established. The surface vortex of the blade is unstable at the valley point of the hydraulic unstable zone. The vortex core and morphological characteristics of the vortex will change in a small range with time. The flow of the best efficiency point and the start point of the hydraulic unstable zone on each turbo surface is relatively stable. At the valley point of the hydraulic unstable zone, the flow and pressure fields are unstable, which causes the flow on each turbo surface to change with time. The hydraulic performance parameters are measured by establishing the double cycle test loop of a waterjet propulsion device compared with numerical simulated data. The verification results show that the numerical simulation method is credible. In this paper, the outcome is helpful to comprehend the unsteady flow mechanism in the pump of waterjet propulsion devices, and improve and benefit their design and comprehensive performance.

Keywords: waterjet propulsion pump; unsteady flow process; test; computational fluid dynamics

---

1. Introduction

Unlike a propeller device, the waterjet propulsion device enables the ship to obtain navigational power by utilizing the reaction force of the water flow ejected by the propulsion pump. The waterjet propulsion device has the advantages of flexible operation, excellent maneuverability, high speed, outstanding anti-cavitation performance, and high efficiency [1]. The propulsion pump is well-protected for being arranged inside. When the ship speed exceeds 25 knots, the total efficiency of the waterjet propulsion device can reach more than 60%. Based on these advantages, the waterjet propulsion device has been widely used in high-speed performance vessels [2,3]. As the core component of the waterjet propulsion device, the performance of the propulsion pump is related to the performance of the entire device, and the nozzle is also an important part. Considering the thrust and layout requirements, the guide vane mixed-flow pump and axial flow pump are generally adopted in current waterjet propulsion vessels. In addition, other types of pump such as screw pump and permanent maglev (shaft less) pump utilized in the waterjet propulsion device are in the experimental research stage [4,5]. The pump employed in the waterjet propulsion device of this research is the guide vane mixed-flow pump. Moreover, pumps are essential for all animals. The heart tube of the embryonic vertebrate has been described as a peristaltic pump before the development of discernable chambers and valves at these early stages [6]. Then the bioinspired valveless pump is designed related to the pumping process of the heart tube and applied in the fields, such as microfluidics, drug delivery, biomedical devices, and cardiovascular pumping systems, becoming an important topic nowadays [7].

In recent years, simulation methods, such as CFD (Computational Fluid Dynamics) and control system simulation, combined with experiment technology, has been applied in numerous fields such as noise [8], vibration [9–13], turbo machinery [14–20], valve [21,22], jet flow [23,24], heat transfer [25], and hydraulic systems [26]. The researchers not only carried out research on waterjet propulsion devices from both theoretical and experimental aspects, but also conducted research works on waterjet propulsion devices by means of CFD technology. By using CFD technology, Ahn predicted the performance of the designed mixed-flow pump. The results are well agreed with the measured data in the cavitation tunnel test [27]. Wu et al. compared and analyzed the cavitation in the tip clearance of propulsion pump at different revolutions utilizing LDV (Laser Doppler Velocimetry) test technology and high-speed photography technology [28]. Tang analyzed the influence of the guide vane on the performance of the axial flow pump, adopting a numerical simulation and model test. A fixed axial clearance value between the impeller and the vane was concluded. When the value was exceeded, the influence on the performance of impeller was negligible [29]. Duerr analyzed the characteristics of the waterjet propulsion device, applying the non-uniform inflow condition [30]. Verbeek and Bulten focused on the uniformity of the flow field in front of the impeller, and analyzed the effects of boundary laminar flow and turbulence intensity on the uniformity of the flow field [31,32]. Brandner and Walker used pressure probes and visual test methods to conduct quantitative and qualitative experimental studies on the waterjet propulsion flush inlet. It was found that cavitation occurred in the lips in a wide range of operating conditions [33,34]. Park conducted a model test on the influent runner model at the wind tunnel laboratory [35,36]. Gong et al. simulated the flow in the entire waterjet propulsion device with unsteady methods and obtained the changes of the free surface in the waterjet propulsion device at different times [37]. Cao et al. compared the effects of uniform inflow and non-uniform inflow on the performance of the waterjet propulsion pump [38]. Zhang et al. analyzed the effect of cavitation on the thrust performance of the nozzle with CFD technology [39]. Cheng and Xia et al. studied the rotation stall existing in the propulsion pump and proposed corresponding suppression measures [40,41]. Recently, works have mainly focused on the property of the waterjet propulsion device, but research on the unsteady flow process between the nozzle and the propulsion pump has not been seen yet. In this study, the hydraulic performance and unsteady flow process of the propulsion pump with nozzle will be analyzed by using CFD technology, and the numerical simulation results are verified compared with the experiment result on the model test.

2. Numerical Simulation Method

2.1. Geometry Model

A computational domain geometric model including the nozzle is established to obtain stable flow field data of the mixed-flow waterjet propulsion device. Along the flow direction, subdomains are the prolonged inlet section, the inlet transition section, the impeller, the guide vane (GV), the nozzle, and the prolonged outlet section, as shown in Figure 1.

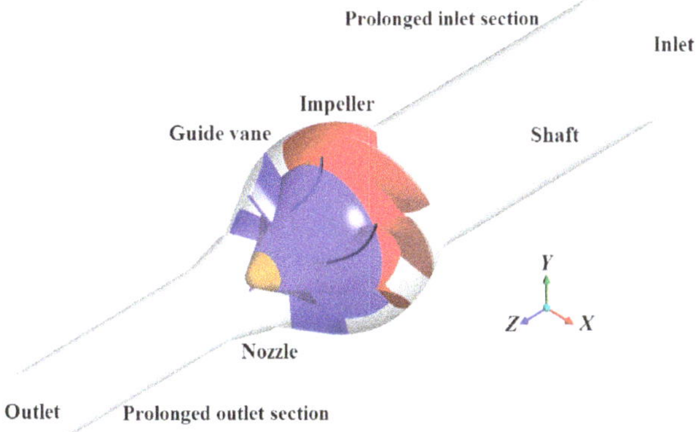

Figure 1. Computational domain of mixed waterjet pump.

2.2. Mesh Generation and Independence Analysis

According to the geometrical characteristics of each sub-domain, different meshing strategies are employed to construct the discrete meshes of each sub-domain. The impeller, the GV and the nozzle adopt the J-type, H-type, and O-type meshing methods, separately. By using ANSYS-ICEM, a reasonable block structure for each sub-domain is created to generate the number of structured meshes by controlling the number of mesh nodes and the growth rate. The accuracy of the calculation results will be poor if the meshes do not meet the calculation requirements. However, if the number of meshes is too large, it will occupy numerous computing resources. Therefore, the mesh independence analysis needs to be performed. The number of impeller meshes is 0.31 million, 0.5 million, 0.7 million, 0.89 million, and 1.1 million, recorded as mesh 1–5, as shown in Figure 2.

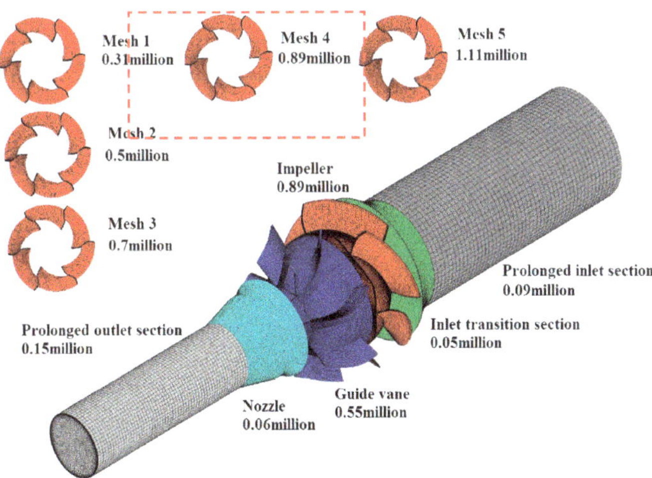

Figure 2. Meshes of computational domain.

The head H and efficiency η is calculated by Equations (1) and (2) when the flow rate is Q_{BEP}, 1.13 Q_{BEP} and 1.33 Q_{BEP}. The relative head H' and relative efficiency η' are calculated by dividing the head and efficiency of mesh 4

$$H = (P_{\text{int}} - P_{outt})/\rho g \tag{1}$$

$$\eta = \frac{\rho g Q H}{P_{shaft}}$$

(2)

where ρ is the density of water and the value is 10^3 kg/m^3, g is the acceleration due to gravity and the value is 9.81 m/s^2, H is the total head of the propulsion pump in m, P_{int} is the total pressure at the entrance of the propulsion pump in Pa, P_{outt} is the total pressure on the outlet of the pump in Pa, P_{shaft} is the shaft power in kW, and η is the efficiency.

The relative head and relative efficiency are drawn in Figure 3. When the flow rate is Q_{BEP}, the relative head H' of each mesh scheme has a certain difference, but the relative efficiency η' is basically consistent. When the flow rate is 1.33 Q_{BEP}, the relative head and the relative efficiency η' has a larger difference, especially mesh 1 and mesh 2. The disparity of mesh 3 tends to decrease. As the number of meshes increases, both the relative head H' and the relative efficiency η' gradually enhance, and this trend exceeds to become clearer as the flow rate increases. When the number of meshes surpass 1.79 million, the relative head H' and the relative efficiency η' tend to be constant. Therefore, mesh 4 is chosen as the final mesh, as shown in Figure 2.

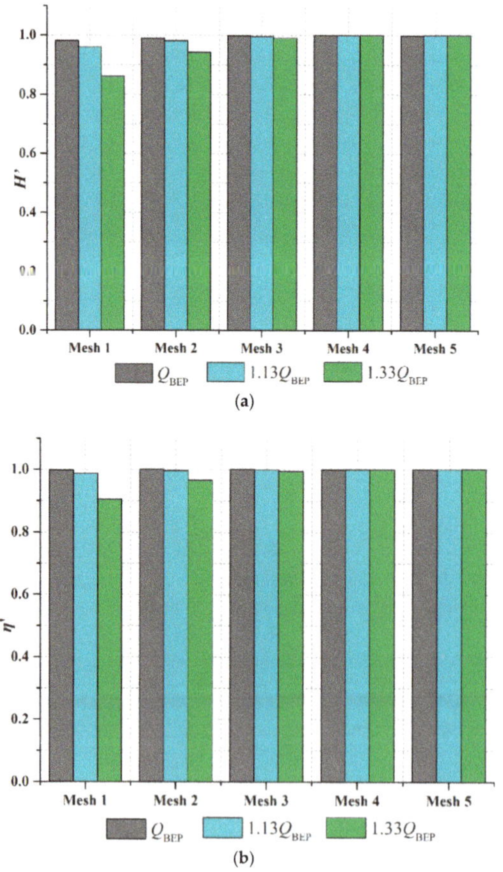

Figure 3. Mesh independence analysis. (**a**) Histogram of H' for each mesh at Q_{BEP}, 1.13 Q_{BEP}, and 1.33 Q_{BEP}. The relative head H' of mesh 4 tends to be constant. (**b**) Histogram of η' for each mesh at Q_{BEP}, 1.13 Q_{BEP}, and 1.33 Q_{BEP}. The relative head η' of mesh 4 tends to be constant.

2.3. Numerical Methodolgy and Boundary Conditions

As heat transfer does not exist in the flow process of waterjet propulsion device, Navier–Stokes equations are utilized as the governing equations to describe the flow, and the finite volume method is also applied to calculate the flow in the computational domains. Considering the incompressible flow, the inlet is set as mass flow. The impeller is the rotating domain and the spinning speed is 700 r/min or 73.27 rad/s. The reference pressure is 1 atm. The outlet is set as average static pressure. In order to guarantee the value transfer, the interfaces between each sub-domain are set, in which interfaces on the inlet and outlet of the impeller are transient rotor stator, and the remaining interfaces are general connection. Scalable wall function is processed on the wall. The standard k-ε turbulence model and the first-order upwind scheme are adopted. The convergence accuracy is 10^{-5}. According to the rotating speed of the waterjet propulsion pump, the time step is 0.00023381 s and the total time is 0.685728 s. The corresponding transient turbulence convergence sample is obtained when the impeller rotates per degree. The transient turbulent convergence sample set of all time steps is obtained when the set time finishes. Usually, the numerous samples will increase the costing time and the amount of data. Therefore, this paper sets 36 samples per period; that is, the data is stored when the impeller rotated per ten degrees. Finally, 288 sample results were saved in 8 periods.

2.4. Pressure Pulsation Monitoring Probe Arrangement

Pressure pulsation can be seen as the difference between the pressure amplitude at different points in time and the average pressure amplitude over the entire time period. Pressure pulsation can usually be classified from pulsation performance and frequency. According to the pulsation performance, pressure pulsation can be divided into turbulent pressure pulsation, which ignores fluid compressibility, and pulse source pressure pulsation, which ignores fluid viscosity. Based on the pulsation frequency, the pressure pulsation can be divided into irregular random pressure pulsation, axial frequency pulsation, and blade frequency pulsation—the latter two pulsations are referred to as regular pressure pulsation. For random pressure pulsation, there are various induced factors, such as cavitation, secondary flow, non-uniform inflow, etc. In terms of the regular pressure pulsation, the main induced factors are the impeller rotation, the pump shaft rotation, and rotor–stator interaction, which are related to the axial frequency and the blade frequency, respectively. Generally, the blade frequency-related pressure pulsation is captured in the impeller. If the pressure pulsation on the upstream and downstream near the impeller shows the discipline, the rotating impeller has an effect on the flow there. The observed pressure pulsation characteristics are distinct for different pumps and for the same pump. The observed pressure pulsation characteristics are diverse under various operating conditions and monitoring positions.

Fifteen monitoring probes in the waterjet propulsion pump are shown in Figure 4, and the locations of each probe are listed in Table 1.

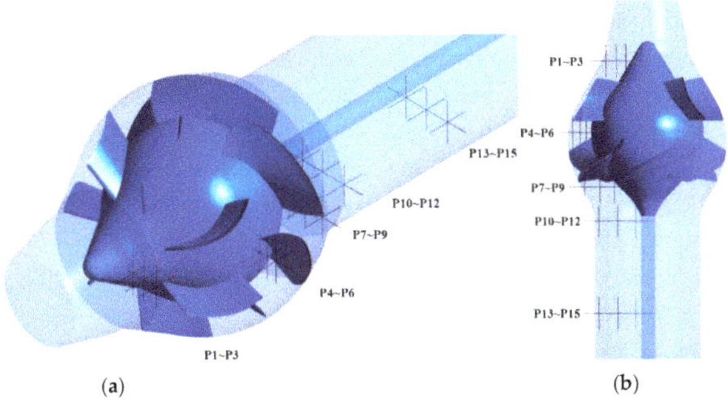

Figure 4. Diagram of pressure pulsation monitoring probe arrangement. (**a**) Lateral view. (**b**) Top view.

Table 1. Locations of each probe.

Number	Location
P1–P3	Outlet of guide vane (GV)
P4–P6	Outlet of impeller
P7–P9	Inlet of impeller
P10–P12	Outlet of the prolonged inlet section
P13–P15	In the prolonged inlet section

Fifteen monitoring probes as above are pre-set in the pre-processing. Data of monitoring parameters are stored when the impeller rotates per degree. 2880 groups of data are obtained. While obvious regular periodic diversification trend starts from the second period in the unsteady calculation process, data of the last six periods, which contains 2160 groups of data, are chosen to execute FFT transformation. The data of the eighth period on each monitoring probes are plotted into the time domain chart of pressure pulsation.

Currently, the time domain analysis method and frequency domain analysis method are the main methods in studying the pressure pulsation. The time domain chart is utilized in the time domain analysis method. In the chart, the abscissa is the time-related parameters, such as the time and period, and the ordinate is the pressure. Frequency domain analysis method converts the irregular pressure pulsation into the superposition of a simple harmonic wave with different frequencies, amplitudes, and phases by performing FFT transformation. These two methods have their own advantages. The time domain chart can intuitively reflect the change of pressure pulsation on the monitoring probe with time. The frequency domain analysis illustrates the main pulsating component of the pressure pulsation and the primary factor affecting the pressure pulsation directly. The pressure pulsation discipline of each monitoring probe on condition A, B, and C will be analyzed by using the methods above.

For the impeller rotating at 700 rev/min, the shaft frequency is 11.67 Hz by using $f_z = n/60$, the blade frequency is 70 Hz by using $f_b = 6f_z$, and multiples of shaft frequency are defined as T_f.

Pressure pulsation coefficient C_p is introduced to analyze the pressure pulsation characteristic of each monitoring probes, and the pressure pulsation coefficient C_p is calculated by applying Equation (3)

$$C_p = \frac{(p - \bar{p})}{0.5\rho V^2} \tag{3}$$

where p is the instantaneous pressure in kPa, $\bar{p}$ is the time-averaged pressure in kPa, ρ is the density of water in kg/m^3, and V is the blade tip speed at the entrance of propulsion pump in m/s.

3. Test Arrangement and Verification

As shown in Figure 5, double cycle waterjet propulsion test bench is established to verify the reliability of the numerical simulation method. The test device consists of two cycles—the main cycle and the secondary cycle. The main cycle, which is applied to provide the navigation speed for the waterjet propulsion pump, consists of the centrifugal auxiliary pump, electromagnetic flow meter, butterfly valve, expansion joint, rectifying device, and piping system. The secondary cycle, which is used to test the hydraulic performance, includes the test zone (mixed pump), electromagnetic flow meter, butterfly valve, and piping system. The flow rate is measured by the flow meter located in the main cycle and secondary cycle. The head is obtained by calculating the pressure measured at the beginning of the inlet flow tube and the outlet pressure measuring tube. The shaft power is calculated from the test data of the torque meter. The comprehensive error of this test bench is ±1.33%.

Numerical simulation and model test are carried out when the rotational speed is 400 rev/min. Dimensionless head H' and efficiency η' are obtained by dividing the head and efficiency of the best efficiency point, plotted in Figure 6. The CFD result shows a consistent trend with the test result. Therefore, the numerical method is reliable and suitable for the following research work.

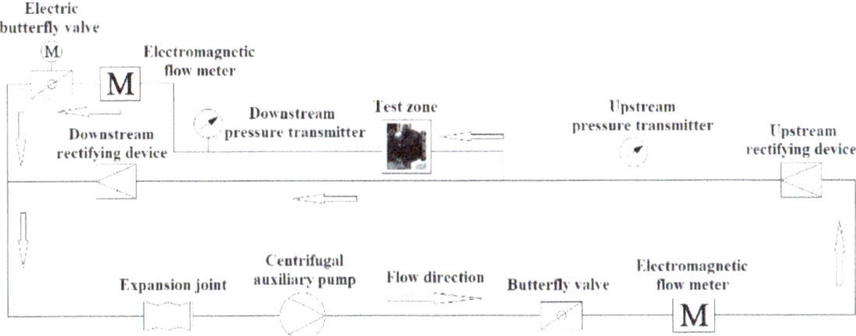

Figure 5. Double cycle waterjet propulsion test bench.

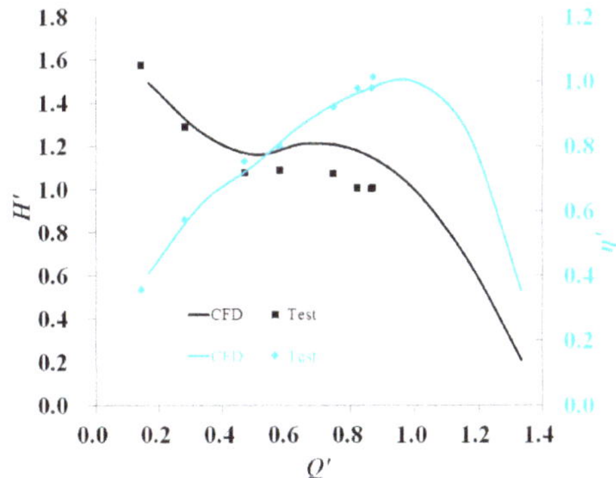

Figure 6. Contrast of CFD result and test result.

4. Results

4.1. Hydraulic Performance Curve

By diving the flow rate, head, and efficiency of the best efficiency point, dimensionless processing is performed for each condition. The dimensionless data is plotted into dimensionless a 'flow rate-head' curve and dimensionless 'flow rate-efficiency' curve, shown in Figure 7, in which the abscissa is Q', the left ordinate, is H' and the right ordinate is η'. The best efficiency point is marked as condition A. Under the condition B, the propulsion pump enters the hydraulic unstable zone. Condition C is the valley point of the hydraulic unstable zone.

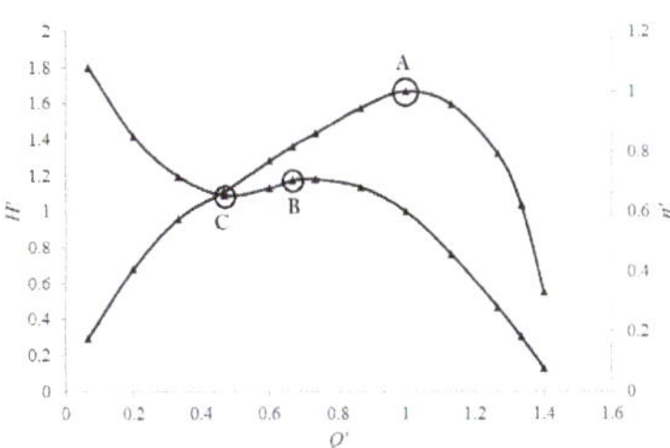

Figure 7. External characteristic curve.

4.2. Unsteady Flow Characteristics

Condition A, condition B, and condition C are chosen to be calculated unsteadily and analyzed. Figures 8–10 list the surface streamline on the blade of three time points (0T, 1/3T, and 2/3T) in the same period for condition A, B, and C. Under the condition A, surface streamlines on the suction side of each blade are smooth and the velocity at the leading edge is high. In the same period, surface streamline on the blade is stable with the lapse of time, which indicates that the pressure field and velocity field near the blade are stable under this condition. Under condition B, the stream at the leading edge flows to the trailing edge and the tip of the blade. The pressure field and velocity field near the blade are still stable. Under condition C, an evident vortex happens on the surface of the blade and the radial location of core vortex is span = 0.65. The size and shape of the vortex change with time, which indicates that the pressure field and velocity field near the blade are unstable.

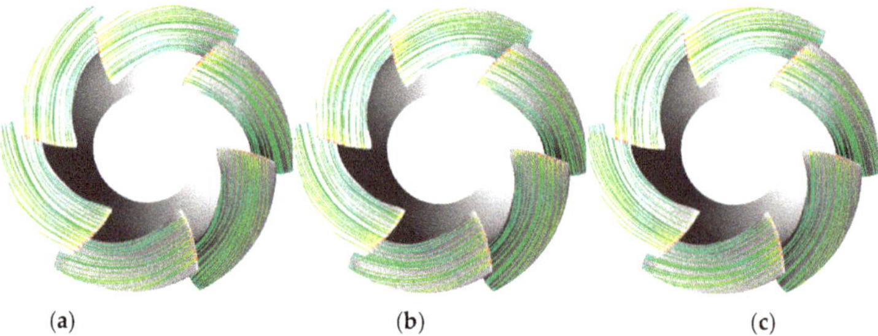

(a) (b) (c)

Figure 8. Streamline on the blade of different times for condition A (best efficiency point). (**a**) 0. (**b**) 1/3T. (**c**) 2/3T.

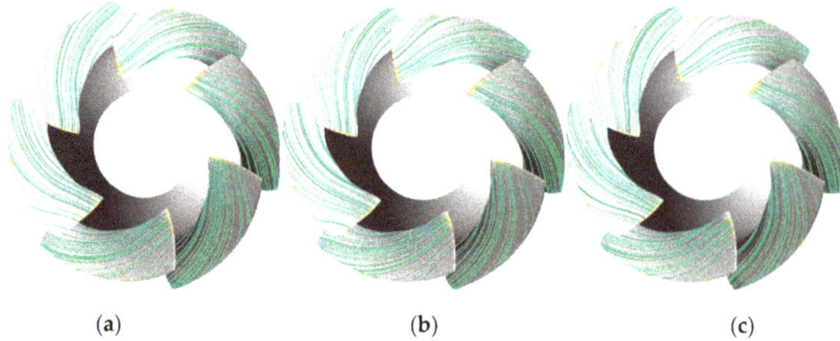

(a) (b) (c)

Figure 9. Streamline on the blade of different times for condition B (start point of hydraulic unstable zone). (**a**) 0. (**b**) 1/3T. (**c**) 2/3T.

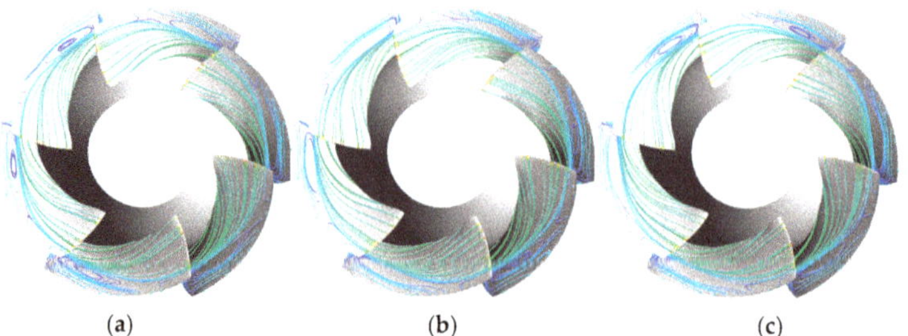

(a) (b) (c)

Figure 10. Streamline on the blade of different times for condition C (valley point of hydraulic unstable zone). (**a**) 0. (**b**) 1/3T. (**c**) 2/3T.

Three turbo surfaces from the hub to the shroud are sliced and recorded as TS1 (span = 0.1) near the hub, TS2 (span = 0.65) near the intermediate surface, and TS3 (span = 0.96) near the shroud, as shown in Figure 11.

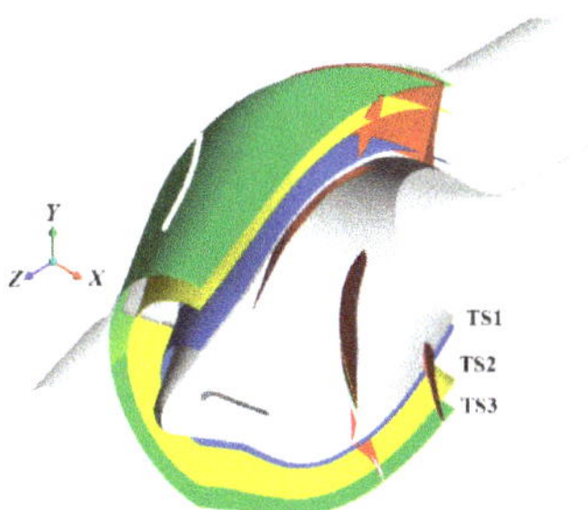

Figure 11. Turbo surfaces of the impeller.

Figures 12–14 are the streamlines on turbo surfaces TS1 under the condition A, B, and C, in which the velocity in the impeller is the relative velocity and the velocity in the GV is the absolute velocity. Under condition A, the streamlines are steady and vary slightly with time on the turbo surfaces of the impeller and GV. The velocity is rapid on the suction side of the impeller. The inlet attack angle is basically identical to the angle of the leading edge of the airfoil, which results in the excellent inflow condition and smooth stream in the blade-to-blade passage. A small-scale spanning vortex occurs at the tailing edge of the suction. Owing to the adjustment of the GV and the location of the vortex away from the impeller, the geometric shape and magnitude of the vortex shows no evident relationship with the time. Under condition B, slight deviation exists between the inlet attack angle and the airfoil angle of the blade in the impeller and the GV. The low-velocity region of the pressure surface of the leading edge of the impeller blade is enlarged. A large-scale spanning vortex occurs in each groove of GV, which extends from the inlet to the outlet in the axial direction, and occupies about 1/3 of the groove in the spanning direction. The streamlines of other parts in the groove are severely skewed and then gather near the outlet of the GV because of the spanning vortex. Under condition C, the smooth streamline in the groove is mildly affected. A distinct spanning vortex in the groove still exists and covers half of the groove on the spanning direction. The status and range of spanning vortex are basically maintained; however, the vortex core migrates in a small scale and the vortex status modifies, meaning the flow characteristic of GV is unstable.

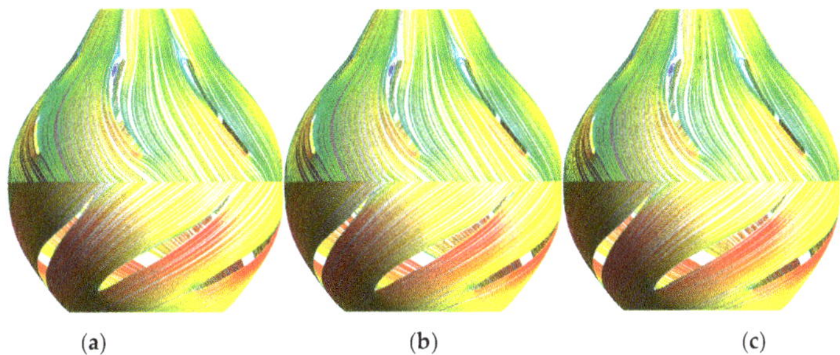

(**a**)   (**b**)   (**c**)

Figure 12. Streamlines on the turbo surface TS1 of condition A (best efficiency point). (**a**) 0. (**b**) 1/3T. (**c**) 2/3T.

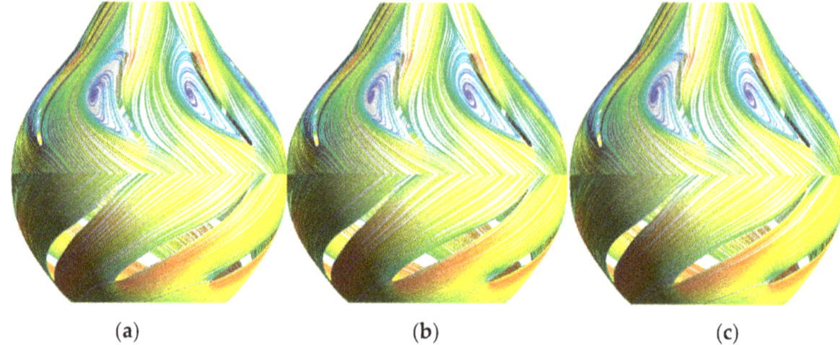

(a)   (b)   (c)

Figure 13. Streamlines on the turbo surface TS1 of condition B (start point of hydraulic unstable zone). (**a**) 0. (**b**) 1/3T. (**c**) 2/3T.

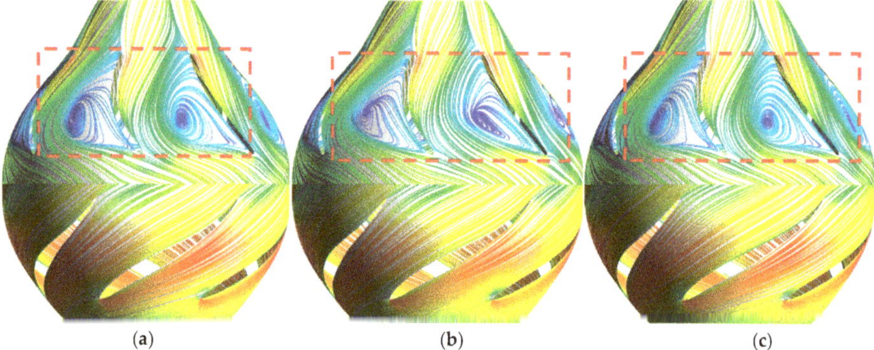

(a)   (b)   (c)

Figure 14. Streamlines on the turbo surface TS1 of condition C (valley point of hydraulic unstable zone). (**a**). 0 (**b**) 1/3T. (**c**) 2/3T.

Figures 15–17 are the streamlines on turbo surfaces TS2 under condition A, B, and C. Under condition A, the streamlines are smooth and no vortex occurs in the blade-to-blade passage of the impeller and GV. Under condition B, a slight deviation also occurs, but the streamlines are smooth in the groove of the impeller. As the arrows indicate in Figure 16, the streamlines deviate slightly near the tailing edge on the suction side of the impeller. The spanning vortex disappears in the groove of GV, but the shedding vortex is observed on the tailing edge of the outlet in the impeller. Part of the streamlines are severely skewed and have little impact on the mainstream. Under condition C, the inlet attack angle is consistent with the airfoil angle of the blade in the impeller. A distinct spanning vortex is observed near the tailing of different blades. The variation between the inlet attack angle of GV and the airfoil angle of the blade is apparent. Three vortexes marked as SV1, SV2, and SV3 occur on the spanning surface of GV. SV1 and SV2, located at the head edge and tailing edge of suction side, are in the channel of groove and rotate in clockwise. The range of SV1 is much larger than SV2. SV3 is the shedding vortex and located at the tailing of GV twirls on the opposite direction of the spanning vortex. As time passes, the shape and range of SV1, SV2, and SV3 varies.

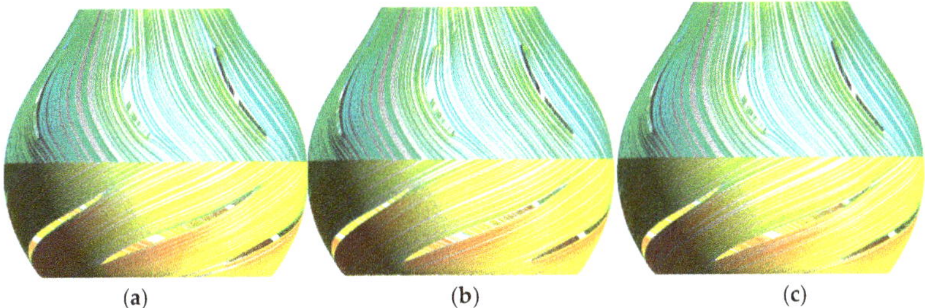

Figure 15. Streamlines on the turbo surface TS2 of condition A (best efficiency point). (**a**) 0. (**b**) 1/3T. (**c**) 2/3T.

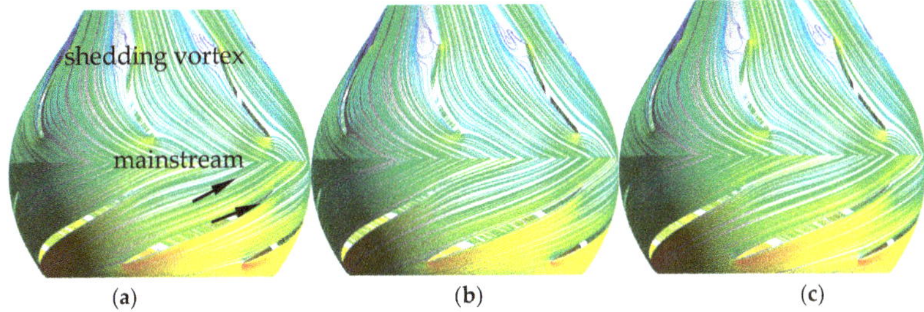

Figure 16. Streamlines on the turbo surface TS2 of condition B (start point of hydraulic unstable zone). (**a**) 0. (**b**) 1/3T. (**c**) 2/3T.

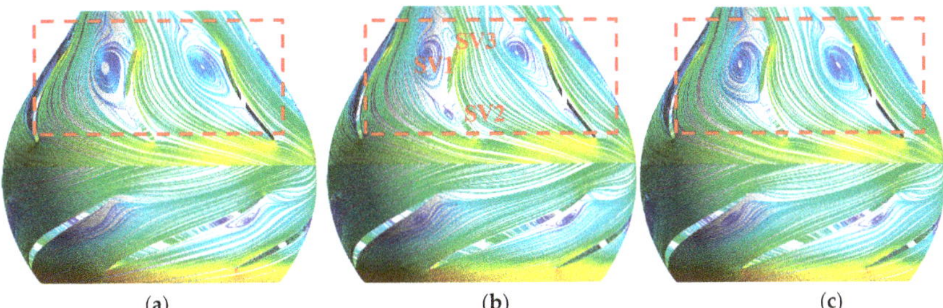

Figure 17. Streamlines on the turbo surface TS2 of condition C (valley point of hydraulic unstable zone). (**a**) 0. (**b**) 1/3T. (**c**) 2/3T.

Figures 18–20 are the streamlines on turbo surfaces TS2 under condition A, B, and C. Under condition A, the inlet attack angle is inconsistent with the angle of the leading edge of the airfoil. In one blade-to-blade passage, part of the stream flows from the suction side to the pressure side, then out of the blade-to-blade passage along the pressure side and the streamlines converge at the tailing edge of airfoil. The flow pattern in the blade-to-blade passage of GV is similar to the impeller. Under condition B, the stream at the entrance of the impeller flows into the neighboring groove on the opposite rotating direction instead of the GV. Part of the stream at the head edge on the suction side flows to the head edge on the pressure side of the neighboring blade and then out of the groove and into the neighboring groove after being dragged by the high-speed steam at the entrance of the groove. A low-velocity zone exists in the groove of the impeller, and the streamline is disordered. Under condition C, the distinction is huge between the inlet attack angle of the impeller and the airfoil angle of the blade, and an obvious spanning vortex appears at the head edge of the pressure side and the tailing edge of the suction side. Both spanning vortexes nearly cover the whole channel of the groove in the spanning direction. The shape of spanning vortex will change with time, but the vortex core will not migrate and the location maintains.

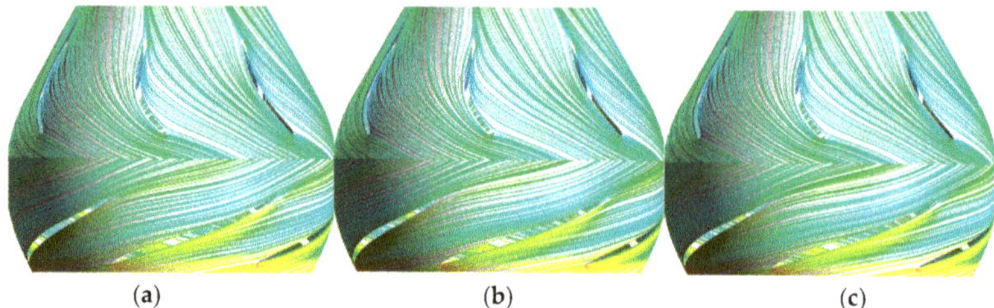

Figure 18. Streamlines on the turbo surface TS3 of condition A (best efficiency point). (**a**) 0. (**b**) 1/3T. (**c**) 2/3T.

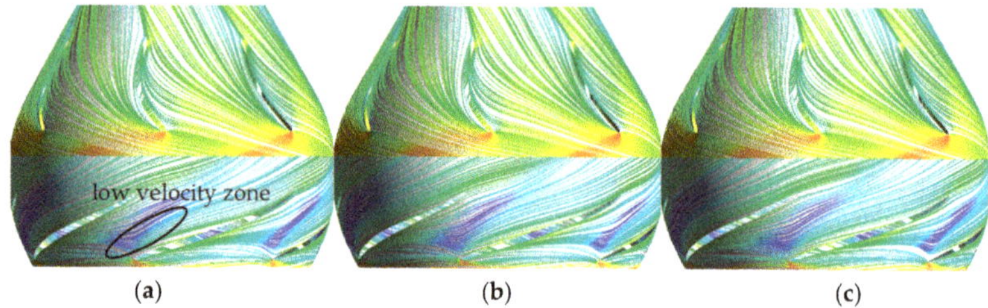

Figure 19. Streamlines on the turbo surface TS3 of condition B (start point of hydraulic unstable zone). (**a**) 0. (**b**) 1/3T. (**c**) 2/3T.

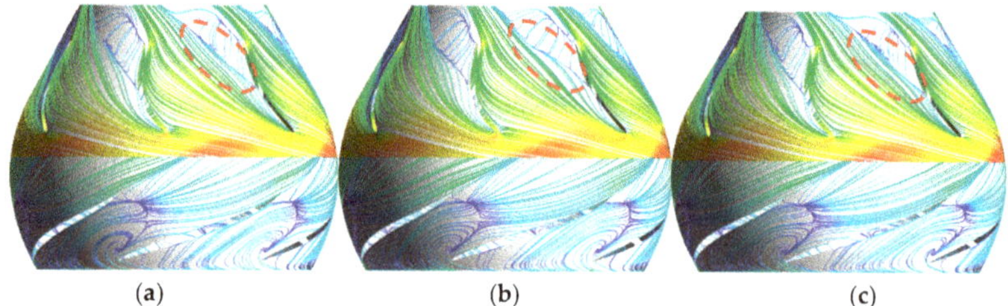

Figure 20. Streamlines on the turbo surface TS3 of condition C (valley point of hydraulic unstable zone). (**a**) 0. (**b**) 1/3T. (**c**) 2/3T.

4.3. Pressure Pulsation

Figure 21 shows the pressure pulsation time domain diagram and comparison of pressure pulsation amplitude of the monitoring probes P13–P15 in a period under condition A, B, and C. The main frequency, the secondary frequency, and the corresponding pressure amplitude of the monitoring probes P13–P15 under condition A, B, and C are listed in Table 2.

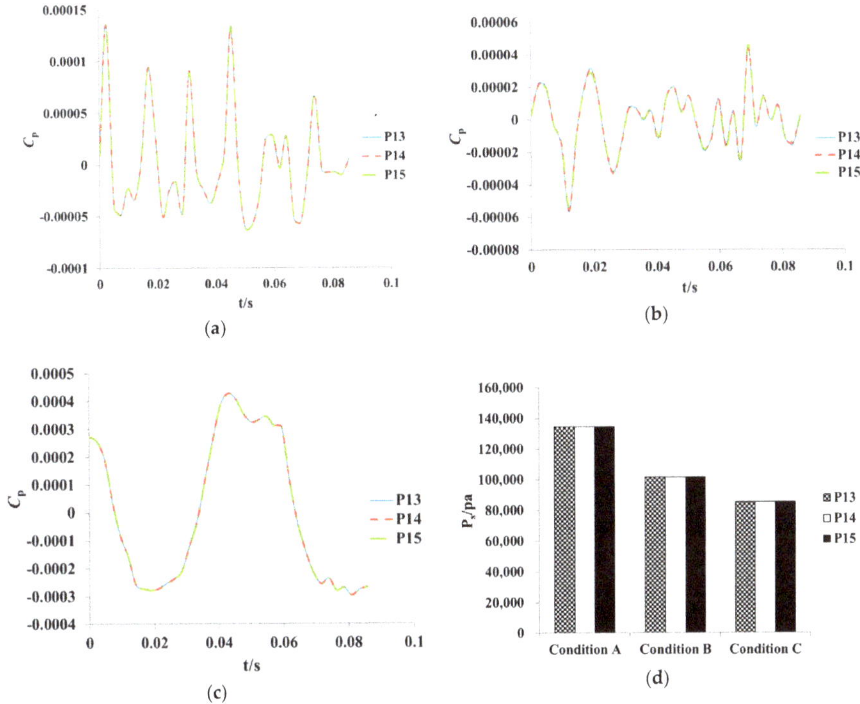

Figure 21. Pressure pulsation time domain diagram and pressure pulsation amplitude (P13–P15). (**a**) Condition A (best efficiency point). (**b**) Condition B (start point of hydraulic unstable zone). (**c**) Condition C (valley point of hydraulic unstable zone). (**d**) Pressure pulsation amplitude.

Table 2. Frequency domains of P13–P15.

Condition	Parameters	P13		P14		P15	
		MF	SF	MF	SF	MF	SF
Condition A	f/Hz	26.25	40.83	26.25	40.83	26.25	40.83
	T_f	2.25	3.50	2.25	3.50	2.25	3.50
	P_s/Pa	584.39	192.31	584.41	192.31	584.43	192.32
Condition B	f/Hz	26.25	40.83	26.25	40.83	26.25	40.83
	T_f	2.25	3.50	2.25	3.50	2.25	3.50
	P_s/Pa	441.86	145.12	441.86	145.13	441.87	145.13
Condition C	f/Hz	26.25	40.83	26.25	40.83	26.25	40.83
	T_f	2.25	3.50	2.25	3.50	2.25	3.50
	P_s/Pa	380.26	120.18	380.26	120.18	380.27	120.18

Under the same condition, the pressure pulsation trend of monitoring probes P13–P15 is completely unanimous, but the amplitudes increase sequentially. This indicates that the farther away from the pump shaft, the smaller the pressure pulsation amplitude, the closer to the pump shaft, and the greater the pressure pulsation amplitude due to the stream disturbance. Under different conditions, the pressure pulsation trend on each monitoring probe is quite different, accompanied by poor periodicity, small flow rate, and pressure pulsation amplitude.

The main frequency and secondary frequency of each monitoring probe are 26.25 Hz and 40.83 Hz, corresponding to the 2.25 times and 3.5 times shaft frequency. For the same monitoring probe, great variation exists between the pressure pulsation amplitude of main frequency (MF) and secondary frequency (SF). The pressure pulsation amplitude of MF is three times that of SF. Under different monitoring probes, the pressure pulsation amplitudes of MF and SF decrease as the flow rate reduces and small variation occurs between the pressure pulsation amplitude of MF and SF.

Figure 22 shows the pressure pulsation time domain diagram and comparison of pressure pulsation amplitude of the monitoring probes P10–P12 in a period under different conditions. Table 3 lists the main frequency, the secondary frequency, and the corresponding pressure amplitude of the monitoring probes P10–P12 under different conditions.

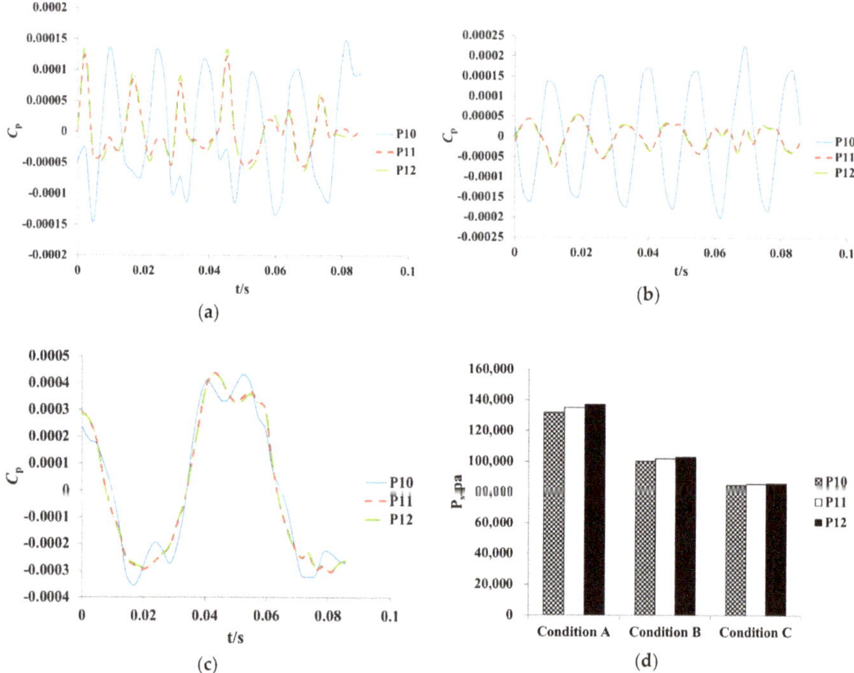

Figure 22. Pressure pulsation time domain diagram and pressure pulsation amplitude (P10–P12). (**a**) Condition A (best efficiency point). (**b**) Condition B (start point of hydraulic unstable zone). (**c**) Condition C (valley point of hydraulic unstable zone). (**d**) Pressure pulsation amplitude.

Table 3. Frequency domains of P10–P12.

Condition	Parameters	P10		P11		P12	
		MF	SF	MF	SF	MF	SF
Condition A	f/Hz	26.25	40.83	26.25	40.83	26.25	40.83
	T_f	2.25	3.50	2.25	3.50	2.25	3.50
	P_s/Pa	572.23	188.36	586.94	193.14	595.22	195.86
Condition B	f/Hz	26.25	40.83	26.25	40.83	26.25	40.83
	T_f	2.25	3.50	2.25	3.50	2.25	3.50
	P_s/Pa	435.41	142.97	443.21	145.57	447.65	147.03
Condition C	f/Hz	26.25	40.83	26.25	40.83	26.25	40.83
	T_f	2.25	3.50	2.25	3.50	2.25	3.50
	P_s/Pa	377.66	119.41	380.65	120.35	382.49	120.94

The pressure pulsation amplitude increases from P10 to P12 under the same condition. The pressure pulsation amplitude at P12 is 1.04, 1.03, and 1.01 times of P10 under condition A, B, and C. The closer to the shaft, the greater pressure pulsation amplitude, meaning the pressure pulsation amplitude is affected by the shaft. Under different conditions, the pressure pulsation amplitude and flow rate of each monitoring probe shows a positive relationship.

Under condition A, B, and C the MF and SF on the outlet of prolonged inlet section is completely the same with those in the prolonged inlet section. The rotating impeller has not yet been able to exert a dominant influence on the flow on the outlet of prolonged inlet section. Such a frequency shows no relationship with the blade frequency. The pressure pulsation amplitudes of MF and SF enlarges from P10 to P12 and the pressure pulsation amplitude of MF is also three times of SF.

The pressure pulsation time domain diagram and comparison of pressure pulsation amplitude of the monitoring probes P7–P9 in a period under different conditions is shown in Figure 23. The main frequency, the secondary frequency and the corresponding pressure amplitude of the monitoring probes P7–P9 under different conditions are listed in Table 4.

In general, the effect of the impeller on the inflow starts from the stream entering into the impeller, but actually the effect of the blade on the inflow begins when the water flow does not enter the impeller, mainly manifested by the pre-spin action on the water flow. Under the same condition, the pressure pulsation amplitude on the monitoring probes P7–P9 is gradually reduced, where P7 is the monitoring probe near the shroud and P9 is the monitoring probe near the hub. Thus, the pressure pulsation amplitude gradually increases from the hub to the shroud. The pressure pulsation amplitude at the monitoring probe P7 is 1.02 times, 1.05 times, and 1.01 times of P9 under condition A, B, and C. Hence, the pressure pulsation amplitude on each monitoring probe is positively correlated with the flow rate under different working conditions.

Table 4. Frequency domains of P7–P9.

Condition	Parameters	P7		P8		P9	
		MF	SF	MF	SF	MF	SF
	f/Hz	70	140	70	26.25	26.25	70
Condition A	T_f	6	12	6	2.25	2.25	6
	P_s/Pa	4600.42	1843.95	2115.09	587.59	593.25	444.94
	f/Hz	70	140	70	140	70	26.25
Condition B	T_f	6	12	6	12	6	2.25
	P_s/Pa	2706.53	1260.64	1358.23	451.00	581.55	414.95
	f/Hz	70	140	70	26.25	70	26.25
Condition C	T_f	6	12	6	2.25	6	2.25
	P_s/Pa	1100.13	370.35	986.14	349.29	385.55	357.89

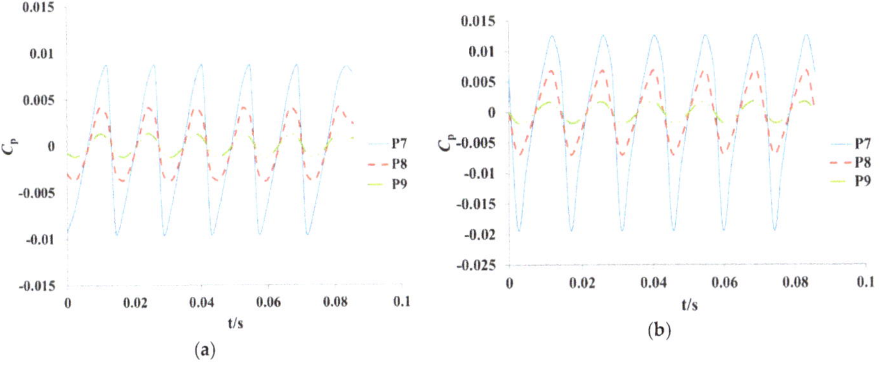

Figure 23. *Cont.*

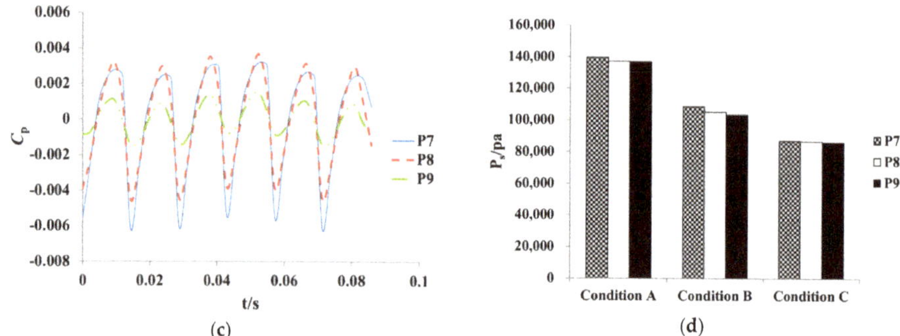

Figure 23. Pressure pulsation time domain diagram and pressure pulsation amplitude (P7–P9). (**a**) Condition A (best efficiency point). (**b**) Condition B (start point of hydraulic unstable zone). (**c**) Condition C (valley point of hydraulic unstable zone). (**d**) Pressure pulsation amplitude.

The frequency domain of monitoring probes set at the inlet of the impeller is observed and analyzed. The MF and SF on the impeller inlet monitoring probe are not certain, wherein the MF is 70 Hz or 26.25 Hz, and the SF is 70 Hz, 140 Hz, or 26.25 Hz. Under condition A, the MF of P7 and P8 is 70 Hz, which is the blade frequency, but the MF of P9 is 2.25 times of the shaft frequency. Under condition B, and C, the MF on each monitoring probe is also the blade frequency, but the SF is 26.25 Hz or 140 Hz. Under all conditions, the pressure pulsation amplitude of MF for P7 and P8 is quite different from the pressure pulsation amplitude of SF. However, the pressure pulsation amplitude of MF of P9 is basically consistent with SF. The monitoring probe P7 is pre-installed near the shroud, and the MF and SF are 70 Hz and 140 Hz, respectively, which is 1 and 2 times the blade frequency. Thus, the blade frequency plays a dominant role in the pressure pulsation near the shroud under condition A, B, and C. The location of monitoring probe P8 is between the shroud and the hub, and the MF is 70 Hz, the SF is 26.25 Hz under condition A and C, but is 140 Hz under condition B. Therefore, the blade frequency is still the dominant factor, but the shaft frequency also begins to play a certain role in it. The monitoring probe P9 is pre-set near the hub. Thus, 26.25 Hz and 70 Hz occur alternately in the MF and the SF of P9, and the pressure pulsation amplitudes of the MF is matched roughly with the SF. This indicates that the pressure pulsation near the hub is affected by both the impeller and the pump shaft.

Figure 24 shows the pressure pulsation time domain diagram and comparison of pressure pulsation amplitude of the monitoring probes P4–P6 in a period under different conditions. Table 5 lists the main frequency, the secondary frequency, and the corresponding pressure amplitude of the monitoring probes P4–P6 under different conditions.

Under the same condition, the pressure pulsation amplitude of P4–P6 decreases successively. Thus, the pressure pulsation amplitude near the shroud of the impeller inlet is higher than near the hub of the impeller outlet. Under condition A, B, and C, the pressure pulsation amplitude of P4 is 1.03 times, 1.04 times, and 1.07 times of P6. Under different conditions, the pressure pulsation amplitude decreases with the reducing flow rate, which indicates that the pressure pulsation amplitude has a positive correlation with the flow rate.

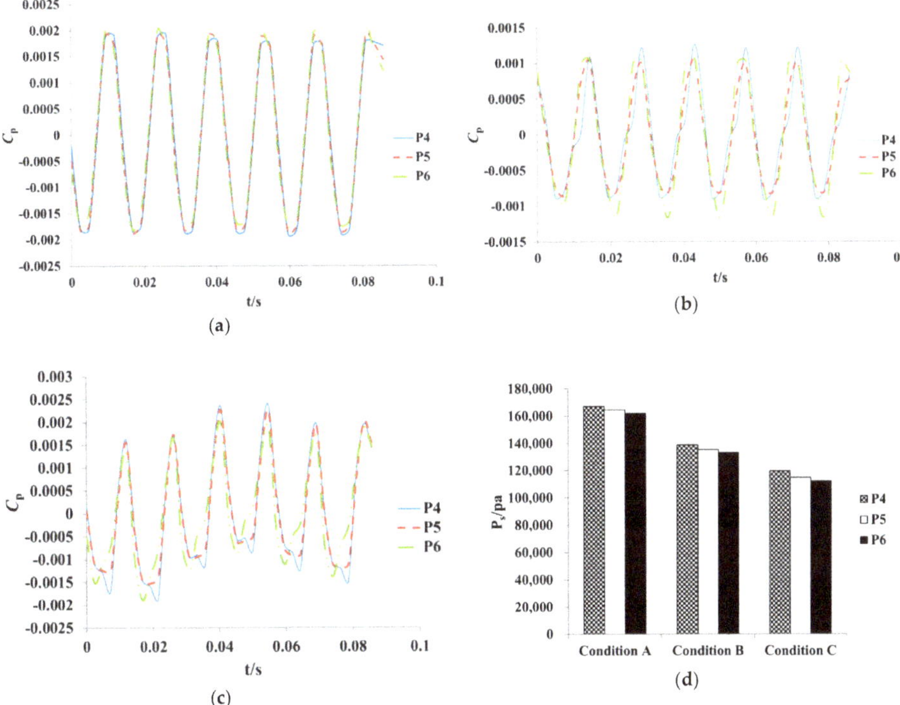

Figure 24. Pressure pulsation time domain diagram and pressure pulsation amplitude (P4–P6). (**a**) Condition A (best efficiency point). (**b**) Condition B (start point of hydraulic unstable zone). (**c**) Condition C (valley point of hydraulic unstable zone). (**d**) Pressure pulsation amplitude.

Table 5. Frequency domains of P4–P6.

Condition	Parameters	P4		P5		P6	
		MF	SF	MF	SF	MF	SF
Condition A	f/Hz	26.25	70	26.25	70	26.25	70
	T_f	2.25	6	2.25	6	2.25	6
	P_s/Pa	720.35	680.70	710.37	680.93	699.73	664.23
Condition B	f/Hz	26.25	70	26.25	70	26.25	70
	T_f	2.25	6	2.25	6	2.25	6
	P_s/Pa	601.99	304.07	587.76	301.68	577.25	372.17
Condition C	f/Hz	70	26.25	26.25	70	26.25	70
	T_f	6	2.25	2.25	6	2.25	6
	P_s/Pa	524.16	520.84	499.13	486.14	482.78	425.16

Under condition A, the MF and SF of each monitoring probe are 26.25 Hz and 70 Hz. The pressure pulsation amplitude corresponding to MF is slightly higher than SF. Under condition B, the MF and the SF of each monitoring probe are also 26.25 Hz and 70 Hz. For monitoring probes P4–P6, the pressure pulsation amplitude of MF is 1.98 times, 1.95 times, and 1.55 times of SF. Under condition C, the MF and SF of the monitoring probe P4 are 70 Hz and 26.25 Hz. The MF and SF of the monitoring probes P5 and P6 are 26.25 Hz and 70 Hz, and the pressure pulsation amplitudes of MF and SF are consistent.

Figure 25 shows the pressure pulsation time domain diagram and comparison of pressure pulsation amplitude of the monitoring probes P1–P3 in a period under different conditions. Table 6 lists the main frequency, the secondary frequency, and the corresponding pressure amplitude of the monitoring probes P1–P3 under different conditions.

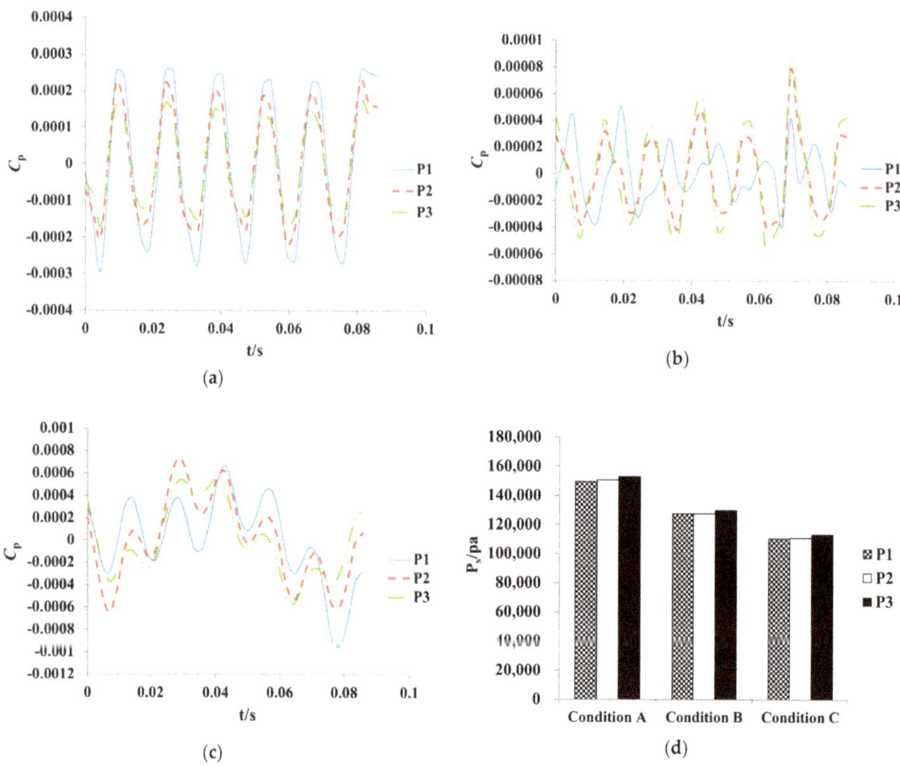

Figure 25. Pressure pulsation time domain diagram and pressure pulsation amplitude (P1–P3). (**a**) Condition A (best efficiency point). (**b**) Condition B (valley point of hydraulic unstable zone). (**c**) Condition C (valley point of hydraulic unstable zone). (**d**) Pressure pulsation amplitude.

Table 6. Frequency domains of P1–P3.

Condition	Parameters	P1		P2		P3	
		MF	SF	MF	SF	MF	SF
Condition A	f/Hz	26.25	40.83	26.25	40.83	26.25	40.83
	T_f	2.25	3.50	2.25	3.50	2.25	3.50
	P_s/Pa	649.35	213.56	654.13	215.19	664.25	218.49
Condition B	f/Hz	26.25	40.83	26.25	40.83	26.25	40.83
	T_f	2.25	3.50	2.25	3.50	2.25	3.50
	P_s/Pa	554.60	182.05	555.11	182.15	565.04	185.43
Condition C	f/Hz	29.62	41.46	29.17	40.83	29.17	40.83
	T_f	2.53	3.55	2.50	3.50	2.50	3.50
	P_s/Pa	488.05	157.02	486.74	156.61	497.70	156.84

Under the same condition, the pressure pulsation amplitudes of the monitoring probes P1–P3 increase sequentially. The pressure pulsation amplitude near the hub of the GVs outlet is higher than it near the shroud of the GVs outlet. Under condition A, B, and C, the pressure pulse amplitude of P1 is 1.02 times of P3. The pressure pulsation amplitude on the same monitoring probe under different conditions is positively correlated with the flow rate.

Under condition A, the MF and the SF of P1–P3 are 26.25 Hz and 40.83 Hz, which are 2.25 times and 3.50 times of the shaft frequency. The pressure pulsation amplitude of the MF is about 3 times of SF. Under condition B, the MF and the SF of P1–P3 are also 26.25 Hz and 40.83 Hz. The pressure pulsation amplitude of the MF is also about 3 times of SF. Under condition C, the MF and SF of P1 are 29.62 Hz and 41.46 Hz. The MF and SF of P2 are 29.17 Hz and 40.83 Hz. The MF and SF of P3 are 29.17 Hz and 40.83 Hz. The MF and SF of P1–P3 are no longer consistent and lacks of regularity. On one hand, the rotation of impeller and shaft affects the flow downstream a little for the outlet of GV is far from the impeller. On the other hand, the flow pattern in the propulsion pump is not stable and the MF and the SF on each monitoring probe are fluctuating in the low frequency range.

5. Conclusions

(1) The mixed-flow waterjet propulsion device is tested by establishing the double circulation test loop of waterjet propulsion system. The test results are consistent with CFD results both in the trend and values. The CFD method is reliable.

(2) Conditions A, B, and C are marked as characteristic conditions by analyzing the hydraulic performance of the propulsion pump, which are the BEP (best efficiency point), start point of hydraulic unstable zone, and the valley point of hydraulic unstable zone. Thus, unsteady calculation is promoted and the unsteady flow process of the propulsion pump at different times of the same period is discussed. The surface vortex on the blade under condition C is unstable, and the vortex core and shape pattern vary on a small scale as time. Three turbo surfaces are sliced to study the flow features on each spanwise under different conditions. The steady flow characteristic of each turbo surface is obtained under condition A and B; however, the flow characteristic of each turbo surface varies as time under condition C, due to the unstable velocity and pressure field.

Author Contributions: Data curation, L.C. and W.J.; Formal analysis, C.W.; Methodology, C.L.; Writing—original draft, H.L.; Writing—review & editing, D.Z.

Funding: This research was funded by [Jiangsu Province Science Foundation for Youths] grant number [BK20170507], [Natural Science Foundation of the Jiangsu Higher Education Institutions] grant number [17KJD580003], [Jiangsu Planned Projects for Postdoctoral Research Funds] grant number [1701189B], [Open Research Subject of Key Laboratory of Fluid and Power Machinery (Xihua University), Ministry of Education] grant number [szjj2019-018], [Science and Technology Innovation and Cultivation Fund of Yangzhou University] grant number [2017CXJ047], [National Natural Science Foundation of China] grant number [51779214], [Priority Academic Program Development of Jiangsu Higher Education Institutions (PAPD)], [Jiangsu Province 333 high level talents training project] grant number [(2018) III-1827], [Peak plan six talents in Jiangsu province], and [Key project of water conservancy in Jiangsu province] [2018042].

Conflicts of Interest: The authors declare no conflict of interest.

References

1. John, A. Marine Waterjet Propulsion. *SNAME Trans.* **1993**, *101*, 275–335.
2. Tang, F. *Design and Turbulence Numerical Analysis of Waterjet Axial-Flow Pump*; Shanghai Jiaotong University: Shanghai, China, 2006.
3. Jin, P. *MarineWater-Jet Propulsion*; National Defence Industry Press: Beijing, China, 1986.
4. Zhou, Z.; Lu, X.; Xia, J. Design of rotor profiles in symmetry twin-screw pump for new water jet propulsion. *J. Nav. Univ. Eng.* **2011**, *23*, 83–85.
5. Cao, P.; Wang, Y.; Qian, K.X.; Li, G. Comparisons of Hydraulic Performance in Permanent Maglev Pump for Water-Jet Propulsion. *Adv. Mech. Eng.* **2014**, 1–16. [CrossRef]
6. Forouhar, A.S.; Liebling, M.; Hickerson, A.; Nasiraei-Moghaddam, A.; Tsai, H.; Hove, J.R.; Fraser, S.E.; Dickinson, M.E.; Gharib, M. The Embryonic Vertebrate Heart Tube Is a Dynamic Suction Pump. *Science* **2006**, *312*, 751–753. [CrossRef] [PubMed]

7. Li, Z.; Seo, Y.; Aydin, O.; Elhebeary, M.; Kamm, R.D.; Kong, H.; Saif, M.T.A. Biohybrid valveless cpump-bot powered by engineered skeletal muscle. *Proc. Natl. Acad. Sci. USA* **2019**, *116*, 1543–1548. [CrossRef] [PubMed]

8. Zhang, J.; Xia, S.; Ye, S.; Xu, B.; Song, W.; Zhu, S.; Xiang, J. Experimental investigation on the noise reduction of an axial piston pump using free-layer damping material treatment. *Appl. Acoust.* **2018**, *139*, 1–7. [CrossRef]

9. Wang, C.; Chen, X.X.; Qiu, N.; Zhu, Y.; Shi, W.D. Numerical and experimental study on the pressure fluctuation, vibration, and noise of multistage pump with radial diffuser. *J. Braz. Soc. Mech. Sci. Eng.* **2018**, *40*, 481. [CrossRef]

10. Ye, S.; Zhang, J.; Xu, B.; Zhu, S.; Xiang, J.; Tang, H. Theoretical investigation of the contributions of the excitation forces to the vibration of an axial piston pump. *Mech. Syst. Signal. Process.* **2019**, *129*, 201–217. [CrossRef]

11. Zhu, Y.; Tang, S.; Quan, L.; Jiang, W.; Zhou, L. Extraction method for signal effective component based on extreme-point symmetric mode decomposition and Kullback-Leibler divergence. *J. Braz. Soc. Mech. Sci. Eng.* **2019**, *41*, 100. [CrossRef]

12. Zhu, Y.; Qian, P.; Tang, S.; Jiang, W.; Li, W.; Zhao, J. Amplitude-frequency characteristics analysis for vertical vibration of hydraulic AGC system under nonlinear action. *AIP Adv.* **2019**, *9*, 035019. [CrossRef]

13. Zhu, Y.; Tang, S.; Wang, C.; Jiang, W.; Yuan, X.; Lei, Y. Bifurcation characteristic research on the load vertical vibration of a hydraulic automatic gauge control system. *Processes* **2019**, *7*, 718. [CrossRef]

14. Wang, C.; Hu, B.; Zhu, Y.; Wang, X.; Luo, C.; Cheng, L. Numerical study on the gas-water two-phase flow in the self-priming process of self-priming centrifugal pump. *Processes* **2019**, *7*, 330. [CrossRef]

15. He, X.; Jiao, W.; Wang, C.; Cao, W. Influence of surface roughness on the pump performance based on Computational Fluid Dynamics. *IEEE Access* **2019**, *7*, 105331–105341. [CrossRef]

16. Wang, C.; He, X.; Zhang, D.; Hu, B. Numerical and experimental study of the self-priming process of a multistage self-priming centrifugal pump. *Int. J. Energy Res.* **2019**, 1–19. [CrossRef]

17. Yang, H.; Zhang, W.; Zhu, Z. Unsteady mixed convection in a square enclosure with an inner cylinder rotating in a bi-directional and time-periodic mode. *Int. J. Heat Mass Transf.* **2019**, *136*, 563–580. [CrossRef]

18. Wang, C.; He, X.; Shi, W.; Wang, X.; Qiu, N. Numerical study on pressure fluctuation of a multistage centrifugal pump based on whole flow field. *AIP Adv.* **2019**, *9*, 035118. [CrossRef]

19. Wang, C.; Shi, W.; Wang, X.; Jiang, X.; Yang, Y.; Li, W.; Zhou, L. Optimal design of multistage centrifugal pump based on the combined energy loss model and computational fluid dynamics. *Appl. Energy* **2017**, *187*, 10–26. [CrossRef]

20. Jiao, W.; Cheng, L.; Xu, J.; Wang, C. Numerical analysis of two-phase flow in the cavitation process of a waterjet propulsion pump system. *Processes* **2019**, *7*, 690. [CrossRef]

21. Qian, J.Y.; Chen, M.R.; Liu, X.L.; Jin, Z.J. A numerical investigation of the flow of nanofluids through a micro Tesla valve. *J. Zhejiang Univ. -Sci. A* **2019**, *20*, 50–60. [CrossRef]

22. Qian, J.Y.; Gao, Z.X.; Liu, B.Z.; Jin, Z.J. Parametric study on fluid dynamics of pilot-control angle globe valve. *ASME J. Fluids Eng.* **2018**, *140*, 111103. [CrossRef]

23. Wang, X.; Su, B.; Li, Y.; Wang, C. Vortex formation and evolution process in an impulsively starting jet from long pipe. *Ocean. Eng.* **2019**, *176*, 134–143. [CrossRef]

24. Chang, H.; Shi, W.; Li, W.; Wang, C.; Zhou, L.; Liu, J.; Yang, Y.; Agarwal, R. Experimental optimization of jet self-priming centrifugal pump based on orthogonal design and grey-correlational method. *J. Therm. Sci.* **2019**, 1–10. [CrossRef]

25. Hu, B.; Li, X.; Fu, Y.; Zhang, F.; Gu, C.; Ren, X.; Wang, C. Experimental investigation on the flow and flow-rotor heat transfer in a rotor-stator spinning disk reactor. *Appl. Therm. Eng.* **2019**, *162*, 114316. [CrossRef]

26. Zhu, Y.; Tang, S.; Wang, C.; Jiang, W.; Zhao, J.; Li, G. Absolute stability condition derivation for position closed-loop system in hydraulic automatic gauge control. *Processes* **2019**, *7*, 766. [CrossRef]

27. Ahn, J.W.; Kim, K.S.; Park, Y.H.; Kim, K.Y.; Oh, H.W. Performance analysis of mixed-flow pump for waterjet. In Proceedings of the Fourth Conference for New Ship and Marine Technology, Shanghai, China, 1 January 2004; pp. 109–116.

28. Wu, H.; Soranna, F.; Micheal, T.; Katz, J.; Jessup, S. Cavitation visualizes the flow structure in the tip region of a waterjet pump rotor blade. In Proceedings of the 27th Symposium on naval hydrodynamics, Seoul, Korea, 5–10 October 2008.

29. Tang, F.-P.; Wang, G.-Q. Influence of Outlet Guide Vanes upon Performances of Waterjet Axial- Flow Pump. *J. Ship Mech.* **2006**, *10*, 19–26.

30. Duerr, P.; von Ellenrieder, K.D. Scaling and Numerical Analysis of Nonuniform Waterjet Pump Inflows. *IEEE J. Ocean. Eng.* **2015**, *40*, 701–709. [CrossRef]

31. Bulten, N.W.H.; Verbeek, R. CFD simulation of the flow through a water jet installation. In Proceedings of the International Conference on Waterjet Propulsion 4, London, UK, 26–27 May 2004.

32. Bulten, N.W.H. Influence of boundary layer ingestion on water jet performance parameters at high ship speeds. In Proceedings of the 5th International Conference on Fast Sea Transportation, Seattle, WA, USA, 31 August–2 September 1999.

33. Brandner, P.A.; Walker, G.J. A waterjet test loop for the tom fink cavitation tunnel. In Proceedings of the International Conference on Waterjet Propulsion III, Amsterdam, The Netherlands, 22–23 October 2001.

34. Brandner, P.A.; Walker, G.J. An experimental investigation into the performance of a flush water-jet inlet. *J. Ship Res.* **2007**, *51*, 1–21.

35. Park, W.G.; Yun, H.S.; Chun, H.H.; Kim, M.C. Numerical flow simulation of flush type intake duct of waterjet. *Ocean. Eng.* **2005**, *32*, 2107–2120. [CrossRef]

36. Park, W.G.; Yun, H.S.; Chun, H.H.; Kim, M.C. Numerical flow and performance analysis of waterjet propulsion system. *Ocean. Eng.* **2005**, *32*, 1740–1761. [CrossRef]

37. Gong, J.; Guo, C.-Y.; Wu, T.-C.; Song, K.-W. Numerical Study on the Unsteady Hydrodynamic Performance of a Waterjet Impeller. *J. Coast. Res.* **2018**, *34*, 151–163. [CrossRef]

38. Cao, P.; Wang, Y.; Kang, C.; Li, G.; Zhang, X. Investigation of the role of non-uniform suction flow in the performance of water-jet pump. *Ocean Eng.* **2017**, *140*, 258–269. [CrossRef]

39. Zhang, Z.; Cao, S.; Luo, X.; Shi, W.; Zhu, Y. Research on the thrust of a high-pressure water jet propulsion system. *Ships Offshore Struct.* **2017**, *13*, 1–9. [CrossRef]

40. Cheng, L.; Qi, W. Rotating stall region of water-jet pump. *Trans. FAMENA* **2014**, *38*, 31–40.

41. Xia, C.; Cheng, L.; Luo, C.; Jiao, W.; Zhang, D. Hydraulic Characteristics and Measurement of Rotating Stall Suppression in a Waterjet Propulsion System. *Trans. FAMENA* **2018**, *42*, 85–100. [CrossRef]

Article

Lattice Boltzmann Simulation on Droplet Flow through 3D Metal Foam

Jian Zhang, Xinhai Yu * and Shan-Tung Tu

Key Laboratory of Pressure Systems and Safety (MOE), School of Mechanical Engineering, East China University of Science and Technology, Shanghai 200237, China; zhangj9@ecust.edu.cn (J.Z.); sttu@ecust.edu.cn (S.-T.T.)
* Correspondence: yxhh@ecust.edu.cn; Tel.: +86-021-64253108

Received: 29 October 2019; Accepted: 18 November 2019; Published: 22 November 2019

Abstract: The hydrodynamics of droplets passing through metal foam is investigated using the lattice Boltzmann method (LBM). The accurate 3D porous structure for the simulation is generated by X-ray micro-computed tomography. The simulated results are in good agreement with the experimental ones using high-speed video. The simulated results show that for droplets passing metal foam, there is a critical capillary number, Ca_c (around 0.061), above which the droplet continues to deform until it breaks up. The simulated results show that the capillary number, droplet size, pores diameter, and thickness of metal foam have the significant effect of droplets deforming and breaking up when the droplets pass through the metal foam. To avoid the calescence of two droplets at the inlet zone of the metal foam, the distance between droplets should be larger than three times the diameter of the droplet.

Keywords: lattice Boltzmann method; metal foam; droplet break; multiphase flow

1. Introduction

Metal foam is a class of materials with a porous structure. A typical property of metal foams is high porosity, and therefore a low density. The thermal and mechanical properties of metal foams remain those of their base metals and sufficiently meet the requirements of light weight, low pressure drop, malleability, improved mixing, and heat transfer. Therefore, metal foams have been applied in many industries involved in enhanced fluid mixing and heat transfer. Metal foams are also used as physical support for catalysts or even as catalyst substrate in chemical processes, such as fuel cells and micro-reactors [1,2]. Metal foams have been used to improve the conversion efficiency in micro-reactors because it can enhance the mixing of the liquid flow [3] and emulsify the immiscible two-phase fluid [4]. However, the flow field and droplet behaviors (e.g., breakup and deformation) are still poorly understood in metal foam reactors.

Various experiments [5] and simulations [6–11] have been carried out to understand the effects of porous structure on fluid hydrodynamics in immiscible binary fluids. However, most of them in this topic rely on empirical correlations of experimental measurements and traditional computational fluid dynamics (CFD). The multiphase flows in the CFD approach are simulated by solving the macroscopic Navier–Stokes equations. Among the approaches of tracking interfaces, the front-tracking method, the volume of fluid (VOF) method, and the level set method are widely used [8–13]. Because the interface must be manually ruptured, the front-tracking methods are not suitable for simulating interface breaking and coalescing [14]. The VOF and level set methods can simulate interface breaking and coalescing; however, to determine the interfacial tension, the force and the flux across the interface is required in the VOF. The level set method uses a signed distance function to represent the interface, which requires a re-initialization procedure to keep the distance property when large topological changes occur around the interface. This process can be time-consuming and not always physically consistent [15]. In addition, the VOF and level set methods will suffer from numerical instability at the interface region when the interfacial tension becomes a dominant factor in complex geometries [16]. For example, it is a challenge to apply the VOF or level set methods to simulate capillary displacement in porous media. Microscopically, the phase segregation and the interfacial dynamics between different phases are due to inter-particle forces or inter-actions [17]. Thus, mesoscopic level models are expected to accurately describe the complex dynamic behavior of multiphase flows [18].

With the advance of computational physics and image technology, the simulation of droplet breakup in metal foams has become feasible. For porous reconstructing, there are two ways of representing the pore scale geometry: an idealized structure and micro X-ray computed tomography (CT). The idealized structure can reconstruct simplified pore geometries by taking into account the structural complexity of the medium [19,20]. A main drawback of the idealized structure is that the pore structure has to be mathematically reasonably simplified to fit the model. This simplification can cause a substantial error in describing the real structure. Micro-CT can accurately regenerate a porous structure. Hundreds of images from various angular views are acquired while the porous object rotates. Then the hundreds of images of virtual cross-section slices are synthesized by a computer to regenerate the porous structure. Montminy et al. [21] reconstructed a 3D metal foam using micro-CT. Carvalho et al. [22] followed their approach to investigate the pressure drop of the single-phase fluid flow passing the porous media.

In recent years, the lattice Boltzmann method (LBM) has emerged as an attractive numerical tool for simulating multiphase flow because of its advantage in dealing with complicated geometry and interfacial dynamics [23,24]. Unlike the methods mentioned before, LBM is based on the solution of macroscopic variables such as velocity and density, and built upon microscopic models and mesoscopic kinetic equations. Compared with traditional numerical methods, LBM is more efficient in dealing with complex boundaries [25–27]. LBM has been successfully applied to investigate the multiphase fluid flow in a porous medium.

Several lattice Boltzmann models such as the Shan-Chen (SC) model, the free energy-based model, and the color gradient-based model [28–30] have been proposed for simulating multiphase flow. Among these models, the SC multiphase model is the simplest. In this model, hydrophobic interaction between fluid phases and additional interaction between the fluid and solid surfaces are taken into account [31–33]. The SC model is capable of simulating the complete range of contact angles and the equilibrium distribution of the phase in a porous medium. Because of these advantages, the SC model has been widely used to study the hydrodynamics of single and multiple phase flows where the interaction between the fluid–fluid and fluid–solid are considered. Li et al. [34] applied the SC model to study the deformation and breakup behavior of liquid droplets past a circular cylinder. Park et al. [35] successfully simulated the motion of liquid droplet flow in a porous medium using the SC model and a reconstructed method. However, the works by Li et al. and Park et al. are limited to two-dimensional simulations. Frank et al. [36] simulated the droplet spreading on a porous surface. Jonas et al. [37] simulated an immiscible binary fluid flow in a porous medium. It should be noted that an ideal porous structure which randomly distributes in the matrix was taken for the simulations by Li et al. and Jonas et al. This ideal porous structure leads to an obvious deviation of simulation results from real results.

In spite of great progress in the simulation of multiphase fluid in porous media using LBM, these porous media are either rocks in geological reservoirs or matrixes of soils. Unlike rock or soil, metal materials have their own properties. The porosity of metal foams is usually larger than rock or soil; therefore, metal foams have a stronger circulation capacity and less capillary action. A few 3D simulations on single phase fluids have been carried out in metal foams using X-ray reconstructed 3D porous structures [38]. However, to the best of our knowledge, no works have been reported on the LBM simulation of droplet breakup and deformation in a metal foam against experimental measurements. Regarding the wide application of metal foams in industries, it is necessary to deepen our understanding of the mechanism of droplet behavior in metal foams.

A primary objective of this paper is to simulate the processes of droplet breakup and deformation in a metal foam generated by X-ray CT using the lattice Boltzmann model with an SC model of multiphase. The Green function ($G_{w,k}$) is obtained by comparing the measured contact angles and the simulation results. The simulation results of the hydrodynamics of droplets passing through metal foam were verified by the results recorded using high-speed video. The effects of several non-dimensional parameters on the hydrodynamics of droplets passing through metal foam are discussed. It should be mentioned that no original point in the numerical method applied, but the hydrodynamics of droplets passing through metal foam revealed by the simulation is crucial to the design of metal foam reactors or mixers.

2. Experimental

2.1. Experimental Setup

The experimental setup for measurements of the hydrodynamics of a droplet passing through a metal foam is illustrated in Figure 1. The metal foams employed in this study were supplied by SiPing AKS Metal Material Technology Co., Ltd. (Siping, China); The width, length, pore density, and porosity of the metal foams were 10 mm, 25 mm, 60 pores per inch (PPI), and 95% porosity, respectively. The height ranged from 2 to 4 mm. The metal foam was placed into a Plexiglass tube with a rectangular shape. The Plexiglas tube was filled with silicon oil (Haishi Co., Shanghai, China). One water droplet above the metal foam was formed by injecting distilled water into the silicon oil with a micro-syringe (HAMILTON Co., Bonaduz, Switzerland). The water droplet diameter was controlled by adjusting the injected water volume. When the water droplet held a steady suspension in silicon oil, the valve connected with the outlet of the Plexiglass tube was opened, and the droplet went through the metal foam. The droplet deformation and breakup after the droplet left the metal foam were recorded by a high-speed video (Integrated Device Technology, Longmont, CO, USA). A ruler was attached to the Plexiglass tube for the calculation of the droplet velocity based on the droplet movement recorded by the high-speed video. All the experiments were carried out at room temperature.

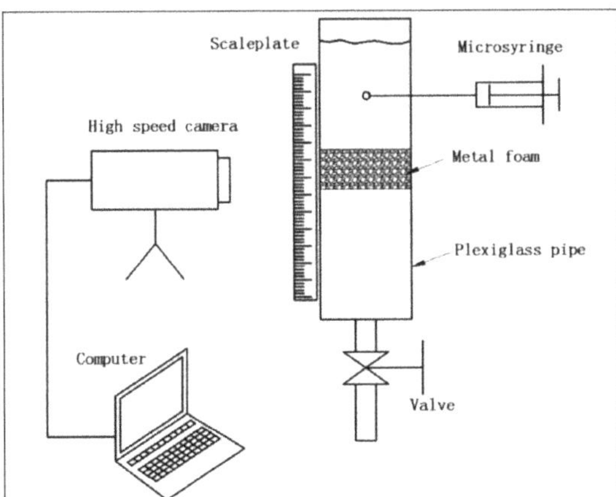

Figure 1. Setup for measurements on the hydrodynamics of a droplet passing through a metal foam.

2.2. Metal Foam Micro-Tomography

The morphology of the metallic foam was generated using the Skyscan high-resolution desktop micro-CT system (Micro Photonics Inc., Allentown, PA, USA). The metal foam sample dimensions were $20 \times 20 \times 2$ mm. The metal foam sample was illuminated by a micro-focus X-ray source at 40 kV with a beam current of 250 μA. A planar X-ray detector collected the magnified projection images with a pixel of 36 μm. The 2D cross-section images acquired from various angular views and the morphology of the metal foam were reconstructed. The Matlab Image Processing Toolbox (Matlab R2006a, MathWorks, Natick, MA, USA, 2007) were used to binarized the 2D cross-section images.

A raw section image of the metal foam structure is shown in Figure 2a. A 3D structure of the restructured metal foam sample is illustrated in Figure 2b. Its porosity was computed to be 93.6%, which is very close to that of 95.0% provided by the metal foam supplier.

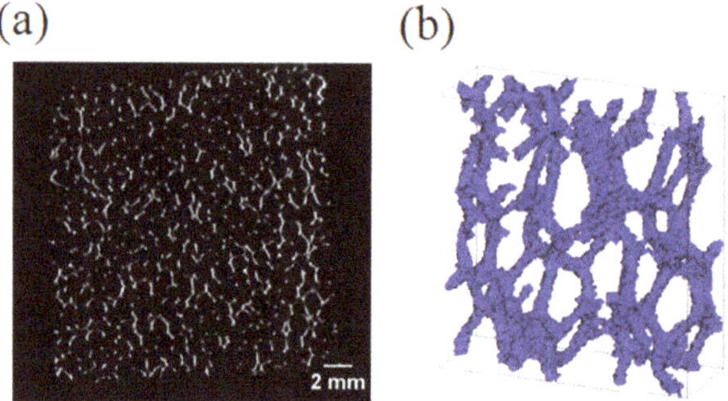

Figure 2. (**a**) Raw section image of the metal foam structure. (**b**) 3D reconstruction of a part of the foam sample of 3.6 × 3.6 × 1.4 mm (mesh number 100 × 100 × 40 of the matrix).

3. Simulation

3.1. Lattice Boltzmann Equations

The hydrodynamics of droplets passing through porous metal foam was simulated using LBM. The lattice Boltzmann simulation includes two steps. First, the collision between fluid particles is calculated, and the second, the particles are streamed. In the case of two immiscible fluids, the fluid distribution function for each fluid is described as the following:

$$f_i^k(\mathbf{x} + \mathbf{e}_i \delta_t, t + \mathbf{e}_i \delta_t) - f_i^k(\mathbf{x}, t) = -\frac{1}{\tau_k}\left[f_i^k(\mathbf{x}, t) - f_i^{k,eq}(\mathbf{x}, t)\right] \tag{1}$$

where k stands for different particles, $\mathbf{x}$ is the discrete lattice location, $\mathbf{e}_i$ is the discrete velocity direction for the D3Q19 lattice structure, δ_t is time step and is chosen as 1, τ_k is relaxation parameter and $f_i^k(\mathbf{x}, t)$ and $f_i^{k,eq}(\mathbf{x}, t)$ are the number density distribution function and local equilibrium function, respectively. $f_i^{k,eq}(\mathbf{x}, t)$ is calculated for the LBM D3Q19 model as

$$f_i^{k,eq}(\mathbf{x}, t) = \omega_i n_k \left[1 + \frac{\mathbf{e}_i \cdot \mathbf{u}_k^{eq}}{c_s^2} + \frac{\left(\mathbf{e}_i \cdot \mathbf{u}_k^{eq}\right)^2}{4c_s^4} + \frac{u_k^2}{4c_s^4}\right] \tag{2}$$

$$\mathbf{e}_i = c\begin{pmatrix} 0 & 1 & -1 & 0 & 0 & 0 & 0 & 1 & -1 & 1 & -1 & 1 & -1 & -1 & 1 & 0 & 0 \\ 0 & 0 & 0 & 1 & -1 & 0 & 0 & 1 & -1 & -1 & 1 & 0 & 0 & 0 & 0 & 1 & -1 \\ 0 & 0 & 0 & 0 & 0 & 1 & -1 & 0 & 0 & 0 & 0 & 1 & -1 & 1 & -1 & 1 & -1 \end{pmatrix} \tag{3}$$

$$\omega_i = \begin{cases} \frac{1}{3}, & \mathbf{e}_i = 0 \\ \frac{1}{18}, & \mathbf{e}_i = c^2 \\ \frac{1}{36}, & \mathbf{e}_i = 2c^2 \end{cases} \tag{4}$$

where n_k is the number density, and c_s^k is the sound velocity of the kth component that satisfies the formula $c_s^k = 1/3$.

The mass density and velocity of the kth component can be obtained:

$$\rho_k = m_k n_k = m_k \sum_i f_i^k \tag{5}$$

$$\rho_k u_k = m_k \sum_i \mathbf{e}_i f_i^k \tag{6}$$

where m_k is the molecular mass of each component. In the Shan-Chen scheme, the interaction force is incorporated into the model by the equal velocity $\mathbf{u}_k^{eq}$ in the equilibrium equation. The $\mathbf{u}_k^{eq}$ is determined by the relation

$$\mathbf{u}_k^{eq} = \mathbf{u}' + \frac{\tau_k F_k}{\rho_k} \tag{7}$$

where $\mathbf{u}'$ is the common velocity, which satisfies the equation

$$\mathbf{u}' = \left(\sum_k \frac{u_k \rho_k}{\tau_k} \right) \Big/ \left(\sum_k \frac{\rho_k}{\tau_k} \right) \tag{8}$$

The total force acting on each component F_k includes the fluid–fluid interaction F_f, the fluid-solid interaction F_{ads} and the external force F_e

$$F_k = F_f + F_{ads} + F_e \tag{9}$$

The SC model assumes that the interaction between particles is nonlocal. This means that the interaction only considers the nearest-neighbors $\mathbf{x}' = \mathbf{x} + \mathbf{e}_i$. The fluid–fluid interaction acting on the kth component at site x is

$$F_{f,k} = -\Psi_k(\mathbf{x}) \sum_{\mathbf{x}'} \sum_k G_{k\bar{k}}(\mathbf{x}, \mathbf{x}') \Psi_{\bar{k}}(\mathbf{x}')(\mathbf{x}' - \mathbf{x}) \tag{10}$$

The effective mass of kth component $\Psi_k(\mathbf{x})$ is related to the number density

$$\rho_k : \Psi_k(\mathbf{x}) = \rho_{k0} \left[1 - exp\left(-\frac{\rho_k}{\rho_{k0}} \right) \right] \tag{11}$$

$G_{k\bar{k}}(\mathbf{x}, \mathbf{x}')$ is Green's function,

$$G_{k\bar{k}}(\mathbf{x}, \mathbf{x}') = \begin{cases} g_{k\bar{k}}, & |\mathbf{x} - \mathbf{x}'| = 1 \\ g_{k\bar{k}}/2, & |\mathbf{x} - \mathbf{x}'| = \sqrt{2} \\ 0, & otherwise \end{cases} \tag{12}$$

where $g_{k\bar{k}}$ controls the interaction strength between different particles, while its sign determines whether the interaction is attraction (negative) or repulsion (positive), and it is sufficient to only involve the nearest neighbor interactions. The interaction F_{ads} for the kth component at the fluid/solid interface is defined as

$$F_{ads,h} = -\Psi_k(\mathbf{x}) \sum_k G_{w,k} s(\mathbf{x}')(\mathbf{x}' - \mathbf{x}) \tag{13}$$

where $s(\mathbf{x}')$ is the switch parameter which is 1 when $(\mathbf{x}')$ is in the pore space of the solid lattice, otherwise it is 0. $G_{w,k}$ determines the interaction's type and magnitude between fluid and solid wall. Different wettability can be obtained by adjusting $G_{w,k}$. Non-wetting fluid and wetting fluid are obtained when $G_{w,k} > 0$ and $G_{w,k} < 0$, respectively. $G_{w,k}$ is obtained by comparing the measured contact angles and the simulation results.

3.2. Computational Domain of Droplet Breakup

The schematic diagram of computational domain and a droplet passing through metal foam is shown in Figure 3, which corresponds to the experimental measurement. A metal foam of 60 PPI was inserted into a rectangular channel of 50 (width) × 50 (height) lattice cells. The physical unit of one lattice is 73 μm. The length of the channel (*L*) is 500 lattice cells, and the distance between the droplet and metal foam (*L1*) is 175 lattice cells. The 3D matrix of the metal foam was restructured by stacking the 2D cross-section images obtained by micro-CT. It should be noted that this reconstructed method perfectly matches the quadrilateral grids of the LBM and has a high computational efficiency, although its accuracy is less than the curve surface mesh. The flow is assumed to be steady, incompressible, and laminar. The rectangular channel is filled with silicon oil, and the water droplet is initially placed at the channel inlet. Constant velocity is imposed on the silicon oil at the inlet boundary of the channel to drive the droplet forward. The half-way bounce back rule is utilized for the solid wall [39]. The velocity inlet boundary condition with uniform velocity and the convective boundary condition are applied at the inlet and outlet of the channel, respectively [40]. It is noted that the conventional SC model suffers several limitations such as the numerical instability at a high-density ratio [41]. In this study, a density ratio of 1.0 is taken for the silicon oil and water system. Such a density ratio of 1.0 is present for the multi-phase related to the micro-reactors and microfluidic devices and is easy to simulate using the SC model. This study does not touch the cases with high-density ratios.

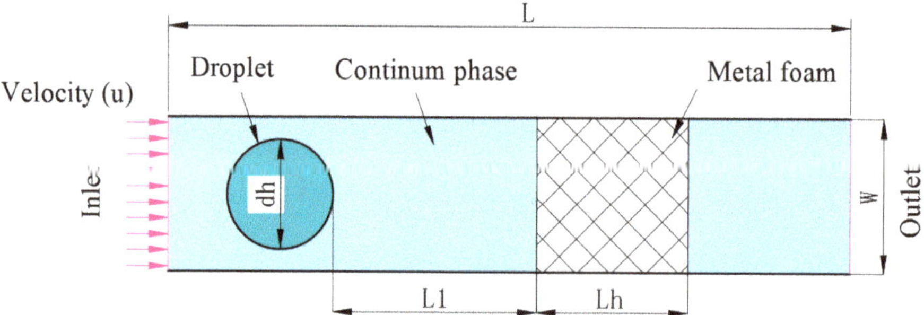

Figure 3. Schematic diagram of a droplet passing through metal foam in a computational domain of 500 (length) × 50 (width) × 50 (height) lattice cell and L1 of 175 lattice cell.

4. Discussion

4.1. Dimensionless Parameters

The physical values are scaled using several dimensionless parameters. One parameter is the capillary number (*Ca*) which describes the relative importance of viscosity and the interfacial tension. Here the *Ca* number is defined as Equation (14) [21]

$$C_a = \frac{u}{\sigma} \tag{14}$$

where u, η, and σ are the average inlet velocity of the silicon oil phase, the dynamic viscosity of the silicon oil, and the interfacial tension between the silicon oil and the water.

The second dimensionless parameter D_d ($D_d = d_h/d_p$) describes the relative size between the pores of the metal foam and the droplet where d_p is the average pore diameter of the metal foam, and d_h is the droplet diameter at the channel inlet. The third parameter, L_d, is defined as the ratio of the thickness of the metal foam (L_h) to d_p. The fourth parameter D_b is defined as $D_b = D/d_h$ where D is the distance between the two successive droplets when the first one reaches the position of L1 (Figure 3).

In this study, the change of droplet superficial area (ΔS) is defined as

$$\Delta S = \frac{S_t}{S_i} \tag{15}$$

where s_i is the droplet's original superficial area before the droplet gets into the metal foam; s_t is the superficial area of the droplet which has entered the metal foam. s_t is obtained by averaging the corresponding values of three simulations with different metal foams (the PPI and porosity are the same). In this way, the generality of the simulation is promoted. Theoretically, when ΔS exceeds 1.0, the droplet definitely deforms or breaks up.

4.2. Modeling Coefficients and Validation of LBM Simulation

The LBM with the SC model requires identifying a correct $G_{w,k}$ before simulating the flow pattern of two immiscible fluids. The water contact angle surrounded by the silicon oil on the surface of a pure nickel material surface was measured using a Drop Shape Analyzer (KRUSS, Hamburg, Germany; see Figure S1a). The relation between the static contact angle and $G_{w,k}$ is investigated in independent LBM simulations by equilibrating a water droplet surrounded by a flat solid surface. The value of the interaction strength between the silicon oil and water ($G_{o,w}$) is fixed and $G_{w,k}$ is altered to examine the wettability of two fluids. As shown in Figure 4b, when $G_{w,k}$ is ±0.06, the simulated contact angle is 135°, which is in good agreement with the experimental value of 132. Therefore, in the following simulation, $G_{w,k}$ is chosen as ±0.06. It is known that $G_{w,k}$ is related to the grid size in the SC model. The physical unit size of one lattice in this study was set as 73 µm, which balanced the simulation accuracy and the calculation time, considering that a high spatial resolution is favorable for improving the simulation accuracy but meanwhile significantly increases the calculation time especially for a 3D simulation. In this study, the accuracy of the grid size value is verified by comparing the simulated and experimental results regarding the contact angle (see Figure S1) and the droplet shape when the droplet leaves the metal foam (see Figure 4).

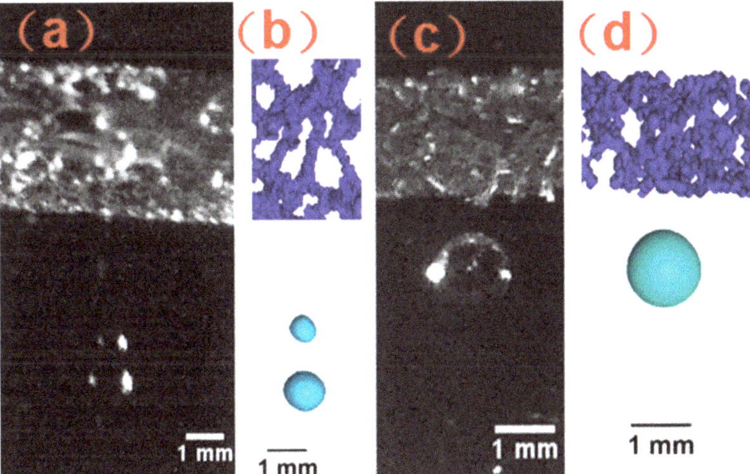

Figure 4. Comparison of the lattice Boltzmann method (LBM) simulation and the experimental results. (**a**) The experimental and (**b**) the simulation results for $Ca = 0.063$, $D_d = 1.06$, and $L_d = 4.88$. (**c**) The experimental and (**d**) the simulation results for $Ca = 0.084$, $D_d = 1.31$, and $L_d = 2.44$.

The experimental results using a Plexiglass tube (Figure 1) were used for the validation of the LBM simulated ones. The droplet deformation and breakup after the droplet left the metal foam were recorded by high-speed video. Correspondingly, the LBM code was used to simulate the hydrodynamics of the droplet. Quantitative comparisons between the simulation and the experiments were performed. When $Ca = 0.063$, $D_d = 1.06$, and $L_d = 4.88$, as shown in Figure 4a, the experimental result showed that the droplet broke into two droplets with the diameters of 0.7 and 1.13 mm, respectively, after it left the foam. The same phenomenon can be observed in the simulation result (see Figure 4b) with the two droplets' diameters of 0.7 and 1.04 mm, respectively. When $Ca = 0.084$, $D_d = 1.31$, and $L_d = 2.44$, only one droplet with the diameter of 1.35 mm was detected after the droplet passed through the metal foam (see Figure 4c). The corresponding simulation result exhibits the same tendency with the droplet's diameter of 1.30 mm (see Figure 4d). Based on the validations above, conclusions can be drawn that the LBM model and code are adequate to simulate the hydrodynamics of droplets passing through metal foam.

In the numerical study, we considered the case of equiviscous droplets with medium viscosity, i.e., $v_w = v_s$ (dynamic viscosity of water and silicon oil), although the experimental setup shows $v_w/v_s = 1/48$. It is known that a large viscosity ratio likely results in an instable LBM simulation. To solve the instability, the multi relaxation time (MRT) approach has been reported to be feasible [42]. However, the MRT method requires a large computer resource that is most likely inapplicable for the 3D LBM simulation on the hydrodynamics of one droplet passing a porous structure like metal foam. Here, despite that a viscosity ratio should be the real value for a more reliable comparison, the droplet dynamics are somehow not so significantly different. From the numerical viewpoint, furthermore, the equiviscous case has a great numerical stability. That is, the efficient calculation compensates for any loss of accuracy.

4.3. The Effect of Ca

For the metal foam reactor with immiscible fluids, smaller droplets produced in the reactor exhibits higher activity. For example, in our previous studies on biodiesel synthesis by transesterification with methanol, methyl ester yield is strongly dependent on droplet size [4]. This is because the mass transfer of triglycerides (TG) from the oil phase to the methanol/oil interface limits the rate of methanolysis reaction and controls the kinetics at the beginning of the reaction [43,44]. The overall volumetric TG mass transfer coefficient increase is attributed to an increase in the specific interfacial area by the decrease in droplet size, which leads to an increase in the TG reaction rate. For the metal foam with two immiscible fluids, when the continuous phase drives the droplets to pass through the porous material, the solid walls stretch the droplets, which deform into an elongated shape and break up. The deformation and breakup increase the superficial area of the droplets. When the distance between two droplets is smaller than one critical valve, they coalesce [45], resulting in a decrease in the superficial area.

Figure 5 shows the effect of Ca on droplet dynamics when the droplet passes through a metal foam. For the Ca values of 0.031 and 0.046, as shown in Figure 5b,c, the droplet is stretched when it collides with the walls of the metal foam. Then, the droplet shape restores at the pores of the metal foam. When the droplet leaves the metal foam, an obvious elongation can be observed. It should be noted that no droplet breakup is detectable for the Ca values of 0.031 and 0.046. Several previous studies [46–49] have indicated, when the viscosity ratio of the droplet to the matrix fluid λ is less than 4, there is a "critical capillary number" Ca_c, above which the droplet continues to deform and finally breaks in the creeping flow regime. The critical capillary number for droplet breakup in shear flow is the lowest for λ, roughly around 0.647, and its value ($Ca_c \approx 0.4$) is slightly less than the case for $\lambda = 1$, where $Ca_c \approx 0.41$ [50]. The droplet hydrodynamics passing through obstructions in confined microchannels were explored by Chung et al. both numerically and experimentally [8]. They found that for the cylinder obstruction, the Ca_c was around 0.1. In our case, as shown in Figure 5d, for a Ca of 0.061, the droplet breaks up and forms one daughter droplet. The Ca_c in our case is around 0.061, which is lower than that of 0.1 for the case of one droplet passing through one cylinder obstruction. It was found that the stretching rate in shear flows may be increased by incorporating periodic reorientations in the flow [51]. Wen et al. [4] also reported that droplet size decreases with increasing reorientations by turns in a zigzag micro-channel for the mixing of ethanol and soybean oil. The smaller Ca_c for the case of metal foam than that for the cylinder obstruction in a micro-channel is mostly likely attributed to the frequent periodic reorientation in the flow. When Ca further increases up to 0.092, three daughter droplets can be observed, as shown in Figure 5f.

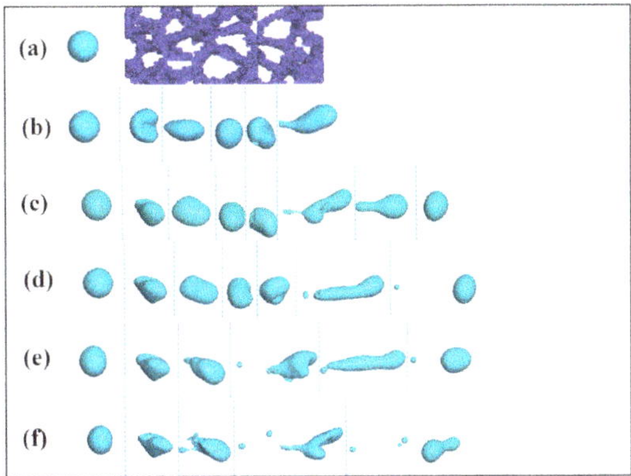

Figure 5. Hydrodynamics of one droplet passing a metal foam with different Ca values (L_d = 7.32, D_d = 1.82): (**a**) schematic diagram, (**b**) Ca = 0.031, (**c**) Ca = 0.046, (**d**) Ca = 0.061, (**e**) Ca = 0.078, and (**f**) Ca = 0.092.

As shown in Figure 6, when Ca increases from 0.031 to 0.092, ΔS rises nonlinearly from 1.0 to 1.2 at the channel length L = 18 mm where the droplet has left the metal foam, and its shape evolves to be stable. The rise in ΔS at L = 18 mm is obvious, with the increase in Ca from 0.031 to 0.061. For Ca = 0.031, the ΔS of 1.0 at L = 18 mm suggests that no daughter droplet is formed when the droplet leaves the porous structure. The high Ca value strengths the droplet breakup and restrains the coalescence.

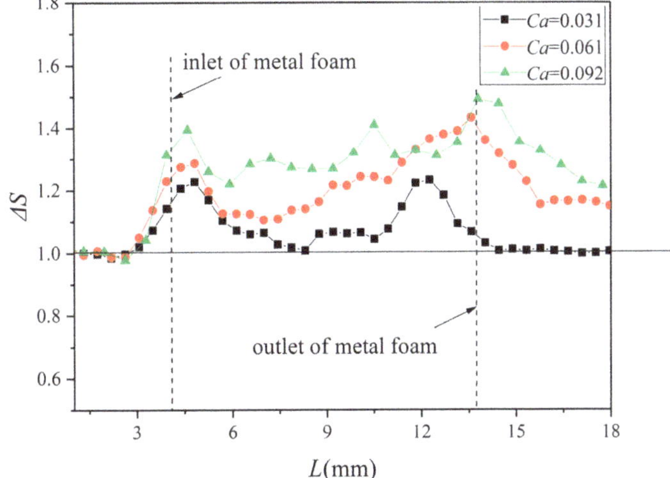

Figure 6. Change of the superficial area of the droplet passing through metal foam with different Ca values (L_d = 7.32; D_d = 1.82).

4.4. The Impact of Dd

PPI and porosity determine the cell diameter of metal foam. The cell average diameter of the metal foam with 60 PPI and the porosity of 95% was calculated to be 0.817 mm. The effect of D_d on droplet dynamics when the droplet passes through a metal foam is show in Figure 7. For the D_d values of 0.89 and 1.07, as shown in Figure 7b,c,the droplet passes though the metal foam without splitting into daughter droplets. For D_d = 0.89, no obvious droplet deformation can be observed. This is because the droplet is small enough for it to pass through the metal foam without a strong interaction with the three-dimensional porous structure of the metal foam. For D_d = 1.07, the droplet is stretched into a linear shape. For $D_d \geq$ 1.25(see Figure 7d–g), a polliwog-shaped droplet is formed, which is the characteristic of the case that an obstacle intrudes into a moving droplet followed by the breakup of droplet. The formation of the tail of the polliwog-shape is attributed to the interfacial area between the droplet and the obstacle. The polliwog-shaped droplets are only present at the outlet of the metal foam for the D_d values of 1.82 and 2.24. In contrast, for 1.25 $\leq D_d \leq$ 1.33, the polliwog-shaped droplets are formed inside the metal foam. This difference is because the polliwog-shaped droplet requires enough space to be stretched until it breaks up. For the big droplet of D_d = 1.82 or 2.24, the tail is so long that the required space for stretching the droplet to break up exceeds the pore size of metal foam, resulting in the formation of polliwog-shaped droplets at the metal foam outlet.

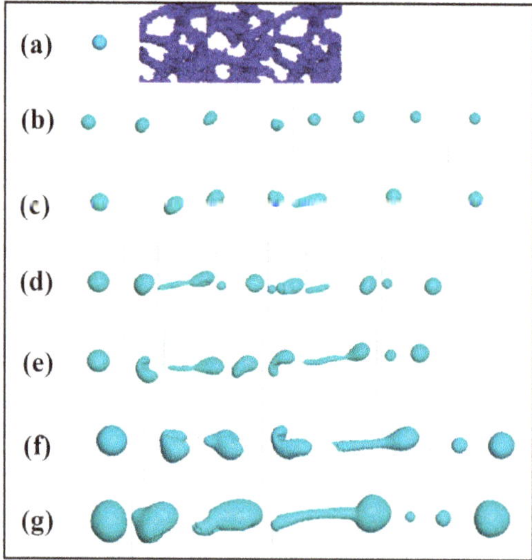

Figure 7. Hydrodynamics of one droplet passing a metal foam with different D_d values (L_d = 7.32, Ca = 0.061): (**a**) schematic diagram, (**b**) D_d = 0.89, (**c**) D_d = 1.07, (**d**) D_d = 1.25, (**e**) D_d = 1.33, (**f**) D_d = 1.82, and (**g**) D_d = 2.24.

To further elucidate the details of droplet breakup evolution in the metal foam, the case of D_d = 1.25 is taken as a representative one (see Figure S2). Droplet splitting in a microfluidic channel by the use of an obstacle was investigated numerically by Lee and Son [9]. The droplet is elongated around a cuboid obstacle and its portion becomes narrow near the front of the obstacle. The elongated portion breaks off at the front corner of the obstacle and then at the rear corner of the obstacle. Chung et al. [8] found that for cylinder obstruction, the thread becomes uniformly thinner around the cylinder, and finally breaks up at the front parts, independent of Ca. They concluded that the thread breakup was attributed to the velocity gradients induced by the geometric effect of the obstructions. Different from the hydrodynamics of droplets passing an obstacle in a micro-channel, for the droplet passing through metal foam, a waist is formed between the head and tail (see Supplementary Figure S2a). The droplet continually lengthens and breaks into two droplets, which are then driven into a spherical shape by surface tension (see Figure S2b,c). That is, the breakup of the droplet occurs at the waist not on the surfaces of the porous obstacles. In this regard, this droplet breakup is to some extent similar to that in a simple shear flow [8].

Figure 8 shows the effect of D_d on ΔS. When D_d increases from 0.89 to 1.33, the ΔS at L = 18 mm rises from 1.0 to 1.2. With a further increase in D_d from 1.33 to 1.82, D_d declines to 1.15 at L = 18 mm. For D_d = 0.89, the ΔS of 1.0 at L = 18 mm suggests that no daughter droplet is formed when the droplet leaves the porous structure.

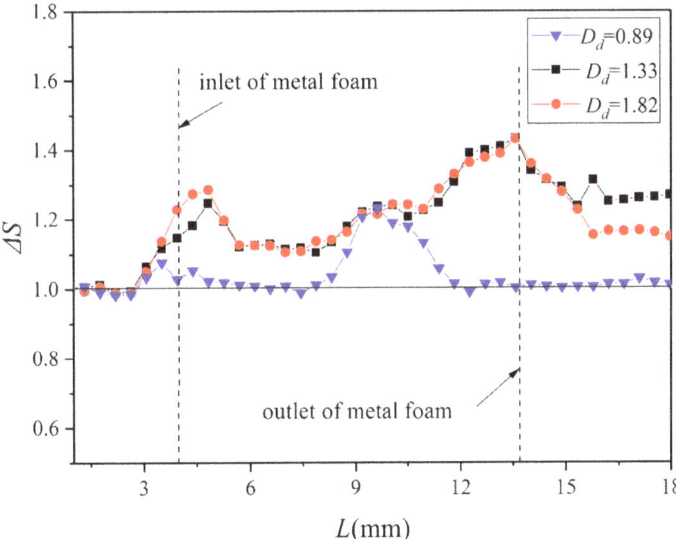

Figure 8. Change of the superficial area of the droplet passing through metal foam with different D_d values (L_d = 7.32; Ca = 0.061).

4.5. The Impact of Ld

The effect of L_d on droplet hydrodynamics is shown in Figure 9. For the case of Ca = 0.061, D_d = 1.25, and L_d = 2.44 (see Figure 9a), the polliwog–shaped droplet is formed at the metal foam outlet and then it breaks up into two droplets. With the increase in L_d from 2.44 to 9.76, the droplet breaks up inside the metal foam and the number of formed droplets is fixed at two. Therefore, no difference in the number of the generated daughter droplets is observed regardless of various L_d. As mentioned in Section 4.4, when D_d < 1.25, the droplet passes though the metal foam without splitting into daughter droplets. For D_d = 1.25, the D_d values of the two broken droplets are both less than 1.25, thus no splitting of the two droplets takes place in the following metal foam part. For the cases of Ca = 0.061, D_d = 1.82, and L_d in the range of 2.44 to 9.76, as shown in Figure 9e–h, the polliwog–shaped droplet is always formed at the metal foam outlet and then it breaks up into two droplets. L_d shows a negligible effect on the number of the generated daughter droplets in this case. In contrast, for a microchannel with obstructions, Chung et al. reported that the number of satellite droplets equals to the number of cylinder obstructions when the penetration of droplet fluid into the cylinder interval occurs [8]. The number of obstructions plays an important role in the number of generated daughter droplets. The difference in the droplet hydrodynamics is mainly because for metal foam, the breakup of the droplet occurs at the waist of the deformed droplet, whereas the breakup takes place on the surfaces of the obstructions in the case of a microchannel with obstructions.

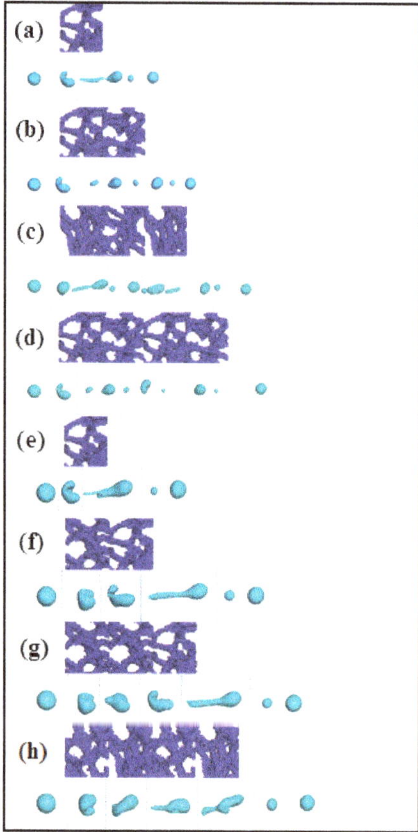

Figure 9. Hydrodynamics of one droplet passing a metal foam with different L_d values ($Ca = 0.061$): (**a**) $L_d = 2.44$, $D_d = 1.25$; (**b**) $L_d = 4.88$, $D_d = 1.25$; (**c**) $L_d = 7.32$, $D_d = 1.25$; (**d**) $L_d = 9.76$, $D_d = 1.25$; (**e**) $L_d = 2.44$, $D_d = 1.82$; (**f**) $L_d = 4.88$, $D_d = 1.82$; (**g**) $L_d = 7.32$, $D_d = 1.82$; and (**h**) $L_d = 9.76$, $D_d = 1.82$.

4.6. The Impact of Db

The above discussions are related to the hydrodynamics of a single droplet passing through metal foam. Actually, metal foam is usually used for the mixing of two immiscible liquids where a series of droplets are formed [4]. Therefore, in this section, the hydrodynamics of two successive droplets passing through metal foam is investigated. The effect of D_b on droplet hydrodynamics is illustrated in Figure 10. As shown in Figure 10, when D_b is below 3.00, the ΔS less than 1 can be found at the inlet of the metal foam. A D_b of 3.00 is the critical value to avoid the coalescence of two droplets at the inlet zone of the metal foam in the case of Ca of 0.061, D_d of 1.33, and L_d of 7.32. To further elucidate the details of two successive droplets hydrodynamics at the inlet zone of the metal foam, the case of D_b of 2.00 is taken as the representative one. A D_b of 2.00 at the inlet of the rectangular channel decreases to 0.35 at the position of 16 mm (1 mm before the inlet of the metal foam). Then, the two droplets merge into a big droplet at the inlet zone of the metal foam (see Figure S3).

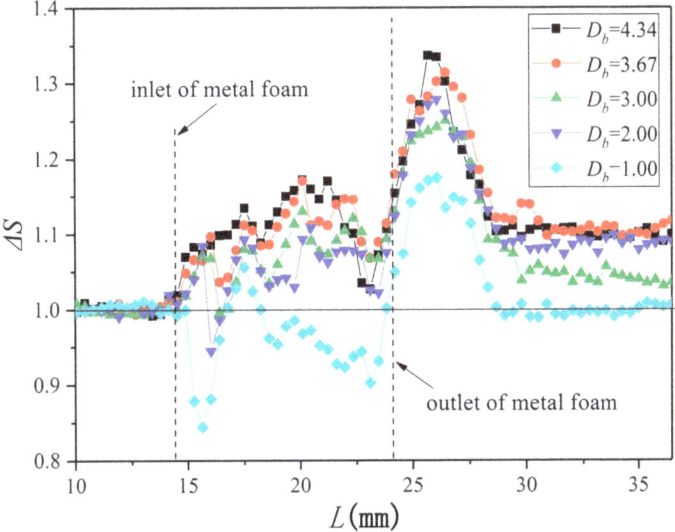

Figure 10. Change of the superficial area of two successive droplets when they pass through a metal foam with different D_b values (L_d = 7.32, Ca = 0.061, and D_d = 1.33).

5. Conclusions

We developed a LBM of droplet deformation and breakup and performed validation experiments. The simulated results arc in good agreement with the experimental ones. We defined the four dimensionless parameters: capillary number (Ca), the relative size between pores of the metal foam and the droplet (D_d), and the metal foam thickness (L_d). The results show that for L_d of 7.32 and D_d of 1.82, the Ca_c for droplet passing metal foam of 60 PPI is around 0.061. This Ca_c is lower than that of 0.1 for the case of one droplet passing through one cylinder obstruction because of the frequent periodic reorientation in the flow of the metal foam. For L_d of 7.32 and Ca of 0.061, when D_d = 0.89, no obvious droplet deform can be observed. For D_d = 1.07, the droplet is stretched into a linear shape. When D_d exceeds 1.25, a polliwog-shaped droplet is formed. The breakup of the droplet occurs at the waist of the deformed droplet and not on the surfaces of the porous obstacles. For Ca of 0.061 and D_d of 1.25, L_d in the range of 2.44 to 9.76 shows a negligible effect on the number of the generated daughter droplets. A D_b of 3.00 is the critical value to avoid the coalescence of two droplets at the inlet zone of the metal foam in the case of Ca of 0.061, D_d of 1.33, and L_d of 7.32.

Supplementary Materials: The following are available online at http://www.mdpi.com/2227-9717/7/12/877/s1, Figure S1: (a) Image of a water droplet surround by silicon oil on pure nickel surface and (b) a droplet simulated by LBM with Gw,k of ± 0.06, Figure S2: Evolution of one droplet breakup: (a) t = 0.09 s, (b) t = 0.096 s, and (c) t =0.102 s, Figure S3: Hydrodynamics of two successive droplets passing a metal foam with Db of 2.00 (Ld = 7.32, Ca = 0.061, Dd = 1.33): (a) t = 0.12 s, (b) t = 0.144 s, and (c) t = 0.168 s.

Author Contributions: Conceptualization, J.Z. and X.Y.; methodology, J.Z.; software, J.Z.; validation, J.Z. and X.Y.; formal analysis, J.Z. and X.Y.; investigation, J.Z.; resources, J.Z.; data curation, J.Z.; writing—original draft preparation, J.Z. and X.Y.; writing—review and editing, J.Z. and X.Y.; visualization, J.Z.; supervision, S.-T.T.; project administration, X.Y.; funding acquisition, X.Y.

Funding: This research was funded by the China Natural Science Foundation, grant number 21476073 and 21176069; and the Sub Topic of Major Science and Technology Project, grant number 2017ZX06002019-003.

Conflicts of Interest: The authors declare no conflict of interest.

References

1. Chen, H.; Yu, H.; Tang, Y.; Pan, M.; Feng, P.; Wang, H.; Peng, F.; Yang, J. Assessment and optimization of the mass-transfer limitation in a metal foam methanol microreformer. *App. Catal. A* **2008**, *337*, 155–162. [CrossRef]
2. Wenmakers, P.W.A.M.; Schaaf, J.V.D.; Kuster, B.F.M.; Schouten, J.C. "Hairy Foam": Carbon nanofibers grown on solid carbon foam. A fully accessible, high surface area, graphitic catalyst support. *J. Mater. Chem.* **2008**, *18*, 2426–2436. [CrossRef]
3. Stemmet, C.P.; Meeuwse, M.; Schaaf, J.V.D.; Kuster, B.F.M.; Schouten, J.C. Gas-liquid mass transfer and axial dispersion in solid foam packing. *Chem. Eng. Sci.* **2007**, *62*, 5444–5450. [CrossRef]
4. Yu, X.; Wen, Z.; Lin, Y.; Tu, S.T.; Wang, Z.; Yan, J. Intensification of biodiesel synthesis using metal foam reactors. *Fuel* **2010**, *89*, 3450–3456. [CrossRef]
5. Stemmet, C.P.; Jongmans, J.N.; Schaaf, J.V.D.; Kuster, B.F.M.; Schouten, J.C. Hydrodynamics of gas–liquid counter-current flow in solid foam packings. *Chem. Eng.* **2005**, *60*, 6422–6429. [CrossRef]
6. Kim, S.M.; Ghiaasiaan, S.M. Numerical Modeling of Laminar Pulsating Flow in Porous Media. *J. Fluids Eng.* **2009**, *131*, 041203. [CrossRef]
7. Koponen, A.; Kataja, M.; Timonen, J.; Kandhai, D. Simulations of Single-Fluid Flow in Porous Media. *Int. J. Mod. Phys. C* **1998**, *9*, 1505–1521. [CrossRef]
8. Chung, C.; Ahn, K.H.; Lee, S.J. Numerical study on the dynamics of droplet passing through a cylinder obstruction in confined microchannel flow. *J. Non-Newton. Fluid* **2009**, *162*, 38–44. [CrossRef]
9. Lee, J.; Lee, W.; Son, G. Numerical study of droplet breakup and merging in a microfluidic channel. *J. Mech. Sci. Technol.* **2013**, *27*, 1693–1699. [CrossRef]
10. Sethian, J.A.; Smereka, P. Level set methods for fluid interfaces. *Annu. Rev. Fluid Mech.* **2003**, *35*, 341–372. [CrossRef]
11. Lee, W.; Son, G. Numerical study of obstacle configuration for droplet splitting in a microchannel. *Comput. Fluids* **2013**, *84*, 351–358. [CrossRef]
12. Qian, J.Y.; Li, Y.J.; Gao, Z.X.; Jin, Z.J. Mixing efficiency and pressure drop analysis of liquid-liquid two phases flow in serpentine microchannels. *J. Flow Chem.* **2019**, *9*, 187–197. [CrossRef]
13. Qian, J.Y.; Li, X.J.; Gao, Z.X.; Jin, Z.J. Mixing efficiency analysis on droplet formation process in microchannels by numerical methods. *Processes* **2019**, *7*, 33. [CrossRef]
14. Juric, D.; Tryggvason, G. A front-tracking method for dendritic solidification. *J. Comput. Phys.* **1996**, *123*, 127–148. [CrossRef]
15. Yang, X.; James, A.J.; Lowengrub, J.; Zheng, X.; Cristini, V. An adaptive coupled level-set/volume-of-fluid interface capturing method for unstructured triangular grids. *J. Comput. Phys.* **2006**, *217*, 364–394. [CrossRef]
16. Scardovelli, R.; Zaleski, S. Direct numerical simulation of free-surface and interfacial flow. *Annu. Rev. Fluid Mech.* **1999**, *31*, 567–603. [CrossRef]
17. He, X.; Chen, S.; Zhang, R. A lattice boltzmann scheme for incompressible multiphase flow and its application in simulation of rayleigh–taylor instability. *J. Comput. Phys.* **1999**, *152*, 642–663. [CrossRef]
18. Qian, J.Y.; Li, X.J.; Wu, Z.; Jin, Z.J.; Sunden, B. A comprehensive review on liquid–liquid two-phase flow in microchannel: Flow pattern and mass transfer. *Microfluid. Nanofluid.* **2019**, *23*, 116. [CrossRef]
19. Boomsma, K.; Poulikakos, D.; Ventikos, Y. Simulations of flow through open cell metal foams using an idealized periodic cell structure. *Int. J. Heat Fluid Flow* **2003**, *24*, 825–834. [CrossRef]
20. Kopanidis, A.; Theodorakakos, A.; Gavaises, E.; Bouris, D. 3d numerical simulation of flow and conjugate heat transfer through a pore scale model of high porosity open cell metal foam. *Int. J. Heat Mass Transf.* **2010**, *53*, 2539–2550. [CrossRef]
21. Montminy, M.D.; Tannenbaum, A.R.; Macosko, C.W. The 3d structure of real polymer foams. *J. Colloid Interf. Sci.* **2004**, *280*, 202–211. [CrossRef] [PubMed]
22. Carvalho, T.P.D.; Morvan, H.P.; Hargreaves, D.; Oun, H.; Kennedy, A. Experimental and Tomography-Based CFD Investigations of the Flow in Open Cell Metal Foams with Application to Aero Engine Separators. In Proceedings of the ASME Turbo Expo 2015: Turbine Technical Conference and Exposition, Montreal, QC, Canada, 15–19 June 2015; p. V05CT15A028.
23. Aidun, C.K.; Clausen, J.R. Lattice-Boltzmann method for complex flows. *Rev. Fluid Mech.* **2010**, *42*, 439–464. [CrossRef]

24. Zhang, J. Lattice boltzmann method for microfluidics: Models and applications. *Microfluid. Nanofluid.* **2011**, *10*, 1–28. [CrossRef]

25. Wu, L.; Tsutahara, M.; Kim, L.S.; Ha, M.Y. Three-dimensional lattice boltzmann simulations of droplet formation in a cross-junction microchannel. *Int. J. Multiphase Flow* **2008**, *34*, 852–864. [CrossRef]

26. Liu, H.; Zhang, Y. Phase-field modeling droplet dynamics with soluble surfactants. *J. Comput. Phys.* **2010**, *229*, 9166–9187. [CrossRef]

27. Wang, W.; Liu, Z.; Jin, Y.; Cheng, Y. Lbm simulation of droplet formation in micro-channels. *Chem. Eng. J.* **2011**, *173*, 828–836. [CrossRef]

28. Rothman, D.H.; Keller, J.M. Immiscible cellular-automaton fluids. *J. Stat. Phys.* **1988**, *52*, 1119–1127. [CrossRef]

29. He, X.; Doolen, G.D. Thermodynamic foundations of kinetic theory and lattice boltzmann models for multiphase flows. *J. Stat. Phys.* **2002**, *107*, 309–328. [CrossRef]

30. Grunau, D.; Chen, S.; Eggert, K. A lattice boltzmann model for multiphase fluid flows. *Phys. Fluids A* **1993**, *5*, 2557–2562. [CrossRef]

31. Shan, X.; Chen, H. Lattice boltzmann model for simulating flows with multiple phases and components. *Phys. Rev. E* **1993**, *47*, 1815–1819. [CrossRef]

32. Shan, X.; Chen, H. Simulation of nonideal gases and liquid-gas phase transitions by the lattice boltzmann equation. *Phys. Rev. E* **1994**, *49*, 2941–2948. [CrossRef] [PubMed]

33. Hou, S.; Shan, X.; Zou, Q.; Doolen, G.D.; Soll, W.E. Evaluation of two lattice boltzmann models for multiphase flows. *J. Comput. Phys.* **1997**, *138*, 695–713. [CrossRef]

34. Li, Q.; Chai, Z.; Shi, B.; Liang, H. Deformation and breakup of a liquid droplet past a solid circular cylinder: A lattice boltzmann study. *Phys. Rev. E* **2014**, *90*, 043015. [CrossRef] [PubMed]

35. Park, J.; Li, X. Multi-phase micro-scale flow simulation in the electrodes of a pem fuel cell by lattice boltzmann method. *J. Power Sources* **2008**, *178*, 248–257. [CrossRef]

36. Frank, X.; Perré, P.; Li, H.Z. Lattice boltzmann investigation of droplet inertial spreading on various porous surfaces. *Phys. Rev. E* **2015**, *91*, 052405. [CrossRef] [PubMed]

37. Tölke, J.; Krafczyk, M.; Schulz, M.; Rank, E. Lattice boltzmann simulations of binary fluid flow through porous media. *Philos. Trans. A Math. Phys. Eng. Sci.* **2002**, *360*, 535–545. [CrossRef]

38. Beugre, D.; Calvo, S.; Dethier, G.; Crine, M.; Toye, D.; Marchot, P. Lattice boltzmann 3d flow simulations on a metallic foam. *J. Comput. Appl. Math.* **2010**, *234*, 2128–2134. [CrossRef]

39. Cornubert, R.; d'Humiè, D.; Levermore, R. A Knudsen layer theory for lattice gases. *Phys. D* **1991**, *47*, 241–259. [CrossRef]

40. Orlanski, I. A simple boundary condition for unbounded hyperbolic flows. *J. Comput. Phys.* **1976**, *21*, 251–269. [CrossRef]

41. Chen, L.; Kang, Q.; Mu, Y.; He, Y.L.; Tao, W.Q. A critical review of the pseudopotential multiphase lattice Boltzmann model: Methods and applications. *Int. J. Heat Mass Transf.* **2014**, *76*, 210–236. [CrossRef]

42. Liu, H.; Ju, Y.; Wang, N.; Xi, G.; Zhang, Y. Lattice boltzmann modeling of contact angle and its hysteresis in two-phase flow with large viscosity difference. *Phys. Rev. E* **2015**, *92*, 033306. [CrossRef] [PubMed]

43. Stamenković, O.S.; Todorović, Z.B.; Lazić, M.L.; Veljković, V.B.; Skala, D.U. Kinetics of sunflower oil methanolysis at low temperatures. *Bioresour. Technol.* **2008**, *99*, 1131–1140.

44. Noureddini, H.; Zhu, D. Kinetics of transesterification of soybean oil. *J. Am. Oil Chem. Soc.* **1997**, *74*, 1457–1463. [CrossRef]

45. Yan, Y.Y.; Zu, Y.Q. Numerical Modelling Based on Lattice Boltzmann Method of the Behaviour of Bubbles Flow and Coalescence in Microchannels. In Proceedings of the 2008 International Conference on Nanochannels, Microchannels, and Minichannels (ASME 2008), Darmstadt, Germany, 23–25 June 2008; pp. 313–319.

46. Li, J.; Renardy, Y.Y.; Renardy, M. Numerical simulation of breakup of a viscous drop in simple shear flow through a volume-of-fluid method. *Phys. Fluids* **2000**, *12*, 269–282. [CrossRef]

47. Grace, H.P. Dispersion phenomena in high viscosity immiscible fluid systems and application of static mixers as dispersion devices in such systems. *Chem. Eng. Commun.* **1982**, *14*, 225–277. [CrossRef]

48. Puyvelde, P.V.; Yang, H.; Mewis, J.; Moldenaers, P. Breakup of filaments in blends during simple shear flow. *J. Rheol.* **2000**, *44*, 1401–1415. [CrossRef]

49. Cristini, V.; Guido, S.; Alfani, A.; Bł(awzdziewicz, J.; Loewenberg, M. Drop breakup and fragment size distribution in shear flow. *J. Rheol.* **2003**, *47*, 1283–1298. [CrossRef]

50. Rallison, J.M. A numerical study of the deformation and burst of a viscous drop in general shear flows. *J. Fluid Mech.* **1981**, *109*, 465–482. [CrossRef]
51. Deroussel, P.; Khakhar, D.V.; Ottino, J.M. Mixing of viscous immiscible liquids. part 1: Computational models for strong–weak and continuous flow systems. *Chem. Eng. Sci.* **2001**, *56*, 5511–5529. [CrossRef]

Article

Design, Simulation, and Experiment of an LTCC-Based Xenon Micro Flow Control Device for an Electric Propulsion System

Chang-Bin Guan *, Yan Shen, Zhao-Pu Yao, Zhao-Li Wang, Mei-Jie Zhang, Ke Nan and Huan-Huan Hui

Advanced Space Propulsion Lab, Beijing Institute of Control Engineering, Beijing 100094, China; shenyan1978@gmail.com (Y.S.); yzp06@sina.com (Z.-P.Y.); wangzhaoli.bionic@gmail.com (Z.-L.W.); zhang-mj15@tsinghua.org.cn (M.-J.Z.); nanke@buaa.edu.cn (K.N.); huihuanhuan@icloud.com (H.-H.H.)
* Correspondence: guanchangbin@163.com; Tel.: +86-010-68112261

Received: 29 October 2019; Accepted: 15 November 2019; Published: 19 November 2019

Abstract: A xenon micro flow control device (XMFCD) is the key component of a xenon feeding system, which controls the required micro flow xenon (µg/s–mg/s) to electric thrusters. Traditional XMFCDs usually have large volume and weight in order to achieve ultra-high fluid resistance and have a long producing cycle and high processing cost. This paper proposes a miniaturized, easy-processing, and inexpensive XMFCD, which is fabricated by low-temperature co-fired ceramic (LTCC) technology. The design of the proposed XMFCD based on complex three-dimensional (3D) microfluidic channels is described, and its fabrication process based on LTCC is illustrated. The microfluidic channels of the fabricated single (9 mm diameter and 1.4 mm thickness) and dual (9 mm diameter and 2.4 mm thickness) XMFCDs were both checked by X-ray, which proved the LTCC method's feasibility. A mathematical model of flow characteristics is established with the help of finite element analysis, and the model is validated by the experimental results of the single and dual XMFCDs. Based on the mathematical model, the influence of the structure parameters (diameter of orifice and width of the groove) on flow characteristics is investigated, which can guide the optimized design of the proposed XMFCD.

Keywords: xenon micro flow control device (XMFCD); low-temperature co-fired ceramic (LTCC); xenon feeding system; electric propulsion system; flow characteristic

---

1. Introduction

Compared with the traditional chemical propulsion system, the electric propulsion system has the advantages of higher specific impulse, simple composition, and no pollution [1]. It has broad application prospects in satellite attitude and orbit control, deep space exploration, interplanetary flight, and other fields. The actuator of the electric propulsion system is an electric thruster, whose principle is to ionize the gas working medium and then accelerate the ion to generate the thrust through an external electric field. Xenon is the ideal working medium for electric thrusters due to its high molecular weight, low ionization energy and easy storage [2].

The xenon feeding system is responsible for the storage, pressure regulation, and flow control of the working medium. It is the key subsystem of the electric propulsion system and belongs to the field of high precision fluid control. The xenon micro flow control device (XMFCD) is the most important component of the xenon feeding system because its control accuracy of the micro xenon flow (µg/s–mg/s) determines the thrust accuracy, working efficiency, and service life of the electric thruster [3]. The core problem of the XMFCD is to achieve ultra-high fluid resistance in a limited volume.

Different XMFCDs have been developed through different implementation methods of ultra-high fluid resistance. The OKB Fakel company of Russia used a capillary type XMFCD to control the micro xenon flow to the electric thruster of SMART-1, which is made of a long-winded capillary tube [4]. The Mott Corporation of USA developed a porous metal type MFC, which is sintered from metal powder [5]. This porous metal type XMFCD was successfully applied in Deep Space 1 [6] and Dawn [7] space detectors. Besides, Northwest Institute for Nonferrous Metal Research of China also produced this type of XMFCD and used it in a hall electric propulsion system [8]. Vacco Industries of USA used chemically etched metal disks as XMFCDs, which realized ultra-high fluid resistance by chemically-etched microfluidic channels [9]. Lee Company invented a stacked-disk containing tangentially machined flow passages, which can be used as an XMFCD [10]. The above XMFCDs all have their shortcomings: (1) The capillary type XMFCD needs very long length to achieve ultra-high fluid resistance, which leads to big weight and volume, (2) the porous metal type XMFCD is heavy and easily produces contaminated particles that causes blockage, (3) the chemically-etched type and the stacked disk type of XMFCDs both have high cost and long production cycle because of their special processing and assembly technologies (such as laser drilling, femtosecond processing, lithography processing, and diffusion welding). Besides, all the above XMFCDs are made of metal, which results in heavy weight and difficulty in constructing microfluidic channels. With the great demand for low-cost and light-weight electric propulsion systems for microsatellites and commercial satellites, the above XMFCDs are no longer able to meet the requirements of the xenon feeding system.

Low temperature co-fired ceramics (LTCC) have been used for almost twenty years to produce a multilayer substrate for packaging integrated circuits [11]. Recently, the multi-layer approach has also been applied to complex three-dimensional (3D) microfluidic structures used as platforms for the fabrication of miniaturized systems for different application fields [12], because LTCC provides a convenient medium for fabricating laminated three-dimensional (3D) micro channel structures due to the easy and inexpensive fabrication process, low sintering temperature, and excellent mechanical, thermal, electrical, and chemical properties [13]. The LTCC microfluidic systems have been applied to many fields such as bioreactors, combustors, mixers, chemical reactors, heat exchangers, and gold nanoparticles generators [14]. The LTCC technology enables the fast and easy fabrication of microfluidic devices and systems. This can both reduce the cost of devices and shorten the development time [15].

Considering the above advantages of LTCC, a miniaturized XMFCD based on LTCC is proposed in this paper. Firstly, the design and fabrication of the LTCC XMFCD, which contains complex 3D micro channels, is illustrated, and the single and dual XMFCD samples are both produced and checked. Secondly, the mathematical model describing the flow characteristics of the XMFCD is established with the help of finite element analysis of the pressure distribution. Thirdly, the experiment setup, which is used to measure the mass flow of the XMFCD, is built. Fourthly, the pressure-flow characteristics of the single and dual XMFCDs are both measured to prove the proposed mathematical model, and the influence of different structural parameters on flow characteristics is discussed. Finally, the conclusion is outlined.

2. Design and Fabrication

The proposed XMFCD in this paper applies a labyrinth type microfluidic passage to realize ultra-high fluid resistance as mentioned in [9]. However, in [9], they used different implementation methods. The XMFCD in [9] is assembled by diffusion welding of several machined metal disks. However, the proposed XMFCD in this paper is produced by LTCC technology, which realizes miniaturization, low cost, and a short production cycle. The proposed XMFCD realizes ultra-high fluid resistance by complex 3D micro channels, which are composed of many chambers, grooves, and orifices in series. The chambers and grooves are embedded in two planes, and the orifices connect with the chambers in two planes.

The production process of the proposed XMFCD based on LTCC is shown in Figure 1. The LTCC technology and material system enable the easy creation of complex three-dimensional microfluidic structures through structuring and assembly before the material is transformed into a rigid glass ceramic device. The starting point in LTCC technology is a green ceramic tape produced by a tape casting method, and various shapes (channels, cavities, vias, etc.) are cut in green LTCC tapes using a laser cutting and drilling machine (SK-MPL50, SANKE, Shanghai, China) in the first step. As shown in Figure 1, there are four shapes of green ceramic tapes (A, B, C, and D). The proposed XMFCD is composed of 14 tape layers (about 0.1 mm thickness each layer), which build the desired complex microfluidic passages. Layers 1, 2, 3, and 4 are A shape tapes which form a central inlet passage; Layer 5 is a B shape tape which hollows out 10 chamber-groove units; Layers 6, 7, 8, and 9 are C shape tapes which form 19 orifices; Layer 10 is a D shape tape which hollows out 10 chamber-groove units; and Layers 11, 12, 13, and 14 are A shape tapes which form a central outlet passage. In the third step, all the stacked LTCC tapes are laminated by a hydraulic press machine (6606-603-400, KISTLER, Beijing, China). Typically, the thermo-compression lamination process is performed at high pressure (up to 20 MPa), elevated temperature (up to 90 °C) for 5 to 30 min. After lamination, the LTCC module is co-fired according to a two-step thermal profile with a maximum temperature of 850 to 900 °C in a muffle furnace (SJL-200, CETC, Changsha, China) [16]. After the co-firing, the multi-layers are sintered into a hard piece. It should be noted that Figure 1 is a simplified schematic diagram for facilitating the disclosure of the manufacturing process which only includes one unit. In the actual processing, the 12 tape layers are all rectangular, which is easy to align. Each layer arrays 20 to 30 same-shape channel units (A, B, C, or D); therefore, 20 to 30 XMFCDs can be produced on one sintering board and then be cut to 20 to 30 independent circle XMFCDs, as shown in Figure 2a, by a laser cutting and drilling machine. This is why the LTCC production is efficient and cheap. The XMFCD made of 14 layers is called "single XMFCD" whose diameter is 9 mm and thickness is 1.4 mm. The mass of the single XMFCD is only 0.2 g, while the existing metal-based XMFCD with the same fluid resistance is at least tens of grams. In order to check the construction effect of the microfluidic channel, X-ray detection of the single XMFCD was performed, as shown in Figure 2b, which indicates that it matches the desired microfluidic channels. The formed complex 3D microfluidic structures of the proposed single XMFCD include 19 small orifices (diameter 0.1 mm), 20 thin grooves (width 0.2 mm, depth 0.1 mm) and 40 chambers (diameter 1 mm, depth 0.1 mm). Fluid enters at the center of the XMFCD and passes through a groove, which is tangential to a spin chamber. Then the fluid discharges through a small center hole into another chamber. This process repeats over and over to realize the ultra-high fluid resistance characteristic.

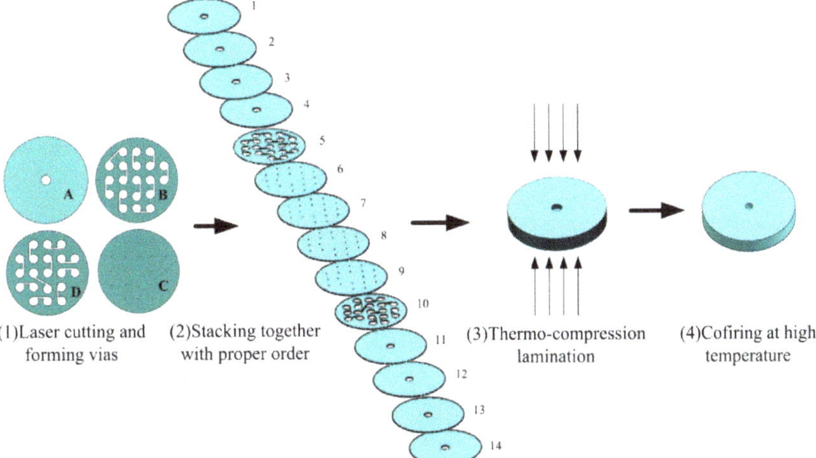

Figure 1. Fabrication process of the proposed xenon micro flow control device (XMFCD) based on low-temperature co-fired ceramics (LTCCs).

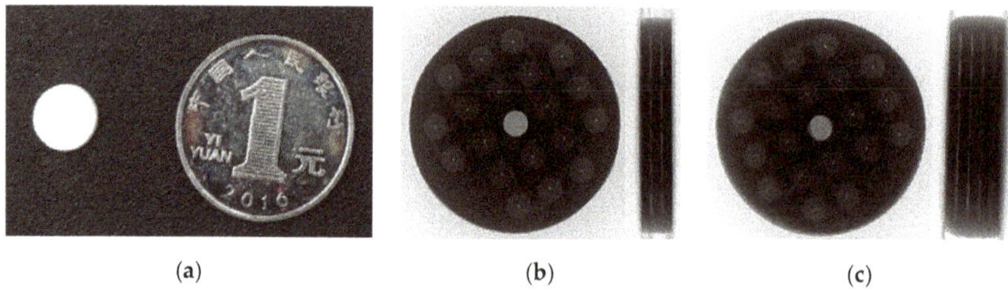

Figure 2. The proposed XMFCD samples and X-ray images. (**a**) Photograph of proposed XMFCD; (**b**) X-ray images of single XMFCD; (**c**) X-ray images of dual XMFCD.

In order to realize bigger fluid resistance, several units can be stacked together by LTCC technology. According to the fabrication process shown in Figure 1, dual XMFCD can be produced by 24 tape layers, which is an increase of 10 layers on the basis of the single XMFCD. The 15th to 24th layers of the dual XMFCD are the same as the Layer 5 to Layer 14 of the single XMFCD. A dual XMFCD was fabricated by LTCC, and its diameter and thickness are, respectively, 9 and 2.4 mm. Its mass is only 0.38 g, while the existing metal-based XMFCD with the same fluid resistance is at least tens of grams. The 3D microfluidic structure of the dual XMFCD includes 38 small orifices (diameter 0.1 mm), 40 thin grooves (width 0.2 mm, depth 0.1 mm) and 80 chambers (diameter 1 mm, depth 0.1 mm). The dual XMFCD is also detected by X-ray, as shown in Figure 2c, which indicates that it matches the designed fluid channel. Compared to the existing metal-based XMFCDs, LTCC-based XMFCDs are smaller, lighter, and cheaper.

3. Modelling and Simulations

For the proposed XMFCD, its mass flow characteristic is the most important performance parameter. In order to investigate the influence of structural parameters on the flow characteristics and guide the design of the proposed XMFCD, its mathematical model should be established. According to 3D microfluidic structures of the proposed XMFCD, the fluid model of the XFCD can be divided into two types of fluid units (Fluid Unit A and Fluid Unit B). Fluid Unit A contains 2 chambers, 1 groove, and 1 orifice, and Fluid Unit B contains 2 chambers and 1 groove. The single XMFCD consists of 19 Fluid Unit As and 1 Fluid Unit B in series, which includes 40 chambers, 19 orifices, and 20 grooves in total. The dual XMFCD consists of 38 Fluid Unit As and 2 Fluid Unit Bs in series which totally includes 80 chambers, 38 orifices, and 40 grooves. According to the actual working condition, the simulation block diagram of the proposed single and dual XMFCDs is given as shown in Figure 3. Therefore, the mathematical model of the proposed XMFCD is composed of 3 types of models: chamber, groove and orifice.

For the groove of the fluid unit shown in Figure 3, the mass flow through the groove can be treated as laminar flow, which can be described as [17]

$$Q_g = \frac{\pi D_g^4 \rho}{256 \mu R T_1 l_g}(P_1^2 - P_2^2) \tag{1}$$

where, ρ and μ are, respectively, the density and viscosity of xenon gas; R is the gas constant; T_1 is the absolute temperature of the gas in the groove; $D_g = \sqrt{4 w_g d_g / \pi}$ and l_g are, respectively, the equivalent cross-sectional diameter and length of the groove; w_g and d_g are, respectively, the width and depth of the groove; and P_1, P_2 are the pressure of chambers at the upstream and downstream of the groove.

For the orifice of Fluid Unit A shown in Figure 3, the mass flow through the orifice can be expressed by [18]

$$Q_o = C_d A \sqrt{k} \frac{P_2}{\sqrt{R T_2}} f(\frac{P_3}{P_2}) \tag{2}$$

$$f\left(\frac{P_3}{P_2}\right) = \begin{cases} \sqrt{\left(\frac{2}{k-1}\right)\left[\left(\frac{P_3}{P_2}\right)^{\frac{2}{k}} - \left(\frac{P_3}{P_2}\right)^{\frac{k+1}{k}}\right]} & \frac{P_3}{P_2} > P_{cr} \quad \text{subsonic flow} \\ \sqrt{\left(\frac{2}{k+1}\right)^{\frac{k+1}{k-1}}} & \frac{P_3}{P_2} \leq P_{cr} \quad \text{Supersonic flow} \end{cases} \tag{3}$$

$$P_{cr} = \left(\frac{2}{k+1}\right)^{\frac{k}{k-1}} \tag{4}$$

where, C_d is the flow coefficient; $A = \pi D_o^2/4$ is the cross-sectional area; D_o is the diameter of the orifice; k is the gas adiabatic index; T_2 is the absolute temperature of the upstream gas of the orifice; P_2, P_3 are, respectively, the upstream and downstream pressure of the orifice; and P_{cr} is the critical pressure ratio.

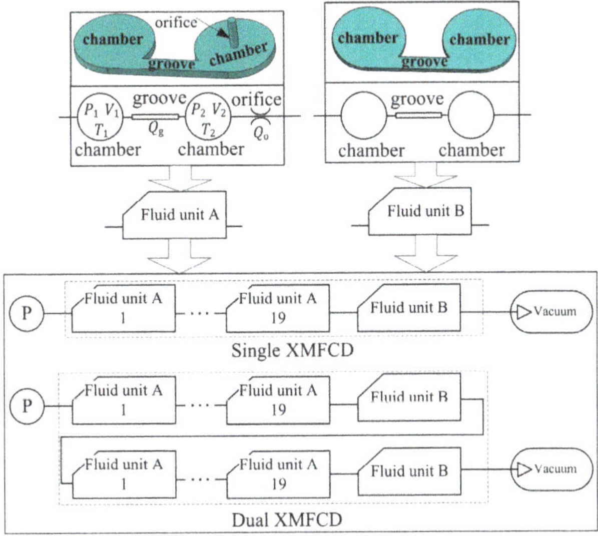

Figure 3. The simulation block diagram of the proposed XMFCD.

The flow characteristic model of the chamber can be expressed by [19] (for example, chamber V_2 shown in Figure 3)

$$\frac{dP_2}{dt} = \frac{R}{V_2}(T_1 Q_g - T_2 Q_o) \tag{5}$$

where, $V_2 = \pi D_c^2 d_g/4$ is the volume of the chamber, and D_c is the diameter of the chamber.

The mass flow of the proposed XMFCD can be iteratively calculated by Equations (1)–(5) when its inlet pressure and outlet pressure are both known as the boundary conditions. However, the flow status through each orifice must be known in order to determine the subsonic flow equation or the supersonic flow equation in Equation (3) [20]. The finite element method by ANSYS is the most effective tool to do numerical analysis on flow and pressure characteristics of microchannels [21,22]. In this work, ANSYS is also applied to simulate the pressure distribution of the single XMFCD. According to its actual working condition in a space electric propulsion system, the inlet pressure and outlet pressure of the single XMFCD are respectively set to 0.15 MPa and 0 because the outlet is directly connected to a vacuum environment. Figure 4 is the pressure distribution map and each chamber's pressure is marked and plotted in Figure 5. Then the ratio of the outlet pressure and inlet pressure of each orifice is calculated and compared with the critical pressure ratio, P_{cr}. In the 40 chambers, only the ratio of the 39th chamber pressure and the 38th chamber pressure is smaller than P_{cr} (for xenon, $k = 1.67$ and $P_{cr} = 0.4867$). Hence, only the flow of the last orifice (19th orifice) is supersonic flow and the other 18 orifices are all subsonic flow. The proposed single and dual XMFCDs both have this feature because their outlets connects to a vacuum.

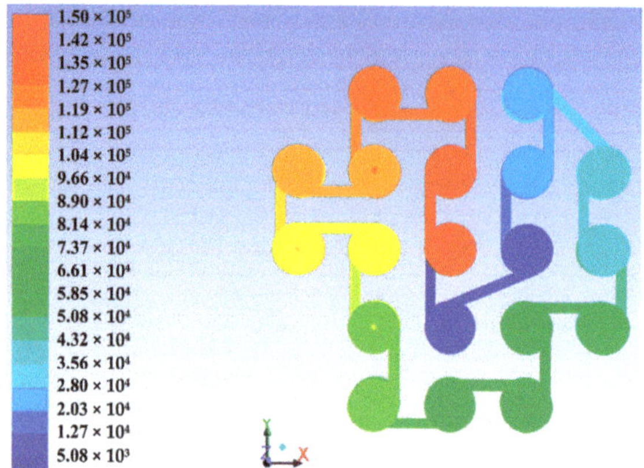

Figure 4. The pressure distribution map of single XMFCD.

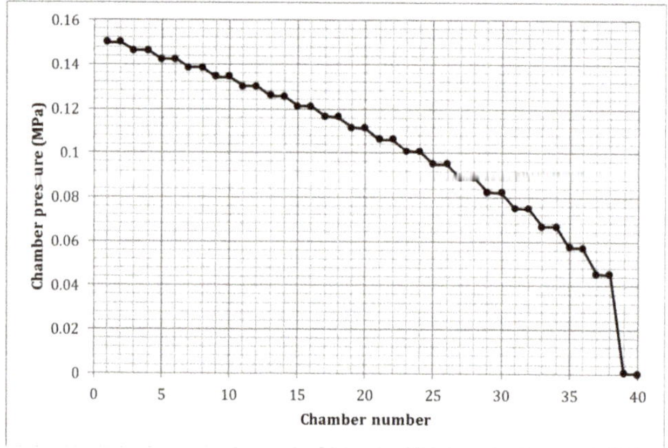

Figure 5. The chamber's pressure values of the single XMFCD.

4. Experimental Setup

Figure 6 presents the experimental setup, which is used to measure the mass flow of the proposed XMFCD. It was built with three manual on/off valves (SS-41GXS1, Swagelok, Solon, OH, USA), three ranges of high-precision mass flow meters (ALICAT SCIENTIFIC, 0–6, 0–20, and 0–100 sccm, Tucson, AZ, USA) and a dry scroll vacuum pump (IDP3A01, Agilent Technologies, Santa Clara, CA, USA). The LTCC XMFCD with a test connecter is installed between the high-precision flow meter and the dry scroll vacuum pump. The manual valves are used to control the on/off of the upstream xenon. The dry scroll vacuum pump vacuums the downstream of the LTCC XMFCD to simulate its actual operating conditions in an electric propulsion system. According to the flow range of the proposed LTCC XMFCD, the 0–20 sccm high-precision flow meter is chosen to measure the micro mass flow. A 5-L xenon tank with 5 MPa pressure supplies xenon for the experiment, which is not shown in Figure 6. The LTCC XMFCD's upstream xenon pressure, which is sampled by a high-precision pressure sensor (P30, 0–0.6 MPa, Wika, Frankfurt, Germany), is precisely adjusted by two pressure regulators (high pressure: KPR1JWA422A20000RD, Swagelok, Solon, OH, USA; low pressure: KLF1FRA411A200000G, Swagelok, Solon, OH, USA) which are not shown in Figure 6.

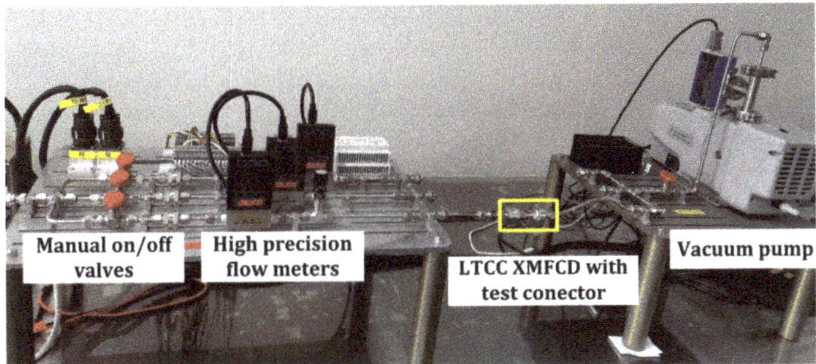

Figure 6. The experiment setup to measure mass flow of the proposed XMFCD.

This experiment setup is used to measure the steady flow of the proposed LTCC XMFCD. At the beginning of the experiment, the vacuum pump is opened and the desired upstream xenon pressure adjusted by two pressure regulators. Then when the value of the flow meter is displayed as 0, which proves the pipeline is already a vacuum, the manual on/off valve is opened and xenon gas is supplied to the LTCC XMFCD. Finally, after the flow meter is stable, the flow value is recorded.

5. Results and Discussions

5.1. Validation of the Mathematical Model

The flow characteristics of the single and dual XMFCDs are both simulated by the proposed mathematical model. The boundary conditions of the simulation are the same as the working conditions in an electric propulsion system. The upstream xenon pressure of the XMFCD is set to 0.1 to 0.2 MPa and the downstream pressure of the XMFCD is set to 0. The temperature is set to 20 °C. The structure parameters of the proposed XMFCD and the physical properties of the xenon are both listed in Table 1.

Table 1. The simulation parameters of the XMFCD.

Parameters	Values
ρ	5.89 kg/m^3
μ	2.11×10^{-5} Pa·s
R	63.29 J/(kg·K)
k	1.67
w_g	0.2 mm
d_g	0.1 mm
l_g	1 mm
C_d	0.7
D_o	0.1 mm
D_c	1 mm

The flows at 11 pressure points between 0.1 to 0.2 MPa are calculated by the proposed mathematical model. In order to validate the mathematical model, the flows at the same pressures are tested. The measured and simulated mass flows of the single and dual XMFCDs are both listed in Table 2. Moreover, its errors are also calculated, which is shown in Table 2. The experimental and simulated flows of the single and dual XMFCDs are both plotted in Figure 7a,b, which show that the simulation flow results are in good agreement with the experiment results. According to Table 2, the error's absolute value of the measured and simulated mass flows of the single XMFCD is 0.00256 to 0.49 mg/s, which is 0.2% to

5.5% of the experimental results. Moreover, the error's absolute value of the measured and simulated mass flows of the dual XMFCD is 0.00318 to 0.03589 mg/s, which is 0.3% to 5.7% of the experimental results. Therefore, the mathematical model proposed in this paper is proven to be very effective for predicting the flow characteristics of the proposed XMFCD. Besides, the experimental results show that the mass flow of the proposed XMFCD is linearly proportional to the inlet pressure.

Table 2. The experimental and simulated mass flow of the proposed XMFCD.

	Inlet Pressure (MPa)	Experimental Mass Flow (mg/s)	Simulated Mass Flow (mg/s)	Error (mg/s)
Single XMFCD	0.1	0.880	0.92900	−0.04900
	0.11	0.993	1.02488	−0.03188
	0.12	1.095	1.12077	−0.02577
	0.13	1.210	1.21667	−0.00667
	0.14	1.310	1.31256	−0.00256
	0.15	1.413	1.40847	0.00453
	0.16	1.520	1.50437	0.01563
	0.17	1.619	1.60027	0.01873
	0.18	1.719	1.69618	0.02282
	0.19	1.817	1.79209	0.02491
	0.20	1.910	1.88800	0.02200
Dual XMFCD	0.1	0.622	0.65789	−0.03589
	0.11	0.702	0.72673	−0.02473
	0.12	0.775	0.79558	−0.02058
	0.13	0.856	0.86445	−0.00845
	0.14	0.927	0.93331	−0.00631
	0.15	0.999	1.00218	−0.00318
	0.16	1.075	1.07106	0.00394
	0.17	1.145	1.13994	0.00506
	0.18	1.215	1.20882	0.00618
	0.19	1.285	1.27771	0.00729
	0.20	1.351	1.34659	0.00441

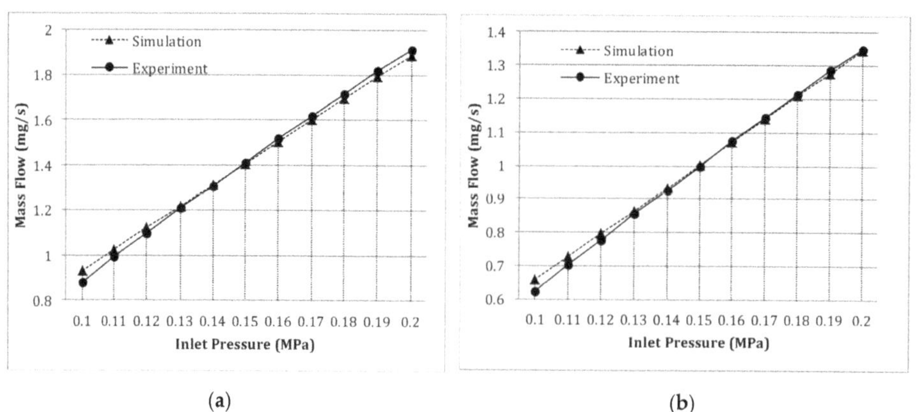

(a) (b)

Figure 7. Comparison of the experimental and simulated mass flow of the proposed XMFCD. (**a**) Single XMFCD; (**b**) Dual XMFCD.

5.2. Influence of Structural Parameters

In order to realize the optimization design of the proposed XMFCD, the influence of structural parameters on flow characteristics should be investigated. According to Figure 3, the diameter of orifice D_o, and the width of groove w_g are the parameters that are easy to adjust under volume limitation.

So the influence of these two parameters on the flow characteristic is focused on analysis using the mathematical model validated above. In the following simulations, the upstream xenon pressure of the XMFCDs is set to 0.15 MPa.

Firstly, the mass flow of the single and dual XMFCDs with different D_o (40, 50, 60, 70, 80, 90, and 100 μm) and other parameters the same as shown in Table 1 are simulated. Moreover, the simulated mass flow results are shown in Table 3 and Figure 8a. According to the data analysis, two conclusions can be obtained as follows: (1) for XMFCDs with the same structural parameters, the mass flow Q_1 of the single XMFCD can be approximately described as 1.4 times of the mass flow Q_2 of the dual XMFCD, that proves the flow of N series orifices is $1/\sqrt{N}$ of a single orifice's flow, and (2) for the proposed XMFCDs, the mass flow is proportional to the square of the orifice diameter.

Table 3. The simulated mass flow of the XMFCD with different diameters of orifice (D_o) and widths of groove (w_g).

Parameters	Value (μm)	Mass Flow Q_1 of Single XMFCD (mg/s)	Mass Flow Q_2 of Dual XMFCD (mg/s)	Q_1/Q_2
D_o	40	0.23	0.165	1.393939
	50	0.358	0.256	1.398438
	60	0.514	0.368	1.396739
	70	0.698	0.499	1.398798
	80	0.908	0.649	1.399076
	90	1.145	0.817	1.401469
	100	1.41	1	1.41
w_g	30	0.5976	0.3289	1.816966
	50	1.021	0.637	1.602826
	80	1.258	0.852	1.476526
	100	1.321	0.914	1.445295
	130	1.368	0.961	1.423517
	150	1.385	0.978	1.416155
	180	1.401	0.995	1.40804
	200	1.41	1	1.41

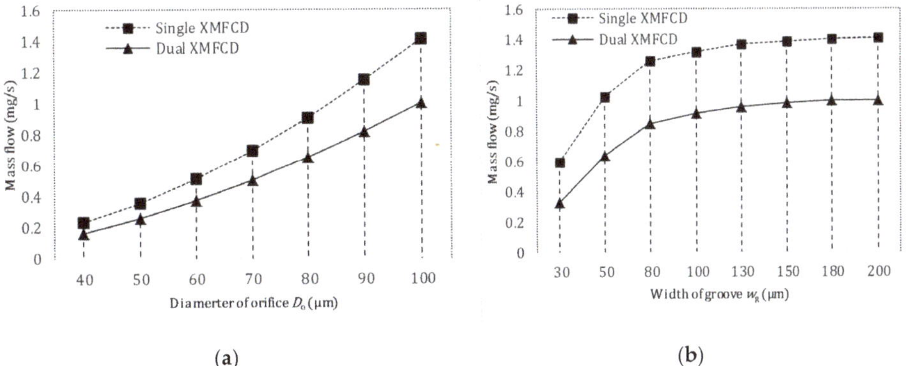

Figure 8. The simulated flow of the XMFCD with different structural parameters. (**a**) Simulated flow with different D_o; (**b**) Simulated flow with different w_g.

Then, the mass flow of the single and dual XMFCDs with different w_g (30, 50, 80, 100, 130, 150, 180, and 200 μm) and other parameters the same as shown in Table 1 are simulated. The simulated mass flow results are listed in Table 3 and drawn in Figure 8b. Referring to the simulated data, we can get conclusions as follows: (1) when $w_g \geq 80$ μm, the flow of the XMFCD is mainly determined by the orifices and the effect of the grooves on the flow is not obvious, and the mass flow in this case conforms to the above flow formula of the series orifices; and (2) when $w_g < 80$ μm, the flow of the XMFCD decreases sharply as w_g decreases because the throttling effect of the grooves becomes increasingly obvious. It is worth noting that when $w_g = 80$ μm, its equivalent diameter of the groove is about 100 μm which is as same as the diameter of the orifices. Therefore, in other words, only when the equivalent diameter of the groove is smaller than the diameter of the orifices, the impact of the groove on the XMFCD's flow will become significant.

6. Conclusions

This work proposed a miniaturized, easily processed, and inexpensive XMFCD, which is based on 3D microfluidic channels and fabricated by LTCC. The single XMFCD (diameter 9 mm, thickness 1.4 mm) and dual XMFCD (diameter 9 mm, thickness 2.4 mm) samples were both produced, and their 3D microfluidic structures were both detected by X-ray which proves the feasibility of this method. Moreover, other XMFCDs with more layers can be fabricated based on LTCC in order to obtain bigger fluid resistance, which can meet smaller flow requirements of space electric thrusters. The flow characteristic mathematical model and the simulation diagram validated by experiment results can be used to analyze and design other gas flow control devices based on microfluidic channels. The influence analysis of the structural parameters on flow characteristics reveals that (1) the diameter of the orifice is the most important parameter, whose square is proportional to the mass flow; (2) the mass flow of N series orifices is approximately $1/\sqrt{N}$ of a single orifice's flow, which can be used to estimate the flow of XMFCDs with different layers; and (3) only when the groove's equivalent diameter is smaller than the orifice diameter, the impact of the groove on the XMFCD's flow will become significant. So this work proposed a novel processing method for realizing the miniaturization of XMFCDs which can also be used in other micro flow control fields, and the mathematic model and simulation method lays a theoretical foundation for research on the flow characteristic of micro gas flow control devices.

Author Contributions: Conceptualization, C.-B.G. and Y.S.; Data curation, K.N. and H.-H.H.; Formal analysis, M.-J.Z. and K.N.; Investigation, H.-H.H. and Z.-L.W.; Methodology, C.-B.G.; Project administration, Y.S. and Z.-P.Y.; Software, M.-J.Z. and Z.-L.W.; Writing—original draft, C.-B.G.; Writing—review & editing, M.-J.Z. and Z.-P.Y.

Funding: This research was funded by National Natural Science Foundation of China, grant number 51805026.

Conflicts of Interest: The authors declare no conflict of interest.

References

1. Jarrige, J.; Elias, P.Q.; Cannat, F.; Packan, D. Performance comparision of an ECR plasma thruster using Argon and Xenon as propellant gas. In Proceedings of the 33rd International Electric Propulsion Conference, Washington, DC, USA, 6–10 October 2013.
2. Nakles, M.R.; Hargus, W.A.; Delgado, J.J.; Corey, R.L. A performance comparison of Xenon and Krypton propellant on an SPT-100 hall thruster. In Proceedings of the 32nd International Electric Propulsion Conference, Wiesbaden, Germany, 11–15 September 2011.
3. Dankanich, J.W.; Cardin, J.; Dien, A.; Kamhawi, H.; Netwall, C.J.; Osborn, M. Advanced xenon feeding system (AXFS) development and hot-fire testing. In Proceedings of the 45th AIAA/ASME/SAE/ASEE Joint Propulsion Conference and Exhibit, Denver, CO, USA, 2–5 August 2009.
4. Koppel, C.R.; Marchandise, F.; Estublier, D. Robust pressure regulation system for the SMART-1 electric propulsion sub-system. In Proceedings of the 40th AIAA/ASME/SAE/ASEE Joint Propulsion Conference and Exhibit, Fort Lauderdale, FL, USA, 11–14 July 2004.
5. Ganapathi, G.B.; Engelbrecht, C.S. Post-launch performance characterization of the xenon system on Deep Space One. In Proceedings of the 35th AIAA/ASME/SAE/ASEE Joint Propulsion Conference and Exhibit, Los Angeles, CA, USA, 20–24 June 1999.
6. Ganapathi, G.B.; Engelbrecht, C.S. Performance of the xenon feed system on Deep Space One. *J. Spacecr. Rocket.* **2000**, *37*, 392–398. [CrossRef]

7. Brophy, J.R.; Ganapathi, G.B.; Garner, C.E.; Lo, J.; Marcucci, M.G.; Nakazono, B. Status of the Dawn ion propulsion system. In Proceedings of the 40th AIAA/ASME/SAE/ASEE Joint Propulsion Conference and Exhibit, Fort Lauderdale, FL, USA, 11–14 July 2004.
8. Tan, P.; Pang, G.Q.; Tang, H.P.; Chen, J.M.; Wang, Q.B.; Kang, X.T.; Chi, Y.D. Microflow controller in hall electric propulsion system. *Funct. Mater.* **2010**, *3*, 385–392.
9. Dyer, K.; Dien, A.; Nishida, E.; Kasai, Y. A xenon propellant management sub-unit for ion propulsion. In Proceedings of the 35th AIAA/ASME/SAE/ASEE Joint Propulsion Conference and Exhibit, Los Angeles, CA, USA, 20–24 June 1999.
10. Toki, K.; Kuninaka, H.; Nishiyama, K. Flight readiness of the microwave ion engine system for MUSES-C mission. In Proceedings of the 28th International Electric Propulsion Conference, Toulouse, France, 17–21 March 2003.
11. Couceiro, P.; Pedro, S.G.; Alonso, J. All-ceramic analytical microsystems with monolithically integrated optical detection microflow cells. *Microfluid. Nanofluid.* **2015**, *18*, 649–656. [CrossRef]
12. Vasudev, A.; Kaushik, A.; Tomizawa, Y.; Norena, N.; Bhansali, S. An LTCC-based microfluidic system for label-free, electrochemical detection of cortisol. *Sens. Actuators B Chem.* **2013**, *182*, 139–146. [CrossRef]
13. Golonka, L.J.; Zawada, T.; Radojewski, J.; Roguszczak, H.; Stefanow, M. LTCC microfluidic system. *Int. J. Appl. Ceram. Technol.* **2006**, *3*, 150–156. [CrossRef]
14. Gomez, H.C.; Cardoso, R.M.; Schianti, J.N.; Oliveria, A.M.; Gongora-Rubio, M.R. Fab on a package: LTCC microfluidic devices applied to chemical process miniaturization. *Micromachines* **2018**, *9*, 285. [CrossRef] [PubMed]
15. Vasudev, A.; Kaushik, A.; Jones, K.; Bhansali, S. Prospects of low temperature co-fired ceramic (LTCC) based microfluidic systems for point-of-care biosensing and environmental sensing. *Microfluid. Nanofluid.* **2013**, *14*, 683–702. [CrossRef]
16. Malecha, K.; Dawgul, M.; Pijanowska, D.G.; Golonka, L.J. LTCC Microfluidic Systems for Biochemical Diagnosis. *Biocybern. Biomed. Eng.* **2011**, *31*, 31–41.
17. Bruschi, P.; Diligenti, A.; Piotto, M. Micromachined gas flow regulator for ion propulsion systems. *IEEE Trans. Aerosp. Electron. Sys.* **2002**, *38*, 982–988. [CrossRef]
18. Jia, G.Z.; Wang, X.Y.; Wu, G.M. Study on extra high pressure and large flowrate pneumatic on-off valve. *Chin. J. Mech. Eng.* **2004**, *40*, 77–81. [CrossRef]
19. Jia, G.Z.; Wang, X.Y.; Wu, G.M.; Tao, G.L.; Chen, Y. Reserch on pricinple and property of classification control pressure reduction by expander in high pressure pneumatic system. *Chin. J. Mech. Eng.* **2005**, *41*, 210–214. [CrossRef]
20. Qian, J.Y.; Li, X.J.; Wu, Z.; Jin, Z.J.; Sunden, B. A comprehensive review on liquid–liquid two-phase flow in microchannel: Flow pattern and mass transfer. *Microfluid. Nanofluid.* **2019**, *23*, 116. [CrossRef]
21. Qian, J.Y.; Li, X.J.; Wu, Z.; Jin, Z.J.; Zhang, J.H.; Sunden, B. Slug formation analysis of liquid-liquid two-phase flow in T-junction microchannels. *J. Therm. Sci. Eng. Appl.* **2019**, *11*, 051017. [CrossRef]
22. Lu, L.; Xie, S.H.; Yin, Y.B.; Ryu, S. Experimental and numerical analysis on the surge instability characteristics of the vortex flow produced large vapor cavity in u-shape notch spool valve. *Int. J. Heat Mass Tran.* **2020**, *146*, 118882. [CrossRef]

Article

Non-Structural Damage Verification of the High Pressure Pump Assembly Ball Valve in the Gasoline Direct Injection Vehicle System

Liang Lu [1,*], Qilong Xue [1], Manyi Zhang [2], Liangliang Liu [2] and Zhongyu Wu [2]

[1]   School of Mechanical Engineering, Tongji University, Shanghai 201804, China; 0xueqilong@tongji.edu.cn
[2]   United Automotive Electronic Systems Co., Ltd., Shanghai 200120, China; manyi.zhang@uaes.com (M.Z.);
     liangliang.liu@uaes.com (L.L.); zhongyu.wu@uaes.com (Z.W.)
*   Correspondence: luliang829@tongji.edu.cn

Received: 28 October 2019; Accepted: 13 November 2019; Published: 16 November 2019

Abstract: The injection pressure of the gasoline direct injection vehicle is currently developing from the low pressure to the high pressure, and the increase of the injection pressure has brought various damage problems to the high pressure pump structure. These problems should be solved urgently. In this paper, the damage problem of the high pressure pump unloading valve ball in a gasoline direct injection vehicle under high pressure conditions is studied. The theoretical calculation of the force of the pressure relief valve is carried out. Firstly, the equivalent friction coefficient is obtained by decoupling analysis of the statically indeterminate model. Based on this, a finite element model is established. The equivalent stress is obtained by numerical simulation. The equivalent stress is compared with the yield strength of the valve ball material to determine that the valve ball damage is a non-static damage. At the same time, the s-N curve of the probability of destruction of one-millionth of the material of the valve ball is given. Then, the fatigue numerical simulation is performed. A safety factor of 3.66 is obtained. In summary, the high pressure relief valve ball in the direct injection high pressure pump should not be a traditional structural damage under high pressure conditions. In the theoretical calculation, the tangential displacement and radial displacement of the ball are all on the micrometer level. It can be presumed that the surface damage of the valve ball is microscopic damage, such as fretting wear.

Keywords: high pressure ball valve; Static friction contact; Static and fatigue analysis; Finite element simulation

---

1. Introduction

Gasoline direct injection is the latest fuel supply technology for gasoline engines. The direct injection gasoline engine directly injects fuel into the cylinder. Compared with the previous carburetor oil supply and single or multi-point electronic fuel injection technology, it has the advantages of accurate oil supply, uniform oil and gas mixing, and high combustion efficiency. Therefore, the technology is used by many car manufacturers [1–3]. Furthermore, the increased pressure of the direct injection oil will make the spray atomization effect better, the combustion efficiency higher, and the generation of discharged particulate matter better suppressed, which make the technology more environmentally friendly [4–6]. More widely, for the process control [7–9] and fluid power control components [10,11], the pressurization technology seems to come into notice more frequently. However, with the increased pressure of the injection gasoline, the high pressure pump will have various types of damage problems under long-term working conditions, which affect normal functions. Li et al. [12] analyzed the pitting corrosion of a high pressure oil pump cam of a direct injection gasoline engine in a cylinder and proposed corrective measures. Wang et al. [13] performed contact stress analysis and fatigue analysis on the high pressure oil pump cam-roller mechanism, which verified the reliability of the mechanism. Lei et al. [14] carried out flow field analysis on the internal fluid of the whole high pressure oil pump and used the obtained oil pressure distribution data to analyze the structural strength of the oil pump structure, which provided research ideas and methods for the design and optimization of the high pressure oil pump. The object of this paper is the pressure relief valve of the high pressure oil pump of the direct injection gasoline vehicle. The surface damage of the valve ball and the valve seat under the high pressure and high frequency working conditions has caused the unloading function to be unachievable. In response to this problem, it is necessary to perform damage verification analysis on the unloading valve structure.

2. Model Analysis

This research object is a spherical unloading valve in the high pressure pump of the direct injection vehicle. Its structure is shown in Figure 1. The unloading valve is shown in Figure 2. The left side of the valve ball is subjected to an alternating oil pressure of 0 to 45 MPa and the right side is subjected to a constant oil pressure of p_2. The left side of the valve ball is initially subjected to the preload force F_t of the spring. When the oil pressure p_2 at the right end of the valve ball increases to a certain value, exceeding the spring force provided at the left end, and the left end oil pressure is changed to 0 MPa, the valve ball will be opened and the right side high pressure oil will be released to the left side, which acts as a safe unloading valve. The ball valve structural parameters and oil pressure parameters are shown in Table 1. The equivalent calculation is carried out according to the pressure, and the contact area between the valve ball and the valve seat is subjected to an equivalent oil pressure of 5–50 MPa.

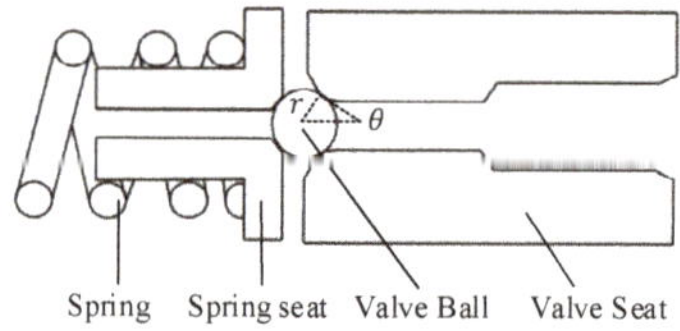

Figure 1. The structure of the pressure relief valve.

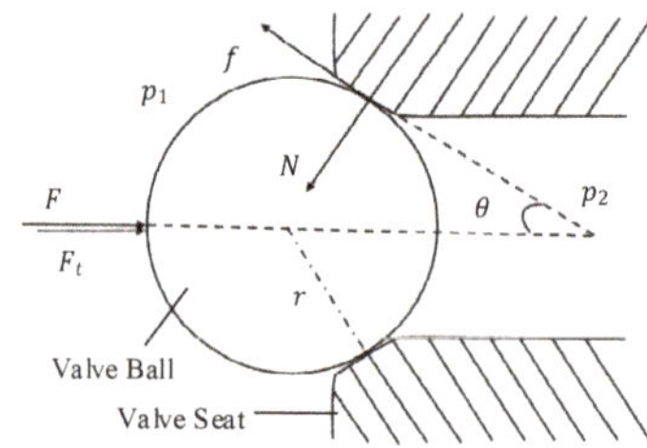

Figure 2. The stress state of the pressure relief valve.

Table 1. The parameters of structure and oil pressure.

r (Ball radius)	0.794 mm
θ (Contact angle)	27.5°
p_1 (Alternating oil pressure)	0–45 MPa
p_2 (High pressure rail constant oil pressure)	35 MPa
F_t (Spring preload force)	63 N

The surface damage problem of the unloading valve ball in 500 h (the number of cycles is 1.728×10^9 times) will occur as shown in Figure 3. The most critical and most important damage mechanism under complex conditions may be hidden and cannot be directly seen in Figure 3, so further analysis and judgment is needed to reach a conclusion. Firstly, it is necessary to consider whether the valve ball will be structurally damaged from the perspective of static force and fatigue.

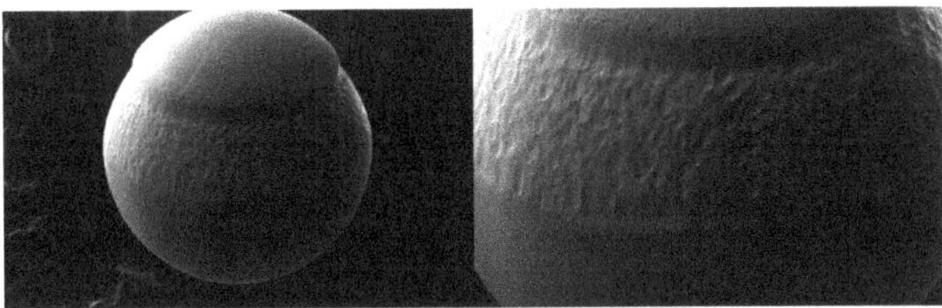

Figure 3. The damage surface topography of the valve ball.

For the verification of the rationality of the structural design, the research can be divided into the following sections:

(1) The contact stress between the valve ball and the valve seat is obtained by theoretical calculation, and the value of the equivalent friction force of the contact part is calculated in this part, which provides a parameter basis for the simulation calculation;

(2) The theoretical calculation results are verified by simulation calculation, and the equivalent stress value and fatigue safety factor are obtained;

(3) The data obtained by the simulation are used to safely check for static and fatigue damage.

3. Theoretical Analysis

3.1. Structural Micro-Division

From the structure of the unloading valve, it can be known that the initial state of contact between the valve ball and the valve seat is a line contact on the circumference. In order to facilitate the analysis of the contact force, the structure is divided into n(n→∞) equal divisions as shown in Figure 4. The force of each micro-element is shown in Figure 5.

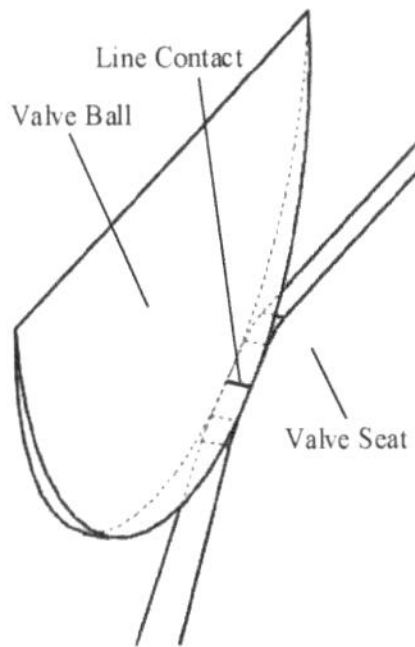

Figure 4. The micro-division of the structure.

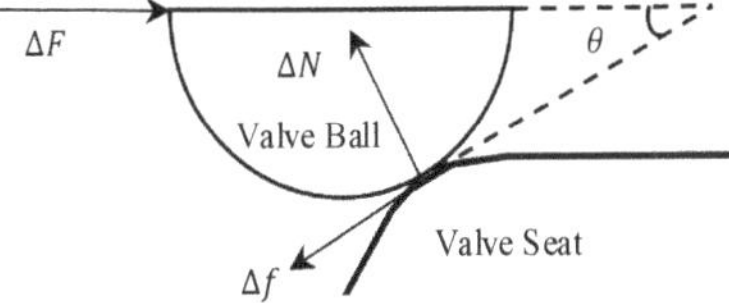

Figure 5. The force of each micro-element.

Under uniform oil pressure and initial spring force, the maximum force that the ball is subjected to can be expressed as:

$$F = F_t + (p_{1max} - p_2) \cdot 2\pi r \cos \theta. \tag{1}$$

The force that each micro-element is subjected to can be expressed as:

$$\Delta F = \frac{F}{n}. \tag{2}$$

It can be obtained from the force model of the micro-element that:

$$\Delta F = \Delta N \cdot \sin \theta + \Delta f \cdot \cos \theta. \tag{3}$$

ΔN and Δf are the positive pressure and friction of the micro-element. From the Equation (3), the relationship between ΔN and Δf is not known; it cannot be solved directly, so it is necessary to obtain the equivalent friction coefficient between the two.

3.2. Equivalent Friction Coefficient Calculation

We can suppose a situation that a block on a horizontal plane is subjected to horizontal thrust F_1 and forward pressure N_1. When the horizontal thrust F_1 is within the range of $0 \leq F_1 \leq f_{max} = \mu_1 N_1$ (μ_1—maximum static friction between the two materials), the slider has no displacement. However, there is a slight slip Δx between the slider and the horizontal plane. It can be considered that the micro slip range in which the slider does not move is $0 \leq \Delta x \leq \Delta x_{max}$. It can be also considered that the equivalent friction coefficient f_x satisfies $f_x \propto \Delta x$ [15].

The meaning of μ_x is the ratio of the value of non-critical equivalent friction to the positive pressure:

$$\mu_Y = \frac{f_x}{N}. \tag{4}$$

The relationship between the equivalent friction coefficient μ_x and the micro-slip Δx is experimentally verified and obtained by Dr. Liu [16]. μ_x and Δx are proportional to each other.

$$\mu_x = k_f \Delta x \tag{5}$$

The valve ball and valve seat structure are steel materials, $k_f = 0.25 \ \mu m^{-1}$ [16].

The micro-element structure of the ball valve is analyzed, and its force deformation is shown in Figure 6. ΔX is the slight slippage of the valve ball under the action of the spring force that can be calculated from the Hertz contact theory in Section 3.3. Here, a is the half width of the contact between the valve ball and the valve seat under the action of positive pressure ΔN. The initial radius of the ball is r. After the force is deformed, the distance between the center point of the contact and the center of the sphere is r_1, which can be expressed as:

$$r_1 = \sqrt{r^2 - a^2}. \tag{6}$$

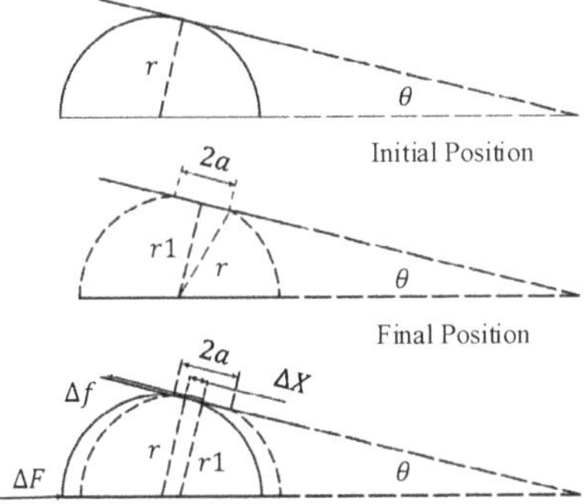

Figure 6. The force model of the ball valve.

The micro-slip distance ΔX can be expressed as:

$$\frac{r - r_1}{\Delta X} = \tan \theta. \tag{7}$$

The relationship between ΔN and Δf in the force model of micro-element can be described as:

$$\Delta f = k_f \Delta X \Delta N. \tag{8}$$

Therefore, the relationship between the equivalent friction and the positive pressure can be linked by the Hertz contact theory.

3.3. Structural Contact Stress Analysis

The contact part of the micro-element structure of the spherical unloading valve can be regarded as a model of contact between the cylinder and the plane. According to the Hertz contact theory [17], the contact half width a satisfies:

$$a = \sqrt{\frac{4 \cdot r \cdot \Delta N}{\pi \cdot E^* \cdot \Delta l}}. \tag{9}$$

The maximum contact stress in the contact portion appears at the contact center, and the maximum value is:

$$p_{max} = \frac{2N}{\pi a L}, \tag{10}$$

where Δl is the length of the micro-element of the contact between the valve seat and the valve ball line, which satisfies:

$$\Delta l = \frac{2\pi r \cos \theta}{n}. \tag{11}$$

E^* is the equivalent elastic modulus, which satisfies:

$$\frac{1}{E^*} = \frac{\left(1 - v_1{}^2\right)}{E_1} + \frac{\left(1 - v_2{}^2\right)}{E_2}. \tag{12}$$

E_1 and E_2 are the elastic modulus values of the valve ball and the valve seat material, respectively. v_1 and v_2 are the Poisson's ratios of the valve ball and the valve seat material, respectively. $E_1 = 200$ GPa, $E_2 = 213$ GPa. $v_1 = 0.3$, $v_1 = 0.29$.

It can be obtained by solving Equations (1)–(3) and (6)–(12) that: $\Delta N = 145.4/n$, $\Delta f = 12.9/n$. The equivalent friction coefficient is:

$$\mu = \frac{\Delta f}{\Delta N} \approx 0.0887. \tag{13}$$

At the same time, the micro-slip distance Δx = 0.355 μm can be obtained; the contact half width a = 17.15 μm.

The maximum contact stress value of the contact between the valve ball and the valve seat is expressed as follows:

$$p_{max} = 1220 \text{ MPa.} \tag{14}$$

4. Simulation Analysis

The valve ball seat structure shown in Figure 7 is added to the ANSYS model. In order to facilitate the addition of force, a plane is selected on the left side of the valve ball. The elastic modulus and Poisson's ratio parameters of the ball and seat are added to the material properties.

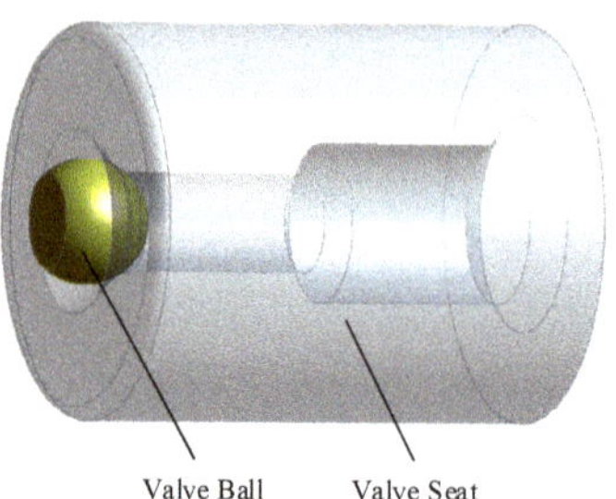

Figure 7. The simulation structure of the pressure relief valve.

The s-N curve of the structural material needs to be known when calculating the fatigue safety factor of the structure. The material used for the valve ball and the valve seat is bearing steel 9Cr18, and its p'-s-N' curve is used to set the material fatigue property parameters [10]. The meaning of the p'-s-N' curve expression is that the life distribution of the material is a normal distribution form under the same stress level [19]. The lifetime of most tested materials is distributed at p' = 0.5, the middle of the normal distribution curve. In this simulation, in order to meet the requirement of one-millionth of the destruction probability of the structure in engineering applications, it is necessary to fit the s-N' curve data in the case of the probability p' = 0.000001 according to the p_1' = 0.01, p_2' = 0.05, p_3' = 0.1 and p_4' = 0.5 data given by the material according to the normal distribution.

The fatigue life curve formula is as shown in Equation (15):

$$N' = C \cdot s^{-m}, \tag{15}$$

where N' is the number of cycles; s (Pa) is the stress size; the values of C and m refer to Table 2.

Table 2. Parameter values under different damage probabilities.

Parameter	$p' = 0.01$	$p' = 0.05$	$p' = 0.10$	$p' = 0.50$
C	9.15×10^{44}	8.66×10^{42}	1.90×10^{41}	2.79×10^{41}
m	10.9314	10.201	9.6618	9.5012

The s-N curve data table with a probability of destruction of one part per million is calculated as shown in Table 3. The meshing parameter is set for the structure, and the local mesh encryption is performed on the key contact portion of the calculation, as shown in Figure 8. The overall mesh size is 0.3 mm and the contact portion mesh is refined to 0.015 mm.

Table 3. The data sheet of fatigue life curve.

Cycles (N)	Stress (MPa)	Cycles (N)	Stress (MPa)
5,000,000	1744.08238	80,000,000	1434.25699
10,000,000	1661.91273	85,000,000	1428.03274
15,000,000	1615.32473	90,000,000	1422.18528
20,000,000	1582.92424	95,000,000	1416.67270
25,000,000	1558.16337	100,000,000	1411.45968
30,000,000	1538.17126	200,000,000	1342.59599
35,000,000	1521.43492	300,000,000	1303.65968
40,000,000	1507.06016	400,000,000	1276.62594
45,000,000	1494.47505	500,000,000	1255.99123
50,000,000	1483.29191	600,000,000	1239.3462
55,000,000	1473.23605	700,000,000	1225.42251
60,000,000	1464.10579	800,000,000	1213.47126
65,000,000	1455.74882	900,000,000	1203.01377
70,000,000	1448.04730	1,000,000,000	1193.72581
75,000,000	1440.90824		

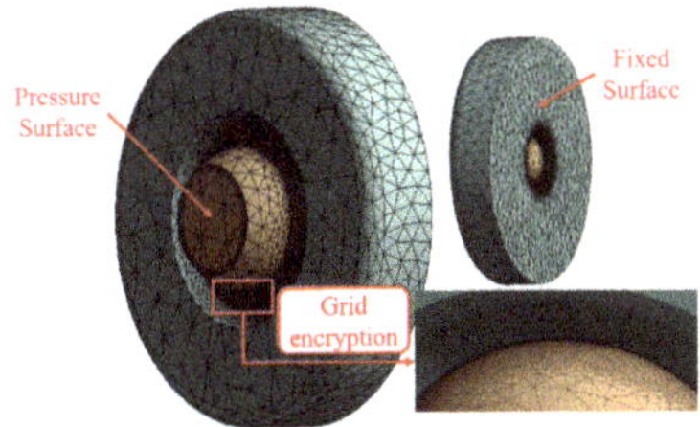

Figure 8. Meshing and simulation boundary conditions.

The structural stress boundary conditions, i.e., the pressure acting surface and the fixed surface, are set. The contact portion was set to frictional contact and the coefficient of friction was defined as 0.0887 as calculated in Section 3.3. The pressure is calculated according to the maximum force under working conditions, that is 50 MPa oil pressure, equivalent to 78.58 N.

The fatigue simulation force is in the form of pulsating circulation. The force application is still the same as the external force in the static simulation. The alternating equivalent oil pressure is 5–50 MPa, and the minimum oil pressure $\sigma_{min} = 0.1 \times 50$ MPa $= 5$ MPa. The pulsating form is shown in Figure 9.

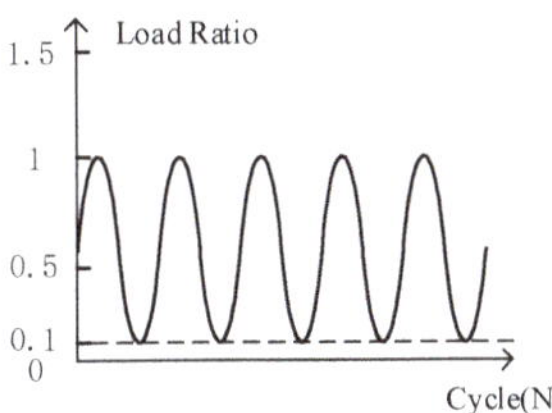

Figure 9. Constant amplitude load ratio.

The contact stress distribution obtained by simulation is shown in Figure 10. The maximum contact stress value calculated by simulation is $p_{max}' = 1107.8$ MPa, and the theoretical calculation result is $p_{max} = 1220$ MPa. The error between the two is about 9.2%, less than 10%, which verifies the correctness of the theoretical model. The overall equivalent stress distribution is obtained as shown in Figure 11. The maximum equivalent stress is 725.63 MPa. The yield strength of the ball and seat material is $\sigma_t = 1900$ MPa, and the tensile yield strength and compressive yield strength of the elastoplastic material are generally considered to be the same, i.e., $\sigma_t = \sigma_c$ (σ_c is the compressive yield strength). The von Mises equivalent stress theoretical shear yield strength satisfies $\tau_s = \sigma_t/\sqrt{3}$, and

$\tau_s \approx 1096$ MPa can be calculated [20]. The check is performed with the minimum shear yield strength, which is much larger than the equivalent stress of 725.63 MPa that is theoretically calculated. It can be judged that the ball valve is safe under static conditions.

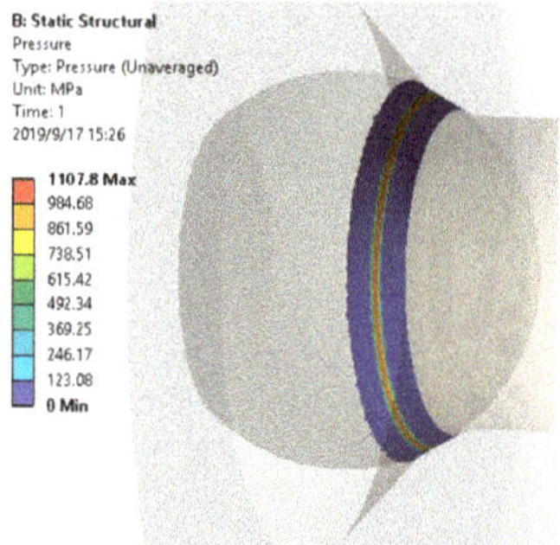

Figure 10. The simulation results of contact stress.

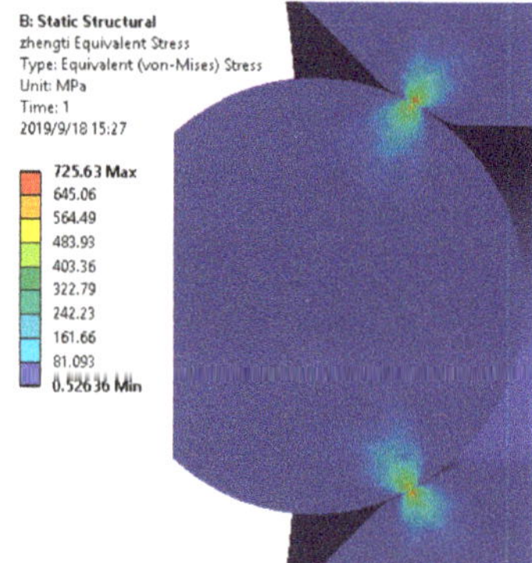

Figure 11. The simulation results of equivalent stress.

The fatigue safety factor distribution of the whole structure under the stress condition of 1.728×10^9 times is shown in Figure 12; the maximum is 3.6558, which indicates that the structural ball will not suffer structural fatigue damage under the action of fatigue alone.

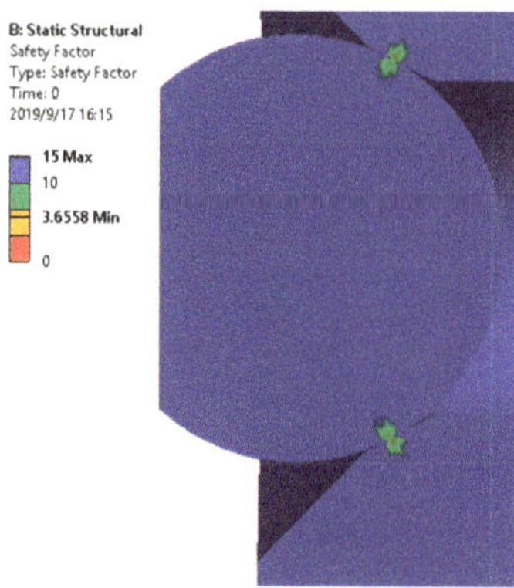

Figure 12. The distribution of fatigue safety factor.

5. Discussion

In this paper, the calculation of the stress of valve ball and valve seat is solved by phenomenological mathematics, obtained by the experiment, and the result is verified by simulation. Through the static and fatigue check, the damage mechanism of the valve ball and the valve seat under working conditions is not static and fatigue damage.

For a further study of the damage mechanism, a high frequency test rig as shown in Figure 13 was designed. The loading force can reach more than 100 N and the operating frequency can reach 500 Hz. A 50-h experiment has been done between the valve ball and the valve seat at 500 Hz and 8–80 N. The damage mechanism needs further research to explain the experimental phenomena or to illustrate the problems in combination with the experiments.

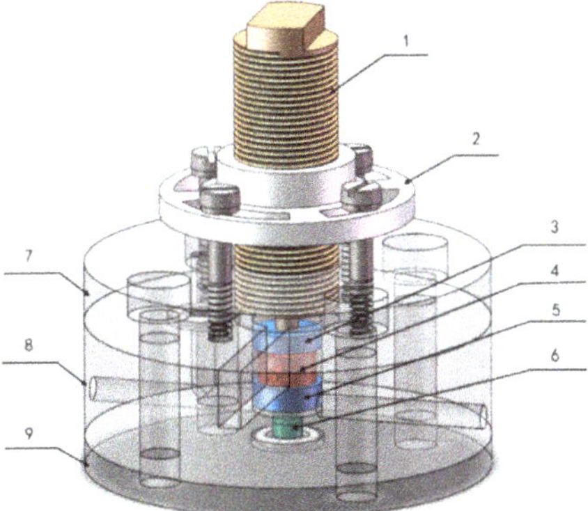

Figure 13. Test bench, (1) piezoelectric actuator; (2) fixed disc; (3) cylinder 1; (4) force sensor; (5) cylinder 2; (6) testing object; (7–9) base.

The ball and seat wear band under the 50-h test is shown in Figure 14. The diameter of the ball is small. It is only a little more than a millimeter. Therefore, the amount of damage cannot be assessed by quality. At present, the best evaluation method is to take the damage surface morphology by microscopy and estimate the damage by width and depth.

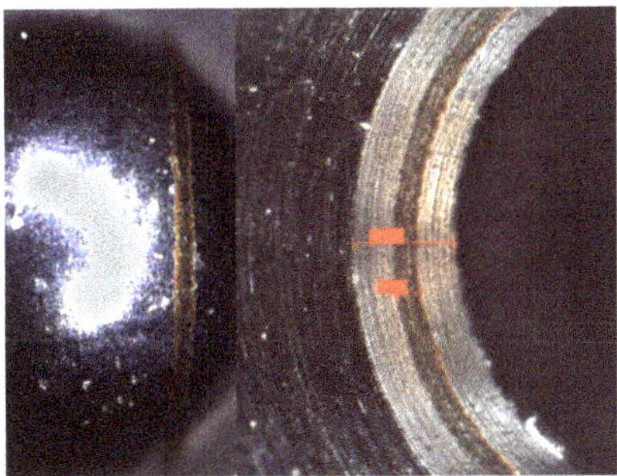

Figure 14. Damage zone (8–80 N, 500 Hz, 50 h).

Although the amount of damage can be estimated by microscopy, this method of evaluation is not particularly accurate. It is also impossible to monitor the damage during the experiment. The test bench can be improved to add better calculation methods. For example, a distance sensor can be added to monitor the depth of the damage zone to determine the depth of damage during the experiment.

6. Conclusions

The injection pressure of the gasoline direct injection vehicle is currently developing from low pressure to high pressure. The structural damage problems brought by the increase of the injection pressure should be solved urgently. Therefore, based on theoretical analysis and numerical analysis, the paper first determines whether there is traditional structural damage.

The theoretical stress calculation and static and fatigue simulation analysis of the unloading valve structure were carried out around the ball damage problem of the high pressure pump unloading valve in a gasoline direct injection vehicle. The following conclusions were obtained:

(1) According to the maximum static force of the valve ball, the theoretical calculation is carried out, the equivalent friction coefficient is obtained by solving the statically indeterminate problem, and the maximum contact stress value of 1220 MPa is obtained by the Hertz contact theory.

(2) Through simulation, the maximum contact stress is 1107.8 MPa and the maximum equivalent stress is 725.63 MPa under maximum static force. The simulated contact stress values are compared with the theoretical calculations and the difference between the two is less than 10%, which verifies the correctness of the theoretical model. At the same time, the equivalent stress is used for static checking, and it is judged that the unloading valve structure will not be damaged under the action of static force.

(3) By the simulation analysis, the fatigue safety factor of the unloading valve is 3.6558 under the condition of one-millionth of the failure probability and 1.728×10^9 cycles. It is verified that the traditional structural fatigue is not the cause of the ball failure of the unloading valve.

It can be seen from the above verification analysis that the damage of the ball valve structure is not caused by static force and fatigue damage; further analysis of the structural damage mechanism is needed. In the theoretical calculation, the tangential displacement (0.355 μm) and radial displacement (17.15 μm) of the valve ball are all in the micron range. The motion state belongs to the fretting category and the surface damage morphology of the valve ball is similar to the fretting damage. It can be preliminarily speculated that the surface damage of the valve ball is a fretting damage. The mechanism still needs further research.

Author Contributions: L.L. (Liang Lu) and Q.X. provided theoretical research methods; L.L. (Liang Lu) and Q.X. conceived and designed the experiments; Q.X. performed the experiments; M.Z. and L.L. (Liangliang Liu) analyzed the data; M.Z., L.L. (Liangliang Liu) and Z.W. contributed to the project supervision and management; Z.W. contributed to the problem research and provided research object and resources. Q.X. provided original draft preparation. L.L. (Liang Lu) provided review and editing of writing.

Funding: The authors are grateful to the National Natural Science Foundation of China (No. 51605333) and the Fundamental Research Funds for the Central Universities (kx0138020173443) for financial support.

Conflicts of Interest: The authors declare no conflict of interest.

References

1. Yang, S.C.; Li, J.; Li, D.G. Analysis on Development for Gasoline Direct Injection Technology. *Veh. Engine* **2007**, *5*, 8–13.
2. Sun, Y.; Wang, Y.J.; Wang, J.X.; Shuai, S.J. New Progress of Gasoline Direct Injection Engines' Research and Technical Problems. *Intern. Combust. Engines* **2002**, *1*, 6–10.
3. Xia, S.M.; Qiu, X.W.; Zhao, X.S.; Feng, H.Q. Survey of Research Situation on Gasoline Direct Injection Technology of Automotive Gasoline Engine. *Trans. Chin. Soc. Agric. Mach.* **2003**, *34*, 168–171.
4. Zhao, F.; Lai, M.C.; Harrington, D.L. Automotive spark-ignited direct-injection gasoline engines. *Prog. Energy Combust Sci.* **1999**, *25*, 437–562. [CrossRef]
5. Pei, Y.Q.; Wang, K.; Zhang, D.; Liu, B.; Hu, T.G.; Ji, S.S. Spray Macroscopic Characteristics of GDI Injector Fueled with Ethanol Using Ultra-High Injection Pressure. *Sci. Technol.* **2018**, *51*, 755–762.
6. Zhong, B.; Hong, W.; Su, Y.; Xie, F.X.; Han, L.P. Effects of Control Parameters on the Particulate Emission Characteristics of Turbocharged Gasoline Direct Injection Engine under Part Load. *J. Xi'an Jiaotong Univer.* **2016**, *50*, 95–100.

7. Qian, J.Y.; Hou, C.W.; Wu, J.Y.; Gao, Z.X.; Jin, Z.J. Aerodynamics analysis of superheated steam flow through multi-stage perforated plates. *Int. J. Heat Mass Transf.* **2019**, *141*, 48–57. [CrossRef]

8. Qian, J.Y.; Chen, M.R.; Gao, Z.X.; Jin, Z.J. Mach number and energy loss analysis inside multi-stage Tesla valves for hydrogen decompression. *Energy* **2019**, *179*, 647–654. [CrossRef]

9. Qian, J.Y.; Wu, J.Y.; Gao, Z.X.; Wu, A.; Jin, Z.J. Hydrogen decompression analysis by multi-stage Tesla valves for hydrogen fuel cell. *Int. J. Hydrog. Energy* **2019**, *44*, 13666–13674. [CrossRef]

10. Lu, L.; Fu, X.; Ryu, S.; Yin, Y.B.; Shen, Z.X. Comprehensive discussions on higher back pressure system performance with cavitation suppression using high back pressure. *Proc. Inst. Mech. Eng. Part J. Mech. Eng. Sci.* **2019**, *233*, 2442–2455. [CrossRef]

11. Lu, L.; Xie, S.H.; Yin, Y.B.; Ryu, S. Experimental and numerical analysis on the surge instability characteristics of the vortex flow produced large vapor cavity in u-shape notch spool valve. *Int. J. Heat Mass Transf.* **2020**, *146*, 118882. [CrossRef]

12. Li, L.F.; Su, Z.G.; Cong, Z.Q. Failure Analysis and Corrective Actions of High Pressure Oil-pump Cam for Direct Injection Gasoline Engine. *Intern. Combust. Engines* **2014**, *4*, 37–39.

13. Wang, J.; Fu, J.; Huang, G.C.; Li, Y.Z.; He, Y.H.; Zhang, Q.H.; Lei, G. Contact analysis and fatigue calculation of the cam mechanism of fuel system. *Mach. Tool Hydraul.* **2017**, *45*, 77–80.

14. Lei, G.; Zhang, X.; Lai, L. Structural strength analysis of high-pressure oil pump under fluid-solid coupling conditions. *Mach. Tool Hydraul.* **2016**, *44*, 74–81.

15. Liu, D. Constitutive Relation of Bodies about Non-Critical Static Friction. *Eng. Mech.* **1999**, *16*, 94–97.

16. Liu, D. The Formula of Non-Critical Static Friction and Its Application. *J. Gansu Sci.* **2006**, *18*, 27–29.

17. Guan, C. Calculation of Critical Parameters for Hertz Contact Between Cylinder and Rigid Plane. *Bearing* **2013**, *8*, 4–7.

18. Zhu, S.D.; Fang, X.W.; Wu, M.D.; Wu, G.C.; Lu, M.J. *Mechanical Engineering Material Performance Data Sheet*; China Machine Press: Beijing, China, 1995.

19. He, C. Fatigue Test Dada Treatment and P-S-N Curve Function. *Automob. Technol. Mater.* **2007**, *4*, 42–44.

20. Niu, J.; Liu, G.; Tian, J.; Zhang, Y.X.; Meng, L.P. Comparison of Yield Strength Theories with Experimental Results. *Eng. Mech.* **2014**, *31*, 181–187.

Article

Pressure Drop and Cavitation Analysis on Sleeve Regulating Valve

Chang Qiu [1], Cheng-Hang Jiang [2], Han Zhang [3], Jia-Yi Wu [1] and Zhi-Jiang Jin [1,*]

[1] Institute of Process Equipment, College of Energy Engineering, Zhejiang University, Hangzhou 310027, China; qiuchang@zju.edu.cn (C.Q.); jiayi-wu@zju.edu.cn (J.-Y.W.)

[2] Hangzhou Special Equipment Inspection & Research Institute, Hangzhou 310051, China; jiangchenghang@aliyun.com

[3] Shanghai Power Equipment Research Institute CO., Ltd., Shanghai 200240, China; zhanghan@speri.com.cn

* Correspondence: jzj@zju.edu.cn; Tel.: +86-0571-8795-1216

Received: 15 October 2019; Accepted: 4 November 2019; Published: 7 November 2019

Abstract: The sleeve regulating valve is widely used in the pipeline systems of process industries to control fluid flow. When flowing through the sleeve regulating valve, the water is easy to reach cavitation because of the pressure drop in the partial region, which may cause serious damage to pipeline system. In this paper, the pressure drop and cavitation characteristics in the sleeve regulating valve for different pressure differences and valve core displacements are investigated using a multiphase cavitation model. The pressure drop, velocity and vapor volume distribution in the regulating valves are obtained and analyzed. The total vapor volumes are also predicted and compared. The results show that the decrease of the valve core displacement induces the enlargement of the vapor distribution region and the increase of the vapor density. The increase of the pressure difference induces a more serious cavitation. The pressure difference has a slight influence on the cavitation intensity and density in the regulating valve when the valve core displacement is 60 mm. With the decrease of the valve core displacement, the effects of the pressure difference on the cavitation intensity are enhanced. This work is of significance for the cavitation control of the sleeve regulating valves.

Keywords: sleeve regulating valve; cavitation; pressure difference; cavitation index

---

1. Introduction

The regulating valve is widely used in the pipeline system of process industries such as power engineering, chemical engineering and petrifaction. A regulating valve is opened with the movement upward of the valve core and the flow rate passing through the regulating valve is changed by adjusting the distance between a stationary valve seat and a movable valve core. For a regulating valve which conveys liquids such as water, cavitation is a serious and destructive problem during its operation because of the pressure drop owing to the variation of velocity. When the local pressure is lower than the corresponding saturated vapor pressure at the same temperature, bubbles can form and then grow until bursting. Longtime cavitation flow can not only induce the waste of energy, but also cause the failure of the piping system. Meanwhile, the lifetime of valves is reduced and noise can also be induced within cavitation flow. Thus, the investigation of cavitation inside the regulating valves is necessary.

In recent years, some meaningful research has been carried out focusing on the cavitation flow. For example, cavitation inside a venturi tube [1], a cone flow channel [2], and pumps [3–7] has been investigated using experimental or numerical methods, and the effects of fluctuating flow [2] and liquid temperature [3,7] or other variables such as nano particles [8] on cavitation distribution have been analyzed. Meanwhile, a variety of meaningful research focusing on the cavitation flow inside the valves has been carried out by experimental and numerical methods. Qian et al. [9] carried out a comprehensive review of cavitation in valves. Hassis [10] conducted the experiment study of the effects of cavitation in butterfly and Monovar valves. Herbertson et al. [11] proposed a novel approach for analyzing the mechanical heart valve cavitation. Gao et al. [12] investigated the cavitation near the orifice of hydraulic valves by numerical method and flow visualization experimental method. Jin et al. [13] also carried out the research on cavitation flow through a micro-orifice. Jia et al. [14] simulated the cavitation flowing through the cylinder valve port and conducted the flow visualization cavitation experiment of cylinder valve. Liu et al. [15] investigated the cavitation flow in the rotary valve of hydraulic power steering gear. Lu et al. [16] researched the acoustic characteristics of cavitation noise in a spool valve with U-notches. Li et al. [17] also researched the cavitation phenomenon of an electrohydraulic servo-valve. The numerical results show a good agreement with experimental observations. Kudzma et al. [18] studied the flow and cavitation in hydraulic lift valve by visualization experiments and acoustic tests. Qu et al. [19] studied the cavitation performance on a pressure-regulating valve with different openings by experiment and numerical simulation. Ulanicki et al. [20] proposed a methodology which is used to evaluate whether a pressure reducing valve is under cavitation. Deng et al. [21] researched the cavitation flow inside the spool valve with large pressure drop using numerical method. Ou et al. [22] simulated the cavitation flow in pressure relief valve with high pressure differentials for different valve openings, inlet pressure and outlet pressure. Yi et al. [23] investigated the interactions between the poppet vibration characteristics and cavitation property in relief valves with the unconfined poppet experimentally. Okita et al. [24] researched the mechanism of noise generation by cavitation in hydraulic relief valve by means of experiment and numerical simulation. Lu et al. [25] analyzed the vortex flow produced large vapor cavity in a u-shape notch spool valve. Jin et al. [26] researched the influence of the structural parameters for globe valves on hydrodynamic cavitation. Pressure loss is an important parameter to influence the cavitation. Some researches focusing on the pressure and energy loss in multi-stage Tesla valves were carried out by Qian et al. [27,28].

To reduce cavitation, there were different designs and methods in a number of published manuscripts. Tao et al. [29] researched the flow loss in a v-port ball valve by experimental and numerical method. Baran et al. [30] proposed a new method of controlling the butterfly valve which operated with cavitation. A position fuzzy controller was used to change the valve opening to suppress cavitation. Zhang et al. [31] proposed a novel approach of suppressing cavitation which induced the pressure back to the orifice to improve the pressure distribution of throttle valves. Shi et al. [32] proposed a modified throttle valve with a drainage device to suppress the cavitation and evaluate the influence of inlet and outlet pressure on the ability of the drainage device to suppress cavitation by experiments. There were also some designs of valve sleeves and perforated plates to suppress cavitation, which used multi-stage sleeves to reduce pressure gradually. For instance, Qian et al. [33] focused on the fluid flow through multi-stage perforated plates. Chern et al. [34] and Yaghoubi et al. [35] researched the influences of throttle sleeve stages on cavitation in a globe valve. Qi et al [36] studied the metrological performance of a swirlmeter affected by flow regulation with a sleeve valve. Furthermore, the parametric analysis on throttle sleeves inside valves were carried out by Qian et al. [37] and Hou et al. [38], respectively.

In this paper, a multiphase cavitation flow model is built to simulate the cavitation inside the sleeve regulating valve. The pressure, velocity and steam volume fraction distribution inside the regulating valves are analyzed and compared for different pressure difference and valve core displacements. The effects of the pressure difference on the flow and cavitation characteristics are revealed. This research is of significance to the cavitation control of the sleeve regulating valve.

2. Numerical Method

2.1. Mathematical Model

Since the actual flow inside the sleeve regulating valve is very complex, the standard k-ε turbulence model was chosen to simulate the turbulence for its advantages in dealing with high Reynolds numbers higher than 10^7. In the cavitation simulations, phase change occurs between the liquid phase and vapor phase. The simulations are conducted in steady state. The governing equations for the cavitation model used in this paper are based on a single-fluid approach, which regarded the mixture as one fluid. Thus, the mixture continuity and momentum equations are shown as

$$L\frac{\partial}{\partial t}\rho_m + \nabla(\rho_m v) = 0 \tag{1}$$

$$\frac{\partial}{\partial t}(\rho_m v) = -\nabla p + \nabla \cdot [(\mu_m + \mu_t)\nabla v] + \frac{1}{3}\nabla[(\mu_m + \mu_t)\nabla \cdot v] - \nabla(\rho_m vv) \tag{2}$$

Here the mixture density ρ_m and μ_m are defined as follows:

$$\rho_m = \alpha\rho_v + (1-\alpha)\rho_l \tag{3}$$

$$\mu_m = \alpha\mu_m + (1-\alpha)\mu_l \tag{4}$$

where ρ_m, ρ_v, and ρ_l represent the mixture, vapor, and liquid densities, respectively; v is the mass average velocity vector; μ_m, μ_v and μ_l represent the mixture, vapor, and liquid dynamic viscosities, respectively, and μ_t is the turbulence viscosity; p is the pressure; α represents the vapor volume fraction.

The cavitation model employed here is based on the Rayleigh–Plesset equation and is developed by Schnerr and Sauer [39]. Although the compressibility of liquid is very important when bubbles break, the liquid density ρ_l is assumed as a constant and so are μ_l, ρ_v, and μ_v. Meanwhile, it is assumed that the bubbles remain spherical and there is no thermal conductivity with tube linked with the valves. The liquid–vapor mass transfer is governed by the vapor transport equation:

$$\frac{\partial}{\partial l}(\alpha\rho_v) + \nabla \cdot (\alpha\rho_v v_v) = R_e - R_c \tag{5}$$

Here, R_e and R_c are defined as follows:

$$R_e = \frac{3\alpha\rho_v(1-\alpha)\rho_l}{\rho_m R_b}\sqrt{\frac{2(p_v - p)}{3\rho_l}}, p_v \geq p, \tag{6}$$

$$R_c = \frac{3\alpha\rho_v(1-\alpha)\rho_l}{\rho_m R_b}\sqrt{\frac{2(p - p_v)}{3\rho_l}}, p_v \leq p, \tag{7}$$

where R_e represents the mass rates of growth of vapor bubbles and R_c represents the mass rates of breaking of vapor bubbles; v_v is the vapor phase velocity, p_v is the saturation pressure of water, R_b is the bubble radius and is defined as follows:

$$R_b = \left(\frac{\alpha}{1-\alpha}\frac{3}{4\pi}\frac{1}{n}\right)^{\frac{1}{3}} \tag{8}$$

where n represents the bubble number density and is usually set as a constant, 1×10^{13}.

2.2. Geometrical Model

Figure 1 demonstrates the schematic structure of the studied sleeve regulating valve. The sleeve regulating valve is comprised of valve body, valve cover, valve core and the sleeve with orifices. When the valve is opening, the valve core is moved upward, driven by the valve rod and driving device. The inlet diameter is 130 mm and the orifice diameter in the sleeve is 4 mm. The maximum valve core displacement is 60 mm. Fully opened state and half opened state are adopted in this study.

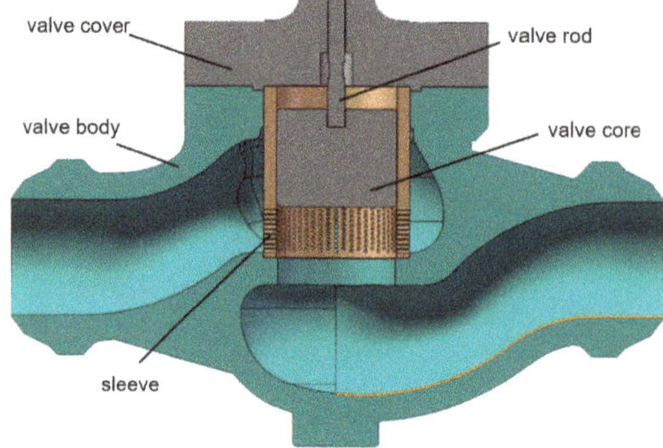

Figure 1. Schematic structure of the studied sleeve regulating valve.

To enable numerical analysis, some simplification is carried out. First, the sleeve regulating valve is assumed as an ideal valve, which means its cutting edge is a right angle exactly with the sharp edges, and the valve core matches the valve seat precisely. Secondly, to save computation time, a 3D axisymmetric geometric model is adopted by considering the symmetry structure. Also, since the cavitation region is partial and the heat transfer due to phase change is little, we take no account of the effects of gravity and heat transfer.

2.3. Mesh and Boundary Conditoms

Figure 2 shows the generated mesh of the sleeve regulating valve with the maximum valve core displacement of 60 mm. To enhance the accuracy of the simulation, the upstream pipe and the downstream pipe length are set as 600 mm which equals to five diameters. Since the flow channel inside the valve is complex, the flow channel is generated using a non-structure mesh partition and the mesh module in ANSYS Workbench 17.2 is employed to generated mesh.

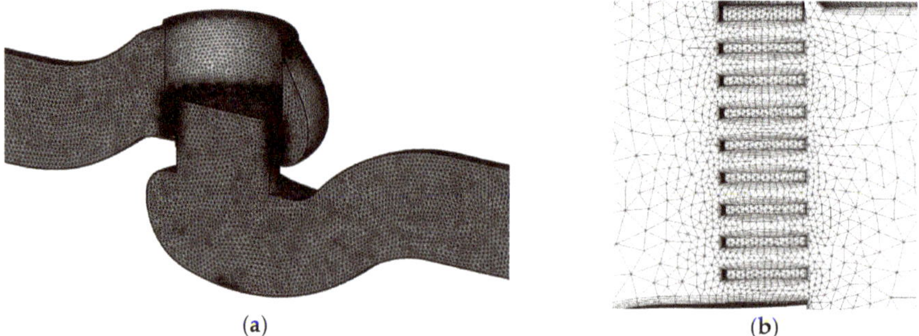

(a)   (b)

Figure 2. Mesh of the sleeve regulating valve with maximum valve core displacement of 60 mm (**a**) overall view; (**b**) mesh in the sleeve.

The mesh independency check is carried out with the maximum valve core displacement. The outlet flow rate and the vapor fraction at the border of orifice are both taken as the judgement parameters with different grid numbers from 1,451,696 to 3,138,669. As is shown in Table 1, when the mesh number ranges between 2,407,840 and 3,138,669, the relative errors of the simulation are kept within 1%, so the mesh number 2,708,138 is chosen.

Table 1. Mesh independency check of various meshes.

Grids	Flow Rate (kg/s)	Volume Fraction
1,451,696	162.72	0.1296
1,705,067	162.14	0.2272
2,055,352	168.31	0.2676
2,407,840	167.74	0.2829
2,708,138	167.37	0.2585
3,138,669	167.01	0. 2323

For the boundary conditions, the inlet condition of the sleeve regulating valve is set as the pressure inlet, and the outlet condition is set as pressure outlet condition of 2 MPa. To analyze the effects of the pressure difference on the flow and cavitation characteristics, the pressure inlet is varied from 10 MPa to 3 MPa and the pressure inlet is kept constant. The initial inlet vapor volume fraction is set as 0. The other faces except the symmetry face are set as wall with no slip condition. The wall function method is adopted in the near wall region by using the finite volume method and first order upwind scheme. Coupling pressure and velocity are based on SIMPLE. In addition, the main medium of the sleeve regulating valve is water at 200 °C. Incompressible liquid water is chosen as the liquid phase while water vapor is chosen as the vapor phase. In the simulation, ρ_l and ρ_v are set as 862.8 kg/m^3 and 7.865 kg/m^3, and μ_l and μ_v are set as 1.357×10^{-4} Pa·s and 1.565×10^{-5} Pa·s, respectively. The vaporization pressure of the liquid phase is set as 1.5 MPa which equals to the saturation vapor pressure of the water at 200 °C. Above operations are all carried out in Fluent 17.2.

3. Results and Discussion

To analyze the effects of the pressure difference on the flow and cavitation characteristics for different valve core displacements, the pressure difference ranges from 1 MPa to 8 MPa by setting the different pressure boundary conditions. In this study, the change of pressure difference is realized by changing the pressure inlet and keeping the outlet pressure as a constant of 2 MPa. The inlet pressure is varied from 3 MPa to 10 MPa. To analyze the effect of the valve openings, the full opening state with the valve core displacement of 60 mm and the half opening state with the valve core displacement of 30 mm are chosen as the analyzed model, respectively.

3.1. Comparsion between Full Opening State and Half Opening State

Figure 3 demonstrates the pressure and velocity distributions of the symmetry cross section inside the sleeve regulating valve with the valve core displacement of 60 mm and 30 mm. The inlet pressure is set as 8 MPa while the outlet pressure is set as 2 MPa.

Although the structure of the sleeve regulating valve is very complex, the pressure and velocity remain stable at the inlet and outlet. As is shown in Figure 3, the pressure for all valve core displacements are kept above 7.5 MPa at the inlet, while the pressure for all valve core displacements remain below 2.5 MPa at the outlet. Meanwhile, the velocity for the valve core displacement of 60 mm is kept above 20 m/s at the inlet while the velocity with the valve core displacement of 30 mm remains below 20 m/s. When the fluid flows through the sleeve which is the main throttling structure, the sudden pressure drop with the velocity increasing appear in the orifices of sleeve for both valve core displacements. Since the throttling cross sections for different valve core displacements are different, the pressure drops induced by the throttling sleeve are totally different. The pressure is decreased below 3.5 MPa for the valve core displacement of 60 mm, while the pressure behind the sleeve is below 2.5 MPa at the half opening state. Accordingly, the region with high velocity in the orifices of sleeve at the half opening state is bigger than these with the valve core displacement of 60 mm. Furthermore, at the outlet of orifices in the sleeve, the pressure is lower than 2 MPa when the valve core displacement is 30 mm. In the left side of the valve chamber, there is always a vortex at both opening states and the pressure for the vortex region is lower than 2 MPa.

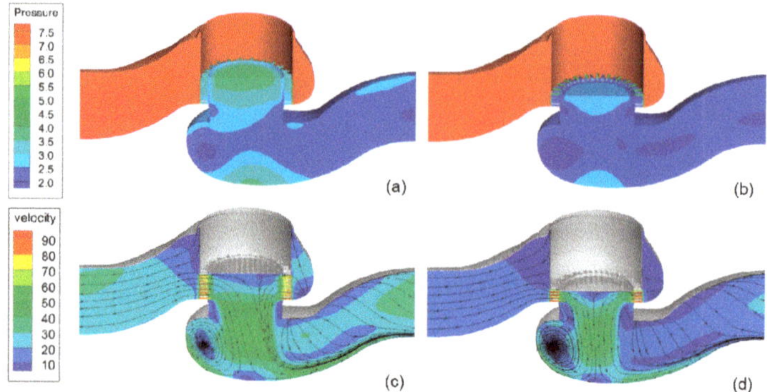

Figure 3. Pressure and velocity contours inside the sleeve regulating valve with valve core displacement of 60 mm and 30 mm (**a**) pressure, 60 mm; (**b**) pressure, 30 mm; (**c**) velocity, 60 mm; (**d**) velocity, 30 mm.

Figure 4 describes the pressure and velocity variation along the horizontal direction with different valve core displacements. Some quantified comparisons can be carried out in Figure 4. Although the inlet pressure for two opening states is set as a constant of 8 MPa, the actual inlet pressure for two opening states is lower than 8 MPa slightly. Specifically, the actual inlet pressure for the full opening state is lower than the inlet pressure for half opening state. When the fluid flows through the sleeve, there is a slight pressure increase for two opening states, which is induced by sudden change of the flow cross section area. The pressure lower than 1.5 MPa can also be visualized at the outlet of orifices in the sleeve for the half opening state. At the center of the sleeve, there is a relative pressure rise for two opening states. The pressure rise for the full opening state is higher than pressure rise for the half opening state.

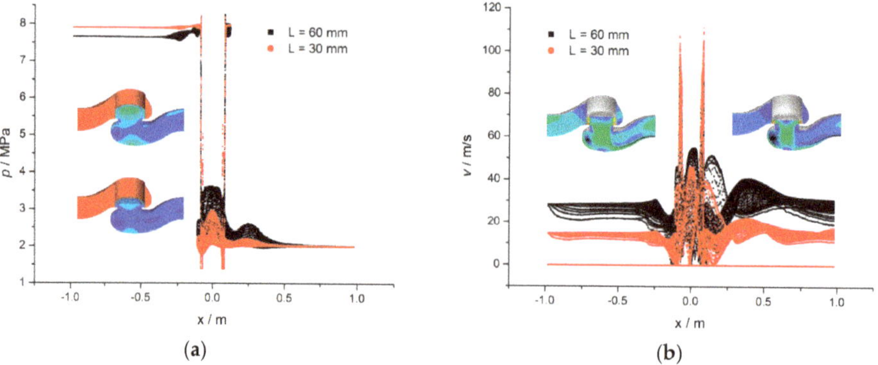

Figure 4. Pressure and velocity variation along the horizontal direction for different valve core displacements (**a**) pressure; (**b**) velocity.

Considering the velocity variation along the horizontal direction, it can be seen that the average velocity at the inlet and outlet with the maximum valve core displacement of 60 mm is higher than the average velocity with the valve core displacement of 30 mm. However, the velocity rise when the fluid flows through the sleeve at the half opening state is higher than the velocity rise at the full opening state. It can be obtained that the energy consume with the valve core displacement of 30 mm is more than the energy consumed with the maximum valve core displacement.

Figure 5 depicts the vapor distributions inside the sleeve regulating valve. It can be seen that the decrease of the valve core displacement induces the enlargement of the vapor distribution region and the increase of the vapor density. The vapor is mainly concentrated in the edge of orifice inlet at the full opening state. When the valve core displacement is decreased from 60 mm to 30 mm, the vapor begins to appear in the orifice outlet and the sealing surface of the valve core.

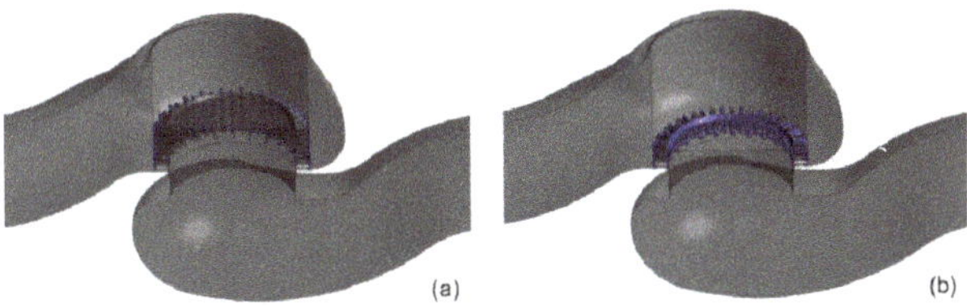

Figure 5. Vapor distributions inside the sleeve regulating valve for different valve core displacements (**a**) 60 mm; (**b**) 30 mm.

3.2. Flow Field Analysis for Different Pressure Difference

Figure 6 demonstrates the pressure distribution of the valve symmetric cross section for different pressure differences with the valve core displacements of 60 mm and 30 mm. When the pressure difference is the lowest of 1 MPa, the pressure distributions for different valve core displacements are shown in Figure 6a,b. The pressure at the inlet and outlet remain stable above 3.5 MPa and below 2.5 MPa, respectively. The pressure drop appears suddenly behind the sleeve, which is the main throttling region. The difference of the pressure distributions induced by the valve core displacement is the region with high pressure at the center of the valve chamber. With the increase in pressure difference, the pressure at the orifice inlet of the sleeve is increased while the pressure drop when flowing through the sleeve is increased. The region with high pressure at the center of the sleeve is enlarged with the increase of the pressure difference while the value of pressure is increased. The outlet pressure for all pressure differences are kept below 2.5 MPa.

To further quantify the pressure variation for different pressure differences, the pressure variations along the horizontal direction for different pressure differences are shown in Figure 7. The actual pressure at the inlet for different pressure differences are all lower than the initial setting pressure and the difference between the actual pressure and setting pressure is increased with the increase of the pressure difference. There is a pressure rise slightly before the fluid flows through the sleeve and the pressure rise is increased with the increase of the pressure difference. In addition, when the fluid flows out of the valve chamber, there is also a pressure rise slightly because of the change of the flow channel structure. The pressure variation along the horizontal direction at the half opening state is similar to the pressure variation at the full opening state, except for the pressure lower than 1.5 MPa at the orifices outlet of the sleeve.

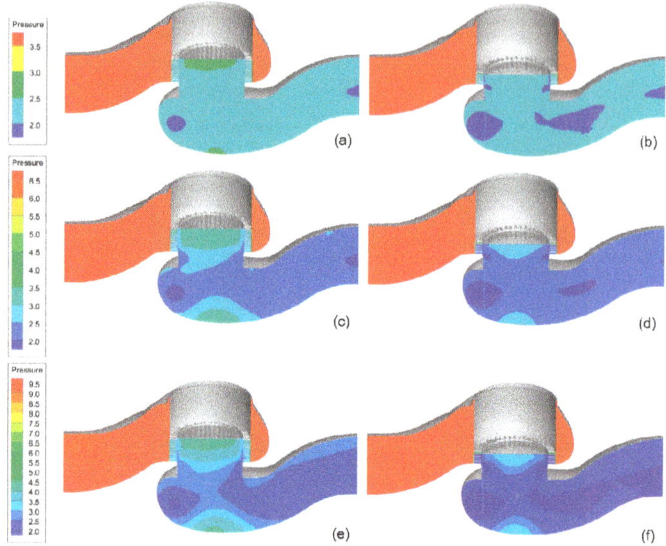

Figure 6. Pressure contours inside the sleeve regulating valve with valve core displacement of 60 mm and 30 mm (MPa) (**a**) 2 MPa, 60 mm; (**b**) 2 MPa, 30 mm; (**c**) 5 MPa, 60 mm; (**d**) 5 MPa, 30 mm; (**e**) 8 MPa, 60 mm; (**f**) 8 MPa, 30 mm.

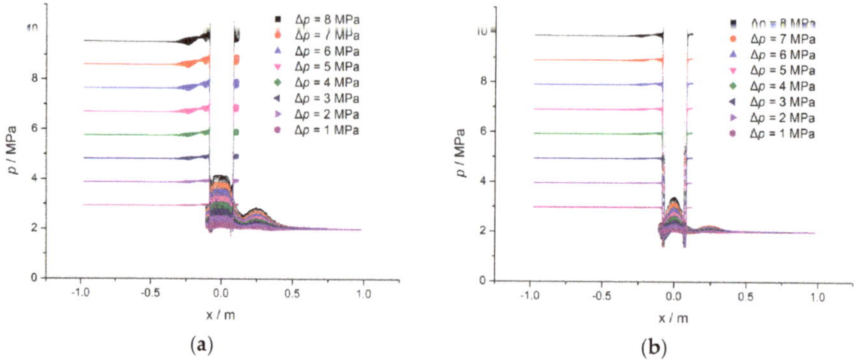

Figure 7. Pressure variation along the horizontal direction for different pressure differences (**a**) L = 60 mm; (**b**) L = 30 mm.

Figure 8 shows the velocity and streamlines distribution of the valve symmetric cross section for different pressure differences with the valve core displacements of 60 mm and 30 mm. It can be found that the inlet velocity is increased with the increase of the pressure difference. For instance, when the valve core displacement is 60 mm, the inlet velocity for the pressure difference of 2 MPa is lower than 20 m/s while the inlet velocity for the pressure difference of 5 MPa is higher than 20 m/s and lower than 30 m/s. The inlet velocity for the maximum pressure difference of 8 MPa is higher than 30 m/s. The velocity rise when flowing through the sleeve is increased while the velocity at the center of valve chamber is increased with the increase of pressure difference. There is no difference in the vortex in the left side of the valve chamber for different pressure difference. The outlet velocity is increased with the increase of the pressure difference.

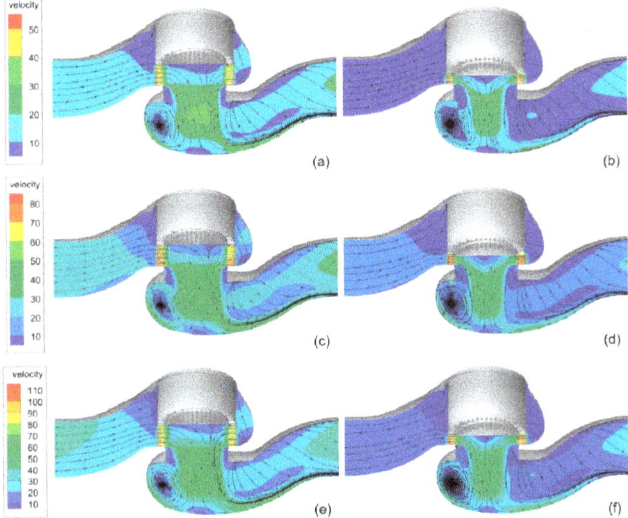

Figure 8. Velocity contours inside the sleeve regulating valve with valve core displacement of 60 mm and 30 mm (m/s) (**a**) 2 MPa, 60 mm; (**b**) 2 MPa, 30 mm; (**c**) 5 MPa, 60 mm; (**d**) 5 MPa, 30 mm; (**e**) 8 MPa, 60 mm; (**f**) 8 MPa, 30 mm.

Figure 9 demonstrates the velocity variation along the horizontal direction for different pressure differences. It can be seen that the velocity at the inlet and outlet when the pressure difference is 1 MPa it is the lowest, while the inlet and outlet velocity for the pressure difference of 8 MPa is the highest. With the increase of the pressure difference, the inlet and outlet velocity are increased. The above phenomenon indicates that the higher initial pressure brings a higher initial energy. The pressure rise in the orifices and the center of the sleeve indicate the same conclusion as the above analysis.

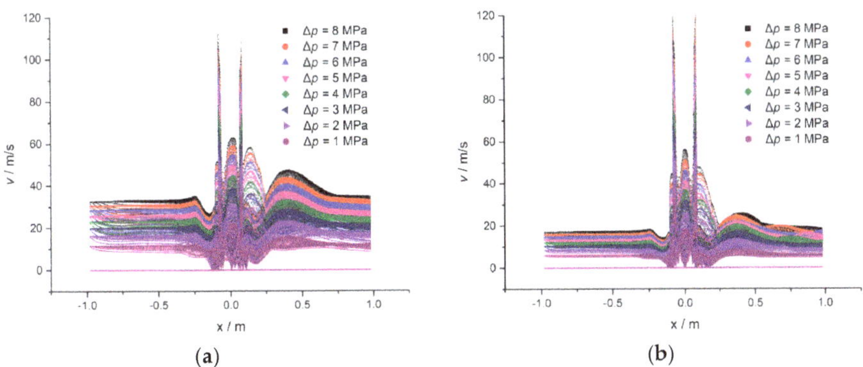

Figure 9. Velocity variation along the horizontal direction for different pressure differences (**a**) L = 60 mm; (**b**) L = 30 mm.

Figure 10 depicts the flow rate variation with the increase of the pressure difference. It can be seen that the mass flow rate is increased with the increase of the pressure difference for two opening states. Totally, the mass flow rates for the full opening state is higher than the flow rates for the half opening state, more than twice flow rates for the half opening state. For example, when the pressure difference is 4 MPa, the mass flow rate with the valve core displacement of 30 mm is 69.07 kg/s while the mass flow rate with the valve core displacement of 60 mm is 131.79 kg/s, which is more than twice 69.07 kg/s. The phenomenon indicates the flux characteristics curve of the studied sleeve regulating valve is convex when comparing with the linear flux characteristic. The increase velocity of the mass flow rate is increased first and then decreased. The increase velocity of the mass flow rate for the full opening state is higher than the increase velocity for the half opening state.

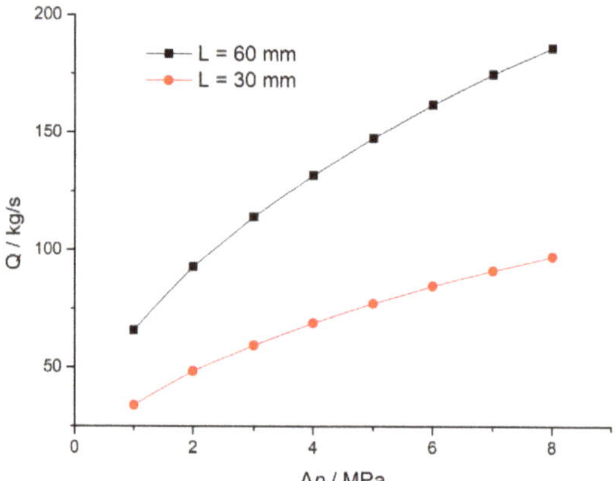

Figure 10. Flow rate variation with the increase of pressure difference for different openings.

3.3. Cavitation Distribution Analysis for Difference Pressure Differences

Figure 11 demonstrates the vapor distributions inside the sleeve regulating valve for different pressure difference with the valve core displacement of 60 mm. In general, when cavitation causes severe valve damages, the vapor volume fraction of each computational cell, α, is higher than 0.5. Figure 11a–c depicts three-dimensional isosurfaces of αlarger than 0.5 in valves with the valve core displacement of 60 mm for different pressure differences.

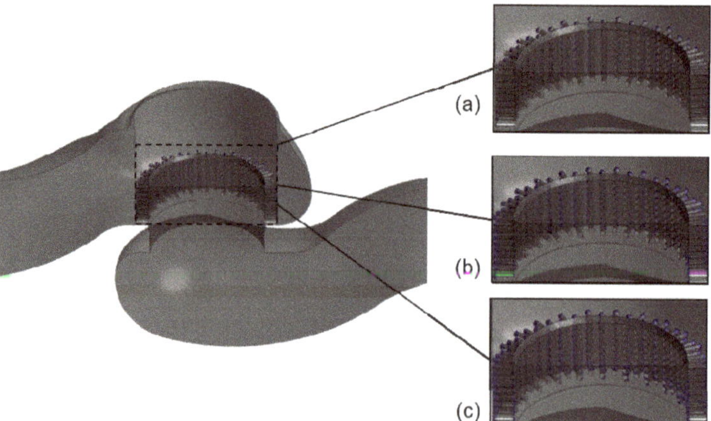

Figure 11. Vapor distributions inside the sleeve regulating valve for different pressure difference with the valve core displacement of 60 mm (**a**) 2 MPa; (**b**) 5 MPa; (**c**) 8 MPa.

For the pressure difference of 2 MPa, the vapor appears at the orifices inlet of the sleeve and the distribution region is very small. When the pressure difference is increased from 2 MPa to 5 MPa, the vapor is still concentrated at the orifice inlet of the sleeve but the distribution region is enlarged slightly. Furthermore, when the pressure difference is 8 MPa, the vapor begins to appear at the outlet of orifices in the sleeve and the distribution region is further enlarged. As a whole, the cavitation intensity and density are tiny when the valve core displacement is 60 mm and the pressure difference has a slight influence on the cavitation intensity and density inside the regulating valve.

When the valve core displacement is decreased from 60 mm to 30 mm, as is shown in Figure 12, the effect of the pressure difference on the cavitation intensity and density is enhanced. The vapor distribution region is still very small when the pressure difference is 2 MPa. However, when the pressure difference is increased from 2 MPa to 4 MPa, the cavitation distribution region is obviously enlarged and the vapor begins to appear on the sealing surface of the valve core. There is no doubt the increase of the pressure difference induces a more serious cavitation. Moreover, when the pressure difference reaches the maximum value of 8 MPa, the vapor almost fills the orifices in the valve and the total sealing surface is filled with vapor. Meanwhile, the vapor distribution region begins to extend downward along the surface of the valve chamber.

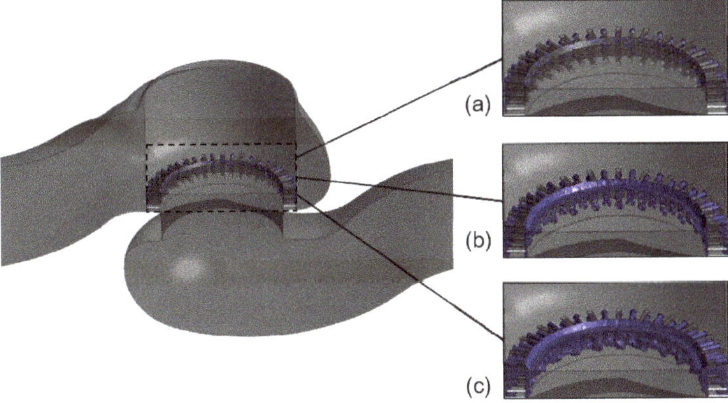

Figure 12. Vapor distributions inside the sleeve regulating valve for different pressure difference with the valve core displacement of 30 mm (**a**) 2 MPa; (**b**) 5 MPa; (**c**) 8 MPa.

To further quantify the influence of the pressure difference on cavitation characteristics of the sleeve regulating valve, the total vapor volume is also calculated using the following equation:

$$V_v = \iiint_{\Omega} \alpha dV \tag{9}$$

where α denotes the vapor volume fraction in an element. The total vapor volume demonstrates the whole vapor caused by cavitation, so the total vapor volume can represent the intensity of the cavitation. The higher the total vapor volume is, the more intense the cavitation intensity is. The total vapor volume quantifies the cavitation intensity.

Figure 13 depicts the total vapor volume variation with the increase of the pressure difference when the valve core displacement is 60 mm and 30 mm. It can be found that the total vapor volumes for two opening state are increased with the increase of the pressure difference. However, the increase trends for two opening states are totally different. When the pressure difference is lower than 3 MPa, the total vapor volumes for two opening states are close to 0. The above phenomenon indicates that the cavitation intensity is still very low for both opening states when the pressure difference is lower than 3 MPa. The low-pressure difference induces the slight cavitation. When the pressure difference is higher than 3 MPa, the increase trend for the half opening state begins to be steep suddenly, while the increase trend for the half full opening state is still very gentle.

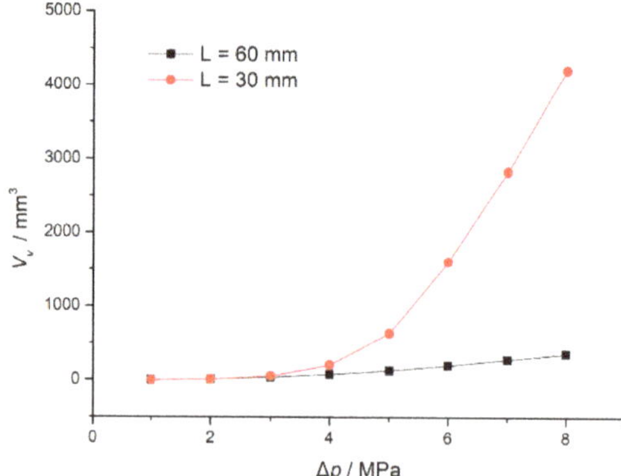

Figure 13. Total vapor volumes inside the valve for different pressure differences with the valve core displacement of 60 mm and 30 mm.

To estimate the probable occurrence of cavitation, the cavitation index σ_v is calculated and denoted as

$$\sigma_v = \frac{p_d - p_v}{p_u - p_d} \tag{10}$$

where p_u is the inlet pressure, p_d is the outlet pressure, and the lower the σ_v is, the higher is the potential. In general, when the σ_v is lower than 1.0, the cavitation will occur. When the σ_v is lower than 0.5, the cavitation phenomenon will be stable.

In this study, the inlet and outlet pressure are determined by the inlet and outlet boundary conditions. The saturation vapor pressure is set as a constant of 1.5 MPa. Therefore, the cavitation index for the valve at specified pressure conditions is certain. To further quantify the actual influence of the cavitation index on the actual cavitation intensity in the sleeve regulating valve, the total vapor volumes variation with the change of the cavitation index is shown in Figure 14. It can be seen that for different opening states, the actual effects of the cavitation index on the total vapor volumes are really different, and the lower the cavitation index is, the higher the difference is for different opening states. As the cavitation index is higher than 0.2, for both two opening state, the variation of the cavitation index has a small influence on the total vapor volume and cavitation intensity. When the cavitation index is lower than 0.2, the total vapor volume is increased with a steep trend at the half opening state, which indicates that a small variation of cavitation index induces a more drastic variation of the total vapor volume. When the valve core displacement is 60 mm, the variation trend of total vapor volume is gentler than the trend at the half opening state when the cavitation is lower than 0.2. As a whole, the relation between actual cavitation intensity and cavitation index is not simply linear and is changed with the change of valve core displacement. When the cavitation index is higher than 0.2, the cavitation index's effects on cavitation intensity are small. As the cavitation index is lower than 0.2, the cavitation index's effects on cavitation intensity are intense.

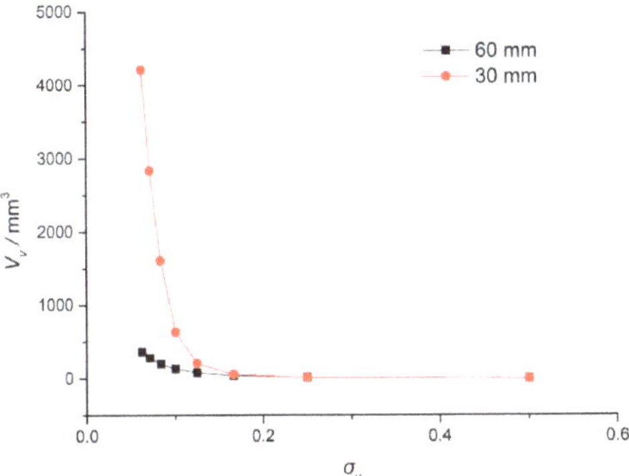

Figure 14. Total vapor volumes inside the valve for different cavitation index with the valve core displacement of 60 mm and 30 mm.

4. Conclusions

The cavitation occurring in a sleeve regulating valve for different pressure differences and valve core displacements has been numerically investigated in this study, and the effects of pressure difference and valve core displacement have been revealed using the proposed numerical model.

First, the flow streamlines, pressure distribution and vapor distribution for different opening state are obtained. A high-velocity and low-pressure region appear behind the valve sleeve because of the sudden decrease of the cross-section area. According to the predicted vapor distribution, the vapor is mainly concentrated in the edge of orifice inlet at the full opening state. The decrease of the valve core displacement induces the enlargement of the vapor distribution region and the increase of the vapor density.

Second, the pressure and flow streamlines for different pressure difference are analyzed. With the increase in pressure difference, the pressure at the orifice inlet of the sleeve is increased while the pressure drop when flowing through the sleeve is increased. The region with high pressure at the center of the sleeve is enlarged with the increase of the pressure difference, while the value of pressure is increased. The inlet velocity is increased with the increase of the pressure difference. The velocity rise when flowing through the sleeve is increased while the velocity at the center of valve chamber is increased with the increase of pressure difference.

Last, the cavitation distributions inside the sleeve regulating valve for different pressure differences are analyzed. It can be seen that the increase of the pressure difference induces a more serious cavitation. The pressure difference has a slight influence on the cavitation intensity and density inside the regulating valve when the valve core displacement is 60 mm. When the valve core displacement is decreased from 60 mm to 30 mm, the effects of the pressure difference on the cavitation intensity are enhanced. The relation between actual cavitation intensity and cavitation index is not simply linear and is changed with the change of valve core displacement. When the cavitation index is higher than 0.2, the cavitation index's effects on cavitation intensity are small. As the cavitation index is lower than 0.2, the cavitation index's effects on cavitation intensity are intense.

Author Contributions: Conceptualization, C.Q. and C.-H.J.; methodology, C.Q. and C.-H.J.; software, C.Q.; validation, C.-H.J., H.Z. and J.-Y.W.; formal analysis, C.Q.; investigation, H.Z.; resources, C.Q.; data curation, C.-H.J.; writing—original draft preparation, C.Q.; writing—review and editing, C.-H.J.; visualization, J.-Y.W.; supervision, Z.-J.J.; project administration, Z.-J.J.; funding acquisition, Z.-J.J.

Funding: This research was funded by the National Natural Science Foundation of China (NSFC), grant number 51875514; the Zhejiang Key Research & Development Project, grant number 2019C01025; and the Zhejiang Quality and Technical Supervision Research Project, grant number 20180117.

Conflicts of Interest: The authors declare no conflict of interest.

Nomenclature

p	pressure (MPa)
p_v	saturation pressure of water (MPa)
t	time (s)
v	mass average velocity (m/s)
V_v	total vapor volume (m^3)
R_b	bubble radius (m)
R_c	rates of breaking of vapor bubbles
α	vapor volume fraction
n	bubble number density
ρ_m	mixture density (kg/m^3)
ρ_l	liquid density (kg/m^3)
ρ_v	vapor density (kg/m^3)
μ_m	mixture dynamic viscosity (Pa·s)
μ_l	liquid dynamic viscosity (Pa·s)
μ_v	vapor dynamic viscosity (Pa·s)
μ_t	turbulent viscosity (Pa·s)
v_v	vapor phase velocity (m/s)
L	valve core displacement (mm)

References

1. Long, X.; Zhang, Y.; Wang, J.; Xu, M.; Lyu, Q.; Ji, B. Experimental investigation of the global cavitation dynamic behavior in a venturi tube with special emphasis on the cavity length variation. *Int. J. Multiph. Flow* **2017**, *89*, 290–298. [CrossRef]
2. Liu, H.; Cao, S.; Luo, X. Study on the effect of inlet fluctuation on cavitation in a cone flow channel. *ASME J. Fluids Eng.* **2015**, *137*, 051301.
3. Kim, J.; Song, S.J. Measurement of temperature effects on cavitation in a turbopump inducer. *ASME J. Fluids Eng.* **2016**, *138*, 011304. [CrossRef]
4. Xiao, L.; Long, X. Cavitating flow in annular jet pumps. *Int. J. Multiph. Flow* **2015**, *71*, 116–132. [CrossRef]
5. Zhang, D.; Shi, W.; Pan, D.; Dubuisson, M. Numerical and experimental investigation of tip leakage vortex cavitation patterns and mechanisms in an axial flow pump. *ASME J. Fluids Eng.* **2015**, *137*, 121103. [CrossRef]
6. Zhu, B.; Hongxun, C. Analysis of the staggered and fixed cavitation phenomenon observed in centrifugal pumps employing a gap drainage impeller. *ASME J. Fluids Eng.* **2016**, *139*, 031301. [CrossRef]
7. Wang, C.; Hu, B.; Zhu, Y.; Luo, C.; Cheng, L. Numerical study on the gas-water two-phase flow in the self-priming process of self-priming centrifugal pump. *Processes* **2019**, *7*, 330. [CrossRef]
8. Qian, J.; Chen, M.; Liu, X.; Jin, Z. A numerical investigation of flow of nanofluids through a micro Tesla valve. *J. Zhejiang Univ. Sci. A* **2019**, *20*, 50–60. [CrossRef]
9. Qian, J.; Gao, Z.; Hou, C.; Jin, Z. A comprehensive review of cavitation in valves: Mechanical heart valves and control valves. *Bio-Design Manuf.* **2019**, *2*, 119–136. [CrossRef]
10. Hassis, H. Noise caused by cavitating butterfly and monovar valve. *J. Sound Vib.* **1999**, *225*, 515–526. [CrossRef]
11. Herbertson, L.H.; Reddy, V.; Manning, K.B.; Welz, J.P.; Deutsch, S. Wavelet transforms in the analysis of mechanical heart valve cavitation. *J. Biomech. Eng.* **2006**, *128*, 217–222. [CrossRef] [PubMed]
12. Gao, H.; Lin, W.; Tsukiji, T. Investigation of cavitation near the orifice of hydraulic valves. *Proc. Inst. Mech. Eng. Part G J. Aerosp. Eng.* **2006**, *220*, 253–265. [CrossRef]
13. Jin, Z.; Gao, Z.; Li, X.; Qian, J. Cavitation flow through a micro-orifice. *Micromachines* **2019**, *10*, 191. [CrossRef] [PubMed]
14. Jia, W.H.; Yin, C.B. Numerical analysis and experiment research of cylinder valve port cavitating flow. In Proceedings of the Fourth International Conference on Experimental Mechanics, Singapore, 18–20 November 2009.

15. Liu, Y.H.; Ji, X.W. Simulation of cavitation in rotary valve of hydraulic power steering gear. *Sci. China* **2009**, *52*, 3142–3148. [CrossRef]

16. Lu, L.; Zou, J.; Fu, X. The acoustics of cavitation in spool valve with U-notches. *Proc. Inst. Mech. Eng. Part G J. Aerosp. Eng.* **2011**, *226*, 540–549. [CrossRef]

17. Li, S.; Aung, N.Z.; Zhang, S.; Cao, J.; Xue, X. Experimental and numerical investigation of cavitation phenomenon in flapper-nozzle pilot stage of an electrohydraulic servo-valve. *Comput. Fluids* **2013**, *88*, 590–598. [CrossRef]

18. Kudźma, Z.; Stosiak, M. Studies of flow and cavitation in hydraulic lift valve. *Arch. Civil Mech. Eng.* **2015**, *15*, 951–961. [CrossRef]

19. Qu, W.S.; Tan, L.; Cao, S.L.; Xu, Y.; Huang, J.; Xu, Q.H. Experiment and numerical simulation of cavitation performance on a pressure-regulating valve with different openings. *IOP Conf. Ser. Mater. Sci. Eng.* **2015**, *72*, 042035. [CrossRef]

20. Ulanicki, B.; Picinali, L.; Janus, T. Measurements and analysis of cavitation in a pressure reducing valve during operation—A Case Study. *Procedia Eng.* **2015**, *119*, 270–279. [CrossRef]

21. Deng, J.; Pan, D.; Xie, F.; Shao, X. Numerical investigation of cavitation flow inside spool valve with large pressure drop. *J. Phys. Conf. Ser.* **2015**, *656*, 012067. [CrossRef]

22. Ou, G.F.; Xu, J.; Li, W.Z.; Chen, B. Investigation on cavitation flow in pressure relief valve with high pressure differentials for coal liquefaction. *Procedia Eng.* **2015**, *130*, 125–134. [CrossRef]

23. Yi, D.L.; Lu, L.; Zou, J.; Fu, X. Interactions between poppet vibration and cavitation in relief valve. *Proc. Inst. Mech. Eng. Part C J. Mech. Eng. Sci.* **2015**, *229*, 1447–1461. [CrossRef]

24. Okita, K.; Miyamoto, Y.; Kataoka, T.; Takagi, S.; Kato, H. Mechanism of noise generation by cavitation in hydraulic relief valve. *J. Phys. Conf. Ser.* **2015**, *656*, 012104. [CrossRef]

25. Lu, L.; Xie, S.; Yin, Y.; Ryu, S. Experimental and numerical analysis on the surge instability characteristics of the vortex flow produced large vapor cavity in u-shape notch valve. *Int. J. Heat Mass Transfer.* **2020**, *146*, 118882. [CrossRef]

26. Jin, Z.; Gao, Z.; Qian, J.; Wu, Z.; Sunden, B. A parametric study of hydrodynamic cavitation inside globe valve. *ASME J. Fluids Eng.* **2018**, *140*, 031208. [CrossRef]

27. Qian, J.; Chen, M.; Gao, Z.; Jin, Z. Mach number and energy loss analysis inside multi-stage Tesla valves for hydrogen decompression. *Energy* **2019**, *179*, 647–654. [CrossRef]

28. Qian, J.; Wu, J.; Gao, Z.; Wu, A.; Jin, Z. Hydrogen decompression analysis by multi-stage Tesla valves for hydrogen fuel cell. *Int. J. Hydrogen Energy* **2019**, *44*, 13666–13674. [CrossRef]

29. Tao, J.; Lin, Z.; Ma, C.; Ye, J.; Zhu, Z.; Li, Y.; Mao, W. An experimental and numerical study of regulating performance and flow loss in a v-port ball valve. *ASME. J. Fluids Eng.* **2020**, *142*, 021207. [CrossRef]

30. Baran, G.; Catana, I.; Magheti, I.; Safta, C.A.; Savu, M. Controlling the cavitation phenomenon of evolution on a butterfly valve. *IOP Conf. Ser. Earth Environ. Sci.* **2010**, *12*, 012100. [CrossRef]

31. Zhang, Z.; Cao, S.; Luo, X.; Shi, W. New approach of suppressing cavitation in water hydraulic components. *Proc. Inst. Mech. Eng. Part C J. Mech. Eng. Sci.* **2016**, *231*, 4022–4034.

32. Shi, W.; Cao, S.; Luo, X.; Zhang, Z.; Zhu, Y. Experimental research on the cavitation suppression in the water hydraulic throttle valve. *J. Press. Vessel Technol.* **2017**, *139*, 051302.

33. Chern, M.J.; Hsu, P.H.; Cheng, Y.J.; Tseng, P.Y.; Hu, C.M. Numerical study on cavitation occurrence in globe valve. *J. Energy Eng.* **2013**, *139*, 25–34. [CrossRef]

34. Qian, J.; Hou, C.; Wu, J.; Gao, Z.; Jin, Z. Aerodynamics analysis of superheated steam flow through multi-stage perforated plates. *Int. J. Heat Mass Transfer* **2019**, *141*, 48–57. [CrossRef]

35. Yaghoubi, H.; Madani, S.A.H.; Alizadeh, M. Numerical study on cavitation in a globe control valve with different numbers of anti-cavitation trims. *J. Cent. South Univ.* **2018**, *25*, 2677–2687. [CrossRef]

36. Liu, Q.; Ye, J.; Zhang, G.; Lin, Z.; Xu, H.; Jin, H.; Zhu, Z. Study on the metrological performance of a swirlmeter affected by flow regulation with a regulating valve. *Flow Meas. Instrum.* **2019**, *67*, 83–94. [CrossRef]

37. Qian, J.; Wei, L.; Jin, Z.; Wang, J.; Zhang, H. CFD analysis on the dynamic flow characteristics of the pilot-control globe valve. *Energy Convers. Manag.* **2014**, *87*, 220–226. [CrossRef]

38. Hou, C.; Qian, J.; Chen, F.; Jiang, W.; Jin, Z. Parametric analysis on throttling components of multi-stage high pressure reducing valve. *Appl. Therm. Eng.* **2018**, *128*, 1238–1248. [CrossRef]
39. Schnerr, G.; Sauer, J. Physical and numerical modeling of unsteady cavitation dynamics. In Proceedings of the 4th International Conference on Multiphase Flow, New Orleans, LA, USA, 27 May–1 June 2001.

Article

Feature Extraction Method for Hydraulic Pump Fault Signal Based on Improved Empirical Wavelet Transform

Zhi Zheng [1], Zhijun Wang [1,*], Yong Zhu [2,*], Shengnan Tang [2] and Baozhong Wang [1]

[1] College of Mechanical Engineering, North China University of Science and Technology, Tangshan 063210, China; zhengzhi@ncst.edu.cn (Z.Z.); wbzhong@ncst.edu.cn (B.W.)

[2] National Research Center of Pumps, Jiangsu University, Zhenjiang 212013, China; tangsn6635@126.com

* Correspondence: zjwang@ncst.edu.cn (Z.W.); zhuyong@ujs.edu.cn (Y.Z.); Tel.: +86-0315-8805440 (Z.W.)

Received: 8 October 2019; Accepted: 31 October 2019; Published: 6 November 2019

Abstract: There are many interference components in Fourier amplitude spectrum of a contaminated fault signal, and thus the segment obtained based on the spectrum can lead to serious over-decomposition of empirical wavelet transform (EWT). Aiming to resolve the above problems, a novel method named improved empirical wavelet transform (IEWT) is proposed. Because the power spectrum is less sensitive to the contaminated interference and manifests the presence of fault feature information, IEWT replaces the Fourier amplitude spectrum of EWT with power spectrum in segment acquirement, and threshold processing is also introduced to eliminate the bad influence on the acquirement, and thus the best decomposition result of IEWT can be obtained based on feature energy ratio (*FER*). The loose slipper fault signal of hydraulic pump is tested and verified. The result demonstrates that the proposed method is superior and can extract the fault feature information accurately.

Keywords: hydraulic pump; fault signal; feature extraction; empirical wavelet decomposition; power spectrum density; feature energy ratio

---

1. Introduction

The hydraulic pump is a power source and supplies pressure energy to hydraulic systems. The pump has been applied in many important industrial fields such as aeronautics, astronautics, metallurgy, petrochemical engineering, and engineering machinery. The hydraulic system in aforementioned fields possess the characteristics such as large-scale continuation, integration and automation, and thus the working condition of pump facing some challenges as high temperature, high pressure and high speed [1,2]. Unfortunately, these challenges accelerate the deterioration of the health status of pumps, so it is very important to diagnose the faults for hydraulic pump [3,4]. The mechanical and fluid impact can cause the vibration of the pump, and vibration severity increases once the pump is broken [5,6]. A lot of fault feature information is contained in the vibration signal, and the signal is also contaminated by many interferences [7,8]. Recently, many scholars have applied vibration signal to diagnose the faults in domestic and foreign. Lan et al. applied wavelet packet transform (WPT), local tangent space alignment (LTSA), empirical mode decomposition (EMD) and local mean decomposition (LMD) to process the fault signal of the pump to extract the eigenvectors, and then the faults could be diagnosed by an extreme learning machine (ELM) [9]. Sun et al. calculated cyclic autocorrelation functions (CAFs) of a fault signal of the pump, and the corresponding slices were extracted from the CAFs and processed by fast Fourier transform (FFT), and then indicators were extracted from the FFT spectra to diagnose the faults [10]. Considering that the early fault signal of the pump is a periodic weak signal, Zhao et al. proposed an intermittent chaos, sliding window symbol sequence statistics-based method, and it was used to detect the early faults [11]. Du et al. presented a new method based on the layered clustering algorithm to diagnose multiple faults of an aircraft pump, and thus the faults could be diagnosed based on risk priority number and their severity layer by layer [12]. Lu et al. applied ensemble empirical mode decomposition (EEMD) to decompose the fault signal of the pump, and eigenvectors were extracted in time, frequency and time-frequency domains as the input, and then the optimized support vector regression (SVR) model was used to diagnose the faults [13].

Wavelet transform (WT) is an effective tool to process a nonlinear and non-stationary signal [14,15]. The time-domain signal can also be decomposed by WT into different frequency bandwidth groups [16,17]. Compared with short-time Fourier transform (STFT), the basic difference between WT and STFT is the basis function. The STFT only uses sine and cosine as basis functions. WT also adopts many kinds of wavelet functions specifying a certain mathematical property. Furthermore, WT fuses the principles of Fourier transform basis function and STFT window function to develop a new oscillation and attenuation wavelet basis function. Thus, WT overcomes the shortcoming of the fixed window, and the new function can change the resolution of time and frequency information with the change of scale factor. Owing to the width adjustment of WT, the signal details can be analyzed [18,19]. Currently, WT has been widely used to diagnose some faults at home and abroad. Aiming at the typical faults of bearing, some researches proposed the method based on WT and utilized it to diagnose the faults successfully [18,19]. Moreover, the method based on WT was used by Kordestani et al. to diagnose the faults of spoiler system effectively [20]. Pointing at the faults of planetary gearboxes, Zhao et al. also proposed a method based on WT, and proved the faults could be diagnosed effectively [21].

Although WT has widely used in some fields, it has some imperfect aspects. Firstly, in matching wavelet basis function with morphological features of a signal, there is no accuracy matching principle and criterion to follow. Secondly, the signal possesses many kinds of features, but the wavelet basis function is not changed when it works, and thus WT is not adaptive. Thirdly, if the basis function and scale factor are selected, the resolution is fixed, and thus it does not have the adaptability. Thirdly, although WT has the capability of multi-scale and multi-resolution, some parts of the signal contained in the window must be approximate stationary (pseudo stationary) [22].

EMD was proposed by Huang in 1998, and it is good at processing the nonlinear and non-stationary signal, and the signal is made up of some multi-component modes [23]. One single-component mode can be separated adaptively from the signal based on its morphological feature information (local maxima points), and the single-component mode is called intrinsic mode function (IMF) [24,25]. Thus far, a large number of studies about EMD have been made by scholars on a global scale. EMD was effectively applied to monitor the healthy condition of wind turbine [19]. Moreover, EMD was also employed to process some fault signals such as gearbox [25], bearings [26–28], and so on.

Although EMD is widely applied, the cubic splines are employed in EMD to fit the lower and upper envelopes in each sift, which can lead to two obvious defects [29]. One is mode-mixing, and another is endpoint effect. In mode mixing, an IMF may be consisted of two or more component modes, and thus it may lead to wrong decomposition. In the endpoint effect, the endpoints of an IMF are divergent, and the divergence is gradually into the inside with the iteration of the decomposition. If the iterative number becomes larger, IMF will be more distorted, then mode mixing and illusive components happen.

Aiming to improve the shortcomings of WT and EMD, a method for processing the nonlinear and non-stationary signal is proposed by Gilles, and the method is called empirical wavelet transform (EWT) [30]. In EWT, Fourier amplitude spectrum of a multi-component mode signal is segmented, and the wavelet orthogonal basis function is established based on each segment. Thus, the signal can be decomposed by EWT into several AM-FM single-component modes, and each mode is compact, supported, and centered on a specific frequency. If the signal is contaminated by interference components, there are also interference components in Fourier amplitude spectrum of the signal, and the segment of EWT is determined by the spectrums of the signal and the interference components together. Thus, the spectrum increase of interference component leads to many segments, which leads to mode mixing and over-decomposition [31]. Dong et al. applied sparsity to improve EWT, and adopted it to diagnose the faults of rolling element bearings [31]. Jiang et al. proposed a novel method based on EWT and ambiguity correlation classifiers, and EWT was used to decompose the fault signal, and ambiguity correlation classifiers was adopted to diagnose the faults of rolling element bearings [32]. Cao et al. used the EWT to decompose the signal of wheel-bearing, and the mode which was rich of fault feature information was selected to analyze the spectrum, and then the faults were diagnosed [33].

In order to resolve the above problems, a new method named improved empirical wavelet transform (IEWT) is proposed. Firstly, the power spectrum of the loose slipper fault signal is obtained. Secondly, different threshold values are adopted to eliminate the power spectrum of the interference components, and the bad influence of interference component on segment acquirement is largely reduced. Then, the best segment number is obtained based on the feature energy ratio (*FER*). Thus, the fault signal of hydraulic pump can be best decomposed by IEWT. The acquired results provide an important basis for the application extension to faults diagnosis study of other rotating machinery.

The rest of the paper is organized as follows. In Section 2, the algorithms of *FER* and IEWT are introduced, and then the flowchart of IEWT is presented. Section 3 depicts the EWT application in a simulated signal and a hydraulic pump fault signal in detail. In Section 4, some conclusions of this investigation are summarized.

2. Methodology

2.1. Feature Energy Ratio

Impact vibration is often caused by the faults of rotating machinery, and the impact vibration energy is generated, and thus the energy is rich in fault feature information. The big value of feature energy ratio (*FER*) signifies that the amount of fault feature information is large.

For the sake of measuring amount of fault features information in a signal, the *FER* is proposed in reference [34]. It can be rewritten as

$$FER = (E_1 + E_2 + \ldots + E_n)/E \tag{1}$$

where $E_1, E_2, \ldots, E_n$ are respectively the energy, and they are respectively presented in fault feature frequency and its harmonics, and E is the total energy of the signal.

2.2. Improved Empirical Wavelet Transform

The algorithm of IEWT is described as follows:

(1) Calculating power spectrum of signals

Different from the Fourier amplitude spectrum of EWT, IEWT is based on power spectrum.

Let $x(t) = (x_1, x_2, \ldots, x_n)$ be a signal, and $X(\omega)$ is its Fourier amplitude spectrum in the frequency domain, and then the power spectrum of the signal is denoted as P, and it is defined as

$$P = \lim_{T \to \infty} \frac{1}{2T} \int_{-T}^{T} x^2(t)dt = \frac{1}{2\pi} \int_{-\infty}^{\infty} \lim_{T \to \infty} \frac{1}{2T} |X_T(\omega)|^2 d\omega \tag{2}$$

(2) Applying the threshold processing

The step is also different from EWT, and the threshold processing is introduced into IEWT. The threshold values are defined as

$$THVA = coefficient \times mean(P) \tag{3}$$

where *coefficient* is an integer and equals to $1, 2, \ldots, L$.

Figure 1 demonstrates how the threshold processing works.

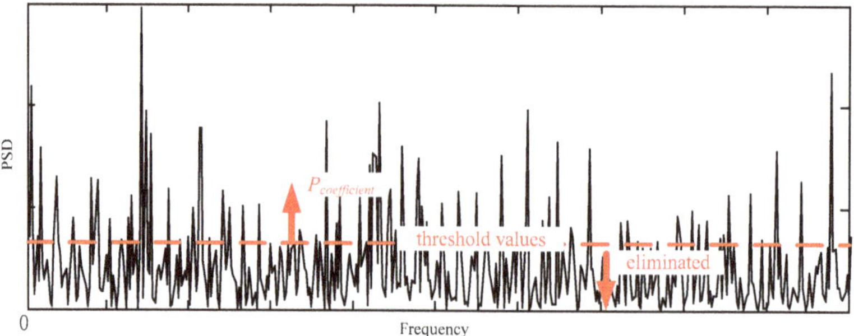

Figure 1. Demonstration of threshold processing.

In Figure 1, the threshold value *THVA* is applied to eliminate the spectrums whose values are smaller than the values (the eliminated spectrum values are set to 0), and thus a new spectrum distribution $P_{coefficient}$ can be obtained. Thus, the bad influence of interference components on adaptive segment acquirement is much reduced.

(3) Decomposing the Signals

Periodicity of a normalized Fourier axis is 2π, and the discussion is restricted to $\omega \in [0, \pi]$ based on $P_{coefficient}$. It is supposed that the Fourier support $[0, \pi]$ is segmented into N contiguous segments, which means that there are $N + 1$ boundaries. However, 0 and π are always used in definition and it is need to be found $N - 1$ extra boundaries. To find the boundaries, local maxima value of power spectrum are selected, and the values are sorted in decreasing order (0 and π are excluded). It is assumed that the algorithm found M maxima, and two cases can appear:

(1) $M \geq N$: enough maxima are selected to define the wanted segment number, and then the first $N - 1$ maxima are adopted.

(2) $M \leq N$: the component mode number of a signal is smaller than expected, and all the selected maxima are kept, and rest N to the appropriate value.

In Figure 2, each segments is defined as $\Lambda_n = [\omega_{n-1}, \omega_n]$, and it can be found that $\mathbf{U}_{n=1}^{N} \Lambda_n = [0, \pi]$. Centered around each ω_n, a transition phase (the gray hatched areas on Figure 2) T_n of width $2\tau_n$ is defined.

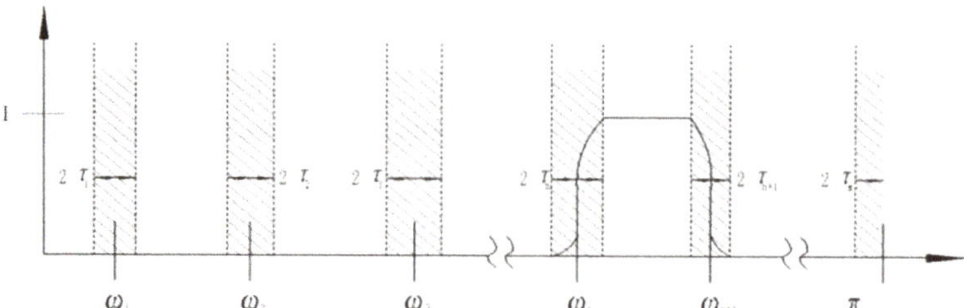

Figure 2. Segments of the Fourier axis.

Based on construction of Littlewood-Paley and Meyer's wavelets, some empirical wavelets of EWT are actually band pass filters on each Λ_n. Then $\forall n > 0$, it can be defined that the empirical scaling function and the empirical wavelets by expressions of Equations (4) and (5) respectively.

$$\hat{\varnothing}_n(\omega) \begin{cases} 1 & \text{if } |\omega| \leq \omega_n - \tau_n \\ \cos\left[\frac{\pi}{2}\beta\left(\frac{1}{2\tau_n}(|\omega| - \omega_n + \tau_n)\right)\right] & \text{if } \omega_n - \tau_n \leq |\omega| \leq \omega_n + \tau_n \\ 0 & \text{otherwise} \end{cases} \tag{4}$$

and

$$\hat{\varnothing}_n(\omega) \begin{cases} 1 & \text{if } \omega_n + \tau_n \leq |\omega| \leq \omega_{n+1} - \tau_{n+1} \\ \cos\left[\frac{\pi}{2}\beta\left(\frac{1}{2\tau_{n+1}}(|\omega| - \omega_{n+1} + \tau_{n+1})\right)\right] & \text{if } \omega_{n+1} - \tau_{n+1} \leq |\omega| \leq \omega_{n+1} + \tau_{n+1} \\ \sin\left[\frac{\pi}{2}\beta\left(\frac{1}{2\tau_n}(|\omega| - \omega_n + \tau_n)\right)\right] & \text{if } \omega_n - \tau_n \leq |\omega| \leq \omega_n + \tau_n \\ 0 & \text{otherwise} \end{cases} \tag{5}$$

where $\gamma < \min_n [\omega_{n+1} - \omega_n/\omega_{n+1} + \omega_n]$, $\beta(x) = 35x^4 - 8x^5 + 70x^6 - 20x^7$.

The definition of IEWT is the same as EWT, and IEWT is defined in the same way as for the classic wavelet transform. Detail coefficient $W_f^\varepsilon(n, t)$ is given by the inner products with the empirical wavelets in Equation (6). Approximation coefficient $W_f^\varepsilon(0, t)$ is given by the inner products with the scaling function in Equation (7).

$$W_f^\varepsilon(n, t) = \langle f, \Psi_n \rangle = \left(\overline{f(\omega)\hat{\Psi}_n(\omega)}\right)^V \tag{6}$$

$$W_f^\varepsilon(0, t) = \langle f, \varnothing_1 \rangle = \left(\overline{f(\omega)\hat{\varnothing}_{1n}(\omega)}\right)^V \tag{7}$$

The signal is reconstructed, and it is defined in Equation (8).

$$f(t) = W_f^\varepsilon(0, t)\varnothing_1(\omega) + \sum_{n=1}^{N} W_f^\varepsilon(n, t)\Psi_n(t) = (\hat{W}_f^\varepsilon(0, \omega)\hat{\varnothing}_1(\omega) + \sum_{n=1}^{N} \hat{W}_f^\varepsilon(n, \omega)\hat{\Psi}_n(\omega))^V \tag{8}$$

Thus the empirical modes are given in Equations (9) and (10)

$$F_0(t) = W_f^\varepsilon(0, t)\varnothing_1(t) \tag{9}$$

$$F_k(t) = W_f^\varepsilon(k, t)\Psi_k(t) \tag{10}$$

(4) Selecting the best decomposition result based on *FER*

The step is very important and also different from EWT, and the purpose of the step is to find the best decomposition.

Because *coefficient* = 1, 2, ... , *L*, there are $P_1, P_2, \ldots, P_L$, and thus *L* decomposition results can be got. Each result is comprised of some component modes, and the mode corresponding to the biggest *FER* value $FER_{coefficient, max}$ is compared with that corresponding to the second biggest *FER* value $FER_{coefficient, max}$, and the comparison result is denoted in Equation (11).

$$A_{coefficient} = (FER_{coefficient, max} - FER_{coefficient, secondmax})/FER_{coefficient, secondmax} \tag{11}$$

Maximum value of $A_{coefficient}$ is denoted as A_{max}, and thus A_{max} corresponds to the best decomposition result, and the corresponding threshold value is the best decomposition value. Therefore, the mode to $FER_{coefficient, max}$ contains the richest fault feature information.

2.3. The Flowchart of IEWT

Firstly, the power spectrum of the loose slipper fault signal is calculated. Secondly, the threshold processing is applied to eliminate the power spectrum of the interference components, and segments can be acquired. Thirdly, the signal can be decomposed by IEWT based on the segments. Lastly, the best decomposition result can be selected by *FER*.

The flowchart of IEWT is shown in Figure 3.

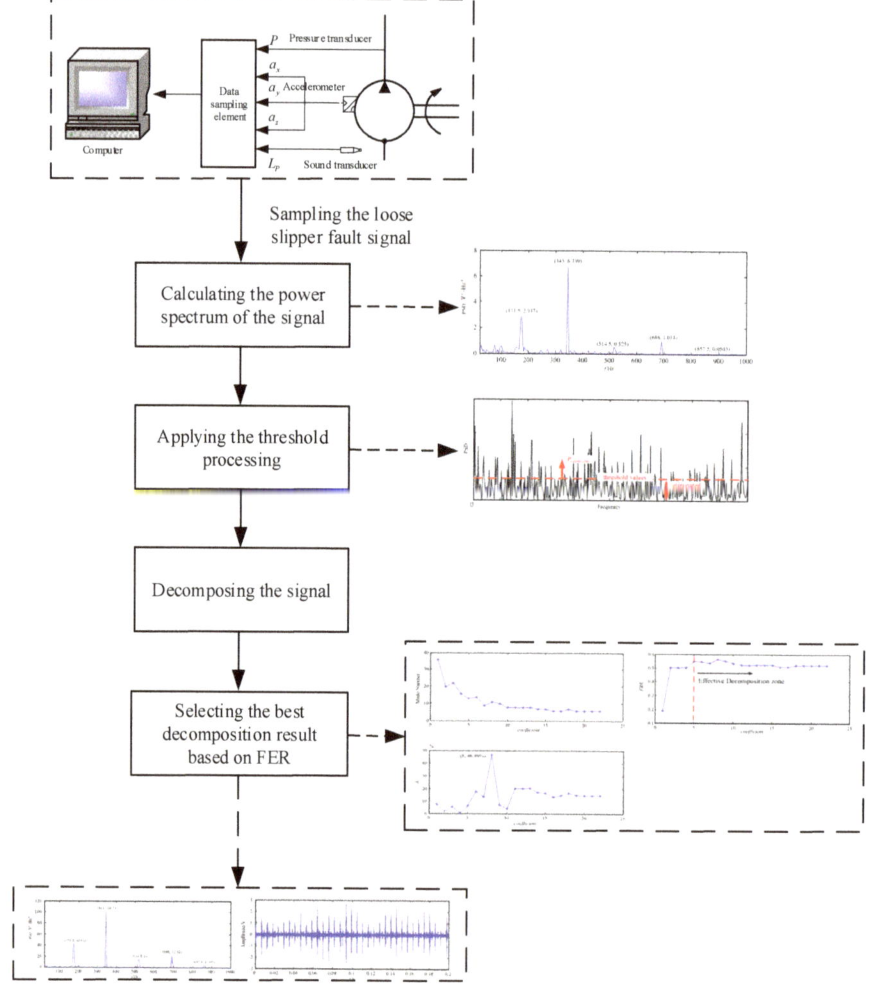

Figure 3. The flowchart of improved empirical wavelet transform (IEWT).

3. Results and Discussion

3.1. Simualtion Study

3.1.1. The Simulated Signal

In order to verify the superiority and effectiveness of IEWT, a simulated signal is defined as

$$x(t) = x_1(t) + x_2(t) \tag{12}$$

The signal consists of two component modes of $x_1(t)$ and $x_2(t)$. $x_1(t)$ is used to simulate an impact signal caused by a fault, and it is an impact signal with periodic exponential attenuation, and its periodicity is 16 Hz, and attenuation function is $e^{-100t}\sin(510\pi t)$ in one periodicity. $x_2(t)$ is a cosine signal, and its periodicity is 20 Hz, and $x_2(t)$ is used to simulate an interference signal of low frequency harmonic. The sampling frequency is 10,240 Hz and sampling time is 1 s.

The signals $x(t)$, $x_1(t)$ and $x_2(t)$ are displayed in Figure 4.

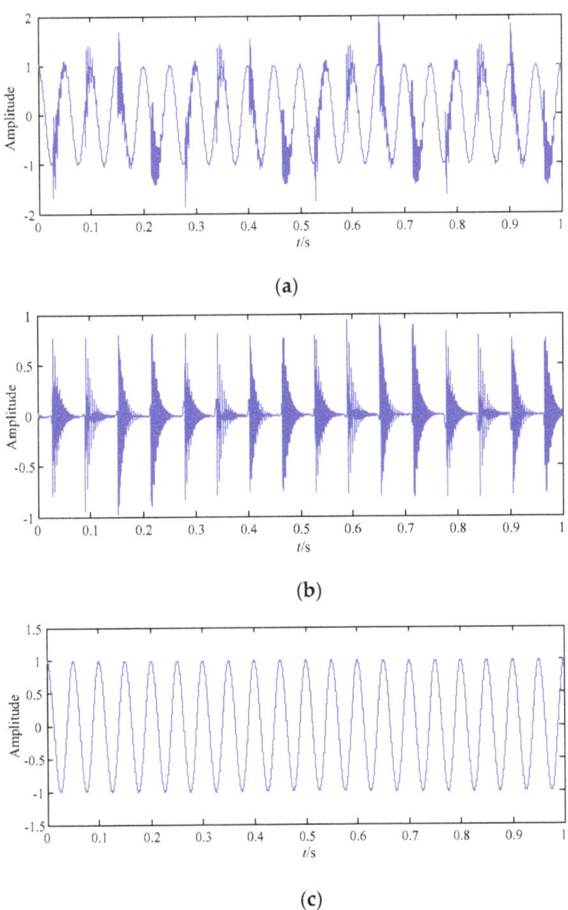

Figure 4. The time domain wave of the simulated signal. (**a**) $x(t)$; (**b**) $x_1(t)$; (**c**) $x_2(t)$.

3.1.2. The Simulated Result Analysis Based on EWT

For demonstrating the superiority and effectiveness of IEWT, the simulated signal is also decomposed by EWT. After the decomposition analysis, three modes can be obtained, which means that there are three contiguous segments and four boundaries. Thus, there is over-decomposition, and the decomposition result is shown in Figures 5 and 6.

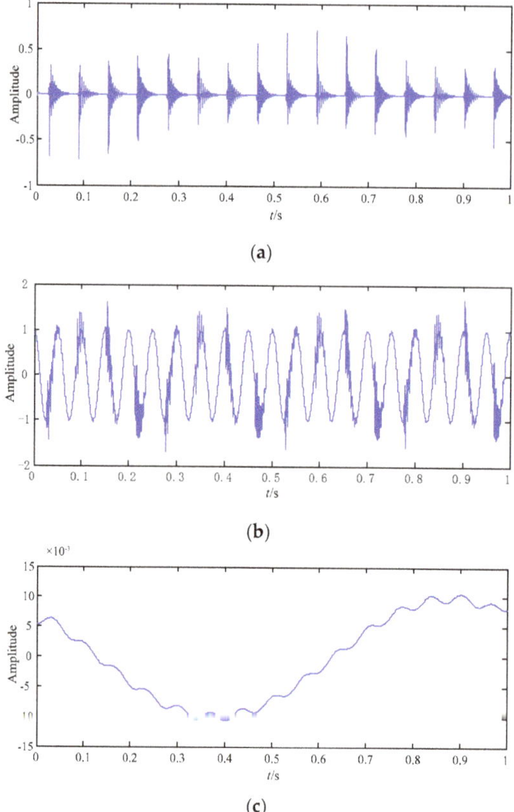

Figure 5. The results of the simulated signal based on empirical wavelet transform (EWT). (**a**) F_3; (**b**) F_2; (**c**) F_1.

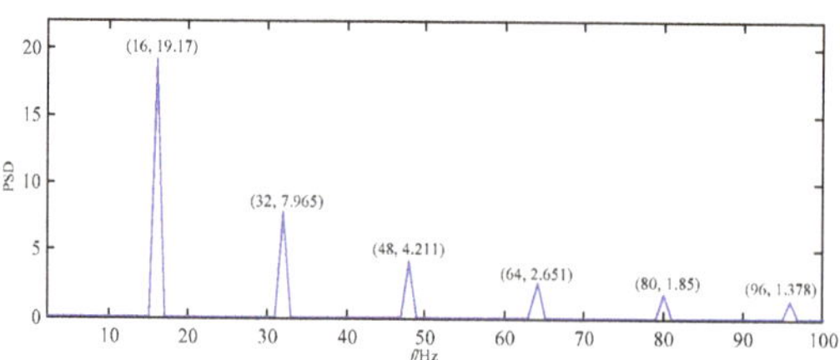

Figure 6. Power spectrum density of F_3.

Figure 5a shows the wave of F_3 in the time domain. Compared with the wave of $x_1(t)$ in Figure 4b, it is shown that the wave of F_3 is distorted. Thus, the periodic impact feature information of F_3 is much different from that of $x_1(t)$.

Figure 5b displays the wave of F_2 in the time domain, and it is seriously interfered with. Compared with the wave of $x(t)$ in Figure 4a, the feature information of F_2 is unexpectedly high similar with that of original signal $x(t)$, but it is known that F_2 is just a mode obtained by EWT, and thus the decomposition is wrong.

Figure 5c depicts the wave of F_1 in the time domain, and its amplitudes are very small. By comparing it with the waves of the three signals of $x(t)$, no signal is the same as the F_1, and its wave is also seriously distorted.

Thus, it is shown that only F_3 is a little similar with the impact signal $x_1(t)$ among the three modes.

Figure 6 shows the spectrum of F_3 in the frequency domain. The fault feature information at fault feature frequency 16 Hz and its harmonics 32 Hz, 48 Hz, 64 Hz, 80 Hz and 96 Hz are all obvious.

It can be concluded from the above analysis that although the F_3 is a little similar with the impact signal $x_1(t)$ in the time domain, there are only two signals in the simulated signal $x(t)$, and there are three modes of F_1, F_2, and F_3 in the decomposition result, and thus $x(t)$ is over-decomposed by EWT.

3.1.3. The Simulated Result Analysis Based on IEWT

The power spectrum of the simulated signal is denoted as $P_{coefficient}$, and threshold value $THVA$ is set as $coefficient \times \mathrm{mean}(P_{coefficient})$, where $coefficient$ is an integer and mean $(P_{coefficient})$ is mean spectrum value of $P_{coefficient}$.

The simulated signal is decomposed by IEWT in [1 122] of the $coefficient$ range, and mode numbers of 122 decomposition results based on different $coefficients$ are displayed in Figure 7.

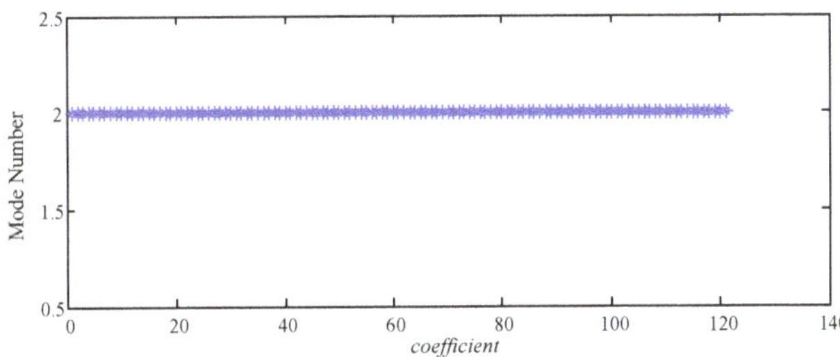

Figure 7. The distribution diagram of mode numbers based on the simulated signal.

It can be seen from Figure 7 that the signal can be effectively and correctly decomposed into only two modes of F_1 and F_2 in each decomposition result, which means that there are two contiguous segments and three boundaries. Thus, there is no over-decomposition and mode mixing, and it is no need to compute comparison between the $FER_{coefficient,\ max}$ and $FER_{coefficient,\ secondmax}$ in Step 4.

In order to get the best decomposition result, FER of each mode is computed based on each of the 122 results. According to the above FER, it can be seen that the FER value of F_2 (the highest-order mode) is the biggest in each result, which means that each F_2 contains the richest fault feature information. FER value of each F_2 in all decomposition results is revealed in Figure 8.

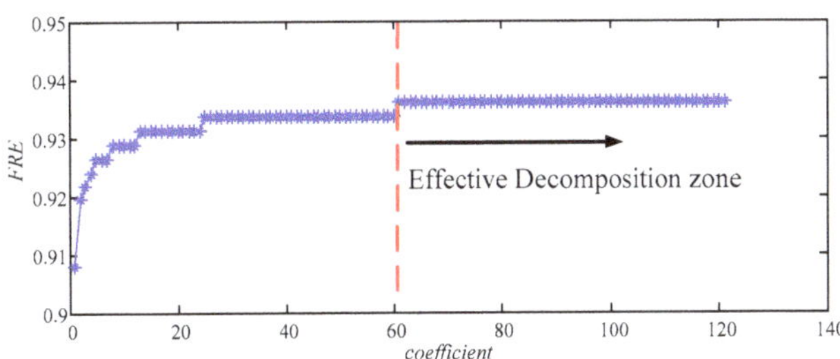

Figure 8. The feature energy ratio (FER) value distribution diagram of all F_2 based on the simulated signal.

It can be known from Figure 8 that FER values in [1 122] of the $coefficient$ range are all above 0.9, and their trend tends to be stable in the range, and all of the biggest values are 0.936 in [61 122]. Thus, the best decomposition result is in [61 122], and all of the results are the same in the range, so only the best result based on $coefficient = 61$ is shown in Figures 9 and 10.

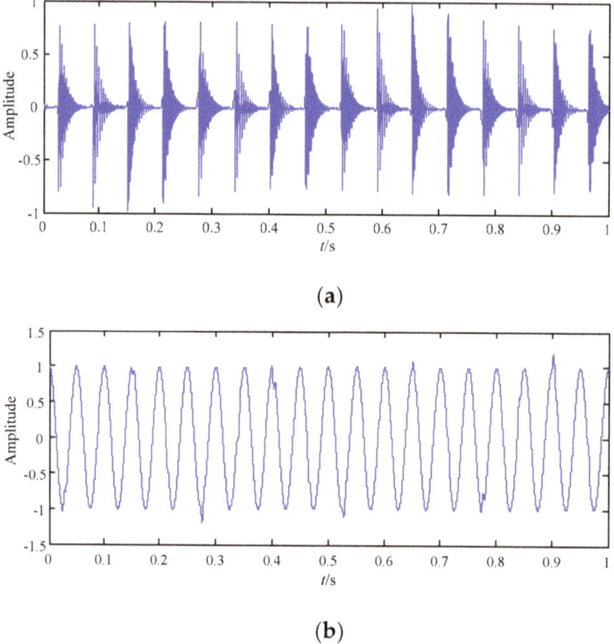

(a)

(b)

Figure 9. The best decomposition results of the simulated signal based on IEWT in the time domain. (a) F_2; (b) F_1.

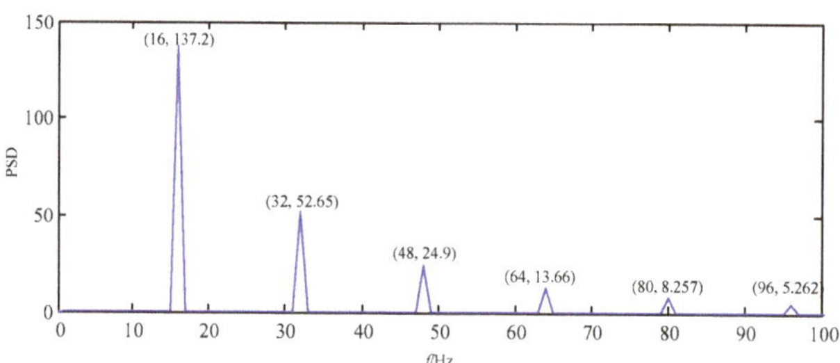

Figure 10. Power spectrum density of F_2.

The wave of F_2 is shown in Figure 9a, and it displays the periodic impact feature information in the time domain. It is compared with the wave of $x_1(t)$ in Figure 4b, and there is nearly no difference between them. Therefore, it is concluded that periodic impact feature information of F_2 is high similar with that of $x_1(t)$ in the time domain.

Figure 9b displays the feature information of cosine curve in F_1, and it is contaminated by little interferences. By comparing the wave of F_1 with that of $x_2(t)$ in Figure 4c, it can be seen that feature information of F_1 is high similar with that of cosine signal $x_2(t)$ in the time domain.

Figure 10 shows the spectrum of F_2 in the frequency domain. The amount of the fault feature information at fault feature frequency 16 Hz and its harmonics 32 Hz, 48 Hz, 64 Hz, 80 Hz and 96 Hz is very large, thus the fault feature information is extracted effectively, and there are nearly no interference components in other frequencies.

It can be known that F_2 got by IEWT and F_3 got by EWT are modes which contain the most of fault feature information, thus the two are compared with each other in the time and frequency domains. The similarity between time waves of F_2 and that of $x_1(t)$ is higher than that between time waves of F_3

and that of $x_1(t)$ in the time domain; the fault feature information amount of F_2 is larger than that of F_3 in the frequency domain.

From the above analysis, it is concluded that the segment can be set according to Fourier power density spectrum in the effective way, and the right mode number can be got, and then one mode which contains the richest feature information can be selected based on *FER*, and thus the simulated signal can be best decomposed by IEWT. Moreover, there is no over-decomposition and mode mixing, so IEWT performs much better than EWT.

3.2. Application to Fault Signals of Hydraulic Pump

3.2.1. Experimental Scheme

An experiment was performed to swash plate axial plunger pump whose type was 10MCY14-1B. Its rotational speed was set as 1470 r/min, and outlet pressure of the pump was set as 15 MPa. The signals of loose slipper fault are sampled by accelerometer a_z at frequency of 50 kHz. The two kinds of fault feature frequency are 171.5 Hz [35]. The experimental system is displayed in Figure 11.

(a)   (b)   (c)

Figure 11. The experiment system of the swash plate axial plunger pump. (**a**) Schematic diagram; (**b**) swash plate axial plunger pump; (**c**) data acquisition equipment.

The length of the loose slipper fault signal is 0.2 s, and the signal is shown in Figure 12.

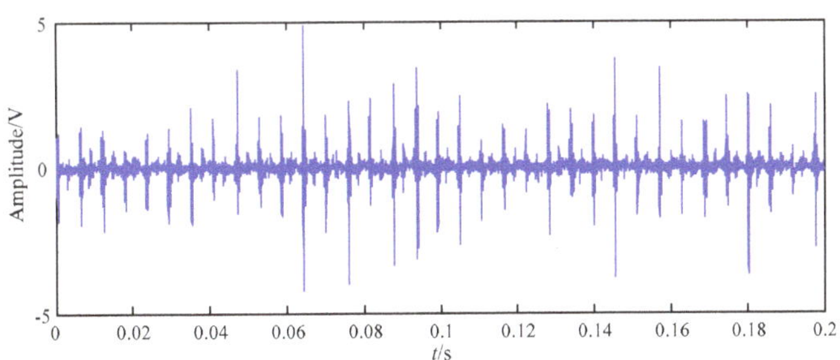

Figure 12. The loose slipper fault signal.

3.2.2. The Application to the Loose Slipper Fault Signal Based on EWT

In order to validate the superiority and effectiveness of IEWT, the loose slipper fault signal is firstly decomposed by EWT, and 58 modes can be obtained, so there is serious over-decomposition. The F_{13} corresponds to maximum *FER* value, thus the mode contains the largest amount of fault feature information, and F_{13} is displayed in Figure 13.

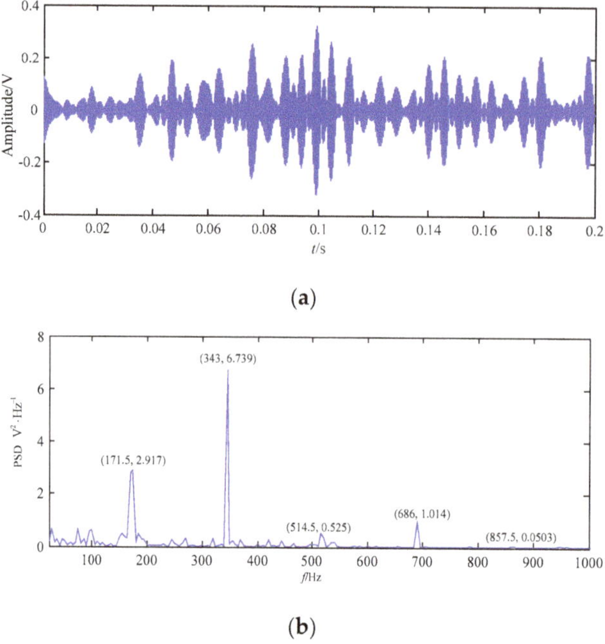

(a)

(b)

Figure 13. F_{13} got based on EWT. (**a**) Time domain wave; (**b**) power spectrum density.

In Figure 13a, the wave of F_{13} is displayed, and the periodic impact feature information is a little obvious, but the periodicity and amplitude information is very irregular. The fault feature information of F_{13} in Figure 13a is low, similar with that of the original loose slipper fault signal in Figure 12.

The spectrum of F_{13} is displayed in Figure 13b. In the frequency domain, the fault feature information at fault feature frequency 171.5 Hz and some of its harmonics is obvious, and there are some noises in the other frequencies.

3.2.3. The Application to the Loose Slipper Fault Signal Based on IEWT

IEWT is also applied to decompose the signal in [1 22] of the *coefficient* range, where the power spectrum of the loose slipper fault signal is denoted as $P_{coefficient}$, the threshold value *THVA* is set as *coefficient* $\times$ mean ($P_{coefficient}$), *coefficient* is an integer, and mean ($P_{coefficient}$) is mean value of $P_{coefficient}$.

The mode number of each decomposition result is obtained based on the different coefficients, as displayed in Figure 14.

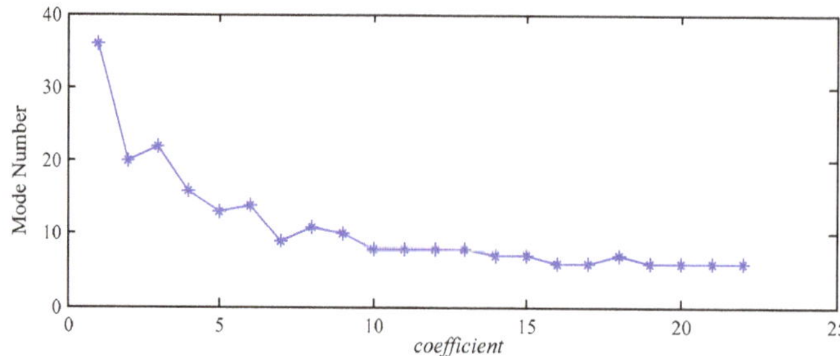

Figure 14. The distribution diagram of mode numbers based on loose slipper fault signal.

From Figure 14, the mode number decreases with the increase of *coefficient* in all results. There are 36 modes in the case of *coefficient* = 1, which means that there are 36 contiguous segments and 37 boundaries. All of the results have six modes in the case of *coefficient* = 19–21. Thus, 36 modes mean that there is over-decomposition, and six modes indicate that there is mode mixing.

For the sake of obtaining the best decomposition result, *FER* of each mode is computed in each of all 22 results and the *FER* value of the highest-order mode is the biggest in each result. The *FER* value of each highest-order mode in all decompositions is displayed in Figure 15.

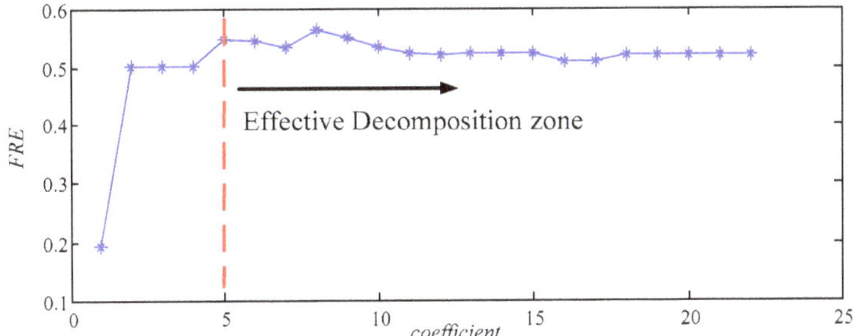

Figure 15. The *FER* value distribution diagram of all highest-order modes based on the loose slipper fault signal.

In the case of *coefficient* ≥ 5, *FER* values change little and maintain at the maximum in Figure 15. Thus, the reasonable decomposition result is in [5 22].

In order to get the best decomposition result, it is necessary to figure out whether there ia over-decomposition and mode mixing in each result. $FER_{coefficient,\,max}$ of the highest-order mode is compared with $FER_{coefficient,\,secondmax}$ of a certain mode in each result. If the above two FERs are very close, there is a real possibility that there is over-decomposition and mode mixing in this result. Comparison result of $FER_{coefficient,\,max}$ and $FER_{coefficient,\,secondmax}$ in each result is demonstrated in Figure 16.

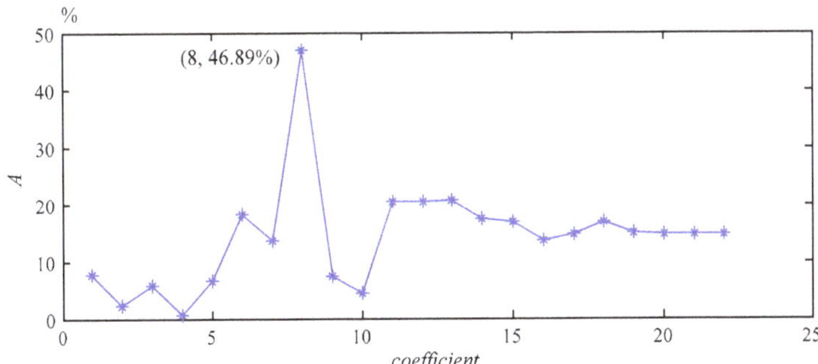

Figure 16. The comparison distribution diagram of FER based on the loose slipper fault signal.

In the case of *coefficient* = 8, it can be found that A_{max} = 46.89% in Figure 16, and it can be also seen that F_{11} is the highest-order mode in Figure 14, which means that there are 11 contiguous segments and 12 boundaries in the result. $FER_{8,\,max}$ corresponding to F_{11} is 0.5644, and $FER_{4,\,max}$ corresponding to F_4 is 0.3842. $FER_{8,\,max}$ is 46.90% bigger than $FER_{4,\,max}$, and thus there is a real possibility that there is no over-decomposition and mode mixing in the case of *coefficient* = 8. The best decomposition result of IEWT can be obtained, as displayed in Figures 17 and 18.

It can be seen from Figure 17 that periodic impact feature information of the highest-order mode F_{11} is more obvious than that of F_1–F_{10}, and that of F_{11} is high similar with that of the original loose slipper fault signal in Figure 12. The spectrum of F_{11} is shown in Figure 18a, the amplitudes at fault feature frequency 171.5 Hz and its harmonics are all obvious; in Figure 18a–k, the amplitudes at the above frequencies are not all extracted in the spectrum of F_1–F_{10}, and there are many interference components in other frequencies. Thus, F_{11} contains the largest amount of fault feature information.

Compared with F_{13} got by EWT in Figure 13a, the periodic impact feature information of F_{11} got by IEWT is very obvious and regular in Figure 17a, and the amplitudes of F_{11} are also much higher than those of F_{13}. By comparing with F_{13} got by EWT in Figure 13b, it can be seen that the amplitudes of F_{11} got by IEWT at the fault feature frequency and its harmonics are all extracted, and the amount of fault feature information is much larger than that of F_{13}.

It can be concluded from the above analysis that the right segment can be obtained according to Fourier power density spectrum, and then the effective mode number is obtained. Based on the mode number, the best decomposition of the loose slipper fault signal can be got by IEWT in the case of *coefficient* = 8, and then the mode which contains the richest fault feature information can be selected based on *FER*. Moreover, it performs better than EWT.

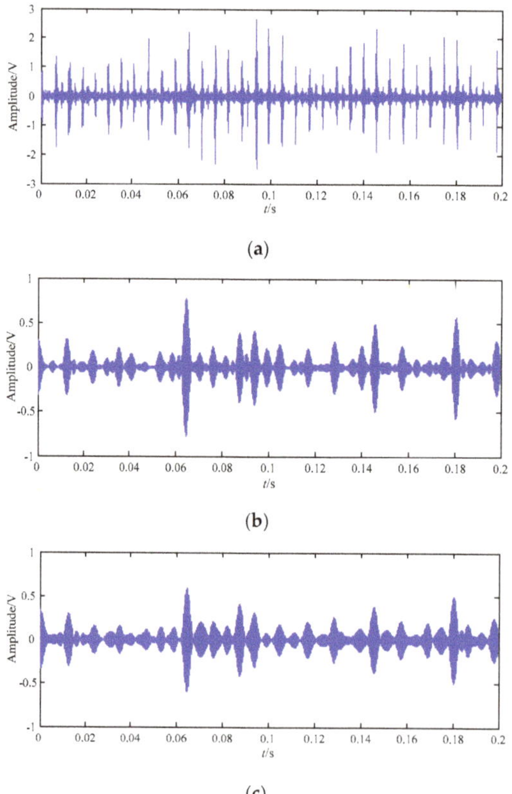

(a)

(b)

(c)

Figure 17. *Cont.*

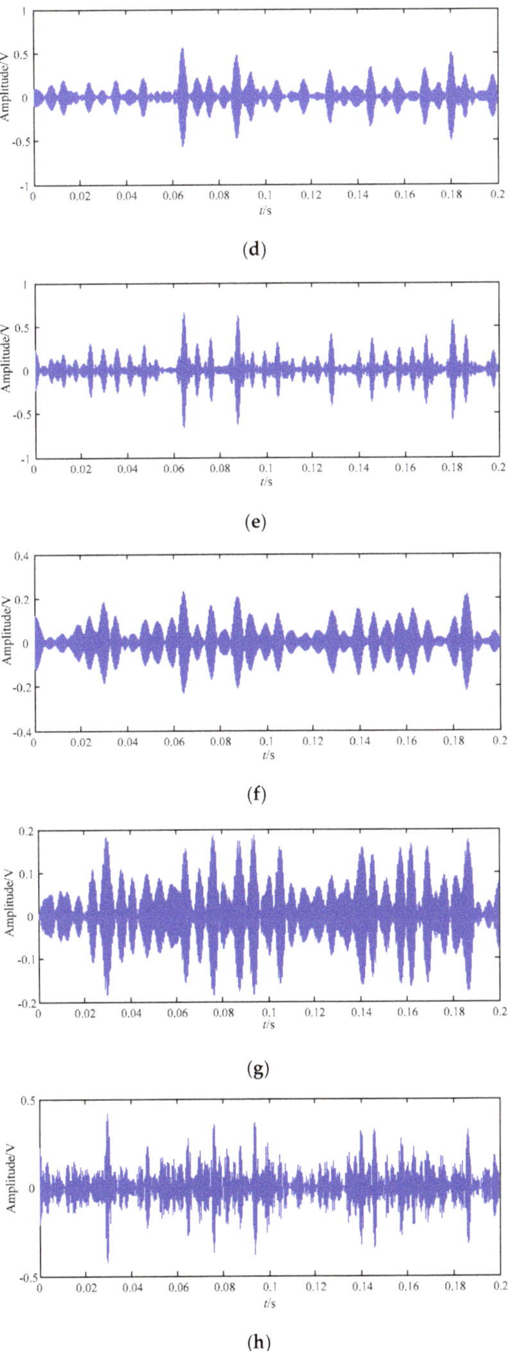

(**d**)

(**e**)

(**f**)

(**g**)

(**h**)

Figure 17. *Cont.*

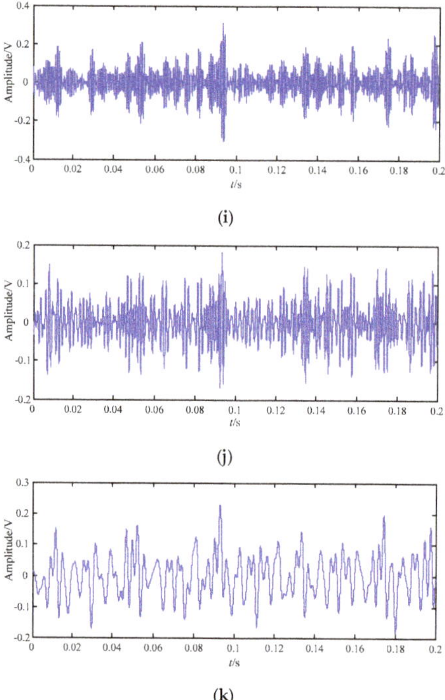

(i)

(j)

(k)

Figure 17. The best decomposition result of loose slipper fault signal based on IEWT in the time domain. (**a**) F_{11}; (**b**) F_{10}; (**c**) F_9; (**d**) F_8; (**e**) F_7; (**f**) F_6; (**g**) F_5; (**h**) F_4; (**i**) F_3; (**j**) F_2; (**k**) F_1.

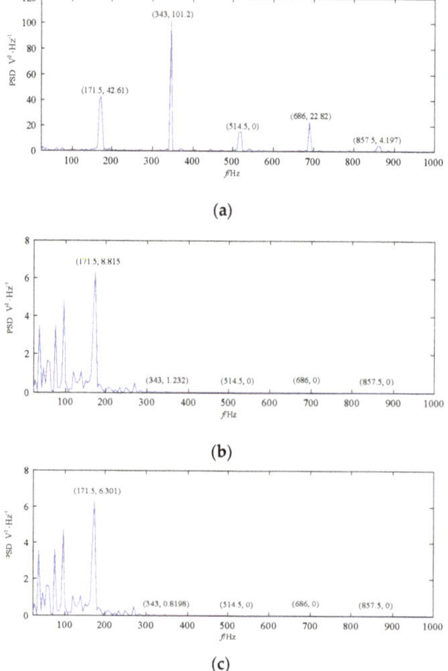

(a)

(b)

(c)

Figure 18. *Cont.*

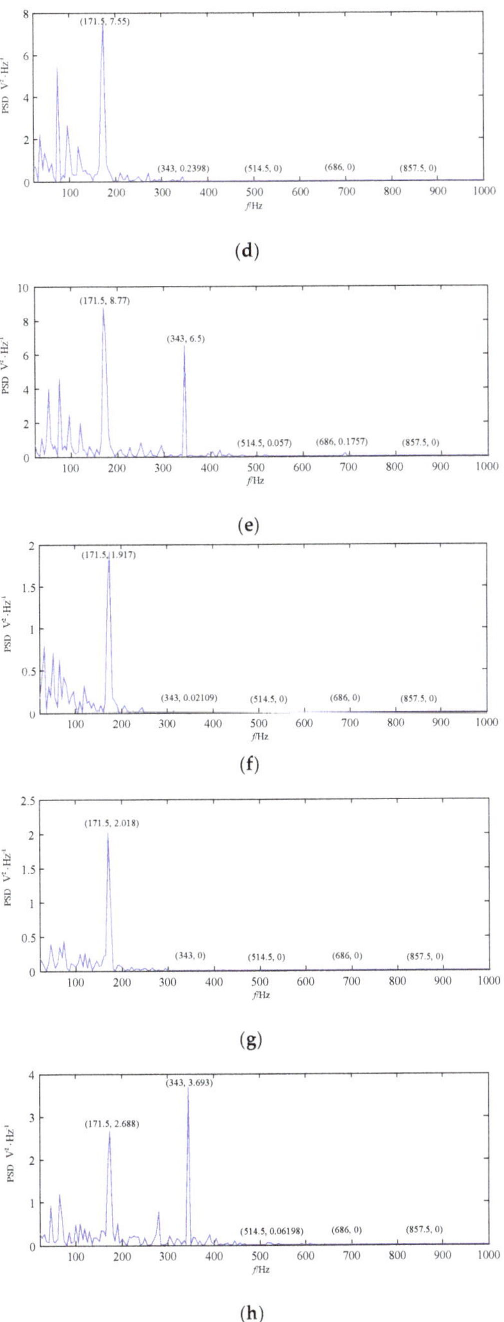

Figure 18. *Cont.*

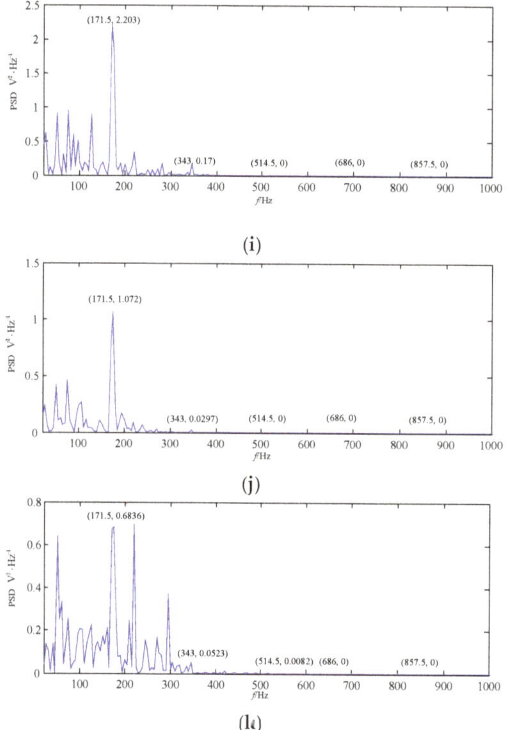

Figure 18. The best decomposition results of the loose slipper fault signal based on IEWT in the frequency domain. (**a**) F_{11}; (**b**) F_{10}; (**c**) F_9; (**d**) F_8; (**e**) F_7; (**f**) F_6; (**g**) F_5; (**h**) F_4; (**i**) F_3; (**j**) F_2; (**k**) F_1.

4. Conclusions

When the hydraulic pump works, it is often faced with of high pressure and high speed working conditions. The vibration of the hydraulic pump is usually caused by mechanical and fluid impact, and the vibration is intensified if it is broken, and thus the fault vibration signal is contaminated by a lot of noises. The Fourier amplitude spectrum is sensitive to the noises, and the segment is got based the above spectrum of the contaminated fault signal in EWT, and thus the signal is decomposed in the wrong way.

Aiming to resolve the shortcomings of EWT, an improved method IEWT is proposed, and IEWT replaced the Fourier amplitude spectrum of EWT with power spectrum in acquiring the segment, and thus the bad influence of the interference on the segment acquirement is much reduced. Based on the right segment, the loose slipper fault signal can be decomposed by IEWT in the best way, and the mode that contains the most amount of the fault feature information can be selected based on *FER*. Therefore, mode-mixing and over-decomposition can be eliminated as much as possible, and IEWT performs much better than EWT.

Author Contributions: Conceptualization, Z.Z., Z.W., Y.Z., and S.T.; Methodology, Z.Z., and Y.Z.; Investigation, Z.Z., Y.Z., and S.T.; Writing-Original Draft Preparation, Z.Z., and Y.Z.; Writing-Review and Editing, Z.W., Z.Z., Y.Z., S.T. and B.W.

Funding: This research was funded by the Startup Foundation for the Doctors of North China University of Science and Technology (0088/28412499), the Cultivation Foundation of North China University of Science and Technology (JP201505), the National Natural Science Foundation of China (51505124, 51805214), and the China Postdoctoral Science Foundation (2019M651722).

Conflicts of Interest: The authors declare no conflict of interest.

Nomenclature

IEWT	improved empirical wavelet transform
EWT	empirical wavelet transform
FER	feature energy ratio
WT	wavelet transform
STFT	short-time Fourier transform
EEMD	ensemble empirical mode decomposition
SVR	support vector regression
CAFs	cyclic autocorrelation functions
WPT	wavelet packet transform
ELM	extreme learning machine
LMD	local mean decomposition
LTSA	local tangent space alignment
FFT	fast Fourier transform
IMF	intrinsic mode function
EMD	empirical mode decomposition
THVA	threshold value
$P_{coefficient}$	a new spectrum distribution got after threshold processing
$FER_{coefficient,\ max}$	the biggest *FER* value in case of the *coefficient* in a decomposition result
$FER_{coefficient,\ secondmax}$	the second biggest *FER* value in case of the *coefficient* in a decomposition result
coefficient	an integer and equals to $1, 2, \ldots, L$
$A_{coefficient}$	comparison result between $FER_{coefficient,\ max}$ and $FER_{coefficient,\ secondmax}$
A_{max}	maximum value of $A_{coefficient}$

References

1. Qian, J.Y.; Chen, M.R.; Liu, X.L.; Jin, Z.J. A numerical investigation of the flow of nanofluids through a micro Tesla valve. *J. Zhejiang Univ. Sci. A* **2019**, *20*, 50–60. [CrossRef]
2. Qian, J.-Y.; Gao, Z.-X.; Liu, B.-Z.; Jin, Z.-J. Parametric Study on Fluid Dynamics of Pilot-Control Angle Globe Valve. *J. Fluids Eng.* **2018**, *140*, 111103. [CrossRef]
3. Zhu, Y.; Qian, P.; Tang, S.; Jiang, W.; Li, W.; Zhao, J. Amplitude-frequency characteristics analysis for vertical vibration of hydraulic AGC system under nonlinear action. *AIP Adv.* **2019**, *9*, 035019. [CrossRef]
4. Zhu, Y.; Tang, S.; Quan, L.; Jiang, W.; Zhou, L. Extraction method for signal effective component based on extreme-point symmetric mode decomposition and Kullback–Leibler divergence. *J. Braz. Soc. Mech. Sci. Eng.* **2019**, *41*, 100. [CrossRef]
5. Wang, C.; Chen, X.; Qiu, N.; Zhu, Y.; Shi, W. Numerical and experimental study on the pressure fluctuation, vibration, and noise of multistage pump with radial diffuser. *J. Braz. Soc. Mech. Sci. Eng.* **2018**, *40*, 481. [CrossRef]
6. Wang, C.; Hu, B.; Zhu, Y.; Wang, X.; Luo, C.; Cheng, L. Numerical Study on the Gas-Water Two-Phase Flow in the Self-Priming Process of Self-Priming Centrifugal Pump. *Processes* **2019**, *7*, 330. [CrossRef]
7. Zhang, J.; Xia, S.; Ye, S.; Xu, B.; Song, W.; Zhu, S.; Tang, H.; Xiang, J. Experimental investigation on the noise reduction of an axial piston pump using free-layer damping material treatment. *Appl. Acoust.* **2018**, *139*, 1–7. [CrossRef]
8. Ye, S.; Zhang, J.; Xu, B.; Zhu, S.; Xiang, J.; Tang, H. Theoretical investigation of the contributions of the excitation forces to the vibration of an axial piston pump. *Mech. Syst. Signal Process.* **2019**, *129*, 201–217. [CrossRef]
9. Lan, Y.; Hu, J.; Huang, J.; Niu, L.; Zeng, X.; Xiong, X.; Wu, B. Fault diagnosis on slipper abrasion of axial piston pump based on Extreme Learning Machine. *Measurement* **2018**, *124*, 378–385. [CrossRef]
10. Sun, H.; Yuan, S.; Luo, Y. Cyclic Spectral Analysis of Vibration Signals for Centrifugal Pump Fault Characterization. *IEEE Sens. J.* **2018**, *18*, 2925–2933. [CrossRef]
11. Zhao, Z.; Jia, M.; Wang, F.; Wang, S. Intermittent chaos and sliding window symbol sequence statistics-based early fault diagnosis for hydraulic pump on hydraulic tube tester. *Mech. Syst. Signal Process.* **2009**, *23*, 1573–1585. [CrossRef]

12. Du, J.; Wang, S.; Zhang, H. Layered clustering multi-fault diagnosis for hydraulic piston pump. *Mech. Syst. Signal Process.* **2013**, *36*, 487–504. [CrossRef]
13. Lu, C.; Wang, S.; Makis, V. Fault severity recognition of aviation piston pump based on feature extraction of EEMD paving and optimized support vector regression model. *Aerosp. Sci. Technol.* **2017**, *67*, 105–117. [CrossRef]
14. Sharma, P.; Newman, K.; Long, C.; Gasiewski, A.; Barnes, F. Use of wavelet transform to detect compensated and decompensated stages in the congestive heart failure patient. *Biosensors* **2017**, *7*, 40. [CrossRef]
15. Zin, Z.M.; Salleh, S.H.; Daliman, S.; Sulaiman, M.D. Analysis of heart sounds based on continuous wavelet transform. In Proceedings of the IEEE Student Conference on Research and Development, Putrajaya, Malaysia, 25–26 August 2003.
16. Tsalaile, T.; Sanei, S. Separation of heart sound signal from lung sound signal by adaptive line enhancement. In Proceedings of the European Signal Processing Conference, Poznan, Poland, 3–7 September 2007.
17. El-Asir, B.; Khadra, L.; Al-Abbasi, A.H.; Mohammed, M.M.J. Time-frequency analysis of heart sounds. In Proceedings of the Digital Processing Applications (TENCON '96), Perth, WA, Australia, 26–29 November 1996.
18. Li, Y.; Liang, X.; Xu, M.; Huang, W. Early fault feature extraction of rolling bearing based on ICD and tunable Q-factor wavelet transform. *Mech. Syst. Signal Process.* **2017**, *86*, 204–223. [CrossRef]
19. Singh, J.; Darpe, A.; Singh, S.; Singh, S. Rolling element bearing fault diagnosis based on Over-Complete rational dilation wavelet transform and auto-correlation of analytic energy operator. *Mech. Syst. Signal Process.* **2018**, *100*, 662–693. [CrossRef]
20. Kordestani, M.; Samadi, M.F.; Saif, M.; Khorasani, K. A New Fault Diagnosis of Multifunctional Spoiler System Using Integrated Artificial Neural Network and Discrete Wavelet Transform Methods. *IEEE Sens. J.* **2018**, *18*, 4990–5001. [CrossRef]
21. Zhao, M.; Kang, M.; Tang, B.; Pecht, M. Deep residual networks with dynamically weighted wavelet coefficients for fault diagnosis of planetary gearboxes. *IEEE Trans. Ind. Electron.* **2018**, *65*, 4290–4300. [CrossRef]
22. Song, L.; Wang, H.; Chen, P. Vibration-Based Intelligent Fault Diagnosis for Roller Bearings in Low-Speed Rotating Machinery. *IEEE Trans. Instrum. Meas.* **2018**, *67*, 1887–1899. [CrossRef]
23. Huang, N.E. The Empirical Mode Decomposition and the Hilbert Spectrum for Nonlinear and Non-stationary Time Series Analysis. *Proc. Math. Phys. Eng. Sci.* **1998**, *454*, 903–995. [CrossRef]
24. Yang, W.; Court, R.; Tavner, P.J.; Crabtree, C.J. Bivariate empirical mode decomposition and its contribution to wind turbine condition monitoring. *J. Sound Vib.* **2011**, *330*, 3766–3782. [CrossRef]
25. Liu, H.; Zhang, J.; Cheng, Y.; Lu, C. Fault diagnosis of gearbox using empirical mode decomposition and multi-fractal detrended cross-correlation analysis. *J. Sound Vib.* **2016**, *385*, 350–371. [CrossRef]
26. Osman, S.; Wang, W. An enhanced Hilbert–Huang transform technique for bearing condition monitoring. *Meas. Sci. Technol.* **2013**, *24*, 085004. [CrossRef]
27. Li, Y.; Xu, M.; Liang, X.; Huang, W. Application of Bandwidth EMD and Adaptive Multiscale Morphology Analysis for Incipient Fault Diagnosis of Rolling Bearings. *IEEE Trans. Ind. Electron.* **2017**, *64*, 6506–6517. [CrossRef]
28. Zhang, J.; Huang, D.; Yang, J.; Liu, H.; Liu, X. Realizing the empirical mode decomposition by the adaptive stochastic resonance in a new periodical model and its application in bearing fault diagnosis. *J. Mech. Sci. Technol.* **2017**, *31*, 4599–4610. [CrossRef]
29. Pan, H.; Yang, Y.; Li, X.; Zheng, J.; Cheng, J. Symplectic geometry mode decomposition and its application to rotating machinery compound fault diagnosis. *Mech. Syst. Signal Process.* **2019**, *114*, 189–211. [CrossRef]
30. Gilles, J. Empirical Wavelet Transform. *IEEE Trans. Signal Process.* **2013**, *61*, 3999–4010. [CrossRef]
31. Wang, D.; Zhao, Y.; Yi, C.; Tsui, K.-L.; Lin, J. Sparsity guided empirical wavelet transform for fault diagnosis of rolling element bearings. *Mech. Syst. Signal Process.* **2018**, *101*, 292–308. [CrossRef]
32. Jiang, X.; Wu, L.; Ge, M. A Novel Faults Diagnosis Method for Rolling Element Bearings Based on EWT and Ambiguity Correlation Classifiers. *Entropy* **2017**, *19*, 231. [CrossRef]
33. Cao, H.; Fan, F.; Zhou, K.; He, Z. Wheel-bearing fault diagnosis of trains using empirical wavelet transform. *Measurement* **2016**, *82*, 439–449. [CrossRef]

34. Shen, L.; Zhou, X.; Zhang, W. Application of morphological demodulation in gear fault feature extraction. *J. Zhejiang Univ. Eng. Sci.* **2010**, *44*, 1514–1519.
35. Jiang, W.; Zheng, Z.; Zhu, Y.; Li, Y. Demodulation for hydraulic pump fault signals based on local mean decomposition and improved adaptive multiscale morphology analysis. *Mech. Syst. Signal Process.* **2015**, *58*, 179–205. [CrossRef]

Article

Natural Frequency Sensitivity Analysis of Fire-Fighting Jet System with Adaptive Gun Head

Xiaoming Yuan [1,2], Xuan Zhu [1,2], Chu Wang [1,2,*], Lijie Zhang [1,2] and Yong Zhu [3,*]

[1] Hebei Key Laboratory of Heavy Machinery Fluid Power Transmission and Control, Yanshan University, Hebei, Qinhuangdao 066004, China; yuanxiaoming@ysu.edu.cn (X.Y.); zx00712@126.com (X.Z.); zhangljys@126.com (L.Z.)

[2] State Key Laboratory of Fluid Power & Mechatronic Systems, Zhejiang University, Zhejiang, Hangzhou 310027, China

[3] National Research Center of Pumps, Jiangsu University, Zhenjiang 212013, China

*   Correspondence: wangchu@stumail.ysu.edu.cn (C.W.); zhuyong@ujs.edu.cn (Y.Z.)

Received: 14 October 2019; Accepted: 28 October 2019; Published: 4 November 2019

Abstract: The gun head is the end effector of the fire-fighting jet system. Compared with a traditional fixed gun head, an adaptive gun head has the advantages of having an adjustable nozzle opening, a wide applicable flow range, and a high fire-extinguishing efficiency. Thus, the adaptive gun head can extinguish large fires quickly and efficiently. The fire-fighting jet system with an adaptive gun head has fluid-structure interaction and discrete-continuous coupling characteristics, and the influence of key design parameters on its natural frequencies needs to be determined by a sensitivity analysis. In this paper, the dynamic model and equations of the jet system were established based on the lumped parameter method, and the sensitivity calculation formulas of the natural frequency of the jet system to typical design parameters were derived. Natural frequencies and mode shapes of the jet system were determined based on a mode analysis. The variation law of the sensitivity of the natural frequency of the jet system to typical design parameters was revealed by the sensitivity analysis. The results show that the fluid mass inside the spray core within a certain initial gas content is the most important factor affecting the natural frequency of the jet system. There was only a 0.51% error between the value of the first-order natural frequency of the jet system determined by the modal experiment and the theoretical one, showing that good agreement with the first-order natural frequency of the jet system was found. This paper provides a theoretical basis for the dynamic optimization design of the adaptive gun head of the fire water monitor.

Keywords: flow control; adaptive gun head; jet system; natural frequency; mode shape; sensitivity analysis

---

1. Introduction

The fire-fighting monitor is an important fire-fighting equipment for long-range fire extinguishing and has become an important part of fire-extinguishing systems in large-scale places [1–3]. According to different spray media, the fire-fighting monitor can be divided into two series: a fire water monitor and a fire foam monitor. Structurally, the fire-fighting monitor is mainly composed of a barrel and a gun head. The barrel mainly includes components such as a motor, a reducer, and an elbow, which can realize horizontal and pitching rotation of the jet direction of the fire-fighting monitor. The gun head is the key component that converts the pressure energy of the fluid into kinetic energy and its working principle is similar to that of a valve component in a hydraulic transmission system [4–6]. Compared with a traditional fixed gun head, the adaptive gun head can automatically adjust the nozzle opening through an adaptive mechanism when the flow and pressure of the fluid change at the entrance of the gun head, so that the performance of the jet system can be better in a wider flow range and fire extinguishing efficiency can be higher.

A sensitivity analysis is a method to study the sensitivity of the system in response to changes in its design parameters, through which the influence degree of design variables on output variables can be sorted. In the process of structural optimization design and parameter identification, the sensitivity analysis can be used to identify important parameters while ignoring unimportant variables, thereby significantly reducing the dimensionality of the design space, providing a basis for identifying weak points in the system [7,8]. When the functional relationship between inputs and outputs is known, the sensitivity of the system can be calculated by the partial derivative of the response function to the design variable. This direct calculation method is generally applicable when the design factor is few, the structure is not complicated, and the sensitivity differential equation is easy to derive. In this case, the method has the characteristics of simplified calculating and can resolve the essential relationship between inputs and outputs. When it is difficult or impossible to determine the analytical relationship between inputs and outputs, the Monte Carlo method, orthogonal test, Kriging, and other modern approximation design methods are generally used to determine the implicit functional relationship between design variables and responses; then the gradient method and the genetic algorithm can be used to calculate the sensitivity of the system. This method of indirect calculation is generally applicable to the case of many design factors, complicated structures, and complex nonlinear relationship between input and output parameters. The method is characterized by a large computational workload, and the error of the determined sensitivity is positively correlated with sample capacity [9–11]. The sensitivity analysis is widely used in various fields such as chemistry [12,13], construction [14,15], environmental protection [16], transportation [17], and machinery [18–21].

The fire-fighting jet system with the adaptive gun head consists of an adaptive gun head, a barrel, a connecting pipe, and its internal fluid. The analysis process of the fluid-structure interaction and discrete-continuous coupling problems in the jet system should be analogized to that of the influencing factors of the dynamic characteristics of valve components [22,23]. Through the sensitivity analysis, the key design parameters affecting the dynamic characteristics of the fire-fighting jet system are discussed. The fire-fighting jet system is subjected to the pressure pulsation of the power element pump in operation, so that the jet system inevitably generates vibration and noise [24–26]. When an adaptive fire-fighting monitor is used instead of a conventional fire-fighting monitor, the stiffness of the end of the jet system is reduced, increasing the tendency of the jet system to resonate. Therefore, the sensitivity analysis can be used to explore the parameters affecting the modal characteristics of the jet system, which can provide theoretical reference for the dynamic optimization design of the adaptive fire-fighting monitor.

In this paper, based on the fluid-structure interaction and discrete-continuous coupling characteristics of the jet system, the dynamic model and equations of the free vibration of the jet system will be established. The sensitivity calculation formulas of the modal characteristics of the jet system to typical design parameters will be derived. The influence law of the modal characteristics of the jet system for each parameter will be analyzed, and the first-order natural frequency of the jet system will be verified according to the modal experiment.

2. Establishment of the Dynamic Model and Equations of the Jet System

The structure of the gun head of the traditional diversion fire-fighting monitor is shown in Figure 1. The structure of the gun head of the new adaptive fire-fighting monitor is shown in Figure 2. The traditional diversion gun head can achieve both straight and spray jets by changing the relative positions of the inner and outer nozzles to meet the fire suppression requirements of different occasions. However, since the internal components are fixedly connected, the opening of the nozzles remains unchanged during the working process. Therefore, when the jet flow changes, the nozzle opening does not match the flow rate, and the fire extinguishing efficiency is low. In contrast, the new adaptive gun head adds an adaptive mechanism consisting of the spray core 9, the end cap 10, the core rod 11, and the spring 12 on the basis of the traditional diversion gun head. When the inlet flow of the gun head increases, the pressure on the left side of the spray core of the adaptive mechanism increases. When the force of the spray core received by the fluid is greater than that of the spring, the spray core moves to the right and the nozzle opening increases. Conversely, when the inlet flow rate is decreased, the spray core moves to the left and the nozzle opening decreases. The adaptive mechanism enables the gun head to automatically adjust the nozzle opening according to changes in inlet flow and pressure, thereby achieving good jet performance under various conditions and extinguishing large fires quickly and effectively.

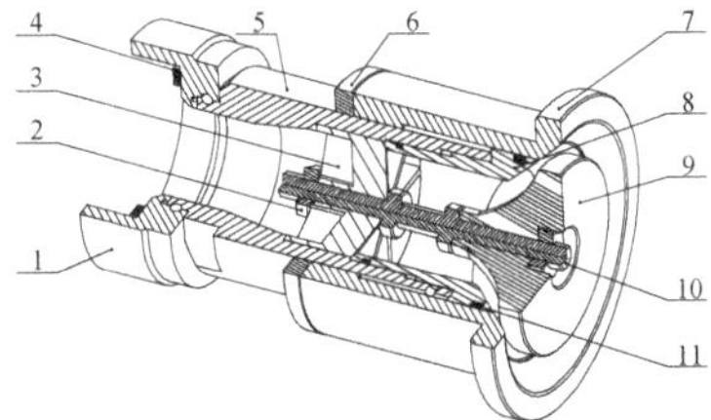

Figure 1. Structure of a traditional diversion gun head. 1. Joint; 2. Nut; 3. Regulator; 4. Gasket; 5. Enclosure; 6. Ring; 7. Outer nozzle; 8. Inner nozzle; 9. Spray core; 10. End cap; 11. Core rod.

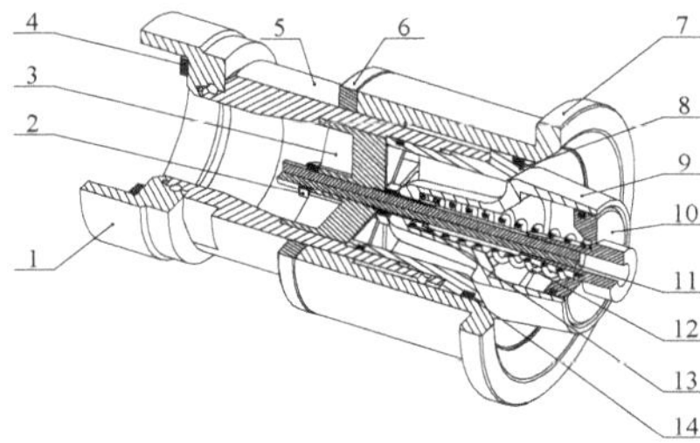

Figure 2. Structure of the new adaptive gun head. 1. Joint; 2. Nut; 3. Regulator; 4. Gasket; 5. Enclosure; 6. Ring; 7. Outer nozzle; 8. Inner nozzle; 9. Spray core; 10. End cap; 11. Core rod; 12. Spring; 13. Core sleeve; 14. Seal ring.

The new adaptive gun head shown in Figure 2 is mounted on the horizontally-rotating and pitch-turning barrel, and combined with the pipes connected to the barrel. The structure above constitutes the fire-fighting jet system with the adaptive gun head studied in this paper, as shown in Figure 3.

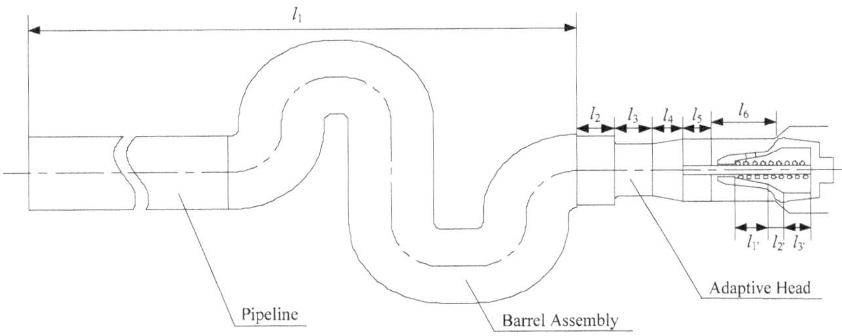

Figure 3. Structure of a fire-fighting jet system with the adaptive gun head.

The fire-fighting jet system with the adaptive gun head has typical fluid-structure interaction characteristics due to the interaction between the fluid, spray core, and spring in the working process.

Moreover, the fluid-structure interaction between the fluid unit and the solid unit in the jet system only occurs at the interface between the two. The interaction between the units in the jet system is shown in Figure 4.

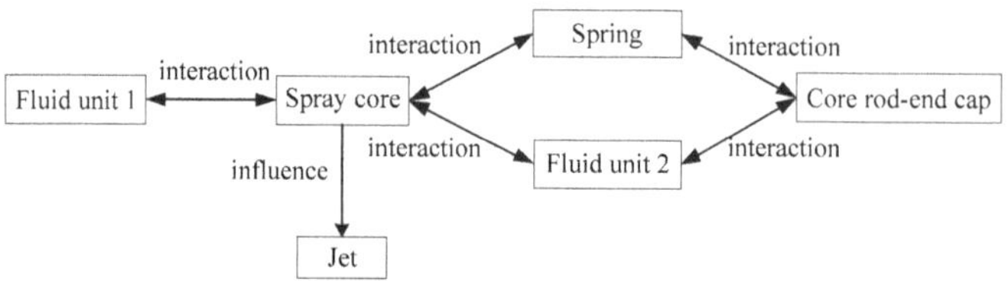

Figure 4. Interaction between the units in the fire-fighting jet system with the adaptive gun head.

In Figure 4, fluid unit 2 is the entire fluid volume enclosed by the inside of the spray core and the core rod-end cap structure. Fluid unit 1 is all fluids in the gun head, the barrel, and the pipes connected to the barrel except fluid unit 2. The fluid units and solid units in the jet system are coupled to each other and interact to make the dynamic modeling and analysis more complicated. To facilitate theoretical analyses, the established overall dynamic model of the jet system follows the following assumptions:

(1) The jet system is modeled based on the lumped parameter method. It is considered that the density, stiffness, pressure, and other attribute parameters of each fluid unit in the jet system are evenly distributed in the control volume and equal everywhere;
(2) Except for the fluid units and the spring, it is considered that the parts such as the spray core, the gun head enclosure, the barrel, and the outer wall of the pipe are rigid bodies, regardless of their deformation;
(3) The spray core, the core rod-end cap structure, and each fluid unit are only affected by the axial force, and the force of the fluid units on the solid units is equivalent to the axial linear spring force;
(4) The damping between the fluid units and the solid units is equivalent to axial linear damping;
(5) Ignore the processing and installation errors of each component.

In the fire-fighting jet system with the adaptive gun head, the inside spray core, fluid unit 1, and fluid unit 2 can be viewed as discrete bodies, and the core rod-end cap structure can be regarded as a continuum due to the particularity of its shape. Therefore, the jet system has not only fluid-structure interaction characteristics, but also discrete-continuous coupling characteristics. The discrete-continuous coupling characteristics in the jet system will increase the difficulty of dynamic modeling and analysis, so the core rod-end cap structure needs to be equally discrete, thus simplifying the problem.

Based on the above analysis and assumptions, the dynamic model of the fire-fighting jet system with adaptive gun head is shown in Figure 5. Among them, the structure consisting of m_1, m_2, m_3, m_4, k_1, k_2, k_3, and k_4 is the equivalent discrete model of the core rod-end cap structure. m_1, m_2, and m_3 are 1/4 of the mass of the core rod, m_4 is the sum of the mass of the 1/4 core rod and the mass of the end cap, and k_1, k_2, k_3, and k_4 are equal in size. m_5, m_6, and m_7 are the mass of fluid unit 2, the spray core, and fluid unit 1, respectively. k_{f1}, k_{f21}, and k_{f22} are fluid stiffnesses. k_{f1} is the stiffness of fluid unit 1. k_{f21} and k_{f22} are numerically equal, and the total stiffness after the parallel connection is the stiffness of fluid unit 2. k_5 is the stiffness of the spring in the adaptive mechanism of the gun head. c_1 is the damping of fluid unit 1. c_2 and c_3 are numerically equal, and the total damping after the parallel connection is the sum of the damping of fluid unit 2 and the damping of the spray core's orifice structure.

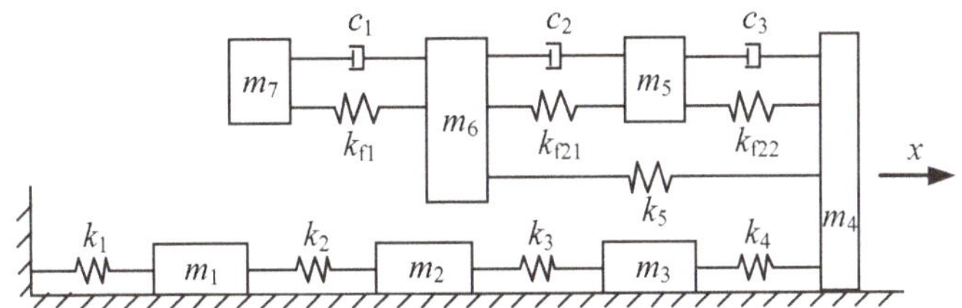

Figure 5. Dynamic model of the fire-fighting jet system with the adaptive gun head.

Based on the dynamic model of the fire-fighting jet system with the adaptive gun head, the free vibration equation of the jet system is established as

$$m\ddot{x} + kx = 0,\qquad(1)$$

where the mass matrix of the jet system is

$$m = \mathrm{diag}(m_1, m_2, m_3, m_4, m_5, m_6, m_7),\qquad(2)$$

where the stiffness matrix of the jet system is

$$k = \begin{bmatrix} k_1 + k_2 & -k_2 & 0 & 0 & 0 & 0 & 0 \\ -k_2 & k_2 + k_3 & -k_3 & 0 & 0 & 0 & 0 \\ 0 & -k_3 & k_3 + k_4 & -k_4 & 0 & 0 & 0 \\ 0 & 0 & -k_4 & k_4 + k_{f22} + k_5 & -k_{f22} & -k_5 & 0 \\ 0 & 0 & 0 & -k_{f22} & k_{f21} + k_{f22} & -k_{f21} & 0 \\ 0 & 0 & 0 & -k_5 & -k_{f21} & k_{f21} + k_5 + k_{f1} & -k_{f1} \\ 0 & 0 & 0 & 0 & 0 & -k_{f1} & k_{f1} \end{bmatrix}.\qquad(3)$$

In the stiffness matrix of the jet system, the stiffness of each fluid unit is calculated equivalently. By the definition of stiffness, there is

$$k_f = -\frac{\Delta F}{\Delta L} = -\frac{S_a \Delta p}{\Delta L},\qquad(4)$$

where ΔF is the amount of load fluctuation of the fluid unit, ΔL is the axial length change of the fluid unit, S_a is the average area of the flow cross section of the fluid unit, Δp is the change of fluid pressure, and L is the axial length of the fluid unit.

The volume of the fluid unit obtained from the average area of the flow cross section of the fluid unit and the axial length of the fluid unit is

$$V = S_a L.\qquad(5)$$

The bulk elastic modulus of the fluid in the fire-fighting jet system with the adaptive gun head is calculated according to the theory of bulk elastic modulus described in [27]. Let the bulk elastic modulus of the fluid be B_f, which is defined as

$$B_f = -\frac{V \Delta p}{\Delta V},\qquad(6)$$

where V is the total volume of the fluid, and ΔV is the volume change of the fluid.

For Equations (4)–(6), the fluid equivalent stiffness can be expressed by the bulk elastic modulus of the fluid:

$$k_f = \frac{B_f S_a}{L}.\qquad(7)$$

As can be seen from Figure 3, the shapes of fluid unit 1 and fluid unit 2 are relatively complicated. In order to improve the calculation accuracy of the fluid equivalent stiffness, fluid unit 1 is divided into six segments and fluid unit 2 is divided into three segments according to the shape of the structure. The first section of fluid unit 1 is a pipe connected to the fire-fighting monitor's barrel and the inlet end of the barrel. Parameters of each section of the two fluid units are shown in Table 1. The equivalent stiffness of each fluid unit subsection can be calculated separately from Equation (7).

Table 1. Data for each segment size of the fluid units.

Fluid Unit	Segmentation	Equivalent Stiffness of Fluid Unit Subsection	Average Area of Flow Cross Section S_{ai}/(mm^2)	Axial Length of the Fluid Domain L_i/(mm)
Fluid unit 1	l_1	k_{f11}	6221	1534.6
	l_2	k_{f12}	6200	38
	l_3	k_{f13}	3200	39
	l_4	k_{f14}	3800	35.5
	l_5	k_{f15}	4400	38
	l_6	k_{f16}	3700	64
Fluid unit 2	$l_{1'}$	$k_{f21'}$	300	46.1
	$l_{2'}$	$k_{f22'}$	800	11
	$l_{3'}$	$k_{f23'}$	2800	40.5

The equivalent stiffness of fluid unit 1 and fluid unit 2 can be obtained from the series relationship of the equivalent fluid springs of the fluid unit subsections, where the equivalent stiffness of fluid unit 1 is

$$\frac{1}{k_{f1}} = \frac{1}{k_{f11}} + \frac{1}{k_{f12}} + \frac{1}{k_{f13}} + \frac{1}{k_{f14}} + \frac{1}{k_{f15}} + \frac{1}{k_{f16}}. \tag{8}$$

The equivalent stiffness of fluid unit 2 is

$$\frac{1}{k_{f2}} = \frac{1}{k_{f21'}} + \frac{1}{k_{f22'}} + \frac{1}{k_{f23'}}. \tag{9}$$

Since k_{f21} and k_{f22} are connected in parallel to be the equivalent stiffness of fluid unit 2 and they are equal in size, then:

$$k_{f21} = k_{f22} = \frac{k_{f2}}{2}. \tag{10}$$

3. Derivation of the Sensitivity Formula of the Jet System

The free vibration differential equation of the fire-fighting jet system with the adaptive gun head is shown in Equation (1), and its corresponding characteristic equation is

$$(k - \omega_{ni}^2 m)\phi_i = 0, \tag{11}$$

where ω_{ni} is natural angular frequency (rad/s), ϕ_i is the principal mode.

Equation (11) left multiplies ϕ_i^T, then

$$\phi_i^T(k - \omega_{ni}^2 m)\phi_i = 0. \tag{12}$$

Assuming that the design variable t, the natural frequency ω_{ni}, the stiffness matrix k, and the mass matrix m are derivable and ω_{ni} is a single eigenvalue, then ω_{ni} is differentiable with respect to design variables. The first-order partial derivatives of both sides of Equation (12) with respect to the design variable t are

$$
\frac{\partial \phi_i^{\mathrm{T}}}{\partial t}\left(k - \omega_{ni}^2 m\right)\phi_i + \phi_i^{\mathrm{T}}\left(\frac{\partial k}{\partial t} - \frac{\partial(\omega_{ni}^2 m)}{\partial t}\right)\phi_i +
$$
$$
\phi_i^{\mathrm{T}}\left(k - \omega_{ni}^2 m\right)\frac{\partial \phi_i}{\partial t} = 0
$$
(13)

Since $\left(k - \omega_{ni}^2 m\right)$ is a real symmetric matrix, ϕ_i^{T} is also the eigenvalue of the characteristic Equation (11), so

$$
\phi_i^{\mathrm{T}}\left(k - \omega_{ni}^2 m\right) = 0.
$$
(14)

Substituting Equations (11) and (14) into Equation (13), then

$$
\phi_i^{\mathrm{T}}\left(\frac{\partial k}{\partial t} - 2\omega_{ni}\frac{\partial \omega_{ni}}{\partial t}m - \omega_{ni}^2\frac{\partial m}{\partial t}\right)\phi_i = 0.
$$
(15)

The formula for calculating the sensitivity of each order natural frequency ω_{ni} to the design variable t from Equation (15) is

$$
\frac{\partial \omega_{ni}}{\partial t} = \frac{1}{2m_{pi}}\left(\frac{1}{\omega_{ni}}\phi_i^{\mathrm{T}}\frac{\partial k}{\partial t}\phi_i - \omega_{ni}\phi_i^{\mathrm{T}}\frac{\partial m}{\partial t}\phi_i\right),
$$
(16)

where $m_{pi} = \phi_i^{\mathrm{T}}m\phi_i$ and $\dfrac{\partial m}{\partial t}$ and $\dfrac{\partial k}{\partial t}$ are determined according to specific design parameters.

In the dynamic model of the fire-fighting jet system with the adaptive gun head, the mass and stiffness of the fluid unit are directly related to its density and bulk elastic modulus. Moreover, the density and bulk elastic modulus of the fluid unit are both functions of fluid pressure and initial gas content of the fluid. Therefore, analyzing the influence of fluid mass and stiffness on the natural frequency sensitivity of the jet system is essential to analyze the influence of fluid pressure and initial gas content on the natural frequency sensitivity of the jet system.

3.1. Sensitivity of the Natural Frequency of the Jet System to the Mass of Fluid Unit 1

The density of fluid unit 1 is equal to that of fluid unit 2, and the density of the fluid can be regarded as a function of fluid pressure and initial gas content in the jet system. Therefore, the change of m_7 indicates that the fluid pressure or initial gas content in the jet system has changed, which leads to the change of m_5. Meanwhile, m_1, m_2, m_3, and m_4 are sub-mass units of the core rod-end cap structure, and their values are independent of m_7. Using the mass matrix of the jet system shown in Equation (2), the partial derivative of the mass matrix of the jet system with respect to the mass of fluid unit 1 m_7 is

$$
\frac{\partial m}{\partial m_7} = \mathrm{diag}\left(0, 0, 0, 0, \frac{\partial m_5}{\partial m_7}, 0, 1\right),
$$
(17)

where

$$
\frac{\partial m_5}{\partial m_7} = \frac{\partial\left(V_{m5}\rho_{\text{water}}\right)}{\partial\left(V_{m7}\rho_{\text{water}}\right)} = \frac{V_{m5}}{V_{m7}}.
$$
(18)

Using the stiffness matrix of the jet system shown in Equation (3), the partial derivative of the stiffness matrix of the jet system with respect to the mass of fluid unit 1 m_7 is

$$\frac{\partial k}{\partial m_7} = \begin{bmatrix} 0 & 0 & 0 & 0 & 0 & 0 & 0 \\ 0 & 0 & 0 & 0 & 0 & 0 & 0 \\ 0 & 0 & 0 & 0 & 0 & 0 & 0 \\ 0 & 0 & 0 & \dfrac{\partial k_{f22}}{\partial m_7} & -\dfrac{\partial k_{f22}}{\partial m_7} & 0 & 0 \\ 0 & 0 & 0 & -\dfrac{\partial k_{f22}}{\partial m_7} & \dfrac{\partial k_{f22}}{\partial m_7}+\dfrac{\partial k_{f21}}{\partial m_7} & -\dfrac{\partial k_{f21}}{\partial m_7} & 0 \\ 0 & 0 & 0 & 0 & -\dfrac{\partial k_{f21}}{\partial m_7} & \dfrac{\partial k_{f21}}{\partial m_7}+\dfrac{\partial k_{f1}}{\partial m_7} & -\dfrac{\partial k_{f1}}{\partial m_7} \\ 0 & 0 & 0 & 0 & 0 & -\dfrac{\partial k_{f1}}{\partial m_7} & \dfrac{\partial k_{f1}}{\partial m_7} \end{bmatrix}. \tag{19}$$

If the initial gas content of the fluid is changed and the fluid pressure remains unchanged, combining Equations (7)–(9) and Equation (10), then

$$\frac{\partial k_{f21}}{\partial m_7} = \frac{\partial k_{f22}}{\partial m_7} = A_1 \frac{\dfrac{\mathrm{d}B_f(p_0, x)}{\mathrm{d}x}}{\dfrac{\mathrm{d}\rho_f(p_0, x)}{\mathrm{d}x}}, \tag{20}$$

$$\frac{\partial k_{f1}}{\partial m_7} = A_2 \frac{\dfrac{\mathrm{d}B_f(p_0, x)}{\mathrm{d}x}}{\dfrac{\mathrm{d}\rho_f(p_0, x)}{\mathrm{d}x}}, \tag{21}$$

where p_0 is the initial fluid pressure, x is the initial gas content of the fluid, and ρ_f is the density of the fluid unit. The expressions of A_1 and A_2 are

$$A_1 = \frac{S_{a1'}S_{a2'}S_{a3'}}{2(L_{1'}S_{a2'}S_{a3'} + L_{2'}S_{a1'}S_{a3'} + L_{3'}S_{a1'}S_{a2'})V_{m7}}, \tag{22}$$

$$A_2 = \frac{S_{a1}S_{a2}S_{a3}S_{a4}S_{a5}S_{a6}}{(L_1 S_{a2}S_{a3}S_{a4}S_{a5}S_{a6} + \ldots + L_6 S_{a1}S_{a2}S_{a3}S_{a4}S_{a5})V_{m7}}. \tag{23}$$

If the fluid pressure is changed and the initial gas content of the fluid remains unchanged, then

$$\frac{\partial k_{f21}}{\partial m_7} = \frac{\partial k_{f22}}{\partial m_7} = A_1 \frac{\dfrac{\mathrm{d}B_f(p, x_0)}{\mathrm{d}p}}{\dfrac{\mathrm{d}\rho_f(p, x_0)}{\mathrm{d}p}}, \tag{24}$$

$$\frac{\partial k_{f1}}{\partial m_7} = A_2 \frac{\dfrac{\mathrm{d}B_f(p, x_0)}{\mathrm{d}p}}{\dfrac{\mathrm{d}\rho_f(p, x_0)}{\mathrm{d}p}}. \tag{25}$$

3.2. Sensitivity of the Natural Frequency of the Jet System to the Stiffness of Fluid Unit 1

Using the mass matrix of the jet system shown in Equation (2), the partial derivative of the mass matrix of the jet system with respect to the stiffness of fluid unit 1 k_{f1} is

$$\frac{\partial m}{\partial k_{f1}} = \mathrm{diag}\left(0, 0, 0, 0, \frac{\partial m_5}{\partial k_{f1}}, 0, \frac{\partial m_7}{\partial k_{f1}}\right). \tag{26}$$

If the initial gas content of the fluid is changed and the fluid pressure remains unchanged, combining Equations (7) and (8), then

$$\frac{\partial m_5}{\partial k_{f1}} = A_3 \frac{\dfrac{d\rho_f(p_0, x)}{dx}}{\dfrac{dB_f(p_0, x)}{dx}}, \tag{27}$$

$$\frac{\partial m_7}{\partial k_{f1}} = A_4 \frac{\dfrac{d\rho_f(p_0, x)}{dx}}{\dfrac{dB_f(p_0, x)}{dx}}. \tag{28}$$

Among them,

$$A_3 = \frac{(L_1 S_{a2} S_{a3} S_{a4} S_{a5} S_{a6} + \ldots + L_6 S_{a1} S_{a2} S_{a3} S_{a4} S_{a5}) V_{m5}}{S_{a1} S_{a2} S_{a3} S_{a4} S_{a5} S_{a6}}, \tag{29}$$

$$A_4 = \frac{(L_1 S_{a2} S_{a3} S_{a4} S_{a5} S_{a6} + \ldots + L_6 S_{a1} S_{a2} S_{a3} S_{a4} S_{a5}) V_{m7}}{S_{a1} S_{a2} S_{a3} S_{a4} S_{a5} S_{a6}}. \tag{30}$$

If the fluid pressure is changed and the initial gas content of the fluid remains unchanged, then

$$\frac{\partial m_5}{\partial k_{f1}} = A_3 \frac{\dfrac{d\rho_f(p, x_0)}{dp}}{\dfrac{dB_f(p, x_0)}{dp}}, \tag{31}$$

$$\frac{\partial m_7}{\partial k_{f1}} = A_3 \frac{\dfrac{d\rho_f(p, x_0)}{dp}}{\dfrac{dB_f(p, x_0)}{dp}}. \tag{32}$$

Using the stiffness matrix of the jet system shown in Equation (3), the partial derivative of the stiffness matrix of the jet system with respect to the stiffness of fluid unit 1 k_{f1} is

$$\frac{\partial k}{\partial k_{f1}} =
\begin{bmatrix}
0 & 0 & 0 & 0 & 0 & 0 & 0 \\
0 & 0 & 0 & 0 & 0 & 0 & 0 \\
0 & 0 & 0 & 0 & 0 & 0 & 0 \\
0 & 0 & 0 & \dfrac{\partial k_{f22}}{\partial k_{f1}} & -\dfrac{\partial k_{f22}}{\partial k_{f1}} & 0 & 0 \\
0 & 0 & 0 & -\dfrac{\partial k_{f22}}{\partial k_{f1}} & \dfrac{\partial k_{f22}}{\partial k_{f1}} + \dfrac{\partial k_{f21}}{\partial k_{f1}} & -\dfrac{\partial k_{f21}}{\partial k_{f1}} & 0 \\
0 & 0 & 0 & 0 & -\dfrac{\partial k_{f21}}{\partial k_{f1}} & \dfrac{\partial k_{f21}}{\partial k_{f1}} + 1 & -1 \\
0 & 0 & 0 & 0 & 0 & -1 & 1
\end{bmatrix}, \tag{33}$$

where

$$\frac{\partial k_{f21}}{\partial k_{f1}} = \frac{\partial k_{f22}}{\partial k_{f1}} = \frac{\dfrac{L_1}{S_{a1}} + \dfrac{L_2}{S_{a2}} + \dfrac{L_3}{S_{a3}} + \dfrac{L_3}{S_{a4}} + \dfrac{L_3}{S_{a5}} + \dfrac{L_3}{S_{a6}}}{2\left(\dfrac{L_{1'}}{S_{a1'}} + \dfrac{L_{2'}}{S_{a2'}} + \dfrac{L_{3'}}{S_{a3'}}\right)}. \tag{34}$$

3.3. Sensitivity of the Natural Frequency of the Jet System to the Mass of Fluid Unit 2

Using the mass matrix of the jet system shown in Equation (2), the partial derivative of the mass matrix of the jet system with respect to the mass of fluid unit 2 m_5 is

$$\frac{\partial m}{\partial m_5} = \mathrm{diag}\!\left(0, 0, 0, 0, 1, 0, \frac{\partial m_7}{\partial m_5}\right), \tag{35}$$

where

$$\frac{\partial m_7}{\partial m_5} = \frac{\partial(V_{m7}\rho_{water})}{\partial(V_{m5}\rho_{water})} = \frac{V_{m7}}{V_{m5}}. \tag{36}$$

Using the stiffness matrix of the jet system shown in Equation (3), the partial derivative of the stiffness matrix of the jet system with respect to the mass of fluid unit 2 m_5 is

$$\frac{\partial k}{\partial m_5} = \begin{bmatrix} 0 & 0 & 0 & 0 & 0 & 0 & 0 \\ 0 & 0 & 0 & 0 & 0 & 0 & 0 \\ 0 & 0 & 0 & 0 & 0 & 0 & 0 \\ 0 & 0 & 0 & \dfrac{\partial k_{f22}}{\partial m_5} & -\dfrac{\partial k_{f22}}{\partial m_5} & 0 & 0 \\ 0 & 0 & 0 & -\dfrac{\partial k_{f22}}{\partial m_5} & \dfrac{\partial k_{f22}}{\partial m_5} + \dfrac{\partial k_{f21}}{\partial m_5} & -\dfrac{\partial k_{f21}}{\partial m_5} & 0 \\ 0 & 0 & 0 & 0 & -\dfrac{\partial k_{f21}}{\partial m_5} & \dfrac{\partial k_{f21}}{\partial m_5} + \dfrac{\partial k_{f1}}{\partial m_5} & -\dfrac{\partial k_{f1}}{\partial m_5} \\ 0 & 0 & 0 & 0 & 0 & -\dfrac{\partial k_{f1}}{\partial m_5} & \dfrac{\partial k_{f1}}{\partial m_5} \end{bmatrix}. \tag{37}$$

If the initial gas content of the fluid is changed and the fluid pressure remains unchanged, combining Equations (7)–(10), then

$$\frac{\partial k_{f21}}{\partial m_5} = \frac{\partial k_{f22}}{\partial m_5} = A_5 \frac{\dfrac{\mathrm{d}B_f(p_0, x)}{\mathrm{d}x}}{\dfrac{\mathrm{d}\rho_f(p_0, x)}{\mathrm{d}x}}, \tag{38}$$

$$\frac{\partial k_{f1}}{\partial m_5} = A_6 \frac{\dfrac{\mathrm{d}B_f(p_0, x)}{\mathrm{d}x}}{\dfrac{\mathrm{d}\rho_f(p_0, x)}{\mathrm{d}x}}. \tag{39}$$

Among them,

$$A_5 = \frac{S_{a1'}S_{a2'}S_{a3'}}{2(L_{1'}S_{a2'}S_{a3'} + L_{2'}S_{a1'}S_{a3'} + L_{3'}S_{a1'}S_{a2'})V_{m5}}, \tag{40}$$

$$A_6 = \frac{S_{a1}S_{a2}S_{a3}S_{a4}S_{a5}S_{a6}}{(L_1 S_{a2}S_{a3}S_{a4}S_{a5}S_{a6} + \ldots + L_6 S_{a1}S_{a2}S_{a3}S_{a4}S_{a5})V_{m5}}. \tag{41}$$

If the fluid pressure is changed and the initial gas content of the fluid remains the same, then

$$\frac{\partial k_{f21}}{\partial m_5} = \frac{\partial k_{f22}}{\partial m_5} = A_5 \frac{\dfrac{\mathrm{d}B_f(p, x_0)}{\mathrm{d}p}}{\dfrac{\mathrm{d}\rho_f(p, x_0)}{\mathrm{d}p}}, \tag{42}$$

$$\frac{\partial k_{f1}}{\partial m_5} = A_6 \frac{\dfrac{\mathrm{d}B_f(p, x_0)}{\mathrm{d}p}}{\dfrac{\mathrm{d}\rho_f(p, x_0)}{\mathrm{d}p}}. \tag{43}$$

3.4. Sensitivity of the Natural Frequency of the Jet System to the Stiffness of Fluid Unit 2

Using the mass matrix of the jet system shown in Equation (2), the partial derivative of the mass matrix of the jet system with respect to the stiffness k_{f21} is

$$\frac{\partial m}{\partial k_{f21}} = \mathrm{diag}\!\left(0, 0, 0, 0, \frac{\partial m_5}{\partial k_{f21}}, 0, \frac{\partial m_7}{\partial k_{f21}}\right). \tag{44}$$

If the initial gas content of the fluid is changed and the fluid pressure remains unchanged, combining Equations (7) and (8), then

$$\frac{\partial m_5}{\partial k_{f21}} = A_7 \frac{\dfrac{\mathrm{d}\rho_f(p_0, x)}{\mathrm{d}x}}{\dfrac{\mathrm{d}B_f(p_0, x)}{\mathrm{d}x}}, \tag{45}$$

$$\frac{\partial m_7}{\partial k_{f21}} = A_8 \frac{\dfrac{\mathrm{d}\rho_f(p_0, x)}{\mathrm{d}x}}{\dfrac{\mathrm{d}B_f(p_0, x)}{\mathrm{d}x}}, \tag{46}$$

where

$$A_7 = \frac{2(L_{1'}S_{a2'}S_{a3'} + L_{2'}S_{a1'}S_{a3'} + L_{3'}S_{a1'}S_{a2'})V_{m5}}{S_{a1'}S_{a2'}S_{a3'}}, \tag{47}$$

$$A_8 = \frac{2(L_{1'}S_{a2'}S_{a3'} + L_{2'}S_{a1'}S_{a3'} + L_{3'}S_{a1'}S_{a2'})V_{m7}}{S_{a1'}S_{a2'}S_{a3'}}. \tag{48}$$

If the fluid pressure is changed and the initial gas content of the fluid remains unchanged, then

$$\frac{\partial m_5}{\partial k_{f21}} = A_7 \frac{\dfrac{\mathrm{d}\rho_f(p, x_0)}{\mathrm{d}p}}{\dfrac{\mathrm{d}B_f(p, x_0)}{\mathrm{d}p}}, \tag{49}$$

$$\frac{\partial m_7}{\partial k_{f21}} = A_8 \frac{\dfrac{\mathrm{d}\rho_f(p, x_0)}{\mathrm{d}p}}{\dfrac{\mathrm{d}B_f(p, x_0)}{\mathrm{d}p}}. \tag{50}$$

Using the stiffness matrix of the jet system shown in Equation (3), the partial derivative of the stiffness matrix of the jet system with respect to the stiffness k_{f21} is

$$\frac{\partial k}{\partial k_{f21}} = \begin{bmatrix} 0 & 0 & 0 & 0 & 0 & 0 & 0 \\ 0 & 0 & 0 & 0 & 0 & 0 & 0 \\ 0 & 0 & 0 & 0 & 0 & 0 & 0 \\ 0 & 0 & 0 & 1 & -1 & 0 & 0 \\ 0 & 0 & 0 & -1 & 2 & -1 & 0 \\ 0 & 0 & 0 & 0 & -1 & 1+\dfrac{\partial k_{f1}}{\partial k_{f21}} & -\dfrac{\partial k_{f1}}{\partial k_{f21}} \\ 0 & 0 & 0 & 0 & 0 & -\dfrac{\partial k_{f1}}{\partial k_{f21}} & \dfrac{\partial k_{f1}}{\partial k_{f21}} \end{bmatrix}, \tag{51}$$

where

$$\frac{\partial k_{f1}}{\partial k_{f21}} = \frac{2\left(\dfrac{L_{1'}}{S_{a1'}} + \dfrac{L_{2'}}{S_{a2'}} + \dfrac{L_{3'}}{S_{a3'}}\right)}{\dfrac{L_1}{S_{a1}} + \dfrac{L_2}{S_{a2}} + \dfrac{L_3}{S_{a3}} + \dfrac{L_3}{S_{a4}} + \dfrac{L_3}{S_{a5}} + \dfrac{L_3}{S_{a6}}}. \tag{52}$$

k_{f21} and k_{f22} are numerically equal and are half of the equivalent stiffness of fluid unit 2, and in the dynamic model of the jet system, they are both located on either side of fluid unit 2. Therefore, k_{f21} and k_{f22} have the same influence on the natural frequency of the jet system, that is, the sensitivity to the jet system is the same. So, this paper only analyzes the influence of k_{f21} on the natural frequency sensitivity of the jet system.

3.5. Sensitivity of the Natural Frequency of the Jet System to the Mass of the Spray Core

Using the mass matrix of the jet system shown in Equation (2), the partial derivative of the mass matrix of the jet system with respect to the mass of the spray core m_6 is

$$\frac{\partial m}{\partial m_6} = \text{diag}(0,0,0,0,0,1,0). \tag{53}$$

The partial derivative of the stiffness matrix of the jet system with respect to the mass of the spray core m_6 in any mode is equal to zero, then

$$\frac{\partial k}{\partial m_6} = 0. \tag{54}$$

3.6. Sensitivity of the Natural Frequency of the Jet System to the Stiffness of the Spring in the Gun Head

The partial derivative of the mass matrix of the jet system with respect to the stiffness of the spring in the gun head k_5 in any mode is equal to zero, then

$$\frac{\partial m}{\partial k_5} = 0. \tag{55}$$

Using the stiffness matrix of the jet system shown in Equation (3), the partial derivative of the stiffness matrix of the jet system with respect to the stiffness of the spring in the gun head k_5 is

$$\frac{\partial k}{\partial k_5} = \begin{bmatrix} 0 & 0 & 0 & 0 & 0 & 0 & 0 \\ 0 & 0 & 0 & 0 & 0 & 0 & 0 \\ 0 & 0 & 0 & 0 & 0 & 0 & 0 \\ 0 & 0 & 0 & 1 & 0 & -1 & 0 \\ 0 & 0 & 0 & 0 & 0 & 0 & 0 \\ 0 & 0 & 0 & -1 & 0 & 1 & 0 \\ 0 & 0 & 0 & 0 & 0 & 0 & 0 \end{bmatrix}. \tag{56}$$

4. Sensitivity Analysis of the Jet System

4.1. Modal Analysis of the Jet System

The design parameters of the jet system are shown in Table 2. The jet fluid is water containing a certain amount of air at 293.15 K and 1013.25 hPa.

Table 2. Design parameters of the jet system.

Parameter Name	Parameter Symbol	Unit	Value
Mass of core rod	m	kg	0.0354
Mass of end cap	m_d	kg	0.1177
Mass of spray core	m_6	kg	0.3163
Mass of Fluid unit 1	m_7	kg	16.4593
Mass of Fluid unit 2	m_5	kg	0.1362
Stiffness of discrete unit of the core rod	k_1	kN/m	333,460
Equivalent stiffness 1 of fluid	k_{f1}	kN/m	354.05
Equivalent stiffness 2 of fluid	k_{f2}	kN/m	424.07
Equivalent stiffness 3 of fluid	k_{f3}	kN/m	424.07
Stiffness of the spring in the gun head	k_5	kN/m	18
Fluid pressure	p	MPa	0.6
Initial gas content of fluid	x	%	2
Bulk elastic modulus of fluid	B_f	MPa	161.1432
Temperature	T	K	293

The parameters shown in Table 2 are substituted into the free vibration equation of the jet system, and the natural frequencies of the jet system and the corresponding mode shapes are shown in Table 3.

Table 3. Natural frequencies and mode shapes of the jet system.

Order	First-Order	Second-Order	Third-Order	Fourth-Order	Fifth-Order	Sixth-Order	Seventh-Order
Natural frequency f_{ni}/Hz	f_{n1} 19.6	f_{n2} 230.8	f_{n3} 427.4	f_{n4} 4040	f_{n5} 24,276.9	f_{n6} 43,906	f_{n7} 57,146.6
Principal mode of each mode Φ_{ni}	0.0004 0.0009 0.0015 0.0019 0.3549 0.7062 1	0.001 0.002 0.003 0.0041 0.7578 1 −0.0252	0.0013 0.0025 0.0038 0.005 1 −0.3205 0.0023	0.261 0.5176 0.7653 1 −0.0049 0.0001 0	0.7231 1 0.6597 −0.0877 0 0 0	1 −0.0186 −0.9997 0.0372 0 0 0	−0.7044 1 −0.7153 0.0155 0 0 0

It can be seen from Table 3 that the first three natural frequencies of the jet system are below 1000 Hz, and it is easy to resonate with the self-pulsation frequency of the incident fluid. Therefore, only the sensitivities of the design parameters to the first three natural frequencies are considered in the sensitivity analysis.

Based on the parameters of the jet system shown in Table 2, combined with the relevant formulas in the second section, the sensitivity of the natural frequency of the jet system to the design parameters can be obtained.

4.2. Sensitivity Analysis of the Natural Frequency of the Jet System to the Mass of Fluid Unit 1

Keeping the fluid pressure unchanged, when the initial gas content changes between 0 and 0.028, the variation law of the first three natural frequencies and their sensitivities of the jet system with m_7 is shown in Figure 6.

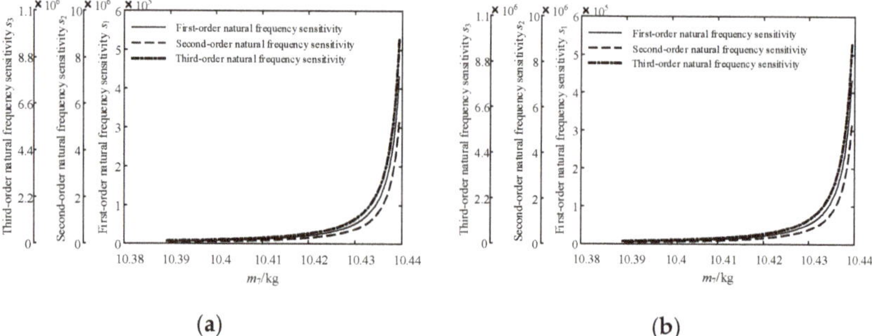

(a)   (b)

Figure 6. The variation law of the first three natural frequencies and their sensitivities of the jet system under a different initial gas content of the fluid with the mass of fluid unit 1. (**a**) Variation law of natural frequency sensitivity. (**b**) Variation law of natural frequency.

It can be seen from Figure 6 that with the increase in the mass of fluid unit 1 under a different initial gas content of the fluid, the sensitivities of the first three natural frequencies of the jet system increase monotonically, and their values are positive. Correspondingly, the first three natural frequencies of the jet system monotonously increase with an increase in the mass of fluid unit 1 and the growth gradually increases.

Keeping the initial gas content of the fluid unchanged, the fluid pressure is changed so that the variation range of m_7 is consistent with that shown in Figure 6. When the pressure is in the range of 0.43–2.36 MPa, the variation law of the first three natural frequencies and their sensitivities of the jet system with m_7 is shown in Figure 7.

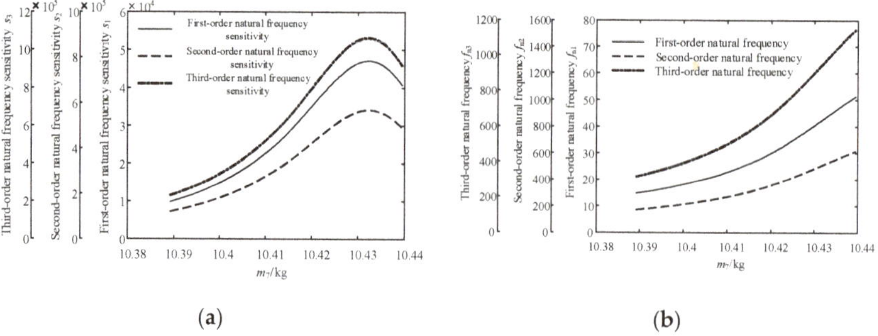

(a)   (b)

Figure 7. The variation law of the first three natural frequencies and their sensitivities of the jet system under a different pressure of the fluid with the mass of fluid unit 1. (**a**) Variation law of natural frequency sensitivity. (**b**) Variation law of natural frequency.

It can be seen from Figure 7 that with the increase in the mass of fluid unit 1 under a different fluid pressure, the sensitivities of the first three natural frequencies of the jet system first rise and then decrease, and their values are positive. Correspondingly, the first three natural frequencies of the jet system monotonically increase with the increase in the mass of fluid unit 1. When the sensitivities are reduced, the growth rate of the natural frequencies of the corresponding mass units of fluid unit 1 decreases in accordance with the increase in the mass of fluid unit 1. However, since the absolute values of the sensitivities are still large, the growth rate of the natural frequency is still high.

It can be seen from the comparison between Figures 6 and 7 that under the current parameter conditions, the initial gas content of the fluid is more sensitive to the first three natural frequencies of the jet system than the fluid pressure within the same mass range of fluid unit 1.

4.3. Sensitivity Analysis of the Natural Frequency of the Jet System to the Stiffness of Fluid Unit 1

Keeping the fluid pressure unchanged, when the initial gas content changes between 0 and 0.028, the variation law of the first three natural frequencies and their sensitivities of the jet system with k_{f1} is shown in Figure 8.

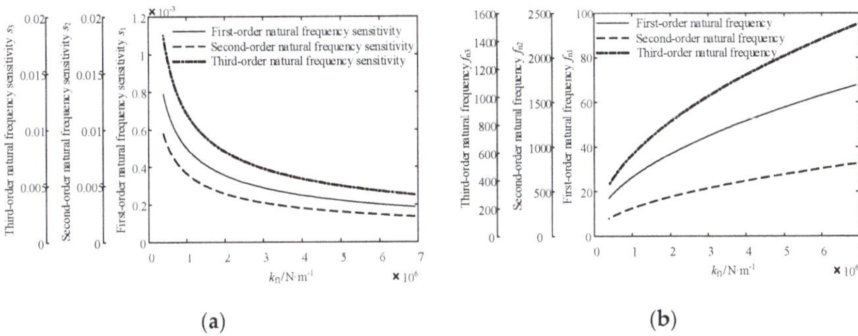

Figure 8. The variation law of the first three natural frequencies and their sensitivities of the jet system under a different initial gas content of the fluid with the stiffness of fluid unit 1. (**a**) Variation law of natural frequency sensitivity. (**b**) Variation law of natural frequency.

It can be seen from Figure 8 that with the increase in the stiffness of fluid unit 1 under a different initial gas content of the fluid, the sensitivities of the first three natural frequencies of the jet system decrease monotonically and the decreasing rate decreases gradually, while the values of sensitivities are small and positive. Correspondingly, the first three natural frequencies of the jet system increase monotonically at a lower rate with the increase in the stiffness of fluid unit 1, and the growth rate gradually decreases.

Keeping the initial gas content of the fluid unchanged, the fluid pressure is changed so that the variation range of k_{f1} is consistent with that shown in Figure 8. When the pressure is in the range of 0.51–35 MPa, the variation law of the first three natural frequencies and their sensitivities of the jet system with k_{f1} is shown in Figure 9.

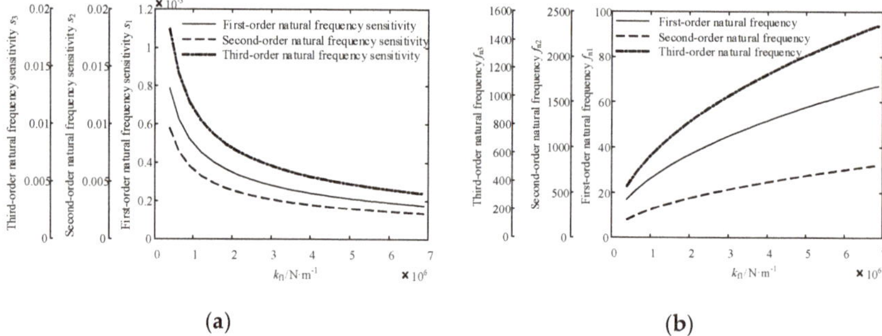

(a) (b)

Figure 9. The variation law of the first three natural frequencies and their sensitivities of the jet system under a different pressure of the fluid with the stiffness of fluid unit 1. (**a**) Variation law of natural frequency sensitivity. (**b**) Variation law of natural frequency.

It can be seen from Figure 9 that with the increase in the stiffness of fluid unit 1, the variation law of the first three natural frequencies and their sensitivities of the jet system under a different fluid pressure is basically the same as that under a different initial gas content of fluid.

It can be seen from Figures 8 and 9 that whether the initial gas content of the fluid changes or the fluid pressure changes, the values of the sensitivities of the first three natural frequencies of the jet system are small within the range of stiffness of fluid unit 1, which means that the change in stiffness of fluid unit 1 has less effect on sensitivity. Among them, the first-order natural frequency sensitivity has the least influence. Under the current parameter conditions, the initial gas content and the fluid pressure have basically the same effect on the sensitivities of the first three natural frequency of the jet system within the same range of stiffness of fluid unit 1.

4.4. Sensitivity Analysis of the Natural Frequency of the Jet System to the Mass of Fluid Unit 2

Keeping the fluid pressure unchanged, when the initial gas content changes between 0 and 0.028, the variation law of the first three natural frequencies and their sensitivities of the jet system with m_5 is shown in Figure 10.

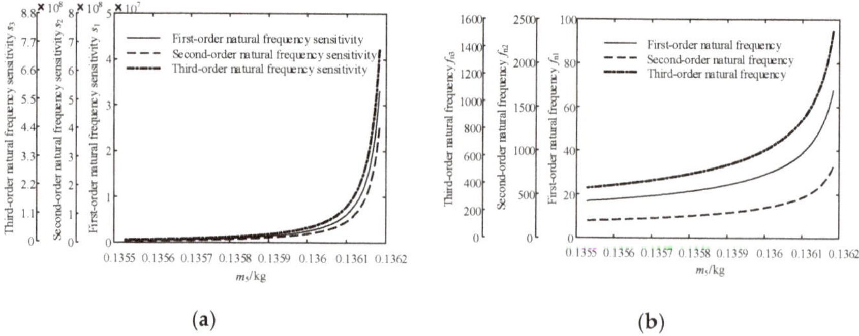

(a) (b)

Figure 10. The variation law of the first three natural frequencies and their sensitivities of the jet system under a different initial gas content of fluid with the mass of fluid unit 2. (**a**) Variation law of natural frequency sensitivity. (**b**) Variation law of natural frequency.

It can be seen from the comparison between Figures 6 and 10 that when the variation range of initial gas content is the same, the variation law of the first three natural frequencies and their sensitivities of the jet system with the mass of fluid unit 2 is basically the same as that with the mass of fluid unit 1. However, with the change of the mass of fluid unit 2 in the same range of initial gas content, the values of the first three natural frequency sensitivities of the jet system are relatively large, indicating that the first three natural frequencies of the jet system are more sensitive to the change in the mass of fluid unit 2, compared with that to the change of the mass of fluid unit 1.

Keeping the initial gas content of the fluid unchanged, the fluid pressure is changed so that the variation range of m_5 is consistent with that shown in Figure 10. When the pressure is in the range of 0.43–2.36 MPa, the variation law of the first three natural frequencies and their sensitivities of the jet system with m_5 is shown in Figure 11.

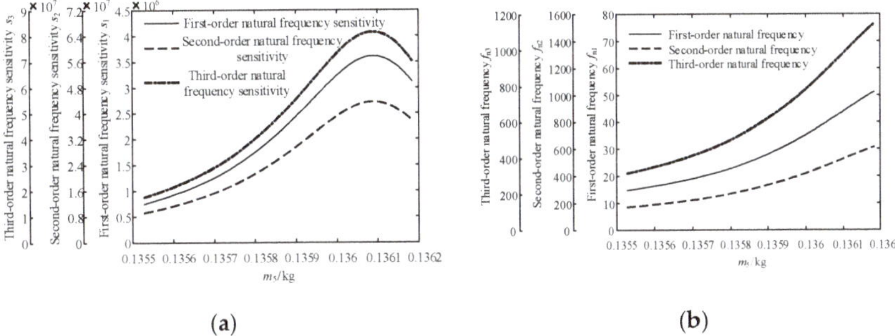

(a)   (b)

Figure 11. The variation law of the first three natural frequencies and their sensitivities of the jet system under a different pressure of the fluid with the mass of fluid unit 2. (**a**) Variation law of natural frequency sensitivity. (**b**) Variation law of natural frequency.

It can be seen from the comparison between Figures 6 and 11 that when the variation range of fluid pressure is the same, the variation law of the first three natural frequencies and their sensitivities of the jet system with the mass of fluid unit 2 is basically the same as that with the mass of fluid unit 1. However, with the change of the mass of fluid unit 2 in the same range of pressure, the values of the first three natural frequency sensitivities of the jet system are relatively large, indicating that the first three natural frequencies of the jet system are more sensitive to the change in the mass of fluid unit 2, compared to the change in the mass of fluid unit 1.

It can be seen from the comparison between Figures 10 and 11 that under the current parameter conditions, the initial gas content of fluid is more sensitive to the first three natural frequencies of the jet system than the fluid pressure within the same mass range of fluid unit 2.

4.5. Sensitivity Analysis of the Natural Frequency of the Jet System to the Stiffness of Fluid Unit 2

Keeping the fluid pressure unchanged, when the initial gas content changes between 0 and 0.028, the variation law of the first three natural frequencies and their sensitivities of the jet system with k_{f21} is shown in Figure 12.

It can be seen from Figure 12 that with the increase in the stiffness of fluid unit 2 under a different initial gas content of fluid, the sensitivities of the first three natural frequencies of the jet system decrease monotonically and the decreasing rate decreases gradually, while the values of sensitivities are small and positive. Correspondingly, the first three natural frequencies of the jet system increase monotonically at a lower rate with the increase in the stiffness of fluid unit 2, and the growth rate gradually decreases.

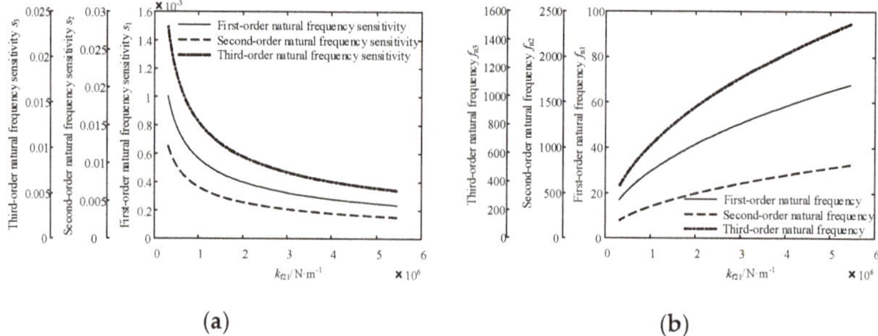

(**a**) (**b**)

Figure 12. The variation law of the first three natural frequencies and their sensitivities of the jet system under a different initial gas content of fluid with the stiffness of fluid unit 2. (**a**) Variation law of natural frequency sensitivity. (**b**) Variation law of natural frequency.

Keeping the initial gas content of the fluid unchanged, the fluid pressure is changed so that the variation range of k_{f21} is consistent with that shown in Figure 12. When the pressure is in the range of 0.51–35 MPa, the variation law of the first three natural frequencies and their sensitivities of the jet system with k_{f21} is shown in Figure 13.

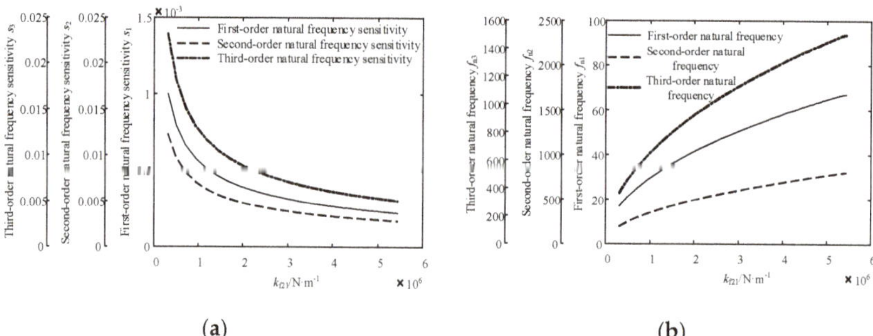

(**a**) (**b**)

Figure 13. The variation law of the first three natural frequencies and their sensitivities of the jet system under a different pressure of the fluid with the stiffness of fluid unit 2. (**a**) Variation law of natural frequency sensitivity. (**b**) Variation law of natural frequency.

It can be seen from Figure 13 that with the increase in the stiffness of fluid unit 1, the variation law of the first three natural frequencies and their sensitivities of the jet system under a different fluid pressure is basically the same as that under a different initial gas content of fluid.

It can be seen from Figures 12 and 13 that whether the initial gas content of the fluid changes or the fluid pressure changes, the values of the sensitivities of the first three natural frequencies of the jet system are small, which means that the change in the stiffness of fluid unit 2 has less effect on sensitivity. Among them, the first-order natural frequency sensitivity has the least influence. Under the current parameter conditions, the initial gas content and fluid pressure have basically the same effect on the sensitivities of the first three natural frequencies of the jet system within the same range of stiffness of fluid unit 2.

4.6. Sensitivity Analysis of the Natural Frequency of the Jet System to the Mass of the Spray Core

Let the initial gas content of the fluid be 0.02 and the fluid pressure be 0.6 MPa. When the mass of the spray core is in the range of 0.1–0.5 kg, the variation law of the first three natural frequencies and their sensitivities of the jet system with m_6 is shown in Figure 14.

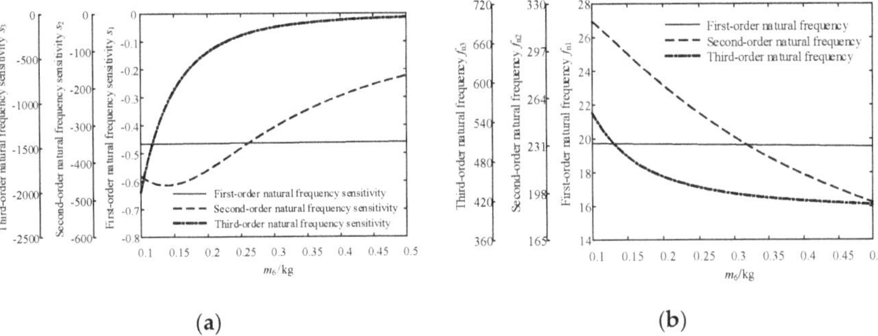

Figure 14. The variation law of the first three natural frequencies and their sensitivities of the jet system with the mass of the spray core. (**a**) Variation law of natural frequency sensitivity. (**b**) Variation law of natural frequency.

From Figure 14, the following laws can be summarized:

(1) With the increase in the mass of the spray core, the first-order natural frequency sensitivity of the jet system increases slightly in the process, but it can be regarded as basically unchanged due to the small increase. The absolute value of the sensitivity is small and negative. In the variation of the natural frequency, the first-order natural frequency decreases slightly from 19.73 Hz to 19.54 Hz with the increase in the mass of the spray core.

(2) With the increase in the mass of the spray core, the second-order natural frequency sensitivity of the jet system is negative, and its absolute value increases first and then decreases. Therefore, the second-order natural frequency of the jet system appears to decrease monotonically with the increase in the mass of the spray core, and the rate of decline increases slightly first and then decreases gradually.

(3) With the increase in the mass of the spray core, the third-order natural frequency sensitivity of the jet system is negative, and its absolute value decreases gradually. Therefore, the third-order natural frequency of the jet system decreases with the increase in the mass of the spray core. The rate of decline decreases gradually and tends to zero eventually.

In summary, the change in the mass of the spray core has a rare effect on the first-order natural frequency of the jet system, but the second-order and third-order natural frequencies of the jet system decrease gradually with the increase in the mass of the spray core.

4.7. Sensitivity Analysis of the Natural Frequency of the Jet System to the Stiffness of the Spring in the Gun Head

Let the initial gas content of the fluid be 0.02 and the fluid pressure be 0.6 MPa. When the stiffness of the spring in the gun head is in the range of 10–30 kN/m, the variation law of the first three natural frequencies and their sensitivities of the jet system with k_5 is shown in Figure 15.

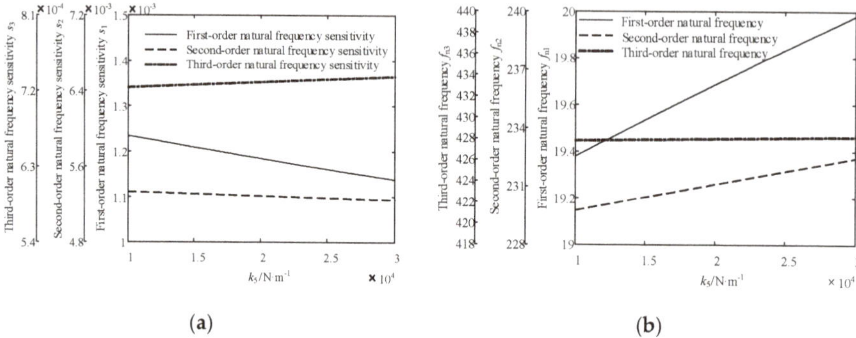

Figure 15. The variation law of the first three natural frequencies and their sensitivities of the jet system with the stiffness of the spring in the gun head. (**a**) Variation law of natural frequency sensitivity. (**b**) Variation law of natural frequency.

It can be seen from Figure 15 that as the stiffness of the spring in the adaptive gun head increases, the first-order and second-order natural frequency sensitivities of the jet system monotonically decrease, the third-order natural frequency sensitivity monotonically increases, and their values are all positive. Corresponding to the change of natural frequency sensitivity of each order, the first three natural frequencies of the jet system gradually increase with the increase in spring stiffness, and the growth rate is small, which indicates that the spring stiffness of the gun head has little effect on the first three natural frequencies of the jet system and their sensitivities.

5. Experimental Verification

In order to verify the accuracy of the modal analysis of the jet system, a platform for the dynamic experiment of the fire-fighting jet system with adaptive gun head was built. The modal test of the working jet system was carried out by the hammering method. The experiment platform is shown in Figure 16. The experimental parameters are consistent with the parameters shown in Table 1.

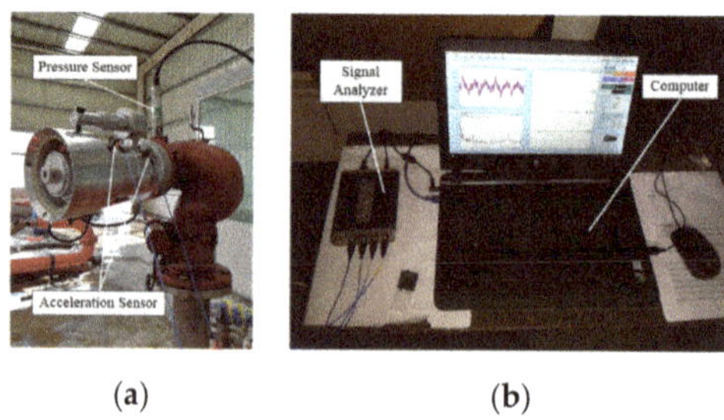

(**a**) (**b**)

Figure 16. Platform for dynamic experiment of the fire-fighting jet system with the adaptive gun head. (**a**) Prototype of gun head. (**b**) Experimental equipment.

The fast Fourier transform was used to analyze the acceleration signal output from the jet system in the frequency domain. The amplitude-frequency characteristics of the acceleration signal of the jet system under the action of the fluid self-pulsation excitation and the hammer step excitation are shown in Figure 17. The coordinate of point A in the Figure 17 is (19.5, 0.1438) and the coordinate of point B is (46.8, 0.1588).

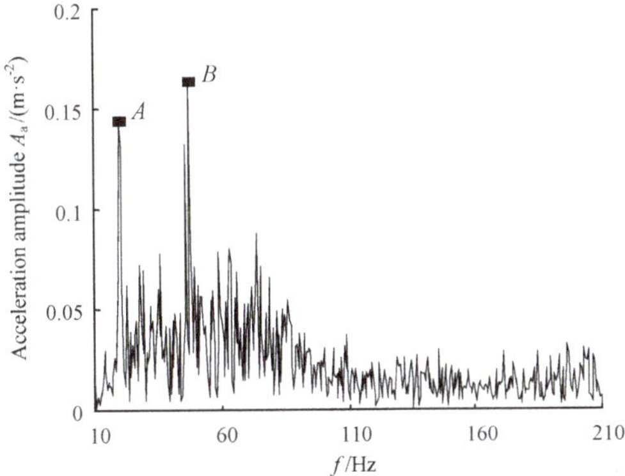

Figure 17. Amplitude-frequency characteristic curve of the fire-fighting jet system with the adaptive gun head.

It can be seen from Figure 17 that the jet system has relatively obvious peaks at the frequencies of 19.5 Hz and 46.8 Hz, of which 46.8 Hz is the axial frequency of the centrifugal pump for the experimental system (2800 RPM) and 19.5 Hz is the first-order natural frequency of the jet system in the experiment. At the same time, due to the influence of fluid turbulence in the fire-fighting monitor, the amplitude-frequency curve superimposes the amplitude-frequency characteristics of white noise, which increases the difficulty of frequency spectrum recognition. Therefore, other order natural frequencies of the jet system cannot be determined experimentally. The experimental data and theoretical value of the first-order natural frequency of the jet system are shown in Table 4.

Table 4. Comparison between experimental data and theoretical value.

Natural Frequency	Experimental Data	Theoretical Value	Error
The first-order natural frequency	19.5 Hz	19.6 Hz	0.51%

It can be seen from Table 4 that the theoretical first-order natural frequency is very close to the experimental data, and the error is only 0.51%, which verifies the accuracy of the proposed dynamic model of the fire-fighting jet system with the adaptive gun head.

6. Conclusions

(1) Considering the fluid-structure interaction and discrete-continuous coupling characteristics of the fire-fighting jet system with the adaptive gun head, the dynamic model and equations of the fire-fighting jet system with adaptive gun head were established based on the lumped parameter method. The sensitivity calculation formulas of the natural frequency of the jet system to typical design parameters were derived.

(2) The modal analysis of the jet system was carried out, and the natural frequencies of the orders and the corresponding modal vectors under given conditions were obtained. The sensitivity analysis of the jet system was carried out, and the variation law of the first three natural frequencies and their sensitivities of the jet system to typical design parameters was revealed. Among the parameters involved, the first three natural frequencies of the jet system were the most sensitive to the change in the mass of fluid unit 2 in the range of a certain initial gas content.

(3) The platform for the dynamic experiment of the fire-fighting jet system with the adaptive gun head was built, and the modal experiment of the jet system was carried out. There was only a 0.51% error between the value of the first-order natural frequency of the jet system determined by

the modal experiment and the theoretical one determined by the analytical method, showing that good agreement with the first-order natural frequency of the jet system was found.

Author Contributions: Conceptualization, X.Y. and C.W.; methodology, X.Z.; investigation, X.Z. and C.W.; writing—original draft preparation, X.Y.; writing—review and editing, C.W. and X.Z.; supervision, L.Z. and Y.Z.

Funding: This work is funded by the National Natural Science Foundation of China (No. 51805468, 51805214), Natural Science Foundation of Hebei Province (No. E2017203129), Open Foundation of the State Key Laboratory of Fluid Power and Mechatronic Systems (No. GZKF-201820), Basic Research Special Funding Project of Yanshan University (No. 16LGB001), and China Postdoctoral Science Foundation (No. 2019M651722). The authors would also like to thank the reviewers for their valuable suggestions and comments.

Conflicts of Interest: The authors declare no conflicts of interest.

References

1. Zhu, J.; Li, W.; Lin, D.; Zhao, G. Study on water jet trajectory model of fire monitor based on simulation and experiment. *Fire Technol.* **2018**, *55*, 773–787. [CrossRef]
2. Hu, G.L.; Chen, W.G.; Gao, Z.G. Flow analysis of spray jet and direct jet nozzle for fire water monitor. *Adv. Mater. Res.* **2010**, *139–141*, 913–916. [CrossRef]
3. Bhende, A.R.; Kadu, P.S. Behavioral analysis of hydraulically driven fire fighting water monitor trailer subjected to type of fire & ways of its fire protection. In Proceedings of the 2009 Second International Conference on Emerging Trends in Engineering & Technology, Nagpur, India, 16–18 December 2009.
4. Berner, O. A volume and pressure controlling spill valve equipped for removal of excess aesthetic gas. *Acta Anaesthesiol. Scand.* **2010**, *16*, 252–258. [CrossRef]
5. Qian, J.Y.; Chen, M.R.; Liu, X.L.; Jin, Z.J. A numerical investigation of the flow of nanofluids through a micro tesla valve. *J. Zhejiang Univ. Sci. A* **2019**, *20*, 50–60. [CrossRef]
6. Qian, J.Y.; Gao, Z.X.; Liu, B.Z.; Jin, Z.J. Parametric study on fluid dynamics of pilot-control angle globe valve. *ASME J. Fluids Eng.* **2018**, *140*, 111102. [CrossRef]
7. Salacinska, K.; El Serafy, G.Y.; Los, F.J.; Blauw, A. Sensitivity analysis of the two dimensional application of the Generic Ecological Model (GEM) to algal bloom prediction in the North Sea. *Ecol. Model.* **2010**, *221*, 178–190. [CrossRef]
8. Gallivan, S. Challenging the role of calibration, validation and sensitivity analysis in relation to models of health care processes. *Health Care Manag. Sci.* **2008**, *11*, 208–213. [CrossRef] [PubMed]
9. Zhu, Y.; Qian, P.; Tang, S.; Jiang, W.; Li, W.; Zhao, J. Amplitude-frequency characteristics analysis for vertical vibration of hydraulic AGC system under nonlinear action. *AIP Adv.* **2019**, *9*, 035019. [CrossRef]
10. Zhu, Y.; Tang, S.; Wang, C.; Jiang, W.; Yuan, X.; Lei, Y. Bifurcation characteristic research on the load vertical vibration of a hydraulic automatic gauge control system. *Processes* **2019**, *7*, 718. [CrossRef]
11. Zhu, Y.; Tang, S.; Quan, L.; Jiang, W.; Zhou, L. Extraction method for signal effective component based on extreme-point symmetric mode decomposition and Kullback-Leibler divergence. *J. Braz. Soc. Mech. Sci. Eng.* **2019**, *41*, 100. [CrossRef]
12. Félix, G.; Ancheyta, J.; Trejo, F. Sensitivity analysis of kinetic parameters for heavy oil hydrocracking. *Fuel* **2019**, *241*, 836–844. [CrossRef]
13. Prieur, C.; Viry, L.; Blayo, E.; Brankart, J.-M. A global sensitivity analysis approach for marine biogeochemical modeling. *Ocean Model.* **2019**, *139*, 101402. [CrossRef]
14. Guo, R.; Hu, Y.; Liu, M.Z.; Heiselberg, P. Influence of design parameters on the night ventilation performance in office buildings based on sensitivity analysis. *Sustain. Cities Soc.* **2019**, *50*, 101661. [CrossRef]
15. Gunay, H.B.; Ouf, M.; Newsham, G.; O'Brien, W. Sensitivity analysis and optimization of building operations. *Energy Build.* **2019**, *199*, 164–175. [CrossRef]
16. Dong, F.; Li, J.Y.; Zhang, S.N.; Wang, Y.; Sun, Z.Y. Sensitivity analysis and spatial-temporal heterogeneity of CO_2 emission intensity: Evidence from China. *Resour. Conserv. Recycl.* **2019**, *150*, 104398. [CrossRef]
17. Rezaie, F.; Bayat, M.A.; Farnam, S.M. Sensitivity analysis of pre-stressed concrete sleepers for longitudinal crack prorogation effective factors. *Eng. Fail. Anal.* **2016**, *66*, 385–397. [CrossRef]

18. Tashakor, S.; Afsari, A.; Hashemi-Tilehnoee, M. Sensitivity analysis of thermal-hydraulic parameters to study the corrosion intensity in nuclear power plant steam generators. *Nucl. Eng. Technol.* **2019**, *51*, 394–401. [CrossRef]

19. Cao, J.K.; Ding, S.T. Sensitivity analysis for safety design verification of general aviation reciprocating aircraft engine. *Chin. J. Aeronaut.* **2012**, *25*, 675–680. [CrossRef]

20. He, X.; Jiao, W.; Wang, C.; Cao, W. Influence of surface roughness on the pump performance based on Computational Fluid Dynamics. *IEEE Access* **2019**, *7*, 105331–105341. [CrossRef]

21. Wang, C.; Hu, B.; Zhu, Y.; Wang, X.; Luo, C.; Cheng, L. Numerical study on the gas-water two-phase flow in the self-priming process of self-priming centrifugal pump. *Processes* **2019**, *7*, 330. [CrossRef]

22. Jia, W.H.; Yin, C.B.; Hao, F.; Li, G.; Fan, X.Z. Dynamic characteristics and stability analysis of conical relief valve. *Mechanika* **2019**, *25*, 25–31. [CrossRef]

23. Zhang, K.Q.; Karney, B.W.; Mcpherson, D.L. Pressure-relief valve selection and transient pressure control. *Am. Water Works Assoc.* **2008**, *100*, 62–69. [CrossRef]

24. Zhang, J.; Xia, S.; Ye, S.; Xu, B.; Song, W.; Zhu, S.; Xiang, J. Experimental investigation on the noise reduction of an axial piston pump using free-layer damping material treatment. *Appl. Acoust.* **2018**, *139*, 1–7. [CrossRef]

25. Ye, S.; Zhang, J.; Xu, B.; Zhu, S. Theoretical investigation of the contributions of the excitation forces to the vibration of an axial piston pump. *Mech. Syst. Signal Process.* **2019**, *129*, 201–217. [CrossRef]

26. Zhu, Y.; Tang, S.; Wang, C.; Jiang, W.; Zhao, J.; Li, G. Absolute stability condition derivation for position closed-loop system in hydraulic automatic gauge control. *Processes* **2019**, *7*, 766. [CrossRef]

27. Imagine, S.A. HYD advanced fluid properties. *Technical Bulletin.* **2007**, *117*, 7.

Article

Flow Characteristics and Stress Analysis of a Parallel Gate Valve

Hui Wu [1], Jun-ye Li [1,*] and Zhi-xin Gao [1,2]

[1] SUFA Technology Industry Co., Ltd, CNNC, Suzhou 215129, China; wuh@chinasufa.com (H.W.); gaozx@chinasufa.com (Z.-x.G.)

[2] Institute of Process Equipment, College of Energy Engineering, Zhejiang University, Hangzhou 310027, China; zhixingao@zju.edu.cn

* Correspondence: lijy@chinasufa.com; Tel.: +86-0512-6662-1921

Received: 29 September 2019; Accepted: 24 October 2019; Published: 3 November 2019

Abstract: Gate valves have been widely used in the piping system and have attracted a lot of attention from researchers. In this paper, a wedge-type double disk parallel gate valve is chosen to be analyzed. The Reynolds number varying from 200 to 500,000, and the valve opening degree varying from 20% to 100%, and the groove depth varying from 2.3 mm to 9 mm are chosen to investigate their effects on the flow and loss coefficients of the gate valve. The results show that the loss coefficient decreases and the flow coefficient increases with the increase of the Reynolds number and the valve opening degree, while with the increase of the groove depth, the loss coefficient barely changes, but the flow coefficient increases if the Reynolds number is larger than 10,000. In addition, the effects of the gaps between the disk and the limit stop on the stress distribution of the bolt are also investigated, and the results show that if the gaps are negative, high stress will act on the bolt at the contact position between the bolt and the limit stop.

Keywords: gate valve; stress analysis; Computational Fluid Dynamics (CFD); flow coefficient; loss coefficient

---

1. Introduction

As a kind of on–off valve, gate valves are widely used in various process industries. When compared with other valves, it needs smaller torque during the open and close processes of the valve, and it also results in smaller flow resistance. However, due to the complexity of the sealing device, more components in gate valves are needed, which might lead to high proneness of gate valve failure [1–3].

Researches about various valves have been focused in the past years. For instance, Qian et al. [4,5] applied Tesla valves in the hydrogen decompression process and investigated the pressure drop and Mach number in multi-stage Tesla valves. Yuan et al. [6] and Jin et al. [7] conducted numerical simulations to study the cavitation inside the poppet valve and the globe valve, and the cavitation characteristics and the structural parameters that affected the cavitation were found. Dasgupta et al. [8] and Zhang et al. [9] and Qian et al. [10] focused on the dynamic valve open and close processes of a proportional valve, a pressure relief valve, and a pilot-control globe valve respectively. Furthermore, Jin et al. [11] and Qian et al. [12,13] investigated the structural parameters of a pilot-control angle valve and a micro Tesla valve, Xu et al. [14] focused on the pulsatile flow of a mechanical heart valve, Chen et al. [15] investigated the thermal stress of a pressure-reducing valve, and Qian et al. [16] analyzed the possibility of the noise distribution in perforated plates connecting pressure-relief valve.

Specific to the gate valves, lots of works have also been done focusing on different issues. There are works regarding the flow resistance and the temperature distributions. Solek and Mika [17] investigated the relationship between the loss coefficient and the Reynolds number for gate valves with ice slurry flow, and their experiments showed that the loss coefficient remained constant in the turbulent regime, but decreased with the increase of the Reynolds number in the laminar regime. Alimonti [18] and Lin et al. [19] studied the flow characteristics and resistance characteristics of a gate valve with different openings and different inlet velocities, and it was found that the flow resistance could be gradually stabilized when the opening was larger than 2/8 [19]. Long and Shurong [20] numerically investigated the flow and temperature distributions in the stem gate valve while using axisymmetric models, and Hu et al. [21] performed experiments and numerical simulations to study the temperature and the convection heat transfer coefficient distributions in a gate valve. In the meantime, there are also works regarding structure optimization [22,23], the corrosion erosion distributions [24–26], and the stress analysis [27–29]. Kolesnikov and Tikhonov [22] focused on the conicity of the output channel of the wedge gate valve and Xu et al. [23] focused on the seal and piston of a subsea gate valve. Babaev and Kerimov [24] found that there was fretting corrosion for a parallel slide gate valve. Lin et al. [25,26] found that the erosion rate in a gate valve was related to the pipe diameter, the cavity width, and the inlet velocity [25], while the open degree had little effects but a large Stokes number could increase the difference of erosion distributions for different valve placements [26]. As for the stress analysis, Liao et al. [27] found that the fatigue of the inlet valve sleeve resulting from collision stress was the main reason for the valve failure. Zakirnichnaya and Kulsharipov [28] analyzed the stress-strain of the wedge gate valves while using fluid-structure interaction technology and they found that the safe operation resources value of the wedge is lower than the valve body. Punitharani et al. [29] applied the finite element analysis (FEA) method to evaluate the residual stresses in a gate valve, and they found that there were large tensile and compressive residual stresses on the circular bead of the gate valve.

Gate valves can be classified into the wedge gate valve, the parallel gate valve, the double disk parallel gate valve, and the double disk wedge gate valve, etc. according to the form of the sealing device. As described above, works regarding parallel gate valves [18,19] and wedge gate valves [22] have been done by researchers. In this paper, a wedge-type double disk parallel gate valve is chosen to be analyzed. The computational fluid dynamics method is used to investigate the flow and loss characteristics under different Reynolds number and different groove depth. Moreover, the structural stress analysis where the valve failure might occur is also done while using a numerical method that is proven by the previous studies [27–31]. This work is helpful for the understanding of the flow characteristics of the gate valve and the judgment of the reason for valve failure in the future.

2. Model Description

2.1. Physic Model

Figure 1 shows the structure of the investigated gate valve and the nominal diameter is 100 mm. Figure 2 shows the detailed illustration of the disk component, which mainly consists of the disk, the disc frame, the nut, the limit stop, and the bolt.

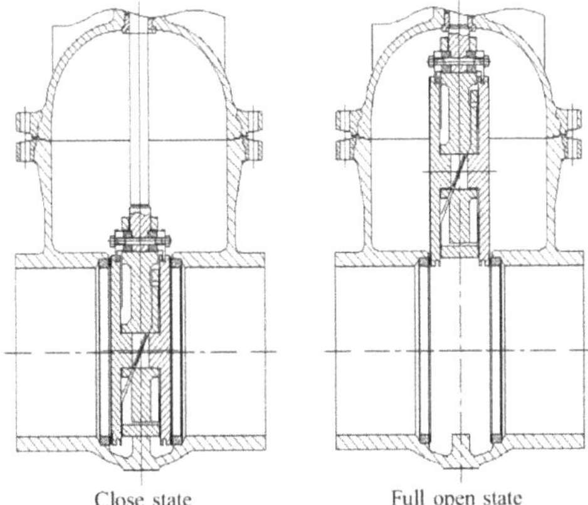

Figure 1. The structure of the wedge-type double disk parallel gate valve.

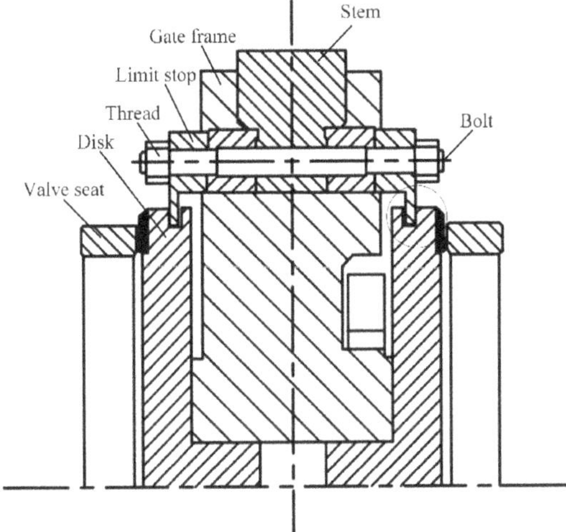

Figure 2. The structure of the disk component.

Figure 3 shows the gaps between the disk and the limit stop. During the mechanical machining process, there might be an asymmetry degree of the opening size of the machine. If the asymmetry degree is too large, gaps C2 and C3 can become negative, which will lead to extra force acting on the bolt.

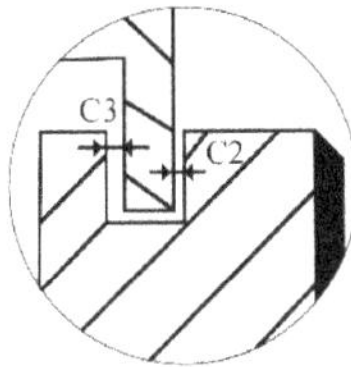

Figure 3. The illustration of the gaps between the disk and the limit stop.

Figure 4 shows the simplified model of the bolt and the limit stop; here, the nominal diameter of the thread is 6 mm, the diameter of the bolt is chosen as 7 mm, and the minimum diameter of the thread is chosen as 4.917 mm.

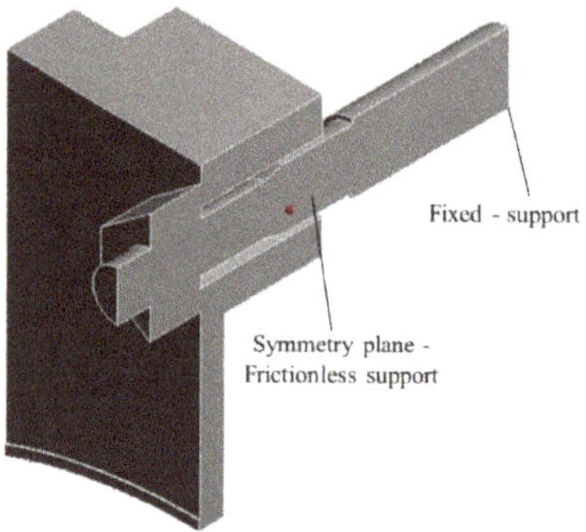

Figure 4. The simplified model of the bolt and the limit stop.

2.2. Numerical Model

The commercial software package ANSYS is used in this paper. The software Fluent that is based on the finite volume method is used to analyze the flow and resistance characteristics, and the software Mechanical based on the finite element method is used to predict the stress of the bolt. To solve the fluid field, the continuity and the momentum equations are used and are shown in the following.

$$\frac{\partial}{\partial x_j}\left(u_j\right) = 0 \tag{1}$$

$$\frac{\partial}{\partial x_j}\left(\rho u_i u_j + p\delta_{ij} - \tau_{ij}\right) = 0 \tag{2}$$

here ρ stands for the density of the fluid, u stands for the velocity of the fluid, p stands for the pressure of the fluid, and τ_{ij} stands for the viscous stress. The Realizable k-ε turbulence model [32] is used to solve the turbulent flow inside the investigated wedge-type double disk parallel gate valve, and the transport equations for turbulence kinetic and turbulence dissipation are shown, as follows

$$\frac{\partial}{\partial x_j}(\rho k u) = \frac{\partial}{\partial x_j}\left[\left(\mu + \frac{\mu_t}{\sigma_k}\right)\frac{\partial k}{\partial x_j}\right] + G_k + G_b - \rho\varepsilon - Y_M + S_K \tag{3}$$

$$\frac{\partial}{\partial x_j}(\rho \varepsilon u) = \frac{\partial}{\partial x_j}\left[\left(\mu + \frac{\mu_t}{\sigma_\varepsilon}\right)\frac{\partial \varepsilon}{\partial x_j}\right] + \rho C_1 S \varepsilon - \rho C_2 \frac{\varepsilon^2}{k + \sqrt{\upsilon\varepsilon}} + C_{1\varepsilon}\frac{\varepsilon}{k}C_{3\varepsilon}G_b + S_\varepsilon \tag{4}$$

here the model constants appear in above equations are σ_k = 1.0, σ_ε = 1.2, $C_{1\varepsilon}$ = 1.44, and C_2 = 1.9.

The hybrid grid is adopted during the computational fluid dynamics simulations and it is shown in Figure 5. A grid independence check is done for the wedge-type double disk parallel gate valve with a different total number of cells, N, and the pressure drop, Δp, between the upstream and the downstream of the gate valve is shown in Table 1. From Table 1, it can be found that, when the cell numbers increase from 13.3×10^5 to 25.3×10^5, the variation of the pressure drop is within 0.9%, so the method that is used to generate the grid 2 is adopted in this study.

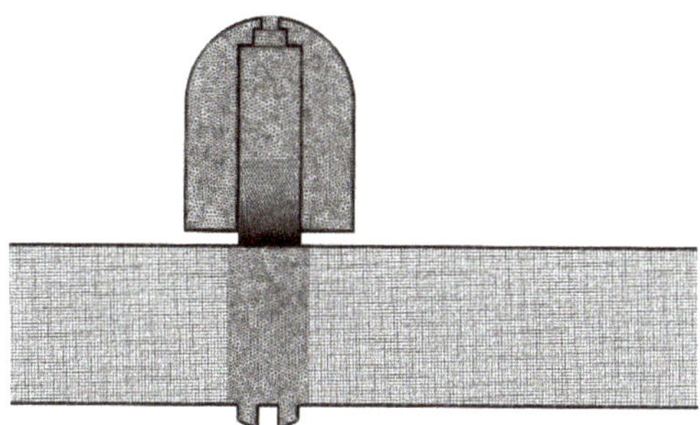

Figure 5. The hybrid grid used in this study.

Table 1. Pressure drop under a different number of cells.

	Grid 1	Grid 2	Grid 3
$N \times 10^5$	6.7	13.3	25.3
Δp (pa)	104.54	102.95	102.08

Water under room temperature is chosen as the working fluid inside the investigated gate valve. During the fluid flowing process, the temperature of water in the upstream of the gate valve and the ambient temperature are the same, which makes a small variation of water temperature. Therefore, the Energy equation is not solved. Symmetrical models are used to save simulation time. The velocity inlet and the pressure outlet boundary conditions are applied, and the outlet pressure is set as the atmosphere in order to investigate the flow and loss coefficients of the wedge-type double disk parallel gate valve. In addition, the no-slip wall is adopted for the wall boundary. For the stress analysis of the bolt, the symmetry plane is set as frictionless support, and the face of the bolt is set as fixed support.

2.3. Model Validation

The numerical results are compared with the experimental results of Lin et al. [19] in order to validate the used methods in this study. Here, the mentioned numerical methods are used to solve the turbulent incompressible gas flow in a parallel gate valve. Table 2 shows the comparison of the loss coefficients, and ξ_E represents the experimental results and ξ_N represents the numerical results. From Table 2, it can be found the relative errors between the experimental results and the numerical results are within 3%, thus the applied methods in this study are appropriate and they can provide results with sufficiently precise.

Table 2. The comparison between the numerical results and the experimental results.

Valve Opening Degree	3/8	5/8	1
ξ_E [17]	10.82	3.69	2.03
ξ_N	11.03	3.74	1.98
Error (%)	1.9	1.3	−2.5

3. Results and Discussion

The flow and loss characteristics of the wedge-type double disk parallel gate valve are numerically investigated, and the effects of the valve opening degree and the groove depth are studied. In addition, the stress analysis is done to uncover the possible reason for the gate valve failure.

3.1. Flow and Loss Characteristics

When water flows through the wedge-type double disk parallel gate valve, there is a pressure drop. The loss coefficient is often used as an indicator of the valve flow capacity, and the greater the loss coefficient, the smaller the valve flow capacity, but the greater the pressure drop. The loss coefficient of a valve can be calculated from the below equation.

$$\xi = \frac{\Delta p}{1/2\rho v^2} \tag{5}$$

Figure 6 shows the pressure drop Δp and the loss coefficient ξ of the investigated gate valve with the Reynolds number varying from 200 to 500,000. It can be found that the pressure drop almost exponentially increases with the increase of the Reynolds number. For the loss coefficient, as the increase of the Reynolds number, it decreases rapidly first and then has very small variation, and there is the largest variation of the loss coefficient when the flow is laminar.

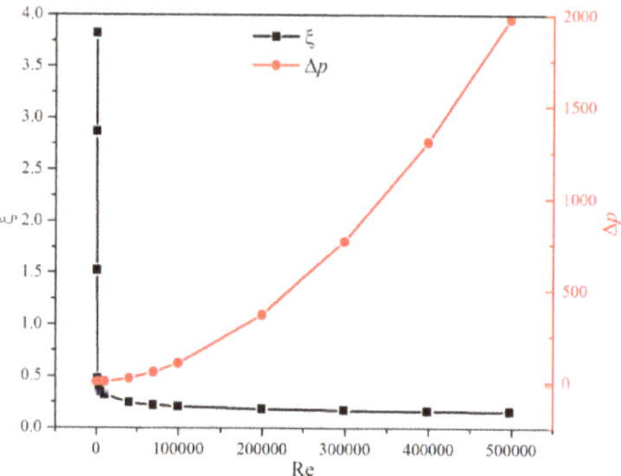

Figure 6. The pressure drop and the loss coefficient with different Reynolds number.

The capacity of water flows through the investigated gate valve can be represented by the flow coefficient, which can be expressed in the following

$$K_v = 10Q\sqrt{\frac{1000\rho}{\Delta p \rho_0}} \tag{6}$$

here Q is the volume flux (m³/h), ρ_0 is the density of water at 15 °C, K_v is the flow coefficient and it represents the flux when the pressure drop of the gate valve is 100 kPa. Figure 7 shows the relationship between the flow coefficient and the Reynolds number. It can be found that the flow coefficient increases with the increase of Reynolds number, and the smaller the Reynolds number, the larger the variation of the flow coefficient.

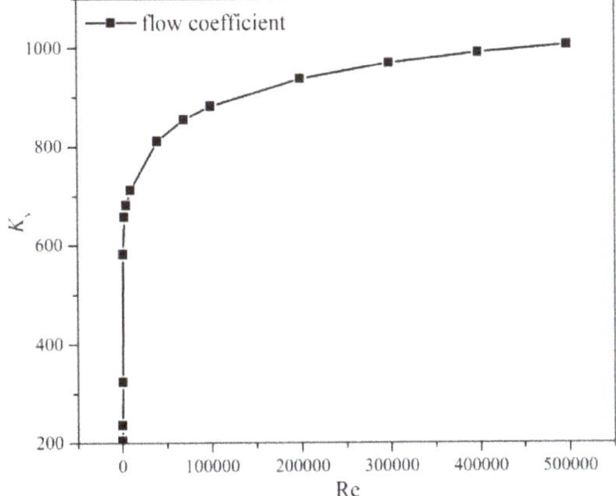

Figure 7. The flow coefficient with different Reynolds number.

To investigate the internal flow and pressure fields, the pressure and velocity distributions on the symmetry plane when the Reynolds number is about 100,000 are shown in Figure 8. It can be found that, when the investigated gate valve is in the fully open position, the whole main pipe in the gate valve is full of water with high speed, and the main resistance of the gate valve is the channel for the movement of the disk. Thus, the pressure drop is relatively small if the gate valve is in the fully open position and the loss coefficient is close to zero.

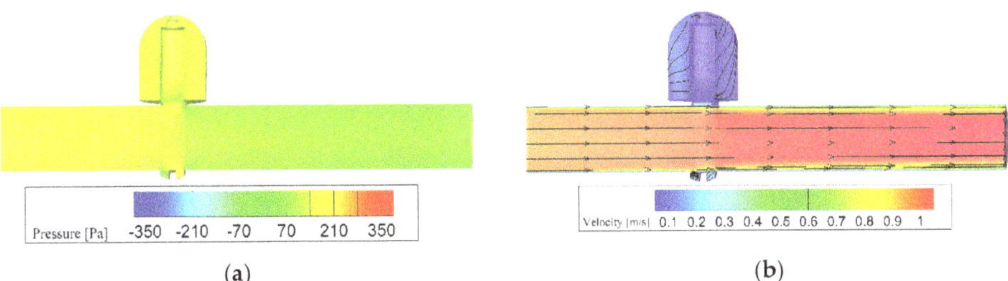

(a) (b)

Figure 8. Pressure and velocity distributions on the symmetry plane: (**a**) Pressure distribution; and, (**b**) Velocity distribution.

3.2. Effects of Valve Opening Degree

Figure 9 shows the internal flow fields and pressure distributions under two different valve opening degrees. Pressure drop mainly appears at the downstream of the gate valve that results from the throttling effect of two disks. With the decrease of the valve opening degree, the flow area decreases and the throttling effect increases, so the upstream pressure increases as the downstream pressure remains constant. Together with Figure 8, it can be found that, when the gate valve is in a 100% opening degree, the water flows directly to the downstream pipe, and there is no swirl in the main pipe of the valve. As the valve opening degree decreases, the maximum velocity increases and it appears under the valve disks. When comparing the gate valve with a 60% opening degree and the gate valve with a 20% opening degree, the maximum velocity increases from 2.2 m/s to 9.3 m/s. From the water streamline under different valve opening degree, it can be found that there is a local recirculation region with low velocity behind the valve disk, and the smaller the valve opening degree, the greater the local recirculation region.

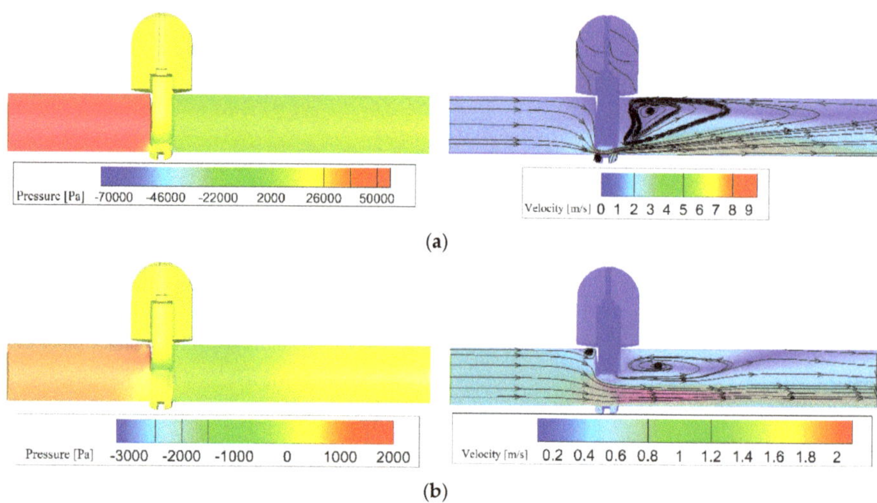

(a)

(b)

Figure 9. Pressure and flow fields under different valve opening degrees: (**a**) 20% opening; and, (**b**) 60% opening.

The loss coefficient and flow coefficient under different valve opening degree are shown in Figures 10 and 11. It can be found that the valve opening degree does not affect the relationship between the loss coefficient, the flow coefficient, and the Reynolds number. If the Reynolds number is smaller than 5000, the loss coefficient decreases with the increase of the Reynolds number rapidly, and the flow coefficient increases sharply as the increase of the Reynolds number. While the Reynolds number is larger than 5000, the loss coefficient barely varies with the increase of the Reynolds number, but the flow coefficient still increases with a reduced increment as the increase of the Reynolds number.

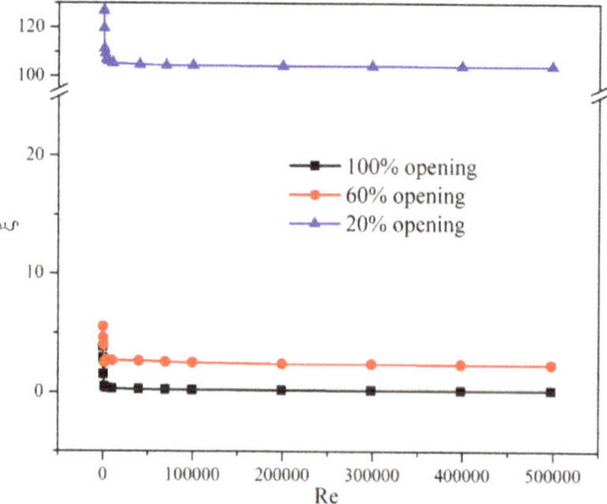

Figure 10. Loss coefficients under different valve opening degree and Reynolds number.

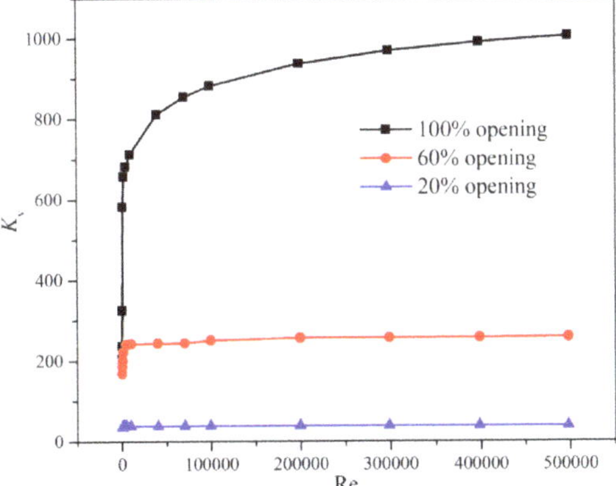

Figure 11. Flow coefficients under different valve opening degree and Reynolds number.

3.3. Effects of the Groove Depth

When the wedge-type double disk parallel gate valve is fully closed, the contact area between the gate valve disks and the valve seat is an annulus, and the size of the annulus can be judged by the groove depth. Three different groove depths are chosen, namely groove depth 1 (2.3 mm), groove depth 2 (4.5 mm), and groove depth 3 (9 mm) to investigate the effects of the groove depth on the flow and loss characteristics of the investigated gate valve.

The pressure and velocity distributions of the investigated gate valve with groove depth 3 are shown in Figure 8, and the pressure and velocity distributions of the investigated gate valve with groove depth 1 and groove depth 2 are shown in Figure 12. Together with Figures 8 and 12, it can be found that the groove depth does not affect the water streamline, which is because the flow channel is similar under different groove depth. While for the water pressure in the investigated gate valve, although the pressure distributions are almost the same, the value of water pressure increases as the groove depth decreases, which results from the resistance loss in the vicinity of the valve seat.

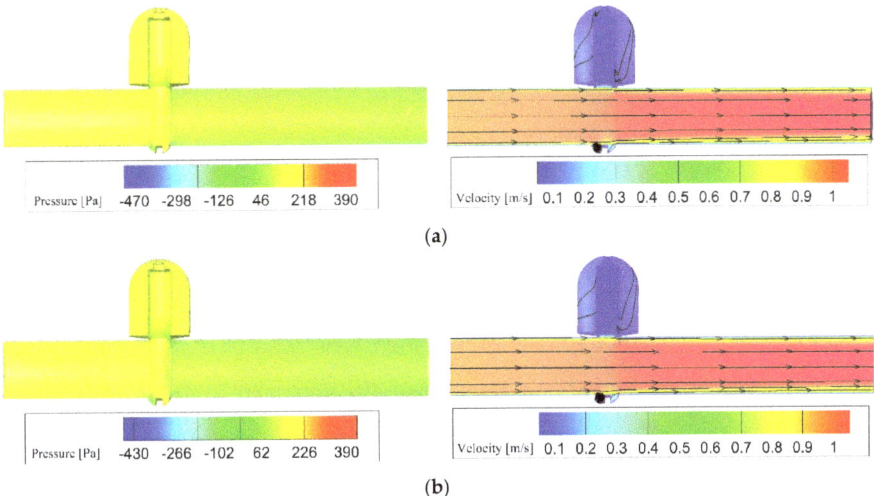

Figure 12. Pressure and flow fields under different groove depth: (**a**) Groove depth 1; and, (**b**) Groove depth 2.

The loss coefficient and flow coefficient of the investigated gate valve with different groove depth are shown in Table 3 and Figure 13. It can be seen from Table 3 that the groove depth does not affect the loss coefficient. Although pressure drop increases with the decrease of the groove depth, the difference can be neglected in the expression of the loss coefficient. When the Reynolds number is smaller than 10,000, the flow coefficient is not affected by the groove depth, while the Reynolds number is larger than 10,000, the flow coefficient increases with the increase of the groove depth, and the larger the Reynolds number, the larger the difference of the flow coefficient. Therefore, a large groove depth should be designed if possible.

Table 3. Loss coefficients under different groove depth.

Re	Groove Depth 1	Groove Depth 2	Groove Depth 3
200	3.763	3.763	3.763
800	1.570	1.569	1.570
5000	0.429	0.428	0.425
40,000	0.293	0.293	0.292
100,000	0.254	0.253	0.252
200,000	0.233	0.233	0.232
300,000	0.223	0.223	0.222
400,000	0.217	0.216	0.215
500,000	0.212	0.212	0.211

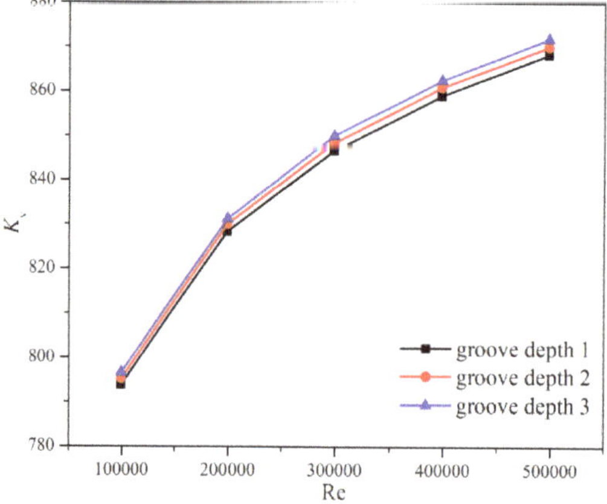

Figure 13. Flow coefficients under different groove depth.

3.4. Stress Analysis of the Bolt

The wedge-type double disk parallel gate valve has a complex actuator component, where there is a high possibility of the occurrence of valve failure. The following part focuses on the stress distribution on the bolt in order to find out one of the reasons for valve failure.

3.4.1. Effects of the Gap C2

If the limit stop shifts to the gate during the mechanical machining process, the gap C2 will become negative. To investigate the effect of gap C2 on the state of the stress distribution of the bolt, the value of gap C2 is assumed to be −0.5 mm.

The stress linearization analysis of the transverse cross-section of the bolt is shown in Figure 14a, and the distribution of the total stress, T_s, which equals to the membrane stress plus the bending stress, is shown in Figure 14b.

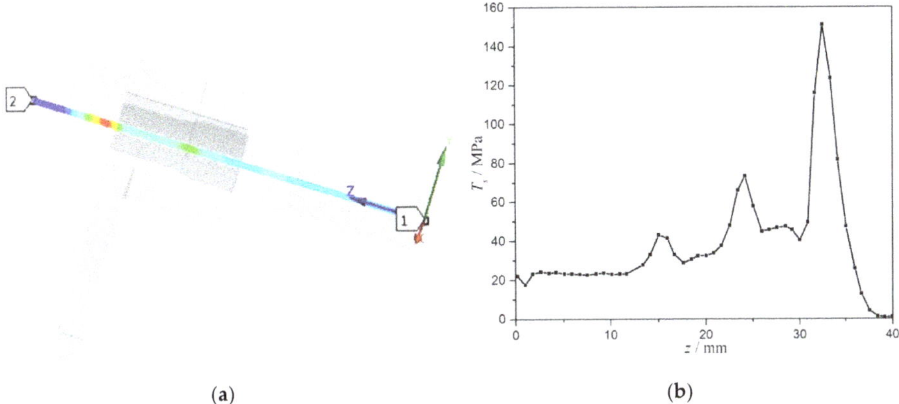

(a)　　　　　　　　　　　　　　　　(b)

Figure 14. The path of the stress linearization analysis and the stress distribution along the path: (**a**) The path of the stress linearization analysis; and, (**b**) The stress distribution along the path.

It can be found in Figure 14 that the maximum of the stress is 152.1 MPa, and the location where there is the maximum stress is 32.5 mm away from the end face of the bolt, where the contact position of the bolt and the limit stop is.

3.4.2. Effects of the Gap C3

If the limit stop shifts away from the gate during the mechanical machining process, the gap C3 will become negative. The value of gap C3 is assumed to be −1 mm to investigate the effect of gap C3 on the state of the stress distribution of the bolt.

Figure 15a shows the stress linearization analysis of the transverse cross-section of the bolt and Figure 15b shows the distribution of the total stress.

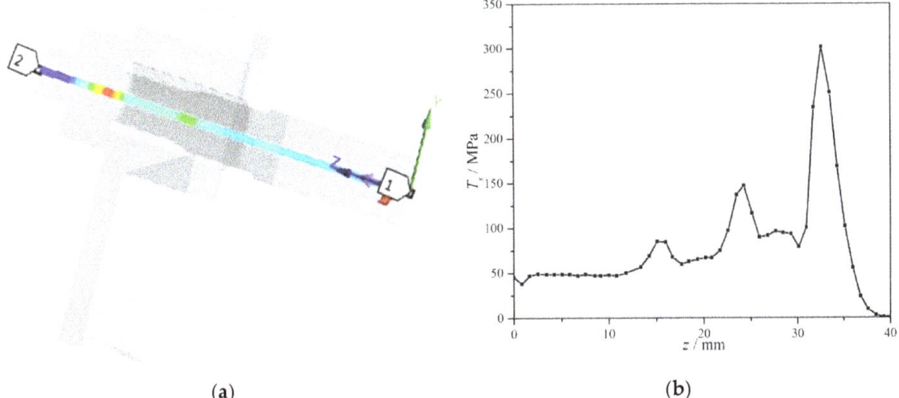

(a)　　　　　　　　　　　　　　　　(b)

Figure 15. The path of the stress linearization analysis and the stress distribution along the path: (**a**) The path of the stress linearization analysis; and, (**b**) The stress distribution along the path.

It can be seen from Figure 15 that the maximum of the stress is 303.1 MPa and the location where there is the maximum stress is 32.5 mm away from the end face of the bolt. Combining Figures 14 and 15, it is found that the stress distribution along the axial direction of the bolt is almost the same when the gap C2 or C3 is negative. Hence, the machining accuracy has great influence on the service life of the valve.

4. Conclusions

Numerical methods are adopted to investigate the flow and loss characteristics and the stress distribution of the wedge-type double disk parallel gate valve. The effects of the Reynolds number, the valve opening degree, and the groove depth are focused. In the meantime, the effects of the gaps between the disk and the limit stop are also discussed. The results show that the loss coefficient decreases, but the flow coefficient increases with the increase of Reynolds number. When Reynolds number is smaller than 5000, the variation of the flow and loss coefficients is very large. While Reynolds number is larger than 5000, the variation of the flow coefficient decreases and the loss coefficient almost remains the same. The valve opening degree has great influence on the flow and loss characteristics, with the decrease of the valve opening degree, the loss coefficient increases rapidly, and the flow coefficient obviously decreases. As for the groove depth, it has negligible effects on the loss coefficient, but the larger the groove depth, the larger the flow coefficient if the Reynolds number is greater than 10,000. The existence of the gaps between the disk and the limit stop is to eliminate the stress acting on the bolt, if the gaps are negative, high stress will appear at the contact position between the bolt and the limit stop. During the design process, a large groove depth should be chosen to provide a large flow coefficient. While, during the machining process, the machining accuracy should be satisfied to avoid stress concentration of the bolt.

Author Contributions: Conceptualization, H.W. and J.-y.L.; Data curation, J.-y.L. and Z.-x.G.; Formal analysis, H.W. and J.-y.L.; Investigation, J.-y.L.; Methodology, H.W.; Project administration, H.W. and J.-y.L.; Resources, H.W. and J.-y.L.; Software, J.-y.L. and Z.-x.G.; Supervision, H.W.; Validation, J.-y.L.; Writing—review & editing, J.-y.L.

Funding: This research received no external funding.

Conflicts of Interest: The authors declare no conflict of interest. The funders had no role in the design of the study; in the collection, analyses, or interpretation of data; in the writing of the manuscript, or in the decision to publish the results.

References

1. Barsoum, I.; Muñoz, A. Failure Analysis of a Large Knife Gate Valve Subjected to Multiaxial Loading. In Proceedings of the ASME 2017 Pressure Vessels and Piping Conference, Waikoloa, HI, USA, 16–20 July 2017; American Society of Mechanical Engineers Digital Collection: New York, NY, USA, 2017.
2. Owen, D.M.; Beavers, J.A. Investigation of Bolt Failures on a Gate Valve from a Natural Gas Pipeline. In Proceedings of the 4th International Pipeline Conference, Calgary, AB, Canada, 29 September–3 October 2002; American Society of Mechanical Engineers Digital Collection: New York, NY, USA, 2002.
3. Ifezue, D.; Tobins, F. Failure Investigation of a Gate Valve Eye Bolt Fracture During Hydrotesting. *J. Fail. Anal. Prev.* **2013**, *13*, 249–256. [CrossRef]
4. Qian, J.Y.; Chen, M.R.; Gao, Z.X.; Jin, Z.J. Mach number and energy loss analysis inside multi-stage Tesla valves for hydrogen decompression. *Energy* **2019**, *179*, 647–654. [CrossRef]
5. Qian, J.Y.; Wu, J.Y.; Gao, Z.X.; Wu, A.; Jin, Z.J. Hydrogen decompression analysis by multi-stage Tesla valves for hydrogen fuel cell. *Int. J. Hydrogen Energy* **2019**, *44*, 13666–13674. [CrossRef]
6. Yuan, C.; Song, J.; Zhu, L.; Liu, M. Numerical investigation on cavitating jet inside a poppet valve with special emphasis on cavitation-vortex interaction. *Int. J. Heat Mass Transfer* **2019**, *141*, 1009–1024. [CrossRef]
7. Jin, Z.J.; Gao, Z.X.; Qian, J.Y.; Wu, Z.; Sunden, B. A parametric study of hydrodynamic cavitation inside globe valves. *ASME J. Fluids Eng.* **2018**, *140*, 031208. [CrossRef]
8. Dasgupta, K.; Ghoshal, S.K.; Kumar, S.; Das, J. Dynamic analysis of an open-loop proportional valve controlled hydrostatic drive. *Proc. Inst. Mech. Eng. Part E* **2019**, 0954408919861247. [CrossRef]
9. Zhang, Z.; Jia, L.; Yang, L. Numerical simulation study on the opening process of the atmospheric relief valve. *Nucl. Eng. Des.* **2019**, *351*, 106–115. [CrossRef]
10. Qian, J.Y.; Wei, L.; Jin, Z.J.; Wang, J.K.; Zhang, H. CFD analysis on the dynamic flow characteristics of the pilot-control globe valve. *Energy Convers. Manage.* **2014**, *87*, 220–226. [CrossRef]

11. Jin, Z.J.; Gao, Z.X.; Zhang, M.; Liu, B.Z.; Qian, J.Y. Computational fluid dynamics analysis on orifice structure inside valve core of pilot-control angle globe valve. *Proc. Inst. Mech. Eng. Part C* **2018**, *232*, 2419–2429. [CrossRef]
12. Qian, J.Y.; Chen, M.R.; Liu, X.L.; Jin, Z.J. A numerical investigation of the flow of nanofluids through a micro Tesla valve. *J. Zhejiang Univ. Sci. A* **2019**, *20*, 50–60. [CrossRef]
13. Qian, J.Y.; Gao, Z.X.; Liu, B.Z.; Jin, Z.J. Parametric study on fluid dynamics of pilot-control angle globe valve. *ASME J. Fluids Eng.* **2018**, *140*, 111103. [CrossRef]
14. Xu, X.G.; Liu, T.Y.; Li, C.; Zhu, L.; Li, S.X. A Numerical Analysis of Pressure Pulsation Characteristics Induced by Unsteady Blood Flow in a Bileaflet Mechanical Heart Valve. *Processes* **2019**, *7*, 232. [CrossRef]
15. Chen, F.Q.; Zhang, M.; Qian, J.Y.; Fei, Y.; Chen, L.L.; Jin, Z.J. Thermo-mechanical stress and fatigue damage analysis on multi-stage high pressure reducing valve. *Ann. Nucl. Energy* **2017**, *110*, 753–767. [CrossRef]
16. Qian, J.Y.; Hou, C.W.; Wu, J.Y.; Gao, Z.X.; Jin, Z.J. Aerodynamics analysis of superheated steam flow through multi-stage perforated plates. *Int. J. Heat Mass Transfer* **2019**, *141*, 48–57. [CrossRef]
17. Solek, M.; Mika, L. Ice slurry flow through gate valves—Local pressure loss coefficient. In Proceedings of the 2nd International Conference on the Sustainable Energy and Environmental Development, Beijing, China, 24–25 March 2019.
18. Alimonti, C. Experimental characterization of globe and gate valves in vertical gas–liquid flows. *Exp. Therm. Fluid Sci.* **2014**, *54*, 259–266. [CrossRef]
19. Lin, Z.; Ma, G.; Cui, B.; Li, Y.; Zhu, Z.; Tong, N. Influence of flashboard location on flow resistance properties and internal features of gate valve under the variable condition. *J. Nat. Gas Sci. Eng.* **2016**, *33*, 108–117. [CrossRef]
20. Yu, L.; Yu, S. High Temperature Flow Characteristics in Non-Full Opening Steam Gate Valve for Thermal Power Plant. In *IOP Conference Series: Materials Science and Engineering*; IOP Publishing: Bristol, UK, 2018.
21. Hu, B.; Zhu, H.; Ding, K.; Zhang, Y.; Yin, B. Numerical investigation of conjugate heat transfer of an underwater gate valve assembly. *Appl Ocean Res.* **2016**, *56*, 1–11. [CrossRef]
22. Kolesnikov, G.N.; Tikhonov, E.A. Influence of the Angle of Taper of Output Channel of Wedge Gate Valve on the Movement of a Liquid. *Chem. Pet. Eng.* **2017**, *52*, 707–709. [CrossRef]
23. Xu, X.; Li, S.; Gong, L.; Wang, H.; Wang, Y. Study on seals of subsea production gate valves. *Int. J. Comput. Appl. Technol.* **2018**, *58*, 29–36. [CrossRef]
24. Babaev, S.G.; Kerimov, V.I. Increase in the Working Capacity of Christmas-Tree Gate Valves Based on Studying the Wear Mechanism of a Gate and Seat Pair. *Chem. Pet. Eng.* **2015**, *51*, 526–529. [CrossRef]
25. Lin, Z.; Ruan, X.D.; Zhu, Z.C.; Fu, X. Three-dimensional numerical investigation of solid particle erosion in gate valves. *Proc. Inst. Mech. Eng. Part C* **2014**, *228*, 1670–1679. [CrossRef]
26. Lin, Z.; Zhu, L.; Cui, B.; Li, Y.; Ruan, X. Effect of placements (horizontal with vertical) on gas-solid flow and particle impact erosion in gate valve. *J. Therm. Sci.* **2014**, *23*, 558–563. [CrossRef]
27. Liao, Y.; Lian, Z.; Long, R.; Yuan, H. Effects of multiple factors on the stress of the electro-hydraulic directional valve used on the hydraulic roof supports. *Int. J. Appl. Electrom.* **2015**, *47*, 199–209. [CrossRef]
28. Zakirnichnaya, M.M.; Kulsharipov, I.M. Wedge gate valves selected during technological pipeline systems designing service life assessment. *Procedia Eng.* **2017**, *206*, 1831–1838. [CrossRef]
29. Punitharani, K.; Murugan, N.; Sivagami, S.M. Finite element analysis of residual stresses and distortion in hard faced gate valve. *J. Sci. Ind. Res.* **2010**, *69*, 129–134.
30. Zhang, C.; Yu, L.; Feng, R.; Zhang, Y.; Zhang, G. A numerical study of stress distribution and fracture development above a protective coal seam in longwall mining. *Processes* **2018**, *6*, 146. [CrossRef]
31. Wang, S.; Li, H.; Li, D. Numerical Simulation of Hydraulic Fracture Propagation in Coal Seams with Discontinuous Natural Fracture Networks. *Processes* **2018**, *6*, 113. [CrossRef]
32. Shih, T.H.; Liou, W.W.; Shabbir, A.; Yang, Z.; Zhu, J. A new k-e eddy viscosity model for high Reynolds number turbulent flows. *Comput. Fluids* **1995**, *24*, 227–238. [CrossRef]

Article

Numerical and Experimental Investigation on Radiated Noise Characteristics of the Multistage Centrifugal Pump

Qiaorui Si [1], Biaobiao Wang [1], Jianping Yuan [1], Kaile Huang [1], Gang Lin [1] and Chuan Wang [1,2,*]

[1] National Research Center of Pumps, Jiangsu University, Zhenjiang 212013, China; siqiaorui@ujs.edu.cn (Q.S.); wbbujs@163.com (B.W.); yh@ujs.edu.cn (J.Y.); huangkaileujs@163.com (K.H.); lingangujs@163.com (G.L.)

[2] College of Hydraulic Science and Engineering, Yangzhou University, Yangzhou 225009, China

* Correspondence: wangchuan@ujs.edu.cn; Tel.: +86-150-508-54196

Received: 4 September 2019; Accepted: 25 October 2019; Published: 2 November 2019

Abstract: The radiated noise of the centrifugal pump acts as a disturbance in many applications. The radiated noise is closely related to the hydraulic design. The hydraulic parameters in the multistage pump are complex and the flow interaction among different stages is very strong, which in turn causes vibration and noise problems because of the strong hydraulic excitation. Hence, the mechanism of radiated noise and its relationship with hydraulics must be studied clearly. In order to find the regular pattern of the radiated noise at different operational conditions, a hybrid numerical method was proposed to obtain the flow-induced noise source based on Lighthill acoustic analogy theory, which divided the computational process into two parts: computational fluid dynamics (CFD) and computational acoustics (CA). The unsteady flow field was solved by detached eddy simulation using the commercial CFD code. The detailed flow information near the surface of the vane diffusers and the calculated flow-induced noise source was extracted as the hydraulic exciting force, both of which were used as acoustic sources for radiated noise simulation. The acoustic simulation employed the finite element method code to get the sound pressure level (SPL), frequency response, directivity, et al. results. The experiment was performed inside a semi-anechoic room with a closed type pump test rig. The pump performance and acoustic parameters of the multistage pump at different flow rates were gathered to verify the numerical methods. The computational and experimental results both reveal that the radiated noise exhibits a typical dipole characteristic behavior and its directivity varies with the flowrate. In addition, the sound pressure level (SPL) of the radiated noise fluctuates with the increment of the flow rate and the lowest SPL is generated at $0.8Q_d$, which corresponds to the maximum efficiency working conditions. Furthermore, the experiment detects that the sound pressure level of the radiated noise in the multistage pump rises linearly with the increase of the rotational speed. Finally, an example of a low noise pump design is processed based on the obtained noise characteristics.

Keywords: multistage centrifugal pump; flow-induced noise; numerical calculation; acoustic analogy

1. Introduction

Multistage pumps with high pressure are widely used in water supply facilities [1–4]. The flow in the multistage pump is highly unsteady and greatly influenced by the interference between the different stages [5,6]. These give rise to the pressure pulsations, mechanical vibrations, and the radiated noise in various pump components [7–10]. Especially, the intensive pressure pulsations are an important source of hydrodynamic excitation force, which in turn produces fluid borne noise and structure-borne noise. Further, fluid-borne noise is a major contributor to radiated noise as well as leading to increased fatigue in the system components. Additionally, the radiated noise is felt as a disturbance in some applications and many conditions are subject to stringent requirements concerning the limitation of noise. Therefore, the investigation of the radiated noises from the centrifugal pump is significant to pump designers.

Computational fluid dynamics (CFD) has been used widely in many engineering fields with the rapid development of computer technology [11–16]. Previous experimental and computational study of the noise generation in the centrifugal pump assured that the noise source in the centrifugal pump is composed of the mechanical noise and flow-induced noise [17–19]. The radiation noise of the pump system can be significantly reduced by increasing the machining accuracy and using components such as low noise motors. However, the low-frequency noise induced by the fluid flow is very difficult to be eliminated. Recently, computational fluid dynamics were widely used to predict the flow-induced noise and analyze its corresponding sound source [20,21]. By employing the Lighthill's acoustic analogy theory, Howe [22] concluded that the dipoles source is the main source of the flow-induced noise in the centrifugal pump. In addition, Kato [23] proposed a one-way coupled simulation method that combines CFD and structure analysis. Ding [24] and Si [25] used the combined CFD/ computational acoustics (CA) method to simulate the hydraulic noise of centrifugal pumps. Keller [26] theoretically and experimentally investigated the effects of the pump-circuit acoustic coupling on the blade-passing frequency perturbation induced by fluid-dynamic interaction between the rotor and the stator, which provides some theoretical supports for the above numerical works. By employing the large eddy simulation (LES) and the Ffowcs Williams and Hawkings (FW-H) acoustic method, Gao [27] concluded that the design operation generates the lowest total sound pressure level. Jiang [28] reported the fluid-induced noise by using the Fluid-solid coupling method and regarded it as a reliable way to predict the fluid-induced noise in the rotating machinery. Sergey [29] simulated the flow-induced noise based on the acoustic-vertex method, which divided the fluid mechanics equation into the sound item and the vortex item. Rual [30] studied the unsteady flow near the tongue and found that the size of the mesh, the time step, and the turbulence model had great influence on the accuracy of the sound simulation. Furthermore, a hybrid method that transforms the information of the unsteady flow into the source of the sound is widely used [31–33]. This method could improve simulation accuracy with lower computational resources [34]. To be specific, the detached eddy simulation (DES) model, which uses the RANS model to simulate the flow near the wall and employs the LES model to simulate the flow far away from the wall, could improve the computational efficiency and produce equal accurate unsteady results [35]. For the acoustic simulation, the finite element method (FEM) method could predict the acoustic distribution at the low frequency well and show more tolerance to the geometric irregularity.

The present paper focuses on the numerical and experimental investigation of the characteristics of the radiated noise of the multi-stage pump. The sound source was extracted from the unsteady computation based on the DES turbulence model, and the simulation code of acoustic finite element method inlaid in software Actran 14 was used to simulate the distribution of the radiated noise of the model pump. In addition, the change of the radiated noise with the variation of the flow rate and the rotation speed is also discussed.

2. Method and Basic Theory

2.1. Theory of the Acoustic Simulation

Acoustic analogies are derived from the Navier–Stokes equations, which governed both flow field and the corresponding acoustic filed. The Navier–Stokes equations expressed as Formula (1) and (2) are rearranged into various forms of the inhomogeneous acoustic wave equation.

$$\frac{\partial \rho}{\partial t} + \frac{\partial}{\partial x_i}(\rho u_i) = 0 \tag{1}$$

$$\frac{\partial(\rho u_i)}{\partial t} + \frac{\partial(\rho u_i u_j)}{\partial x_j} = -\frac{\partial p}{\partial x_i} + \frac{\partial e_{ij}}{\partial x_j} \tag{2}$$

where e_{ij} is the viscous stress tensor, and the last Lighthill function is expressed as:

$$\frac{\partial^2(\rho - \rho_0)}{\partial t^2} - c_0{}^2 \frac{\partial^2}{\partial x_i \partial x_j}(\rho - \rho_0) = \frac{\partial^2 T_{ij}}{\partial x_i \partial x_j} \tag{3}$$

where T_{ij} is the stress tensor, and $T_{ij} = \rho u_i u_j - e_{ij} + \delta_{ij}[(p - p_0) - c_0{}^2(\rho - \rho_0)]$; δ_{ij} is the Kronecker function, ρ is the fluid density, ρ_0 is the undisturbed density, c_0 is the sound velocity, t is the time, x is the space coordinates, while the i, j represent the direction of the coordinate axis.

Furthermore, by employing several mathematical manipulations, the above function could be transformed as:

$$\int_\Omega \left(\frac{\partial^2}{\partial t^2}(\rho - \rho_0)\delta \rho + c_0^2 \frac{\partial}{\partial x_i}(\rho - \rho_0)\frac{\partial(\delta \rho)}{\partial x_i} \right) dx = -\int_\Omega \frac{\partial T_{ij}}{\partial x_j}\frac{\partial(\delta \rho)}{\partial x_i} dx + \int_\Gamma \frac{\partial \Sigma_{ij}}{\partial x_j} n_i \delta \rho d\Gamma(x) \tag{4}$$

The first term of the right is the volume source, and the second term is the surface source. In the acoustic simulation, the solution of the Navier-Stokes equation firstly assumes the water as incompressible to calculate the flow-induced acoustic source. Then, the compressibility of the water is to be considered to solve the acoustic wave propagation.

In addition, the description of the radiated noise is related to time and space. Mathematically, this relationship could be expressed as an acoustic wave equation. The process of the sound radiation meets the basic physical law, which is expressed as:

$$(\rho_0 + \rho')\left(\frac{\partial}{\partial t} + v\nabla\right)v = -\nabla(p_0 + p') \tag{5}$$

$$\frac{\partial(\rho_0 + \rho')}{\partial t} = (\rho_0 + \rho')q' - (\rho_0 + \rho')\nabla v \tag{6}$$

$$p' = \frac{\gamma p_0}{\rho_0}\rho' + \frac{\gamma(\gamma - 1)}{2\rho_0{}^2}(\rho')^2 \tag{7}$$

where, ρ_0, v_0 and p_0 respectively represent the static density, static velocity, and static sound pressure. And ρ', v', p', and q', respectively represent the increment of the density, velocity, sound pressure, and the mass. With some mathematical treatment, the acoustic wave equation is expressed as:

$$\nabla^2 p' = \frac{1}{c^2}\frac{\partial^2 p'}{\partial t^2} - \rho_0 \frac{\partial q'}{\partial t} \tag{8}$$

where $\nabla^2 = \frac{\partial^2}{\partial x^2} + \frac{\partial^2}{\partial y^2} + \frac{\partial^2}{\partial z^2}$ and $c = \sqrt{\gamma \frac{p_0}{\rho_0}}$, which is the sound velocity in the fluid.

The vibration at any time could be decomposed as several simple harmonic vibrations expressed as:

$$p' = p(x, y, z)e^{j\omega t} \tag{9}$$

$$q' = q(x, y, z)e^{j\omega t} \tag{10}$$

A combination of the above three equations could deduct the acoustic wave equation expressed as;

$$\nabla^2 p(x, y, z) - k^2 p(x, y, z) = -j\rho_0 \omega q(x, y, z) \tag{11}$$

where $k = \frac{2\pi f}{c}$ and this value represents the number of the waves, ω is the angular velocity.

2.2. DES Method

The DES method is a modification of a RANS model in which the model switches to a subgrid-scale formulation in regions fine enough for LES calculations, which calculates the sound source information and simultaneously cuts down the cost of the computation. It is initially formulated for the Spalart–Allmaras model, which is expressed as:

$$\frac{D\tilde{v}}{Dt} = c_{b1}\tilde{S}\tilde{v} + \frac{1}{\sigma}[\nabla\cdot((v+\tilde{v})\cdot\nabla\tilde{v} + c_{b2}(\nabla\tilde{v})]^2 - c_{w1}f_w[\frac{\tilde{v}}{d_w}]^2 \tag{12}$$

where the variables that represent turbulent motion are the quantities directly solved by the S-A equation. The relationship with the motion viscosity coefficient is defined as:

$$v_t = \frac{\mu_t}{\rho} = \tilde{v}f_{v1} \tag{13}$$

where f_{v1} is a dimensionless function defined as:

$$f_{v1} = \frac{\chi^3}{\chi^3 + c_{v1}^3}, \chi = \frac{\tilde{v}}{v} \tag{14}$$

where v is an expression of the molecular viscosity generation term expressed as:

$$\tilde{S} = f_{v3}S + \frac{\tilde{v}}{\kappa^2 d_w^2}f_{v2} \tag{15}$$

where S is the absolute value of vorticity, and f_{v2} and f_{v3} express dimensionless functions, which is respectively expressed as:

$$f_{v2} = (1 + \frac{\chi}{c_{v2}})^{-3}, f_{v3} = \frac{(1+\chi f_{v1})(1-f_{v2})}{\chi} \tag{16}$$

f_w is expressed as:

$$f_w = g(\frac{1+c_{w3}^6}{g^6 + c_{w3}^6})^{\frac{1}{6}}, g = r + c_{w2}(r^6 - r), r = \frac{\tilde{v}}{\tilde{S}\kappa^2 d_w^2} \tag{17}$$

Constant value is evaluated as:

$$c_{b1} = 0.1355, c_{b2} = 0.622, \sigma = 2/3, \kappa = 0.41,$$
$$c_{w1} = \frac{c_{b1}}{\kappa^2} + \frac{1+c_{b2}}{\sigma}, c_{w1} = 0.3, c_{w3} = 2, c_{v1} = 71 \tag{18}$$

DES method is based on the S-A model which replaces the feature-length d_w as:

$$\tilde{d} = \min(d_w, C_{DES}\Delta) \tag{19}$$

where $\Delta = \max(\Delta x, \Delta y, \Delta z)$ represents the maximum grid distance, $C_{DES} = 0.6$.

The default value of the software is adopted in the article. The size of the characteristic length is related to the scale of the grid, and the DES method is an S-A model of the LES simulation.

2.3. Simulation Procedure

Considering that the flow field inside the multistage pump is complex and the acoustic simulation domain is irregular, the finite element method is used in this acoustic simulation which could avoid the inaccuracy during the transformation of the sound source. Firstly, the computational flow domain and corresponding structured grids were prepared to process the steady simulation of the pump flow field based on the experimental inlet and outlet boundary conditions. Simultaneously, the test rig was built to use the obtained pump performance data such as the pump head to verify or modify the numerical calculation model. The shear stress transport (SST) turbulence model embedded in ANSYS CFX 14.1 software (ANSYS, Inc., Commonwealth of Pennsylvania, USA) was repeatedly used at this step. Afterwards, the correct fluid calculation domain and the appropriate grid were obtained. Secondly, the unsteady simulation of the pump flow field was obtained using the DES method and the detailed flow information, such as flow velocity, pressure, density etc, was extracted and transformed as the sound source in the acoustic simulation. Meanwhile, the acoustic calculation domains were built and the acoustic meshes were generated containing an interface setting. Finally, the radiated noise calculation was completed by the acoustic finite element method using Actran12.0 software (Free Field Technologies MSC Software Company, Mont-Saint-Guibert, Belgium). The acoustic characteristics of the model pump were analyzed after experimental verification. The flow chart of the simulation work is shown in Figure 1.

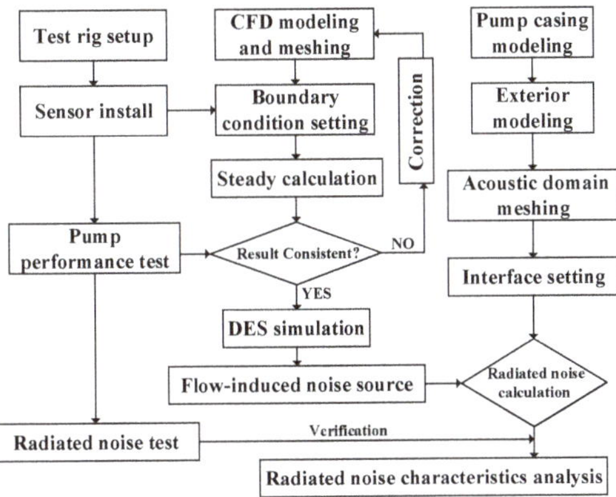

Figure 1. The flow chart of the radiated noise calculation of the multi-stage centrifugal pump.

3. Numerical Method

3.1. Study Object

The study object is focused on a self-priming multistage pump type, which is widely used to feed water to the high floors of a building. The field measurement revealed that this type of multistage pump generated a higher sound pressure level than the technical requirements. The basic analysis of this problem showed the noise is caused by the hydraulic design. The geometry features of the example multistage pump with five stages in this study is shown in Figure 2a. Its inflow is axial and its outflow is along the radial direction. Each stage of this pump is composed of one impeller and one vane diffuser. The impeller constitutes seven backswept blades, and the vane diffuser is two-dimensional with 12 guiding vanes and 12 returning vanes. The design impeller rotation speed of the pump is 2800 r/min and the design flow rate of the pump is 8 m^3/h. The single-stage head of the pump is 10 m and the total head is a superposition of the stages. The detailed hydraulic parameters of each stage are presented in Table 1. In order to simplify the model, this study only takes the three-stages hydraulic part of the pump and makes a prototype as shown in Figure 2b for the simulation and experimental test.

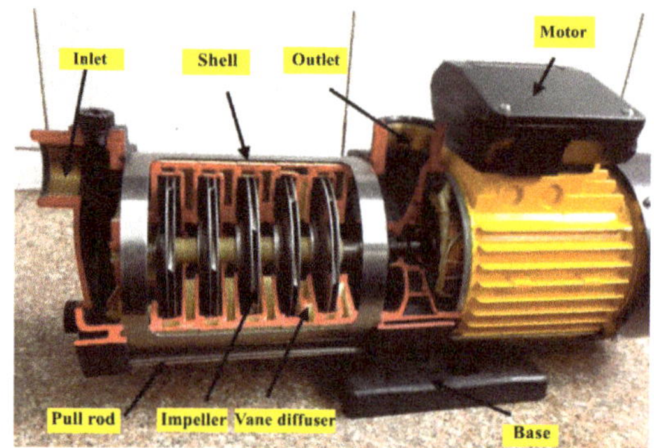

(**a**) The cutaway view of the multi-stage centrifugal pump prototype.

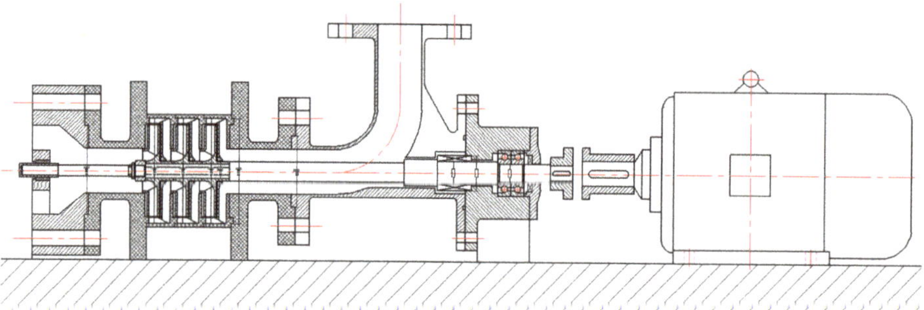

(**b**) Test prototype of the simplified model pump.

Figure 2. Pump model.

Table 1. Main characteristics of the model pump.

Geometry	Values	Geometry	Values
Impeller inlet diameter, D_j	45/mm	Blade outlet angle, β_2	37/°
Impeller outlet diameter, D_2	103/mm	Impeller blades number, Z_a	7
Impeller outlet width, b_2	10/mm	Diffuser vane number, Z_d	12
Diffuser inlet diameter, D_3	104.5/mm	Specific speed, n_s	86

3.2. Fluid Field Simulation

3.2.1. Computational Domain

In general, the multistage pump is composed of the first stage, the last stage, and several middle stages. To save the computational resources, the number of the middle stages was reduced to one stage and the first and last stages were still kept, as presented in Figure 3. Such simplification is appropriate to reflect the physical problems of the study object because the flow in the middle stages is similar except the initial pressure. Each computational stage includes one impeller, one vane diffuser, and the gap between the impeller and vane diffuser. Apart from the three stages, the computational stage is composed of the inlet section, the wear-ring section, pump cavity, impeller, vane diffuser, and the outlet section, which are consistent with the flow path of the experimental model pump. Both the inlet and outlet suction are extended appropriately to ensure the numerical stability, simultaneously to comply with the experimental boundary conditions.

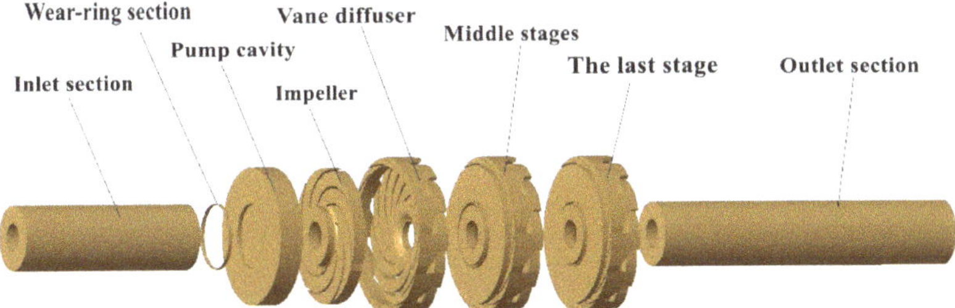

Figure 3. Domains of the flow field calculation.

3.2.2. Mesh Generation and Boundary Conditions

The structured mesh is applied for the computational domains by using ANSYS ICEM version 14.1 (ANSYS, Inc., Commonwealth of Pennsylvania, USA). As shown in Figure 4, a special refinement is applied around the blades and vane diffusers to improve the accuracy of the simulation. In addition, six groups of computational grids have been chosen to analyze the influence of the mesh size on the prediction of the hydraulic performance of the model pump. Figure 5 presents the analysis of mesh independence verification, the total element number is chosen as 6,680,830 by considering the simulation accuracy and efficiency. In addition, boundary conditions concern a bar at the inlet and the mass flow rate at the outlet of the computational domain. All the other walls are treated as non-slip boundaries. The maximum nondimensional wall distance, $y+ < 10$, was obtained in the complete flow field, which could satisfy the requirement of all turbulence modeling methods used in this paper. The pre-converged steady flow field (based on the SST turbulence model) obtained is accepted as the initial condition followed by the unsteady simulation (based on the DES model). The time step for the unsteady simulation is set as 1.7857×10^{-4} s.

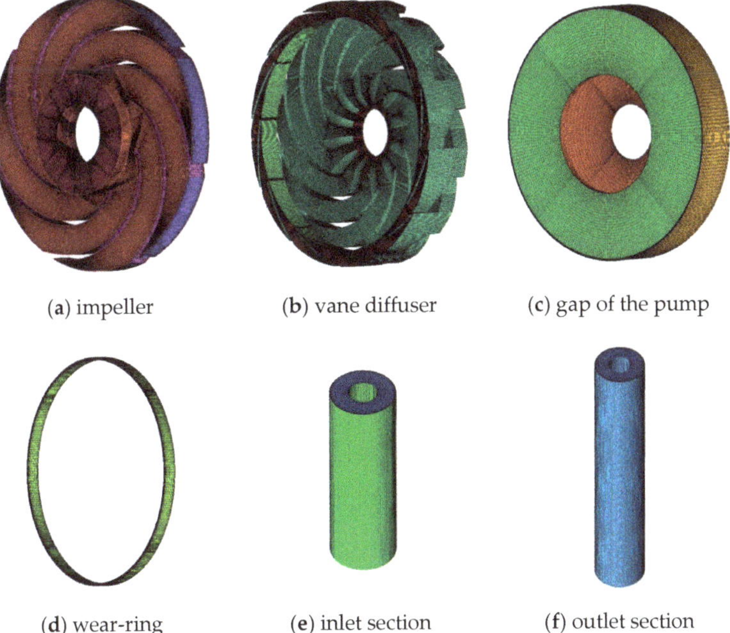

Figure 4. Mesh of the multi-stage centrifugal pump.

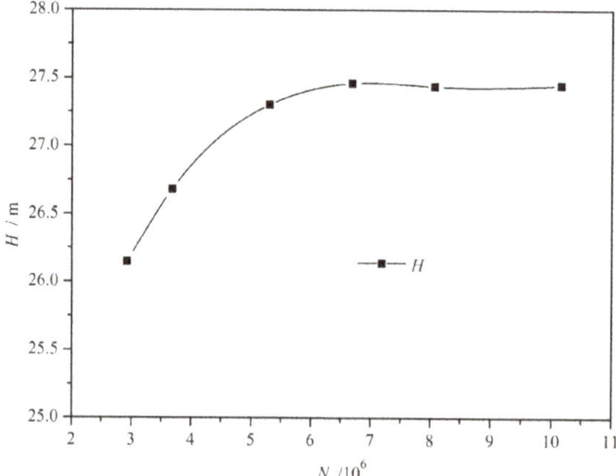

Figure 5. Analysis of mesh independence.

3.2.3. Flow Field Results

The global performance of the model pump is shown in Figure 6. The experimental and simulated data of the total head agree well with nearby design flow rates. The calculated head is slightly different from the measured value at the part-load and overload condition. The value of calculated efficiency is lower than the measured one and the differences become bigger with the increment of the flow rate. While the error of the head between the experiment and the simulation is lower than 4.5% and the error of the efficiency is lower than 3.8%, the established calculation model and the selected number of grids could fully support the next acoustic calculations. By the way, the highest efficiency working condition of the model pump appears in the $0.8Q_d$. This might be related to subsequent acoustic characterization.

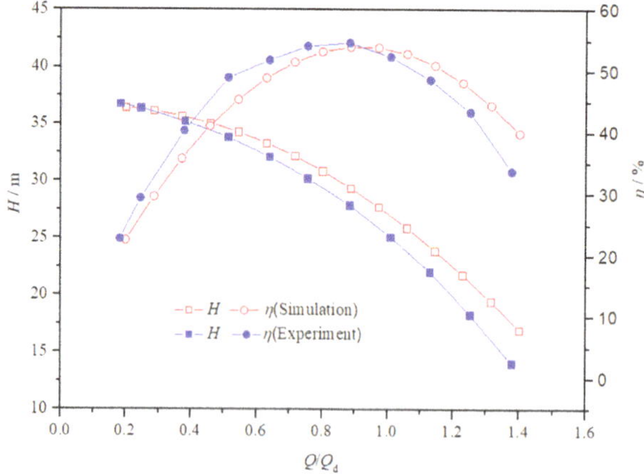

Figure 6. External characteristics of numerical calculation and experiment.

Figure 7a–d describes the velocity distribution on the mid-span of the impeller and the diffuser. At the $0.6Q_d$ condition, the velocity distribution in the impeller is not uniform and the separation is detected at the inlet of the diffuser. With the increase of the flow rate, the flow field becomes better distributed while the diffusers are impugned by the high-velocity flow. The wake flow induced by the impingement can generate higher pressure pulsations. As pointed out by Gülish [36], the wake flow and the separation in the centrifugal pump gives rise to the pressure pulsations and the subsequent radiated noise.

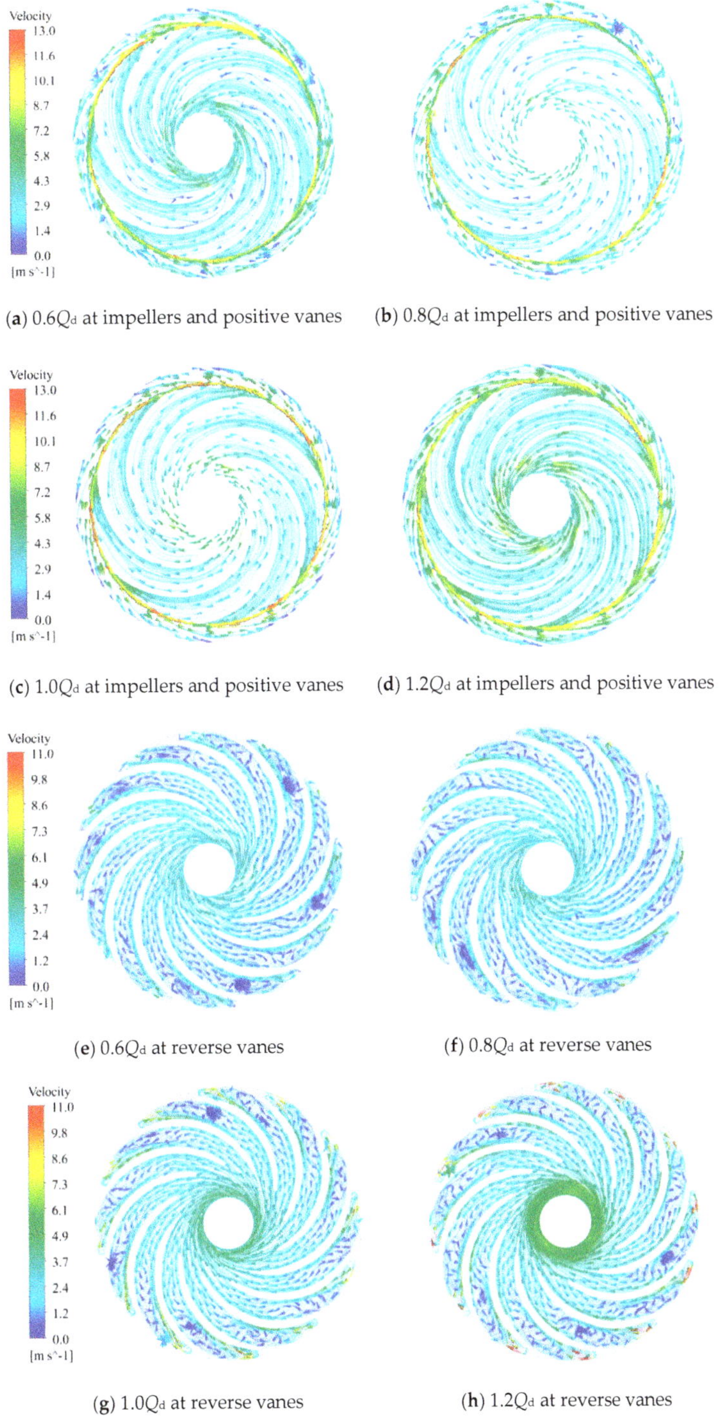

(**a**) $0.6Q_d$ at impellers and positive vanes

(**b**) $0.8Q_d$ at impellers and positive vanes

(**c**) $1.0Q_d$ at impellers and positive vanes

(**d**) $1.2Q_d$ at impellers and positive vanes

(**e**) $0.6Q_d$ at reverse vanes

(**f**) $0.8Q_d$ at reverse vanes

(**g**) $1.0Q_d$ at reverse vanes

(**h**) $1.2Q_d$ at reverse vanes

Figure 7. Velocity distribution in the mid-scan of the pump's first stage at different flowrates.

Considering the influence of the upstream flow, the flow in the returning vane diffuser passages is more complex, as shown in Figure 7e–h. At the part-load condition, several separations block several passages. Additionally, the flow is greatly deaccelerated in the diffuser passages at the overload condition and the separation flow is detected near the suction side of the diffuser. When the flow reaches $0.8Q_d$, the number of separations is the fewest. This might indicate that minor noise would produce at this flow condition.

In order to understand the pressure pulsation characteristics of the flow in the model pump at various stages of the multistage centrifugal pump, 12 monitoring points at every stage, totaling 36 points, are set on the cross-section of the impeller, the positive vane, and the reverse guide vane. As shown in Figure 8, the first letter in each monitoring point name is: F means the first level, S means the second level, and T means the third level. Moreover, the second letter in each monitoring point name is: Y means the impeller, D means the positive vane, and F means the reverse guide vane. The dimensionless pressure pulsation coefficient $C_p{}^*$ as shown in Formula (20) is used for further data reduction.

$$C_P{}^* = \frac{(p - \bar{p})}{0.5\rho v_2{}^2} \tag{20}$$

where p is the static pressure of the monitoring point, $\bar{p}$ is the average value of the static pressure, ρ is the fluid density, and v_2 is the circumferential velocity at the impeller outlet. Afterward, standard deviation C_{PS} was used to characterize the pressure pulsation intensity of each monitoring point, and fast Fourier transform processing was used to obtain the frequency spectrum of the flow pressure fluctuation.

As shown in Figure 8b, the pressure pulsation intensity of the monitoring points in the multi-stage centrifugal pump shows a periodic change, and it always became larger around the positive guide vane flow channel, indicating that the flow between the stages is similar and the biggest hydraulic exciting force appears at the positive guide vane. The maximum value of the pressure pulsation intensity appears on monitoring points near the throat of the positive guide vane, indicating that the rotor-stator interaction between impeller and positive guide vane makes the greatest contribution to the pressure pulsation intensity. Other flow conditions show the same results. For more detail, the maximum pressure pulsation intensity appears in the first-stage positive guide vane flow channel, and the secondary and final guide vanes make less difference. Then, frequency analysis of the pressure fluctuation at different flow rates is processed on the monitoring points of the first positive guide vane, as shown in Figure 8c–f. Since the sampling time step of the numerical calculation is 1.7857×10^{-4} s, the sampling frequency is 5600 Hz. According to the Nyquist sampling theorem, the maximum frequency that can be analyzed is 2800 Hz.

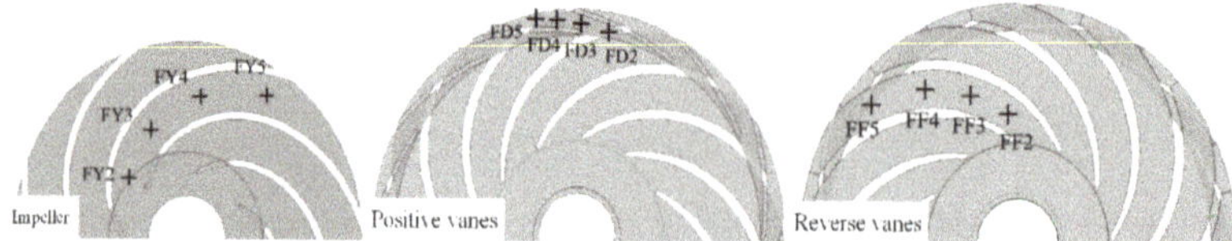

(**a**) monitoring points at the first stage of the model pump

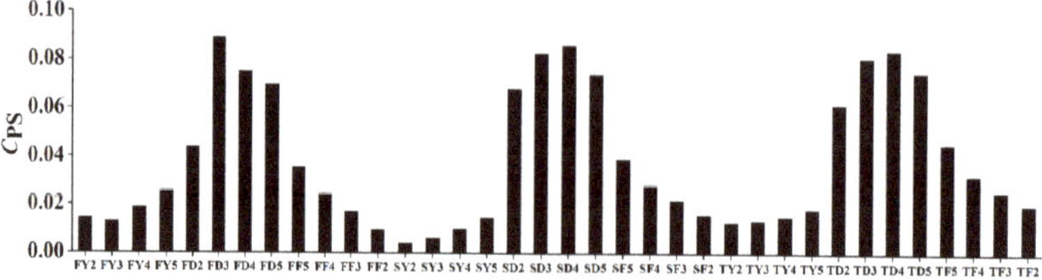

(**b**) amplitude distribution of pressure fluctuation at $1.0Q_d$

Figure 8. *Cont.*

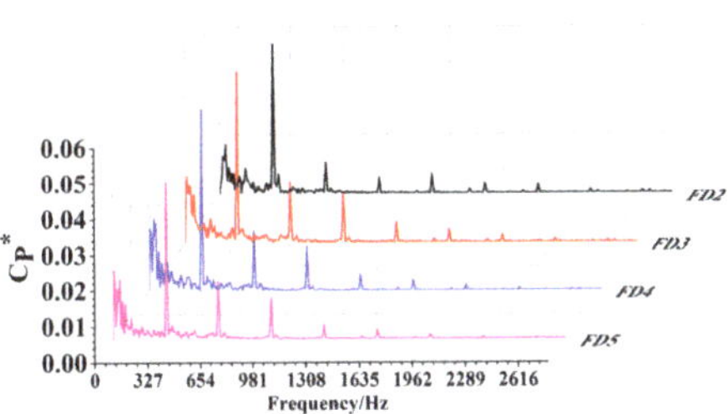

(**c**) spectrum of pressure fluctuation at 0.6Q_d

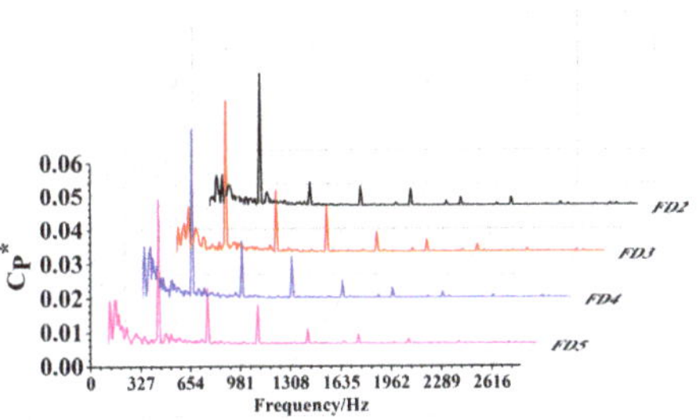

(**d**) spectrum of pressure fluctuation at 0.8Q_d

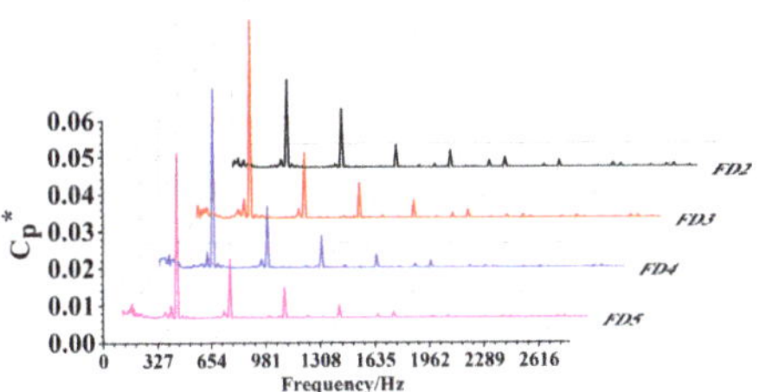

(**e**) spectrum of pressure fluctuation at 1.0Q_d

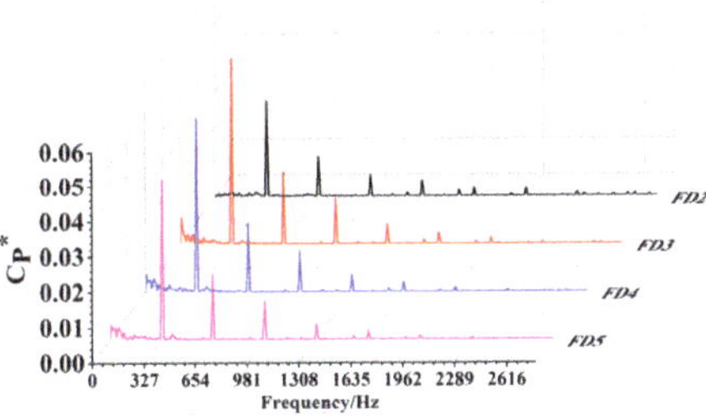

(**f**) spectrum of pressure fluctuation at 1.2Q_d

Figure 8. Pressure fluctuation of monitoring points.

The results of the frequency spectrum show that the pressure pulsation is mainly concentrated in the low-frequency region within 1000 Hz, and the main frequency is the blade passing frequency (327 Hz) and its multiple. The monitoring point with a large peak is also located near the throat of the positive guide vane at all flow rates, which corresponds to the pressure pulsation intensity results. Under the small flow conditions such as $0.6Q_d$ and $0.8Q_d$, there are many low-frequency pulsations smaller than the blade passing frequency for all monitoring points. It means the periodic pressure pulsation caused by the rotor-stator interaction between impeller and positive guide vane is the major source of the pump unstable operation. The $0.8Q_d$ has the lowest amplitude of the frequency spectrum, which corresponds to the phenomenon that this flowrate present the highest efficiency.

3.3. Radiated Noise Analysis

3.3.1. Computational Domain

The computational domain for the acoustic simulation includes the structural domain and air-borne domain, as shown in Figure 9. The inner surface of the vane diffuser of the structural domain is loaded with information of the unsteady flow to get the sound source in the subsequent simulation, and the definition of the air-borne domain is to get the distribution of the radiated noise in the subsequent simulation.

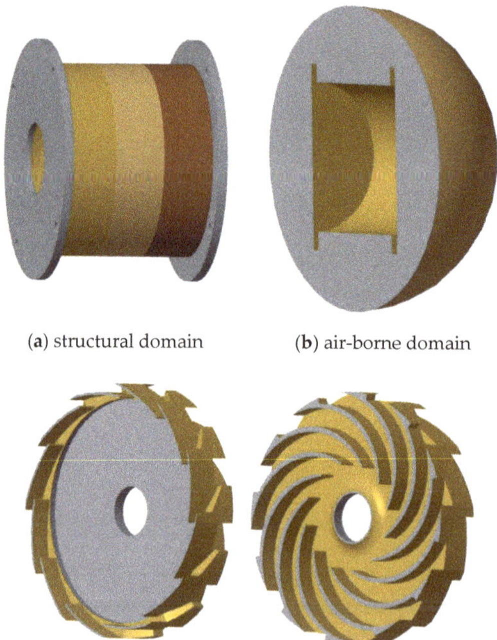

(**a**) structural domain (**b**) air-borne domain

(**c**) the structure of the vane diffuser

Figure 9. Calculation domains of radiated noise simulation.

3.3.2. Mesh Generation and Boundary Condition

Unstructured mesh, which has better adaptability to the geometry, is applied in the acoustic simulation. To guarantee the precision of the acoustic computation, the maximum mesh size should meet the Formula (21). In this study, the maximum mesh size is 0.008 m. And the mesh for acoustic simulation is shown in Figure 10.

$$L_{max} < \frac{c}{6 f_{max}} \tag{21}$$

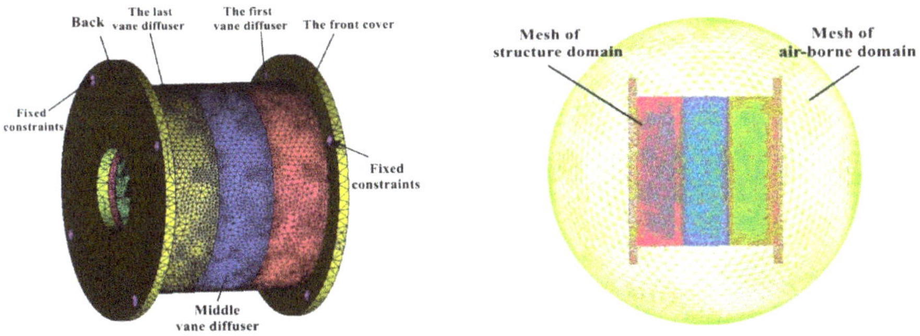

(**a**) mesh for the structural domain   (**b**) acoustic simulation mesh

Figure 10. The sound field calculation mesh of multistage centrifugal pump.

The detailed information of the unsteady flow is extracted as the sound source near the surface of the vane diffusers. In addition, the material properties of the structure domain are shown in Table 2. Considering the total time of the unsteady CFD simulation and the time step, the frequency range of the acoustic simulation is set from 0 Hz to 2800 Hz, and the resolution is set as 5.3 Hz. The data transmission of the interface between the structural domain and the air-borne domain is finished with the integral interpolation method. In order to analyze the properties of the radiated noise, 60 monitor points are mounted equally in the mid-span surface of the second impeller and distances between the diffuser surfaces of these points are 1 m, as shown in Figure 11.

Table 2. Material properties of the structure domain.

Material	Density/(kg/m^3)	Young's Modulus/GPa	Poisson's Ratio
ABS	1040	200	0.394

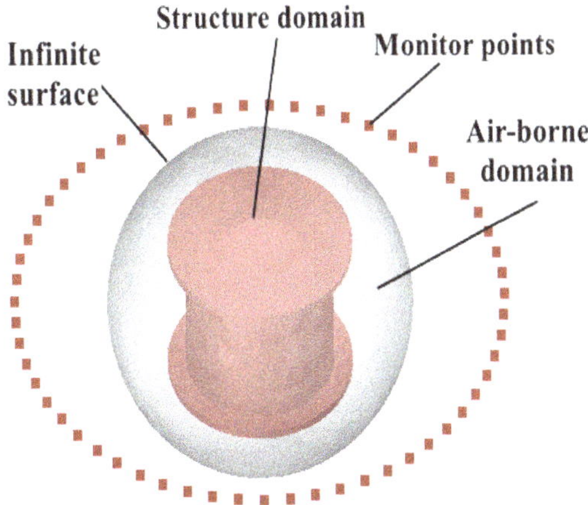

Figure 11. The pre-setting of radiated noise calculation.

3.3.3. Acoustic Field Results

Figure 12 shows the predicted sound pressure spectra at different flow rates. As can be seen, the dominant frequency of the multi-stage pump is the blade passing frequency (327 Hz). Furthermore, the sound pressure level (SPL) of the radiated noise at the low frequency increases with the increase of the flow rate. In contrast, the SPL above 2700 Hz has the opposite trend. Additionally, the SPL at the blade passing frequency decreases slightly before $0.8Q_d$, and increases afterward. In addition, there is a slight increase in the SPL around 1500 Hz, which is assumed to be linked with the vibration mode of the pump system.

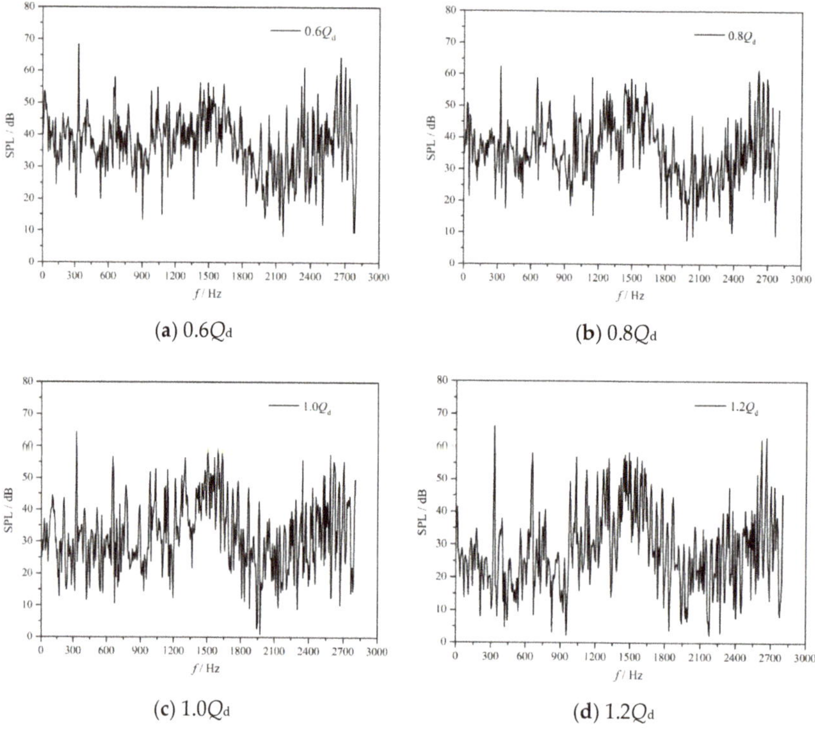

(a) $0.6Q_d$ (b) $0.8Q_d$

(c) $1.0Q_d$ (d) $1.2Q_d$

Figure 12. Frequency response curves of radiated noise.

Figure 13 shows the sound pressure level contour at blade passing frequency. The sound pressure level reaches the highest at the $0.6Q_d$, while the sound pressure level reaches the highest near the vane diffuser of the first stage. This is due to the fact that the intensity of the pressure fluctuation at the first stage is highest and the structural strength is relatively lower around the first stage.

The profiles of the SPL of the directivity field shown in Figure 14 demonstrate the dipole characteristic behavior. It is found that two SPL valleys appear around at $\theta = 120°$ and $300°$ and fluctuate slightly with the change of the flowrate. This phenomenon illustrates that the rotor/stator interaction is the main source of the fluid-induced noise.

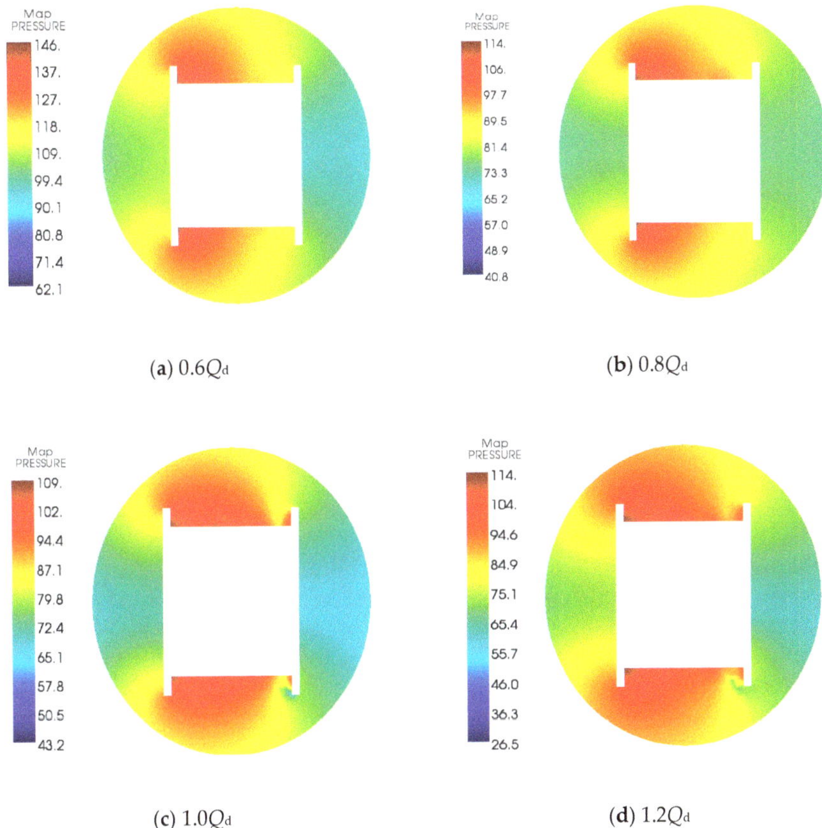

(**a**) $0.6Q_d$

(**b**) $0.8Q_d$

(**c**) $1.0Q_d$

(**d**) $1.2Q_d$

Figure 13. The sound pressure level contour at blade passing frequency.

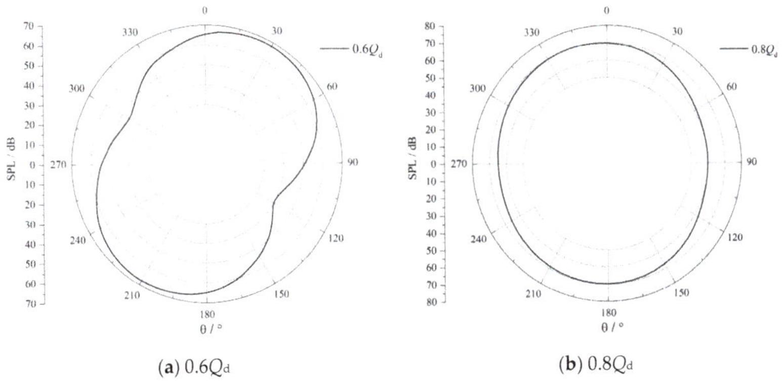

(**a**) $0.6Q_d$

(**b**) $0.8Q_d$

Figure 14. *Cont.*

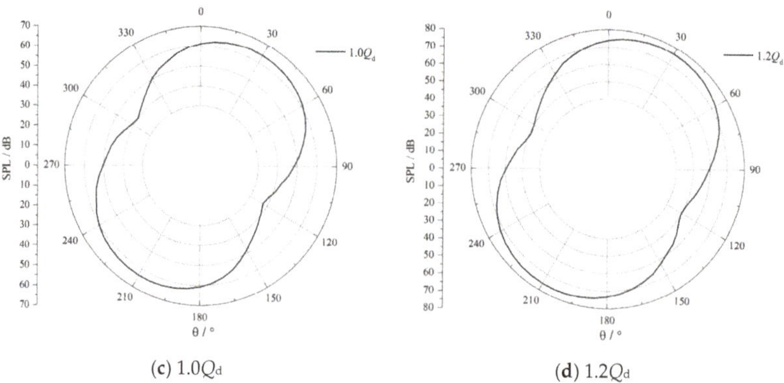

(**c**) 1.0Q_d (**d**) 1.2Q_d

Figure 14. The directivity of the radiated noise field at blade passing frequency.

4. Experimental Verification

The experimental research was done in a closed-loop test rig beside the semi-anechoic room of the National Research Center of Pumps, China. The test rig shown in Figure 15 for pump performance measurement meets the Grade 2 accuracy based on the ISO 9906.2012 standard. The accuracy of the flow rate measurements is ±2.5%, the head is ±3%, the torque is ±2.5%, and the rotation is ±1%. In addition, to verify the validity of the acoustic simulation results, the radiated noise of the multi-stage pump is measured in the semi-anechoic room built with 15 dB background noise and 50 Hz cut-off frequency. As shown in the above figure, the five walls of this semi-anechoic room are equipped with an anechoic wedge that could form a semi-free acoustic field. In order to improve the accuracy of the measurement, the multi-stage pump is installed on a damping base and the pipelines are supported with the damping disc. In addition, an anechoic tank is arranged between the water tanks and the multi-stage pump to reduce the flow-induced noise inside the pipelines. All the data collection equipment is put in the monitoring room to minimize the influence of the accuracy of the measurement.

The data collection system is composed of the module to measure the static characteristic and the other module to capture the radiated noise of the model pump. The magnetic inductive flowmeter performs the measurements of the volume flow, pressure sensors capture the static pressure at the inlet and outlet of the multi-stage pump and monitor the torque value of the model pump. The noise is measured with the PCB 14043 type microphone (PCB Piezotronics Inc., New York, USA) and processed based on the LMS test Lab platform. The accuracy of the SPL of the radiated noise is ±1 dB. The acquisition sampling frequency is set to 6000 Hz, and the resolution frequency is set to 0.5 Hz. Every test used 120 s for the signal acquisition and treated frequency resolution by the Hanning window in general.

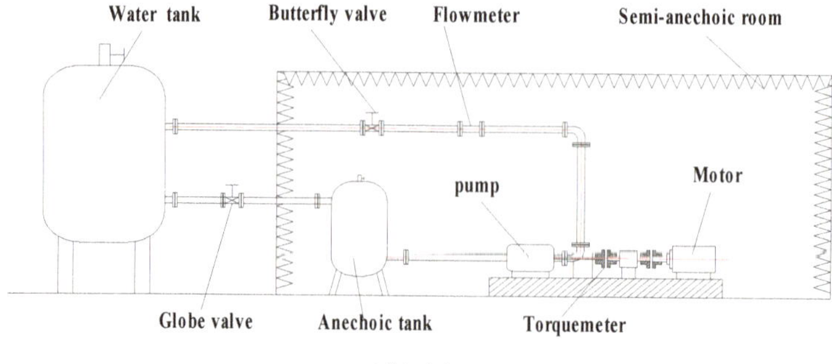

(a) test rig

Figure 15. *Cont.*

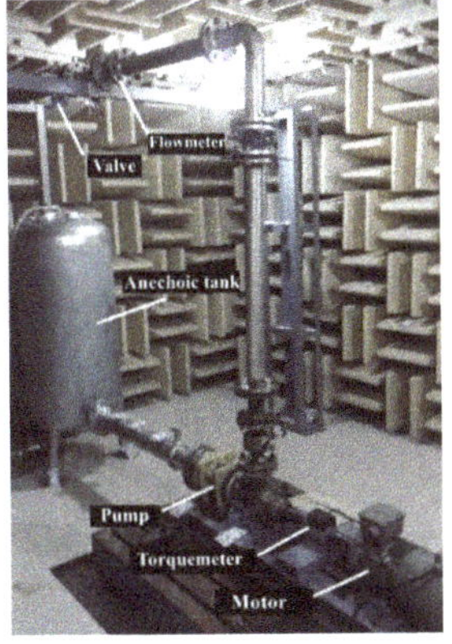 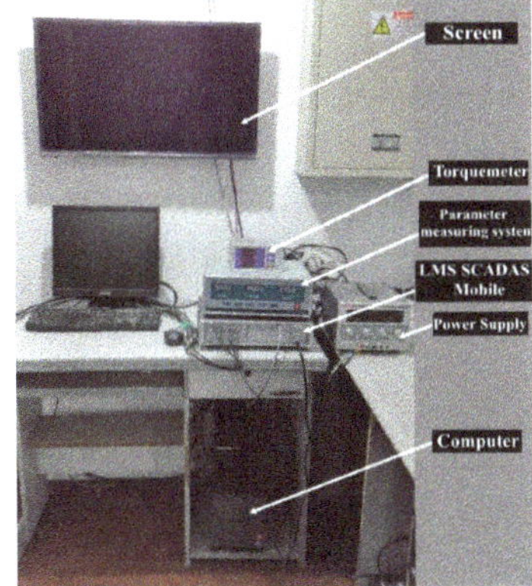

<table>
<tr><td align="center">(b) model pump and the layout of the pipelines</td><td align="center">(c) monitoring room</td></tr>
</table>

Figure 15. Test of the multi-stage centrifugal pump.

4.1. Radiated Noise at the Different Flow Rate

To investigate the relationship between the radiated noise and flow rate, the flow rate is regulated by the butterfly valve to detect the radiated noise between 3 m³/h–11.6 m³/h ($0.375Q_d$–$1.45Q_d$). Five positions set for the microphones followed the Chinese standard GB/T 29529-2013 [37], as shown in Figure 16a, consistent with the position of the monitoring point set in the numerical simulation. Figure 16b shows the setup with enclosures. Figure 17 shows the radiated noise of the model pump at different flowrates concerning the above two conditions and the comparison with the numerical simulation results. In order to analyze the variation of the total sound pressure level of the radiated noise of the multi-stage centrifugal pump, the average total sound pressure level $\overline{L}_P$ is expressed as follows. Where N represents the number of sound monitoring points and L_{pi} is the total sound pressure level of each sound monitoring point.

$$\overline{L}_P = 10 \cdot \lg\left[\frac{1}{N}\sum_{i=1}^{N} 10^{0.1L_{Pi}}\right] \tag{22}$$

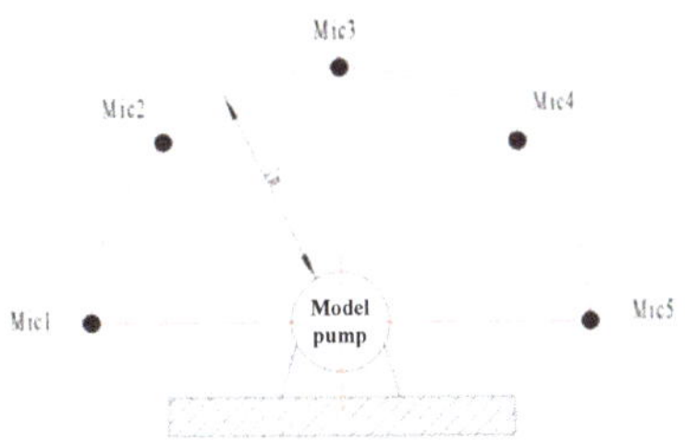

<table>
<tr><td align="center">(a) location of the microphone</td><td align="center">(b) with acoustic enclosure</td></tr>
</table>

Figure 16. Experiment method.

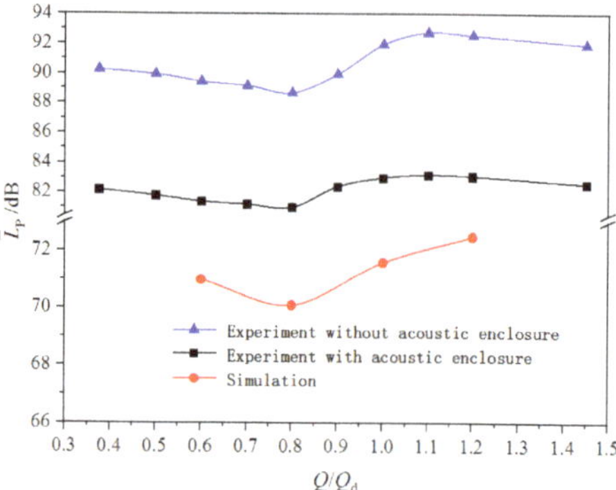

Figure 17. The radiated noise of the multi-stage centrifugal pump at different flowrates.

As seen in Figure 17, the radiated noise of the model pump at all flow rates with acoustic enclosure is 8 dB lower than the case without the enclosure. This fact proves that the enclosure is necessary for the experiment. On the other hand, the simulated total sound pressure level is consistent with the experiment that the total sound pressure level fluctuates with the increment of the flow rate and the minimal value emerges around $0.8Q_d$ where the efficiency is the highest. As pointed out by Gülish [24], the sound pressure level of the induced noise is in the inverse relationship with the efficiency. The differences between the simulation and experiment are within the order of 10 dB because the background noise inside the anechoic chamber still provides some disturbance. Although we used sound elimination materials on the pipe system, the motor still emits a strong radiated noise even when we used an enclosure.

4.2. Radiated Noise at the Different Rotational Speed

The rotational speed of the centrifugal pump has the direct influence of the pressure distribution in the model pump and meantime influence on the radiated noise of the model pump. A frequency converter allows changing the rotational speed of the pump. As seen in Figure 18, the sound pressure level increases almost linearly with the increase of the rotational speed. Processed by the fitting instruments, their relation meets the following formula:

$$y = 0.00545x + 66.7 \tag{23}$$

In this formula, y represents the sound pressure level, x represents rotational speed, and the linear dependence between the fitted curve and the measured data is 0.97894.

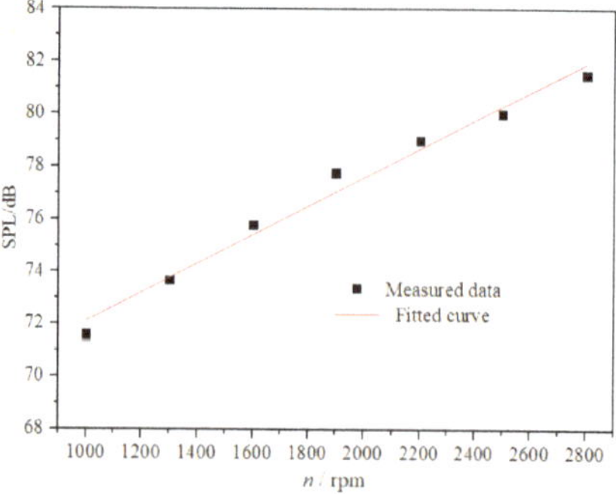

Figure 18. The radiated noise of the multi-stage centrifugal pump at different speeds.

4.3. Directivity of Radiated Noise at Different Flowrates

To analyze the contribution of the different types of sound sources, the sound pressure level in the hemispherical surface around the source is measured. Microphones are mounted equally in the mid-span surface of the model pump and these five microphones are 1 m away from the surface of the second diffuser.

Considering the number of the microphones is limited, an interpolation method is used to get the directivity of the radiated noise at different flow rates, as shown in Figure 19. The experimental and simulated data of directivity agree well, considering the effect of background noise. It is obvious that the dipole is the main sound source of the multi-stage centrifugal pump at the different flowrates. Additionally, directivity at the $0.8Q_d$ is vertical, which explains that the directivity varies with the change of the flowrate.

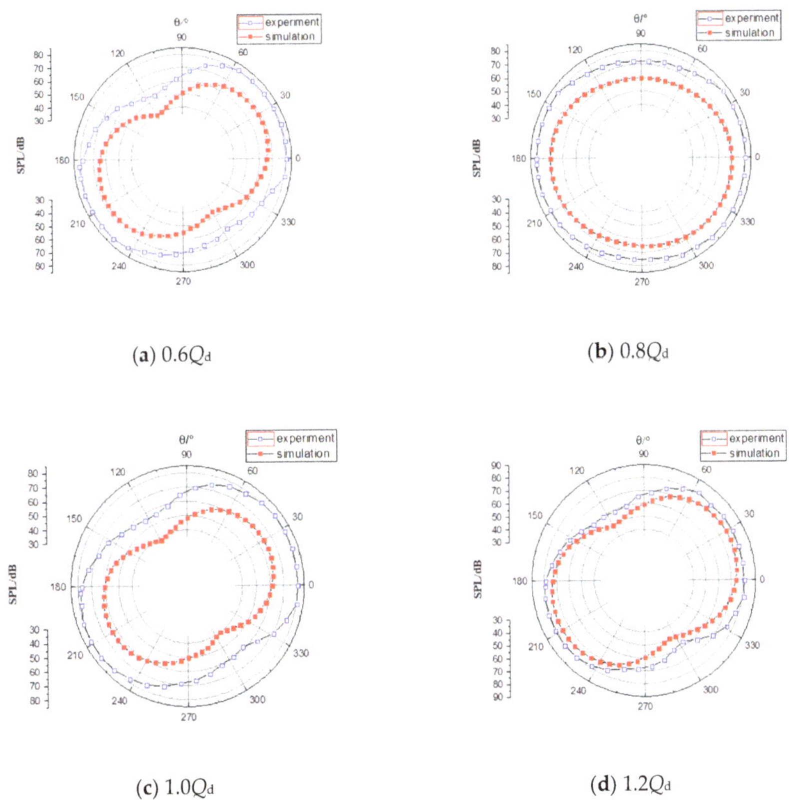

(a) $0.6Q_d$ (b) $0.8Q_d$

(c) $1.0Q_d$ (d) $1.2Q_d$

Figure 19. Directivity of radiated noise at different flowrates.

5. Hydraulic Optimization Design

From the above experimental and numerical analysis, the main frequency of pressure pulsation and radiation noise of the prototype model pump is the blade passing frequency, and the amplitude under it is the largest. The flow passage is too narrow due to the unreasonable design of the number of blades of the impeller and the vane, which leads to poor fluid permeability. There are many vortices in the flow passage of the prototype pump model, and the intensity of its pressure pulsation is large, thereby generating a large radiated noise.

In general, the main measures for reducing the radiated noise of the multi-stage pump include changing the rotational speed and process hydraulic optimization. Optimizing the effect of rotor-stator interaction is the most important region for the hydraulic design of a low noise multi-stage pump because the rotating speed is generally determined by the matching motor. Since the matching between the impeller and the guide vane of the multi-stage centrifugal pump also has a great influence on the pump performance, this study starts with the internal flow improvement and carries out a multi-objective design for the head, efficiency, and noise. The best match between the impeller and the guide vane is derived according to Formula (24) and (25). Finally, the number of impeller blades is six and the vane blade number is nine through mathematical calculation. Other geometric parameters were obtained by the velocity coefficient design method of the centrifugal pump and orthogonal optimization design [1].

$$H_{impeller} = \frac{u_2}{g}\left(\sigma u_2 - \frac{Q\cot\beta_2}{\pi D_2 b_2 \psi_2}\right) \tag{24}$$

$$H_{vane} = \frac{\omega}{g}\frac{Q}{a_3 b_3 Z_{vane}}\frac{D_3}{2} \tag{25}$$

The numerical calculation process including fluid calculation domain modeling, meshing, boundary condition setting, and flow field calculation is the same as before. Figure 20 gives the simulation results of the optimized pump. Seen from the calculated pump performance, the head and efficiency are higher than the original one after optimization. The efficiency after optimization increased by 12.9% under design flow. At the same time, the optimized multi-stage centrifugal pump achieves the highest efficiency around $1.1Q_d$, and the highest efficiency zone is widened.

The optimized multi-stage centrifugal pump pressure pulsation intensity shown in Figure 20c has the same characteristics as the original pump, that is, the pressure pulsation intensity in the impeller is the smallest, the pressure pulsation intensity in the positive guide vane is the largest, and the maximum value still appears in the positive guide vane passage near the throat at all stages. The pressure pulsation intensity of the optimized multi-stage centrifugal pump is reduced correspondingly to the same position as the original pump. Figure 20d revels that the dominant frequency of the radiated noise of the optimized multi-stage centrifugal pump is still blade passing frequency, but the sound pressure level under it is reduced by 2 dB compared with the original one shown in Figure 12c, which indicates that matching the relationship between the impeller and the guide vane is better after optimization. Reasonably, comparing the spectrum before and after optimization, it can be found that the magnitude of the radiated noise of the optimized multi-stage centrifugal pump in the low-frequency region is significantly reduced.

Finally, the optimized design was manufactured and sent to the product quality inspection center for a pump performance test. The multi-stage centrifugal pump after optimization has improved the head by 4.62 m, the efficiency by 11.57%, and reduced the average total sound pressure level by 2.6 dB at the design flow rate, which indicates that the simulation process and optimization method proposed in this paper is suitable for the pump designer.

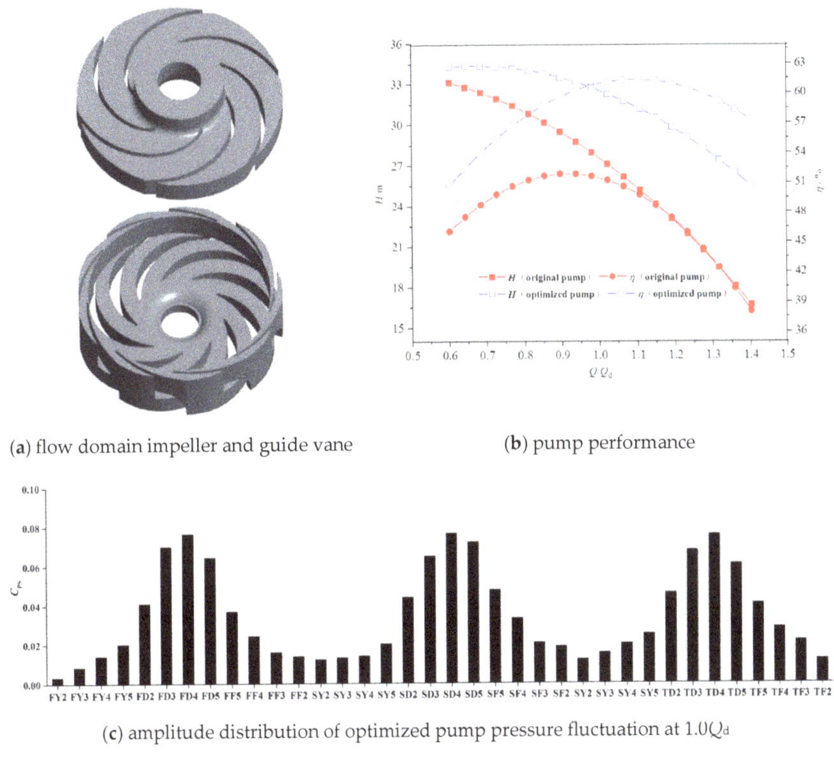

(**a**) flow domain impeller and guide vane   (**b**) pump performance

(**c**) amplitude distribution of optimized pump pressure fluctuation at $1.0Q_d$

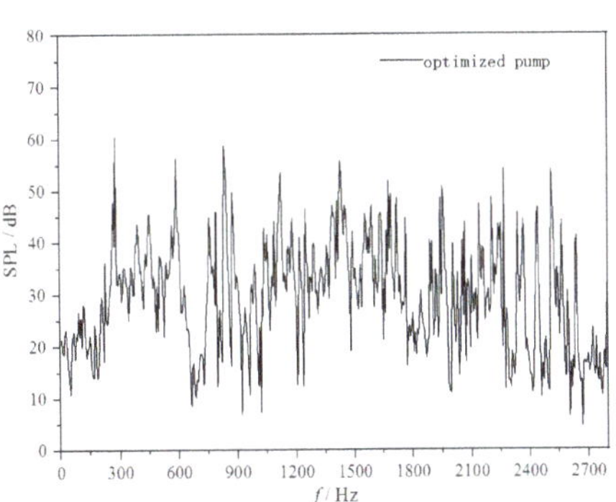

(**d**) frequency response curves of the optimized pump noise at $1.0Q_d$

Figure 20. Simulation results of the optimized pump.

6. Conclusions

The study presents the acoustic simulation and experiment of radiated noise in a multi-stage pump. The following results were obtained:

(1) The sound pressure level in the multi-stage pump increases first and decreases afterward with the increment of the flowrate. The sound pressure level reaches its lowest value at $0.8Q_d$, which corresponds to maximum efficiency working conditions. The sound pressure level of the radiated noise in the multistage pump rises linearly with the increase of the rotational speed.

(2) The radiated noise of the multi-stage pump is characterized by dipoles. Furthermore, the main frequency of the radiated noise is the blade passing frequency (327 Hz). This fact proves that the rotor-stator interaction between impeller and diffuser is still the main hydraulic exciting force and sound source. In addition, the directivity of the sound source changes with the variation of the flowrate.

(3) Flow-induced radiated noise of the multi-stage centrifugal pump was calculated by the combined CFD(DES)/CA(FEM) method based on the Lighthill acoustic analogy theory. Comparisons of numerical predictions with the measured/analytical results reveal that the model can yield good results on the noise and the flow field. The most important aspect of the hydraulic design of a low-noise multi-stage centrifugal pump is to select the appropriate number of impeller blades and its matched guide vane.

Author Contributions: Q.S. and C.W. conceived and designed the experiments; G.L. performed the experiments and simulation; B.W. and K.H. analyzed the data; Q.S. wrote the paper; J.Y. funding acquisition.

Funding: This research was funded by National Key R&D Program of China (2018YFC0810505), National Natural Science Foundation of China (51976079, 51779107), Senior Talent Foundation of Jiangsu University (15JDG048), Open Foundation of National Research Center of Pumps, Jiangsu University (NRCP201604).

Conflicts of Interest: The authors declare no conflict of interest.

References

1. Wang, C.; Shi, W.; Wang, X.; Jiang, X.; Yang, Y.; Li, W.; Zhou, L. Optimal design of multistage centrifugal pump based on the combined energy loss model and computational fluid dynamics. *Appl. Energy* **2017**, *187*, 10–26. [CrossRef]

2. Wang, C.; He, X.; Zhang, D.; Hu, B.; Shi, W. Numerical and experimental study of the self-priming process of a multistage self-priming centrifugal pump. *Int. J. Energy Res.* **2019**, *43*, 4074–4092. [CrossRef]

3. Huang, K.; Yuan, J.; Si, Q.; Lin, G. Numerical simulation on pressure pulsation in multistage centrifugal pump under several working conditions. *J. Drain. Irrig. Mach. Eng.* **2019**, *37*, 387–392.

4. Wang, C.; Hu, B.; Zhu, Y.; Wang, X.; Luo, C.; Cheng, L. Numerical study on the gas-water two-phase flow in the self-priming process of self-priming centrifugal pump. *Processes* **2019**, *7*, 330. [CrossRef]

5. Rochecarrier, N.L.; Ngoma, G.D.; Ghie, W. Numerical Investigation of a first stage of a multistage centrifugal pump: Impeller, diffuser with return vanes, and casing. *Isrn Mech. Eng.* **2015**, *2013*, 578072.

6. Wang, C.; He, X.; Shi, W.; Wang, X.; Wang, X.; Qiu, N. Numerical study on pressure fluctuation of a multistage centrifugal pump based on whole flow field. *Aip Adv.* **2019**, *9*, 035118. [CrossRef]

7. Zhang, J.; Xia, S.; Ye, S.; Xu, B.; Song, W.; Zhu, S.; Xiang, J. Experimental investigation on the noise reduction of an axial piston pump using free-layer damping material treatment. *Appl. Acoust.* **2018**, *139*, 1–7. [CrossRef]

8. Ye, S.; Zhang, J.; Xu, B.; Zhu, S. Theoretical investigation of the contributions of the excitation forces to the vibration of an axial piston pump. *Mech. Syst. Signal Process.* **2019**, *129*, 201–217. [CrossRef]

9. Wang, J.; Feng, T.; Liu, K.; Zhou, Q.-J. Experimental Research on the Relationship between the Flow-induced Noise and the Hyaraulic Parameters in Centrifugal Pump. *Fluid Mach.* **2007**, *6*, 1–13.

10. Brennen, C.E. A review of the dynamics of cavitating pumps. *J. Fluids Eng.* **2013**, *135*, 061301. [CrossRef]

11. He, X.; Jiao, W.; Wang, C.; Cao, W. Influence of surface roughness on the pump performance based on Computational Fluid Dynamics. *IEEE Access* **2019**, *7*, 105331–105341. [CrossRef]

12. Qian, J.Y.; Chen, M.R.; Liu, X.L.; Jin, Z.J. A numerical investigation of the flow of nanofluids through a micro Tesla valve. *J. Zhejiang Univ. Sci. A* **2019**, *20*, 50–60. [CrossRef]

13. Qian, J.Y.; Gao, Z.X.; Liu, B.Z.; Jin, Z.J. Parametric study on fluid dynamics of pilot-control angle globe valve. *ASME J. Fluids Eng.* **2018**, *140*, 111103. [CrossRef]

14. Zhu, Y.; Tang, S.; Wang, C.; Jiang, W.; Yuan, X.; Lei, Y. Bifurcation Characteristic research on the load vertical vibration of a hydraulic automatic gauge control system. *Processes* **2019**, *7*, 718. [CrossRef]

15. Zhu, Y.; Tang, S.; Quan, L.; Jiang, W.; Zhou, L. Extraction method for signal effective component based on extreme-point symmetric mode decomposition and Kullback-Leibler divergence. *J. Braz. Soc. Mech. Sci. Eng.* **2019**, *41*, 100. [CrossRef]

16. Zhu, Y.; Qian, P.; Tang, S.; Jiang, W.; Li, W.; Zhao, J. Amplitude-frequency characteristics analysis for vertical vibration of hydraulic AGC system under nonlinear action. *Aip Adv.* **2019**, *9*, 035019. [CrossRef]

17. Chu, S.; Dong, R.; Katz, J. Relationship Between Unsteady Flow, Pressure Fluctuations, and Noise in a Centrifugal Pump—Part A: Use of PDV Data to Compute the Pressure Field. *J. Fluids Eng.* **1995**, *117*, 24–29. [CrossRef]

18. Chu, S.; Dong, R.; Katz, J. Relationship between Unsteady Flow, Pressure Fluctuations, and Noise in a Centrifugal Pump—Part B: Effects of Blade-Tongue Interactions. *J. Fluids Eng.* **1995**, *117*, 30–35. [CrossRef]

19. Khalifa, A.E.; Al-Qutub, A.M.; Ben-Mansour, R. Study of Pressure Fluctuations and Induced Vibration at Blade-Passing Frequencies of a Double Volute Pump. *Arab. J. Sci. Eng.* **2011**, *36*, 1333–1345. [CrossRef]

20. Parrondo, J.; Pérez, J.; Barrio, R.; Gonzales, J. A simple acoustic model to characterize the internal low frequency sound field in centrifugal pumps. *Appl. Acoust.* **2011**, *72*, 59–64. [CrossRef]

21. Liu, H.L.; Dai, H.W.; Ding, J. Numerical and experimental studies of hydraulic noise induced by surface dipole sources in a centrifugal pump. *J. Hydrodyn. Ser. B* **2016**, *28*, 43–51. [CrossRef]

22. Howe, M.S. On the Estimation of Sound Produced by Complex Fluid-Structure Interactions, with Application to a Vortex Interacting with a Shrouded Rotor. *Proc. Math. Phys. Sci.* **1991**, *433*, 573–598. [CrossRef]

23. Kato, C.; Yoshimura, S.; Yamade, Y.; Jiang, Y.Y.; Wang, H.; Imai, R.; Katsura, H.; Yoshida, T.; Takano, Y. Prediction of the Noise from a Multi-Stage Centrifugal Pump. In Proceedings of the ASME 2005 Fluids Engineering Division Summer Meeting, Houston, TX, USA, 19–23 June 2005; pp. 1273–1280.

24. Ding, J.; Liu, H.L.; Wang, Y.; Tan, M.G.; Cui, J.B. Numerical study on the effect of blade outlet angle on centrifugal pump noise. *J. Vib. Shock* **2014**, *33*, 122–127.

25. Si, Q.R.; Ali, A.; Yuan, J.P.; Fall, I.; Muhammad, Y.F. Flow-induced noises in a centrifugal pump: A review. *Sci. Adv. Mater.* **2019**, *11*, 909–924. [CrossRef]

26. Keller, J.; Barrio, R.; Parrondo, J.; Barrio, R.; Fernandez, J.; Blanco, E. Effects of the Pump-Circuit Acoustic Coupling on the Blade-Passing Frequency Perturbations. *Appl. Acoust.* **2014**, *76*, 150–156. [CrossRef]

27. Gao, M.; Dong, P.; Lei, S.; Turan, A. Computational Study of the Noise Radiation in a Centrifugal Pump When Flow Rate Changes. *Energies* **2017**, *10*, 221. [CrossRef]

28. Jiang, Y.Y.; Yoshimura, S.; Imai, R.; Katsura, H.; Yoshida, T.; Kato, C. Quantitative evaluation of flow-induced structural vibration and noise in turbomachinery by full-scale weakly coupled simulation. *J. Fluids Struct.* **2007**, *23*, 531–544. [CrossRef]

29. Timushev, S. Development and Experimental Validation of 3D Acoustic-Vortex Numerical Procedure for Centrifugal Pump Noise Prediction. In Proceedings of the ASME 2009 Fluids Engineering Division Summer Meeting, Vail, CO, USA, 2–6 August 2009; pp. 389–398.

30. Barrio, R.; Parrondo, J.; Blanco, E. Numerical analysis of the unsteady flow in the near-tongue region in a volute-type centrifugal pump for different operating points. *Comput. Fluids* **2010**, *39*, 859–870. [CrossRef]

31. Dong, L.; Dai, C.; Kong, F.; Fu, L.; Bai, Y. Impact of blade outlet angle on acoustic of centrifugal pump as turbine. *Trans. Chin. Soc. Agric. Eng.* **2015**, *31*, 69–75.

32. Rui, X.P.; Zhao, Y. Numerical simulation and experimental research of flow-induced noise for centrifugal pumps. *J. Vibroeng.* **2016**, *18*, 622–636.

33. Hao, Z. Vibration Noises and Acoustic Optimization Design of the Centrifugal Pump. In Proceedings of the International Conference on Machinery, Materials and Computing Technology, Hangzhou, China, 23–24 January 2016.

34. Heng, Y.; Yuan, S.; Hong, F.; Yuan, J.; Si, Q.; Hu, B. A Hybrid Method for Flow-Induced Noise in Centrifugal Pumps Based on LES and FEM. In Proceedings of the ASME 2013 Fluids Engineering Division Summer Meeting, Vail, CO, USA, 2–6 August 2013; p. V01BT10A034.

35. Tu, S.; Shahrouz, A.; Reena, P.; Watts, M. An implementation of the Spalart-Allmaras DES model in an implicit unstructured hybrid finite volume/element solver for incompressible turbulent flow. *Int. J. Numer. Methods Fluids* **2010**, *59*, 1051–1062. [CrossRef]

36. Gülish, J.F. *Centrifugal Pumps*; Springer: Berlin/Heidelberg, Germany; New York, NY, USA, 2008.

37. *Methods of Measuring and Evaluating Noise of Pumps*; China Standard Press: Beijing, China, 2013; GB/T 29529-2013.

Article

Research on the Vertical Vibration Characteristics of Hydraulic Screw Down System of Rolling Mill under Nonlinear Friction

Yongshun Zhang [1,2], Wanlu Jiang [1,3,*], Yong Zhu [4,*] and Zhenbao Li [1]

[1] Hebei Provincial Key Laboratory of Heavy Machinery Fluid Power Transmission and Control, Yanshan University, Qinhuangdao 066004, China; zhangyongshun@ysu.edu.cn (Y.Z.); lizhenbao_ysu@163.com (Z.L.)

[2] National Engineering Research Center for Equipment and Technology of Cold Strip Rolling, Yanshan University, Qinhuangdao 066004, China

[3] Heavy-duty Intelligent Manufacturing Equipment Innovation Center of Hebei Province, Yanshan University, Qinhuangdao 066004, China

[4] Research Center of Fluid Machinery Engineering and Technology, Jiangsu University, Zhenjiang 212013, China

* Correspondence: wljiang@ysu.edu.cn (W.J.); zhuyong@ujs.edu.cn (Y.Z.); Tel.: +86-0335-806-1729 (W.J.); +86-0511-8879-9918 (Y.Z.)

Received: 31 August 2019; Accepted: 15 October 2019; Published: 2 November 2019

Abstract: The rolling mill with hydraulic system is widely used in the production of strip steel. For the problem of vertical vibration of the rolling mill, the effects of different equivalent damping coefficient, leakage coefficient, and proportional coefficient of the controller on the hydraulic screw down system of the rolling mill are studied, respectively. First, a vertical vibration model of a hydraulic screw down system was established, considering the nonlinear friction and parameter uncertainty of the press cylinder. Second, the correlation between different equivalent damping coefficient, internal leakage coefficient, proportional coefficient, vertical vibration was analyzed. The simulation results show that, in the closed-loop state, when Proportional-Integral-Derivative (PID) controller parameters are fixed, due to the change of the equivalent damping coefficient and internal leakage coefficient, the system will have parameter uncertainty, which may lead to the failure of the PID controller and the vertical vibration of the system. This study has theoretical and practical significance for analyzing the mechanism of vertical vibration of the rolling mill.

Keywords: rolling mill; hydraulic screw down system; vertical vibration; LuGre model; friction-induced vibration; flow control; PID controller

---

1. Introduction

With the continuous improvement of strip material properties, high requirements for rolling equipment appears. The increase of rolling speed and rolling force weakens the reliability of original equipment, and vertical vibration of the rolling mill appears in more and more fields. This kind of vibration will cause the thickness difference of the strip steel, which is no longer stable, or the surface of the strip steel will produce vibration marks, the surface quality will become poor [1], and even cause serious production accidents, such as strip breaking. Therefore, vibration detection systems based on acceleration sensors are added in many production lines to avoid production accidents by reconstructing the real-time signal [2] to analyze the vertical vibration and reducing the speed artificially [1]. However, it is difficult to deal with this kind of monitoring in the first time, and, consequently, it is hard to avoid vibration.

It is found that vertical vibration of the rolling mill has many inducements, but it is more related to the hydraulic system [3,4], because the hydraulic screw down system is the power source in the vertical direction of the mill [5]. Figure 1 shows that the rolling force is generated by applying the screw down cylinder to the roll system. Moreover, the hydraulic system itself is composed of many nonlinear links [6–8]; the hydraulic pump, as an oil-source system, is a vibration source [9], and the working mechanism of the plunger pump will make the system oil-source pressure pulsation [10–13]. The valve orifice characteristics of the hydraulic system are also nonlinear [14,15]. As the actuator of the hydraulic screw down system, the hydraulic cylinder has a large diameter and long stroke due to its own structure. Due to processing or sealing, the hydraulic cylinder will be affected by obvious nonlinear friction [16,17] and oil leakage. Moreover, the hydraulic cylinder oil also has nonlinear spring-force action [18]. All these factors result in making the hydraulic screw down system work unstable. Therefore, the comprehensive consideration of the nonlinear factors of the hydraulic screw down system and roll system to establish a precise model [19–21] is very important and of theoretical significance.

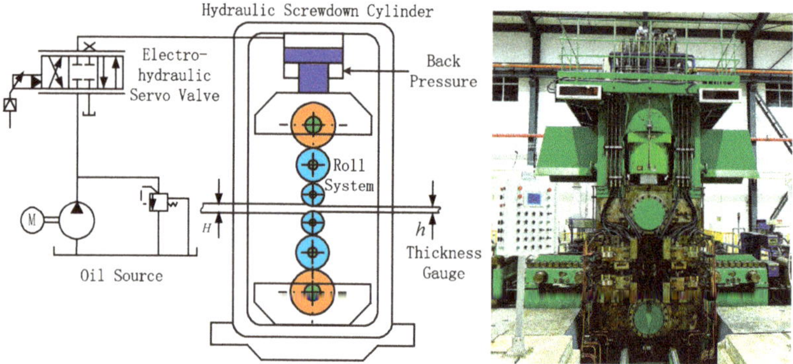

Figure 1. Structural diagram of rolling mill.

Among those inducements, nonlinear friction widely exists in mass-spring model sliding on the transmission belt [22], pin sliding on the rotating disc, turbine blade, water-lubricated bearing, wheel-rail system, disc braking system, and machine tool system [23]. By establishing nonlinear dynamics models to explain friction phenomena such as stick-slip [24], flutter, and chaos, it can promote the understanding of various friction mechanisms to minimize the deterioration effect of friction. Therefore, it is of theoretical and engineering significance to explore the influence of the hydraulic screw down system's response to the vertical vibration of the rolling mill under the action of nonlinear friction.

In this paper, the nonlinear vibration model of the hydraulic screw down system is established, considering the influence of nonlinear friction with the LuGre model [16,25]. The closed-loop control system is built by using the Proportional-Integral-Derivative (PID) controller. The influence of the damping coefficient, leakage coefficient, and control proportional coefficient on the vibration characteristics of the system is revealed. The purpose of this work is to clarify the influence of process parameters on the vertical vibration of the rolling mill, with nonlinear friction force taken into account, and to evaluate the influence of vertical vibration by introducing the vibration severity [26,27], which makes the vibration monitoring of the rolling mill in industrial fields more intuitive and can effectively provide the safe and stable operation of equipment.

2. Principle of Hydraulic Screw Down System

The structural diagram of the vertical vibration of hydraulic screw down system is shown in Figure 2. x_d is the given displacement signal, and x is the cylinder displacement obtained by the displacement sensor. In this paper, in order to study the influence of nonlinear friction $F_f(v)$ on the vertical vibration of the load roll system, a vertical vibration model of the load roll system was established under the action of nonlinear friction force of the hydraulic screw down system.

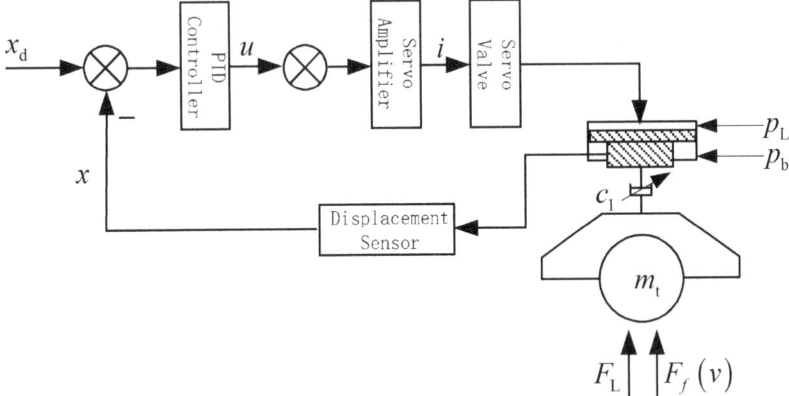

Figure 2. Structural diagram of the vertical vibration of hydraulic screw down system.

The schematic diagram of the hydraulic screw down system of the rolling mill is displayed in Figure 3. p_s is the oil-supply pressure, p_t is the return pressure, and p_b is the back pressure of the rod cavity of hydraulic cylinder. The servo valve is a three-way valve formed by a standard four-way spool valve blocking a control port. When the main spool of the servo valve works in the right position, the high-pressure oil enters into the rod-less cavity of the cylinder, and the piston rod extends to realize the pressing action. When the main spool of the servo valve works in the left position, the high-pressure oil of the rod-less cavity flows back to the oil tank, and the piston rod retracting action is realized under the constant back pressure of the cavity of pole.

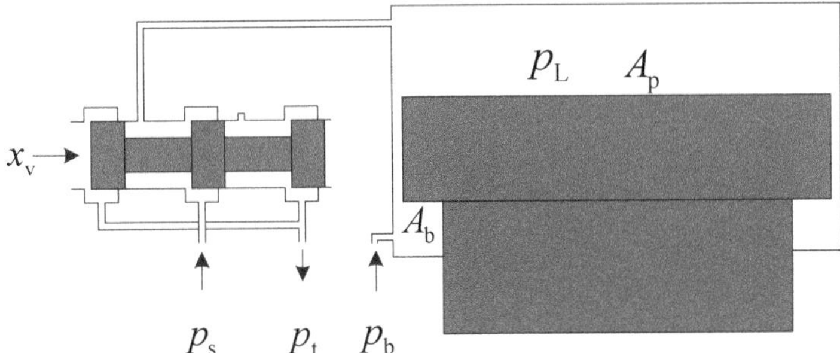

Figure 3. Hydraulic principle diagram of hydraulic screw down system.

3. Mathematical Model

3.1. Nonlinear Flow Equation of Four-Way Valve

The flow equation of the servo valve can be expressed by the following formula [28]:

$$Q_{\mathrm{L}} = C_{\mathrm{d}} W x_{\mathrm{v}} \sqrt{\frac{2(p_{\mathrm{s}} - p_{\mathrm{L}})}{\rho}} \quad x_{\mathrm{v}} \geq 0$$

$$Q_{\mathrm{L}} = C_{\mathrm{d}} W x_{\mathrm{v}} \sqrt{\frac{2(p_{\mathrm{L}} - p_{\mathrm{t}})}{\rho}} \quad x_{\mathrm{v}} < 0 \tag{1}$$

C_{d}: port flow coefficient of four-way valve;
W: area grades of four-way valve;
x_{v}: spool displacement of four-way valve;
p_{L}: rod-less cavity pressure of hydraulic cylinder;
ρ: oil density.

3.2. Model of Servo Amplifier and Servo Valve

The servo amplifier and servo valve are equivalent to the proportional component, so Equation (2) is applied:

$$K_{\mathrm{p1}} = \frac{i}{u} \tag{2}$$

$$K_{\mathrm{sv}} = \frac{x_{\mathrm{v}}}{i} \tag{3}$$

u: output of vibration controller;
K_{p1}: gain of the servo amplifier;
i: output current of servo amplifier;
K_{sv}: gain of servo valve.

3.3. Flow Equation of Hydraulic Cylinder

Ignoring the external leakage characteristics of the hydraulic cylinder has Equation (4) [29]:

$$Q_{\mathrm{L}} = A_{\mathrm{p}} \dot{x} + C_{\mathrm{t}}(p_{\mathrm{L}} - p_{\mathrm{b}}) + \frac{V}{\beta_{\mathrm{e}}} \dot{p}_{\mathrm{L}} \tag{4}$$

where

$$V = A_{\mathrm{p}} L_1 + A_{\mathrm{p}} x;$$

A_{p}: cross-sectional area of rod-less cavity of cylinder;
x: vibration displacement of cylinder piston rod;
C_{t}: leakage coefficient of hydraulic cylinder;
V: rod-less cavity volume of cylinder;
L_1: the initial stroke of the cylinder piston;
β_{e}: volume elastic modulus of oil.

3.4. Nonlinear Friction

There are static and dynamic friction models. The LuGre friction model is a kind of dynamic model, which is more reasonable for practical mill system dynamic changes. Moreover, the model can be expressed in simple mathematical form. The parameter identification theory based on this model is mature and more conducive to realize.

The nonlinear friction force of a hydraulic cylinder based on the LuGre model can be expressed by the following equations [30]:

$$\frac{dz}{dt} = v - \frac{\sigma_0 z}{g(v)}|v| \tag{5}$$

$$F_f(v) = \sigma_0 z + \sigma_1 \frac{dz}{dt} + \sigma_2 v \tag{6}$$

$$g(v) = f_c + (f_s - f_c)e^{-\left(\frac{v}{v_s}\right)^2} \tag{7}$$

where

z: average deformation of the sideburns;
σ_0: stiffness coefficient of sideburns;
σ_1: microscopic damping coefficient;
σ_2: viscous damping coefficient;
f_c: coulomb force of friction;
f_s: maximum static friction force;
v_s: Stribeck speed.

The LuGre friction model is composed of Equations (5)–(7), where f_c, f_s, v_s, and σ_2 are static parameters, and σ_0 and σ_1 are dynamic parameters.

(1) When velocity $v = 0$, $\frac{dz}{dt} = 0$, the friction force is $F_f(0) = \sigma_0 z$ and is a constant. For the convenience of analysis, the constant is defined as f_s, and $\dot{F}_f(0) = 0$.

(2) When the velocity $v \neq 0$, $F_f(v)$ can be expressed by Equation (6), and $\dot{F}_f(v)$ is continuously differentiable.

3.5. Balance Equation of Cylinder Piston Force

Considering the influence of nonlinear friction on the piston rod, as shown in Equations (5)–(7), the force balance equation of piston rod is established as follows [31]:

$$A_p p_L - A_b p_b = m_t \ddot{x} + c_1 \dot{x} + k_1 x + F_f(v) + F_L \tag{8}$$

where

A_b: the effective area of rod cavity of cylinder;
m_t: equivalent total mass of piston and load;
c_1: viscous damping coefficient of piston;
k_1: spring stiffness of load;
$F_f(v)$: nonlinear friction;
F_L: external force;
v: speed of hydraulic cylinder.

3.6. System State Equation

State variables are defined as follows:

$$\begin{cases} x_1 = x \\ x_2 = \dot{x} \\ x_3 = \ddot{x} \end{cases} \tag{9}$$

Then, the system equations can be obtained as follows:

$$\begin{cases} \dot{x}_1 = x_2 \\ \dot{x}_2 = x_3 \\ \dot{x}_3 = -\alpha_0 x_3 - \alpha_1 x_2 - \alpha_2 x_1 - \alpha_3 f_x + \alpha_4 u \end{cases} \tag{10}$$

Among them are the following equations:

$$\alpha_0 = \frac{C_t \beta_e}{A_p L_1 + A_p x_1} + \frac{c_1}{m_t},$$

$$\alpha_1 = \frac{\beta_e A_p{}^2}{(A_p L_1 + A_p x_1) m_t} + \frac{\beta_e C_t c_1}{(A_p L_1 + A_p x_1) m_t} + \frac{k_1}{m_t},$$

$$\alpha_2 = \frac{C_t k_1 \beta_e}{(A_p L_1 + A_p x_1) m_t},$$

$$\alpha_3 = \frac{\beta_e A_p}{(A_p L_1 + A_p x_1) m_t},$$

$$f_x = \frac{A_p L_1 + A_p x_1}{\beta_e A_p} \dot{F}_f(v) + \frac{C_t}{A_p} F_f(v) + \frac{C_t}{A_p} F_{Lb} + \frac{C_t A_b}{A_p} p - C_t p_b,$$

$$\alpha_4 = \frac{C_d W K_{p1} K_{sv} \beta_e A_p g(x_v) \sqrt{\frac{2}{\rho}}}{(A_p L_1 + A_p x_1) m_t},$$

$$g(x_v) = \sqrt{p_s - p_L}\, x_v \geq 0,$$

$$g(x_v) = \sqrt{p_L - p_t}\, x_v < 0.$$

Due to the influence of different working states and ambient temperature, the equivalent damping coefficient of the system and leakage coefficient in the hydraulic cylinder will be changed during the operation of the system and bring parameter uncertainty of the coefficients of Equation (10), which will affect the working characteristics of the hydraulic system. This phenomenon is discussed in detail in the following.

3.7. Vibration Controller

The vibration controller adopts the classical PID control, which can be expressed as Equation (11):

$$u = K_P (x_d - x) + K_I \int (x_d - x)\,dt + K_D \frac{d(x_d - x)}{dt}, \tag{11}$$

where

K_P: proportionality coefficient;
K_I: integral coefficient;
K_D: differential coefficient.

4. Numerical Simulation

4.1. The Parameters Selection of Simulation Model

The MATLAB is adopted for system modeling and simulation, and ode45 algorithm is employed. Considering that the pressure pulsation caused by the oil pump is 12%, the speed of plunger pump is 990 rpm, the number of plungers is seven, so the pulsation frequency is 116 Hz. The nominal parameters of the hydraulic screw down system of the rolling mill used in the simulation are shown in Table 1.

Table 1. Nominal value of main parameters of hydraulic screw down system.

Parameters	Value	Parameters	Value
A_p	0.19635 m^2	K_{p1}	0.0125 A/V
c_1	2.25×10^6 N·s/m	k_1	2.5×10^9 N/m
m_t	1200 kg	F_L	2×10^6 N
C_t	5.0×10^{-16} m^3/s·Pa	p_s	1.8×10^7 Pa
β_e	7×10^8 Pa	ρ	850 kg/m^3
C_d	0.62	W	0.025 m
K_{sv}	0.01 m/A	L_1	0.15 m
L	0.26 m	A_b	0.030159 m^2
p_b	1.0×10^6 Pa	p_t	0.0×10^5 Pa
σ_0	4.4178×10^9 N/m	σ_1	1×10^5 N·s/m
σ_2	1×10^5 N·s/m	f_c	$16{,}000$ N
f_s	$22{,}000$ N	v_s	0.1 m/s

4.2. Simulation Results and Analysis

4.2.1. Influence of Different Equivalent Damping Coefficients on Nonlinear Dynamic Behavior

It is found that different equivalent damping coefficients have a great influence on the nonlinear system. At the initial value $X_{00} = [0.001, 0, 0]$, the system damping coefficients are respectively defined as $c_1 = 2.25 \times 10^6$ N·s/m, $c_1 = 2.25 \times 10^5$ N·s/m, and $c_1 = 1.25 \times 10^5$ N·s/m. The time–domain diagrams of the vibration displacement of the cylinder are displayed in Figure 4a,b. The phase diagrams of the vibration displacement–velocity are shown in Figure 4c. An improved frequency–domain integral method [22,23] based on low-frequency filtering of vibration intensity is adopted to obtain the vibration intensity of the system under different damping coefficients. The curve is shown in Figure 5.

Figure 4 shows the vibration simulations that result from the hydraulic cylinder of the rolling mill with different equivalent damping coefficients under the same conditions. Through the vibration displacement curve, it can be clearly found that the vibration amplitude of vibration displacement significantly increases with the damping coefficient decreasing. Through the phase diagram, it can be found that the system moves from a stable state to a periodic vibration state and becomes unstable with the damping coefficient decreasing. From the perspective of vibration intensity, as shown in Figure 5, it can clearly be found that the vibration intensity of the system can be effectively suppressed by increasing the damping.

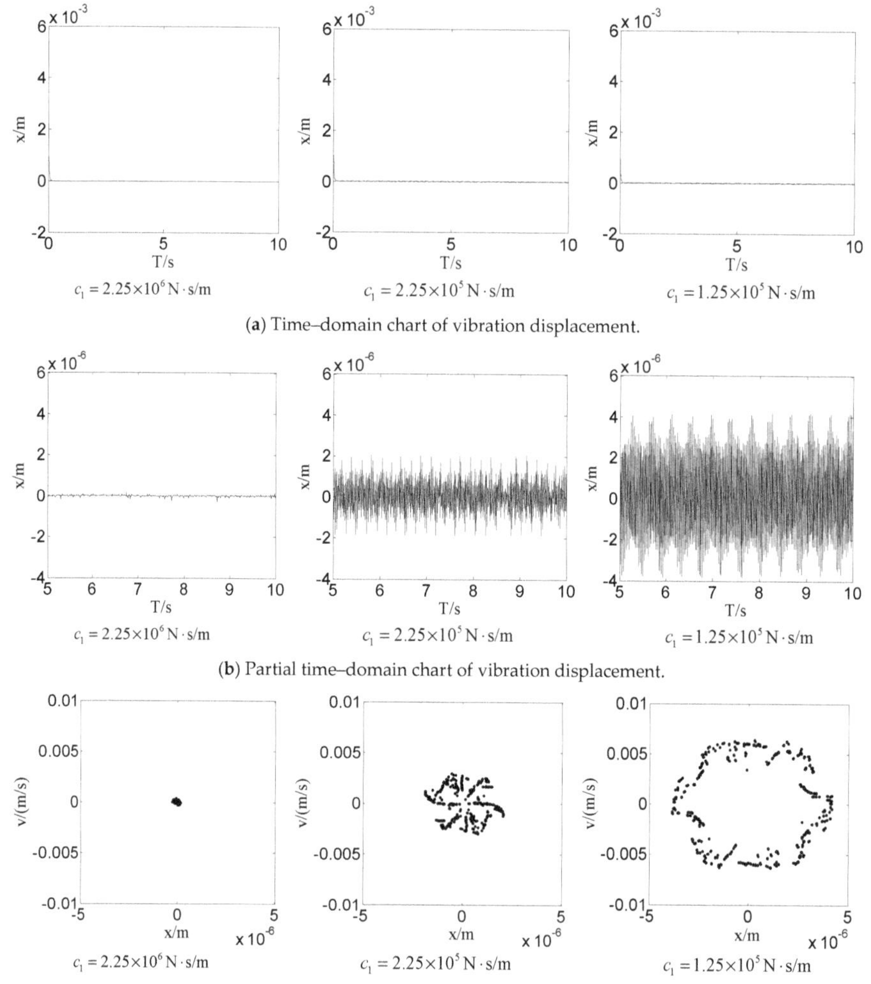

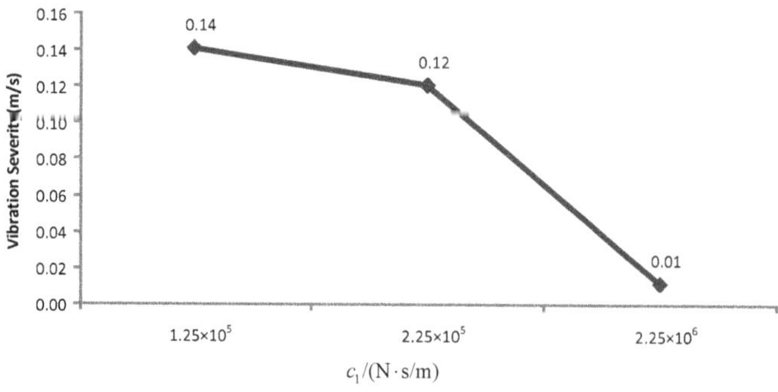

(a) Time–domain chart of vibration displacement.

(b) Partial time–domain chart of vibration displacement.

(c) Phase diagram of vibration displacement and speed.

Figure 4. Effect of equivalent damping on hydraulic cylinder vibration response.

Figure 5. Effect of different equivalent damping on vibration severity of hydraulic cylinder.

4.2.2. Influence of Different Leakage Coefficients on Nonlinear Dynamic Behavior

The hydraulic system is sealed with a rubber ring. Due to the influence of oil temperature and oil cleanliness, the seal of the hydraulic cylinder will deteriorate with time, resulting in different levels of oil leakage. By studying the influence of different internal leakage coefficients on the system, the vibration mechanism of the hydraulic cylinder caused by the leakage can be analyzed.

The leak coefficients of the hydraulic cylinder are defined as $C_t = 5.0 \times 10^{-16}$ m^3/s·Pa, $C_t = 5.0 \times 10^{-10}$ m^3/s·Pa, and $C_t = 5.0 \times 10^{-9}$ m^3/s·Pa. If you take the initial value $X_{00} = [0.001, 0, 0]$, the time–domain diagram of the vibration displacement of the cylinder and the phase diagram of the vibration displacement–speed are shown in Figure 6. The vibration intensities of the system under different leakage coefficients are displayed in Figure 7.

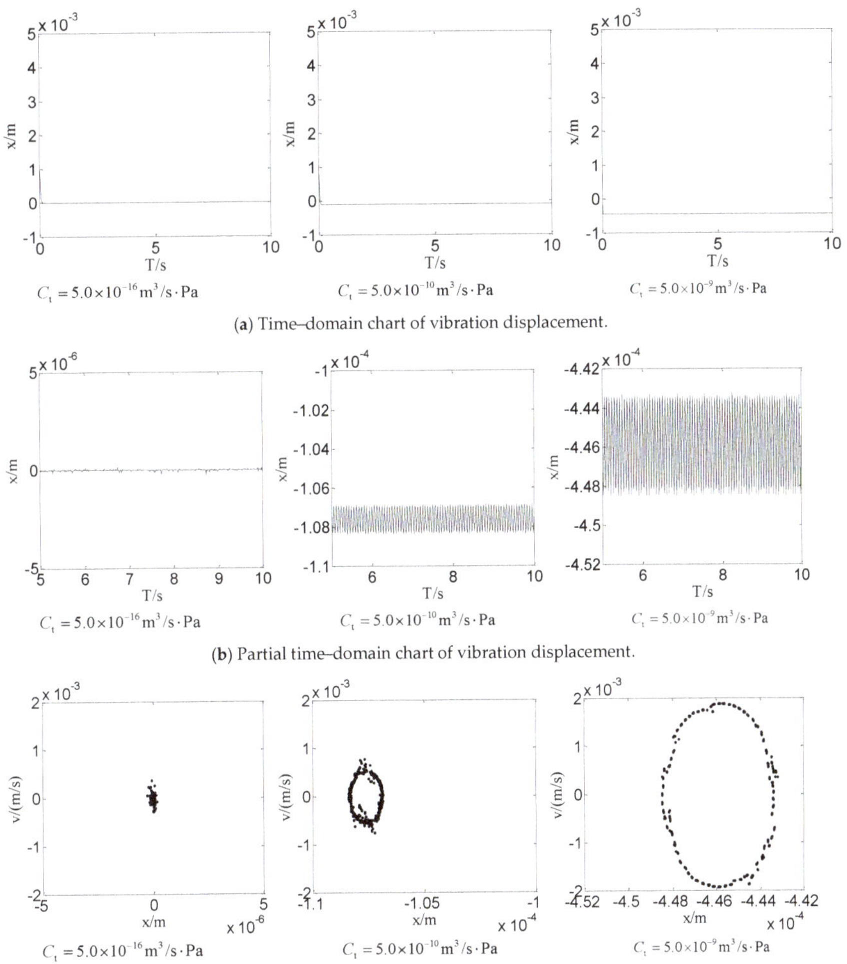

Figure 6. Effect of leakage coefficient on hydraulic cylinder vibration response.

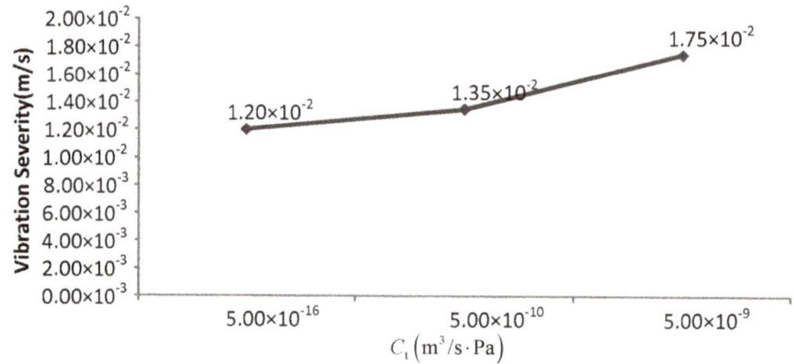

Figure 7. Effect of leakage coefficient on vibration severity of hydraulic cylinder.

Figure 6 shows the simulation results of the different leakage coefficients of the hydraulic cylinder under the same other conditions. It can be seen from Figure 6 that the vibration displacement of the hydraulic cylinder is small when the leakage coefficient is small. And from the phase diagram, it can be observed that the system is convergence. By increasing the leakage coefficient of the hydraulic cylinder, the amplitude of the vibration displacement increases. From the phase diagram, it is difficult to stabilize from the convergence to the stage of multi-period vibration. From the variation of vibration intensity in Figure 7, it can be found that the vibration intensity increases with the increase of the leakage coefficient. Therefore, the increase of system leakage will increase the intensity of vertical vibration.

4.2.3. The Influences of Proportional Parameter of Vibration Controller on Nonlinear Dynamic Behavior

The proportional parameters of the controller affect the control effect of vibration suppression. As certain process parameters or structural parameters of the system are changed, fixed parameters K_P are bound to have adverse effects. When other parameters are the same, the proportional coefficients of the controller are set as $K_P = 100,000$, $K_P = 3,500,000$, and $K_P = 3,900,000$. Integral coefficient $K_I = 0.001$, and differential coefficient $K_D = 0$. The initial value is set as $X_{00} = [0.001, 0, 0]$. The time–domain diagram and the vibration displacement–speed phase diagram of the cylinder are shown in Figure 8. With the proportional coefficient increasing, the vibration intensity of the cylinder is reveled in Figure 9.

The PID control algorithm, with the characteristics of simple structure and convenient implementation, can be well applied to practical engineering. In the PID control algorithm, the proportion coefficient determines the speed of system response, and the appropriate integral coefficient can effectively eliminate the system error. Therefore, increasing the proportional coefficient can improve the response speed of the system, but a big proportional coefficient may cause the system instability.

As shown in Figure 8, the amplitude of the vibration displacement of the cylinder increases significantly with the proportional coefficient of the controller increasing. It can be seen from the phase diagram that the system migrates from the convergence state to the periodic and quasi-periodic states. From the variation of vibration intensity in Figure 9, it can be found that the vibration severity increases with the increase of the proportion coefficient. The results show that reasonable control parameters K_P can effectively suppress system vibration. As the process parameters or system's structural parameters change, the control parameters K_P become relatively large, which will increase the vibration and cause the system to become unstable.

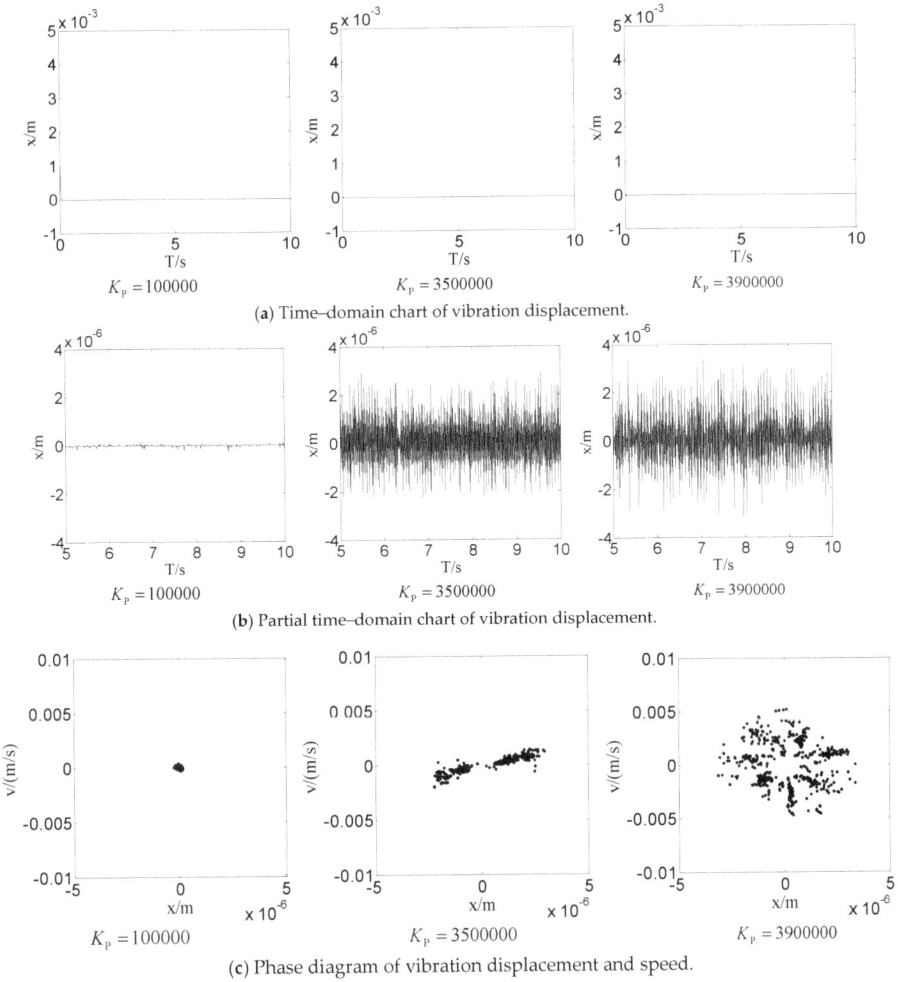

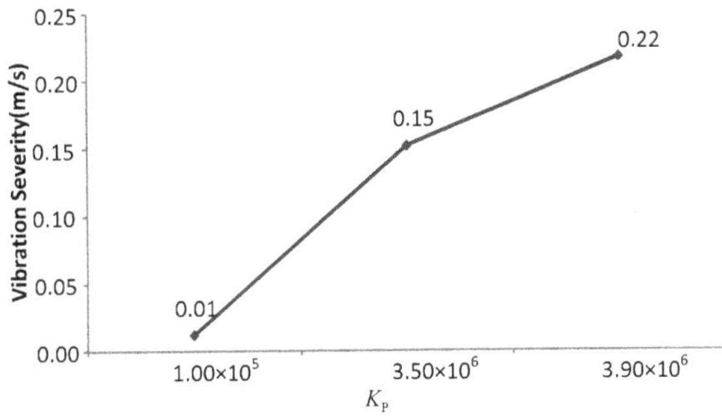

$K_\mathrm{p} = 100000$ $K_\mathrm{p} = 3500000$ $K_\mathrm{p} = 3900000$

(**a**) Time–domain chart of vibration displacement.

$K_\mathrm{p} = 100000$ $K_\mathrm{p} = 3500000$ $K_\mathrm{p} = 3900000$

(**b**) Partial time–domain chart of vibration displacement.

$K_\mathrm{p} = 100000$ $K_\mathrm{p} = 3500000$ $K_\mathrm{p} = 3900000$

(**c**) Phase diagram of vibration displacement and speed.

Figure 8. Effect of proportional parameter on hydraulic cylinder vibration response.

Figure 9. Effect of proportional parameter on the vibration severity of hydraulic cylinder.

5. Conclusions

In this paper, the effects of different equivalent damping coefficients, leakage coefficients in the hydraulic cylinder, and proportional coefficients of the PID controller on the vertical vibration of the rolling mill were studied. The following conclusions are obtained:

(1) The friction force applied to the cylinder is nonlinear. The equivalent damping coefficient and internal leakage coefficient change with the working state and ambient temperature. The vibration model of the rolling mill is established by considering the influence of nonlinear friction and parameter uncertainty on the system characteristics.

(2) When the vibration response of the hydraulic cylinder is analyzed by using a different damping coefficient, it is found that the vibration can be suppressed effectively by increasing the damping coefficient.

(3) The hydraulic system will inevitably bring leakage problems with the aging of the sealing device. Through the analysis of the leakage coefficient, it is found that the vibration attenuation becomes slower and even periodic vibration appears with the increase of the leakage coefficient.

(4) The controller is an effective mean to ensure accurate position control. Reasonable controller parameters can effectively suppress vibration, but fixed controller parameters will make negative effects on vibration suppression.

Author Contributions: Conceptualization, W.J. and Y.Z. (Yong Zhu); methodology, Y.Z. (Yongshun Zhang); investigation, Y.Z. (Yongshun Zhang); writing—original draft preparation, Y.Z. (Yongshun Zhang); writing—review and editing, Y.Z. (Yong Zhu) and Z.L.

Funding: This research was funded by [the National Natural Science Foundation of China] grant number [No. 51875498, 51805214], [Key Program of Hebei Natural Science Foundation] grant number [No. E2018203339], [China Postdoctoral Science Foundation] grant number [No. 2019M651722], and [Young issues in the special project of basic research of Yanshan University] grant number [No. 15LGB005].

Conflicts of Interest: The authors declare no conflict of interest.

References

1. Prihod'ko, I.Y.; Krot, P.V.; Solov'yov, K.V.; Parsenyuk, E.A.; Chernov, P.P.; Pimenov, V.A.; Dolmatov, A.P.; Kharin, A.V. Vibration monitoring system and the new methods of chatter early diagnostics for tandem mill control. In Proceedings of the International Conference on Vibration in Rolling Mills, London, UK, 2006; pp. 87–106.

2. Zhu, Y.; Tang, S.; Quan, L.; Jiang, W.; Zhou, L. Extraction method for signal effective component based on extreme-point symmetric mode decomposition and Kullback-Leibler divergence. *J. Braz. Soc. Mech. Sci. Eng.* **2019**, *41*, 100. [CrossRef]

3. Zhu, Y.; Tang, S.; Wang, C.; Jiang, W.; Yuan, X.; Lei, Y. Bifurcation characteristic research on the load vertical vibration of a hydraulic automatic gauge control system. *Processes* **2019**, *7*, 718. [CrossRef]

4. Zhu, Y.; Tang, S.; Wang, C.; Jiang, W.; Zhao, J.; Li, G. Absolute stability condition derivation for position closed-loop system in hydraulic automatic gauge control. *Processes* **2019**, *7*, 766.

5. Yan, X.Q. Machinery-electric-hydraulic Coupling Vibration Control of Hot Continuous Rolling Mills. *J. Mech. Eng.* **2011**, *47*, 61–65. [CrossRef]

6. Zhang, W.; Wang, Y.Q.; Gao, Y.J. Modeling and Simulation of Hydraulic Screw-down System in a Strip Mill. *Chin. Hydraul. Pneum.* **2004**, *1*, 21–24.

7. Jiang, W.L.; Zhu, Y.; Zhang, Z.; Zhang, S. Nonlinear Vibration Mechanism of Electro-hydraulic Servo System and Its Experimental Verification. *J. Mech. Eng.* **2015**, *51*, 175–184. [CrossRef]

8. Zhu, Y.; Qian, P.F.; Tang, S.G.; Jiang, W.L.; Li, W.; Zhao, J.H. Amplitude-frequency Characteristics Analysis for Vertical Vibration of Hydraulic AGC System under Nonlinear Action. *AIP Adv.* **2019**, *9*, 035019. [CrossRef]

9. Zhang, J.; Xia, S.; Ye, S.; Xu, B.; Song, W.; Zhu, S.; Xiang, J. Experimental investigation on the noise reduction of an axial piston pump using free-layer damping material treatment. *Appl. Acoust.* **2018**, *139*, 1–7. [CrossRef]

10. Ye, S.G.; Zhang, J.H.; Xu, B.; Zhu, S.Q. Theoretical investigation of the contributions of the excitation forces to the vibration of an axial piston pump. *Mech. Syst. Signal Proccess.* **2019**, *129*, 201–217. [CrossRef]

11. Wang, C.; Chen, X.X.; Qiu, N.; Zhu, Y.; Shi, W.D. Numerical and experimental study on the pressure fluctuation, vibration, and noise of multistage pump with radial diffuser. *J. Braz. Soc. Mech. Sci. Eng.* **2018**, *40*, 481. [CrossRef]

12. He, X.; Jiao, W.; Wang, C.; Cao, W. Influence of surface roughness on the pump performance based on Computational Fluid Dynamics. *IEEE Access* **2019**, *7*, 105331–105341. [CrossRef]

13. Wang, C.; Hu, B.; Zhu, Y.; Wang, X.; Luo, C.; Cheng, L. Numerical study on the gas-water two-phase flow in the self-priming process of self-priming centrifugal pump. *Processes* **2019**, *7*, 330. [CrossRef]

14. Qian, J.Y.; Gao, Z.X.; Liu, B.Z.; Jin, Z.J. Parametric study on fluid dynamics of pilot-control angle globe valve. *ASME J. Fluids Eng.* **2018**, *114*, 111103. [CrossRef]

15. Qian, J.Y.; Chen, M.R.; Liu, X.L.; Jin, Z.J. A numerical investigation of the flow of nanofluids through a micro Tesla valve. *J. Zhejiang Univ.-Sci. A* **2019**, *20*, 50–60. [CrossRef]

16. Xuan, B.T.; Nur, H.; Hideki, Y. Modeling of dynamic friction behaviors of hydraulic cylinders. *Mechatronics* **2012**, *22*, 65–75.

17. Yanada, H.; Khaing, W.; Endo, H.; Tran, X. Effect of friction model on simulation of hydraulic actuator. *Proccess. J. Syst. Control Eng.* **2014**, *228*, 690–698.

18. Liu, B.; Li, P.; Liu, F.; Liu, H.; Jiang, J. Chaotic Prediction Control of Hydraulic Cylinder Nonlinear Stiffness Constrain System. *China Mech. Eng.* **2017**, *28*, 559–564.

19. Yang, X.; Tong, C.N.; Yue, G.F.; Meng, J.J. Dynamic model of chatter for cold rolling. *J. Iron Steel Res. Int.* **2010**, *17*, 30–34. [CrossRef]

20. Yang, X.; Tong, C.N. Coupling dynamic model and control of chatter in cold rolling. *J. Dyn. Syst. Meas. Control* **2012**, *4*, 0141001. [CrossRef]

21. Bar, A.; Bar, O. Types of mid-frequency vibrations appearing during the rolling mill operation. *J. Mater. Proccess. Technol.* **2005**, *162*, 461–464. [CrossRef]

22. Alborz, N.; Farhang, K. Vibration instability in a large motion bistable compliant mechanism due to stribeck friction. *J. Vib. Acoust.* **2018**, *140*, 061017.

23. Ibrahim, R.A. Friction-induced vibration, chatter, squeal, and chaos-part II: Dynamics and modeling. *Appl. Mech. Rev.* **1994**, *47*, 227–253. [CrossRef]

24. Leine, R.I.; Vancampen, D.H.; Kraker, A.D.E. Strick-slip vibrations induced by alternate friction models. *Nonlinear Dyn.* **1998**, *16*, 41–54. [CrossRef]

25. Zhao, W.W.; Qiu, S.X.; Feng, C.Z. Study on the stick-slip phenomenon of hydraulic cylinder based on the LuGre friction model. *Des. Res.* **2017**, *12*, 88–92.

26. Zhu, Y.; Jiang, W.L.; Hu, H.S.; Kang, H.Z. A Precise Frequency-domain Integral Method of Vibration Acceleration Signal. *ICIC Control Express Lett.* **2015**, *9*, 1617–1624.

27. Zhu, Y.; Jiang, W.L.; Kong, X.D.; Zheng, Z.; Hu, H.S. An Accurate Integral Method for Vibration Signal Based on Feature Information Extraction. *Shock Vib.* **2015**, *2015*, 962793. [CrossRef]

28. Khan, H.; Seraphin, C.A.; Sepehri, N. Nolinear observer-based fault detection technique for electro-hydraulic servo-positioning systems. *Mechatronics* **2005**, *15*, 1040–1041. [CrossRef]

29. Liu, J.X.; Qiao, B.J.; Zhang, X.W. Adaptive vibration control on electrohydraulic shaking table system with an expanded frequency range: Theory analysis and experimental study. *Mech. Syst. Signal Process.* **2019**, *132*, 125. [CrossRef]

30. Su, X. Modeling method and control performance of hydraulic AGC system. *QinHuangdao* **2018**, *11*, 37–43.

31. Guan, C.; Pan, S.X. Adaptive sliding mode control of electro-hydraulic system with nonlinear unknown parameters. *Control Eng. Pract.* **2008**, *16*, 1276. [CrossRef]

Article

Absolute Stability Condition Derivation for Position Closed-Loop System in Hydraulic Automatic Gauge Control

Yong Zhu [1], Shengnan Tang [1,*], Chuan Wang [1,2,*], Wanlu Jiang [3], Jianhua Zhao [3,4] and Guangpeng Li [1]

[1] National Research Center of Pumps, Jiangsu University, Zhenjiang 212013, China; zhuyong@ujs.edu.cn (Y.Z.); cherish957@126.com (G.L.)
[2] School of Hydraulic Energy and Power Engineering, Yangzhou University, Yangzhou 225002, China
[3] Hebei Provincial Key Laboratory of Heavy Machinery Fluid Power Transmission and Control, Yanshan University, Qinhuangdao 066004, China; wljiang@ysu.edu.cn (W.J.); zhaojianhua@ysu.edu.cn (J.Z.)
[4] State Key Laboratory of Fluid Power and Mechatronic Systems, Zhejiang University, Hangzhou 310027, China
* Correspondence: 2111811013@stmail.ujs.edu.cn (S.T.); wangchuan@ujs.edu.cn (C.W.); Tel.: +86-0511-88799918 (S.T. & C.W.)

Received: 18 September 2019; Accepted: 12 October 2019; Published: 18 October 2019

Abstract: In the metallurgical industry, hydraulic automatic gauge control (HAGC) is a core mechanism for thickness control of plates used in the rolling process. The stability of the HAGC system's kernel position closed-loop is key to ensuring a process with high precision, speed and reliability. However, the closed-loop position control system is typically nonlinear, and its stability is affected by several factors, making it difficult to analyze instability in the system. This paper describes in detail the functioning of the position closed-loop system. A mathematical model of each component was established using theoretical analysis. An incremental transfer model of the position closed-loop system was also derived by studying the connections between each component. In addition, based on the derived information transfer relationship, a transfer block diagram of disturbance quantity of the system was established. Furthermore, the Popov frequency criterion method was introduced to ascertain its absolute stability. The results indicate that the absolute stability conditions of the position closed-loop system are derived in two situations: when spool displacement is positive or negative. This study lays a theoretical foundation for research on the instability mechanism of an HAGC system.

Keywords: rolling mill; hydraulic automatic gauge control system; position closed-loop system; absolute stability condition; Popov frequency criterion; flow control

---

1. Introduction

The development of "intelligent" and "green" manufacturing equipment has propelled the metallurgical industry to pursue intelligence in their rolling equipment, and to ensure high quality of plates and strips used in the industry [1]. However, it has been demonstrated that mass production often results in instability in the rolling process, hindering high-precision and intelligent development. The hydraulic automatic gauge control (HAGC) system is a core mechanism for thickness control of plates used in the rolling process. Its function is to automatically adjust the roll gap of a rolling mill when external disturbance factors change, so as to ensure that the target thickness of the strip is within the index range. The stability of the system's closed-loop kernel position is key to ensuring a process with high precision, speed and reliability.

The HAGC system is complex with multiple links—it is nonlinear and has several parameters that influence its functioning. Because the system's working mechanism is multifaceted, it is difficult to analyze its instability, a problem that researchers in the engineering field have been trying to solve [2–4]. Scholars are currently studying the dynamic characteristics of the HAGC system. Roman et al. [5] researched the thickness control of cold-rolled strips and proposed a system that compensates for errors caused by the hydraulic servo-system used for positioning of the rolls. Hu et al. [6] analyzed the rolling characteristics of the tandem cold-rolling process and proposed an innovative multivariable optimization strategy for thickness and tension based on inverse linear quadratic optimal control. Sun et al. [7] proposed a dynamic model of a cold rolling mill based on strip flatness and thickness integrated control. They conducted dynamic simulation of the rolling process, obtaining information on thickness and flatness. Prinz et al. [8] compared two different AGC setups and developed a feed forward approach for lateral asymmetry of entry thickness. They also developed a new feed forward control approach for the thickness profile of strips in a tandem hot rolling mill [9]. Kovari [10] studied the effect of internal leakage in a hydraulic actuator on dynamic behaviors of the hydraulic positioning system. Li et al. [11] presented a robust output-feedback control algorithm with an unknown input observer for the hydraulic servo-position system in a cold-strip rolling mill with uncertain parameters, immeasurable states and unknown external load forces. Sun et al. [12] introduced key technical features and new technology of the improved cold strip mill process control system: system architecture, hardware configuration and new control algorithms. Yi et al. [13] analyzed HAGC's step response test process: they simulated and established a transfer function model of the test using matrix and laboratory (MATLAB). Liu et al. [14] built a vibration system dynamic model with hydraulic-machinery coupling for four-stand tandem cold rolling mills. The model integrated MATLAB software with automatic dynamic analysis of mechanical systems (ADAMS). Wang et al. [15] established a mathematical model for position–pressure master–slave control of a hydraulic servo system, then simulated the system with AMESim and MATLAB. Hua et al. [16] provided rigorous proof of the exponential stability of the HAGC system by implementing the Lyapunov stability theory. Zhang et al. [17] studied the control strategy of a hydraulic shaking table based on its structural flexibility. Qian and Wang et al. [18,19] researched the effects of important elements, such as valves [20–22], pumps [23–26] and rotors [27], on stability. The influence of excitation forces on the vibration of a pump and measures of noise reduction were studied by Ye et al. [28,29]. Bai et al. [30–32] studied the vibration in a pump under varied conditions. These researchers have had great results with their experiments, providing the basis for further study. However, theoretical derivation and research on the instability mechanism of an HAGC system is still relatively rare.

Scholars have previously applied the Lyapunov method to study the absolute stability of a nonlinear closed-loop control system [33,34]. However, this method has certain reservations, and application of the required Lyapunov function is difficult [35,36]. In 1960, V. M. Popov created a frequency criterion method to determine absolute stability of a nonlinear closed-loop control system. It relied on a classical transfer function and eliminated the dilemma of reconstructing a decision function. This method is of great applicatory value and has been widely recognized by scholars worldwide [37,38]. However, there are still rare results via applying the Popov frequency criterion method to the stability of the HAGC system. The HAGC system is a typical nonlinear closed-loop control system with many influence parameters, and its dynamic characteristics are complex and changeable. When the system is in certain working states, the nonlinear vibration may be induced [39,40]. If the instability mechanism cannot be effectively mastered and controlled in time, a major vibration accident may occur in the system [41–43]. Therefore, it is very important and urgent to explore the instability mechanism of the HAGC system and solve the problem of dynamic instability and inhibition from the source. Conducting the theoretical derivation and in-depth study of the instability mechanism of the HAGC system by using Popov frequency criterion method, is a new technique which needs to be further explored.

In this paper, the Popov frequency criterion method is introduced to theoretically deduce the absolute stability condition for key position closed-loop system in HAGC. The purpose is to lay a theoretical foundation for the study on the instability mechanism of the HAGC system. In Section 2, the mathematical model of the position closed-loop system is established. In Section 3, the incremental transfer model of the position closed-loop system is deduced. In Section 4, the absolute stability condition for the position closed-loop system is deduced. In Section 5, some conclusions are provided.

2. Mathematical Model of Position Closed-Loop System

The HAGC system is mainly controlled by electro-hydraulic servo valve and oil cylinder to realize the setting and adjustment of roll gap or rolling pressure. In terms of control function, a complete HAGC system is composed of several automatic control systems. The most important three control loops are as follows: cylinder position closed loop, rolling pressure closed loop and thickness gauge monitoring closed loop, as shown in Figure 1.

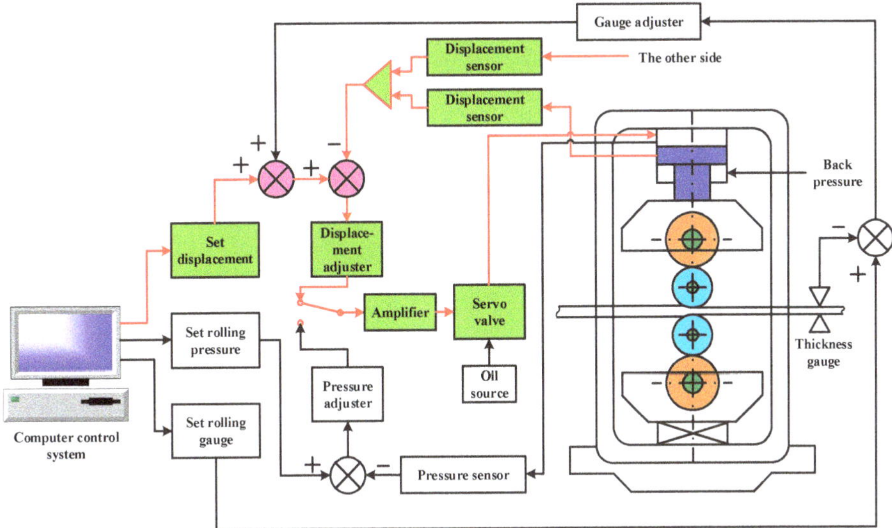

Figure 1. Function diagram of the hydraulic automatic gauge control (HAGC) system.

As the basis of the whole thickness control, cylinder position closed loop is used to control the displacement in a timely and accurate manner with the change of rolling conditions, so as to achieve the setting and controlling of the roll gap. In the position closed-loop system, the measured displacement value is negatively fed back to the signal input end and compared with the given displacement value. If there is a deviation, it will be adjusted by the displacement adjuster and converted into current signal by the power amplifier and further sent to the electro-hydraulic servo valve. After the servo valve obtains the current signal, it will control the flow into the working chamber of the cylinder through the movement of valve spool and then adjust the piston displacement of the cylinder until the feedback value is equal to the set value.

2.1. Mathematical Model of Controller

The controller generally adopts proportion-integration-differentiation (PID) adjuster and its dynamic transfer function can be expressed as:

$$G_c(s) = K_p\left(1 + \frac{1}{T_i s} + T_d s\right), \tag{1}$$

where K_p is proportionality coefficient, T_i is integral time constant, T_d is differential time constant and s is the Laplace operator.

2.2. Mathematical Model of Servo Amplifier

The function of the servo amplifier is to convert voltage signal into current signal and then control the servo valve to realize flow regulation. Since the response time of the servo amplifier is extremely short, it can be treated as a proportional component and its dynamic transfer function is:

$$K_a = \frac{I}{U} \tag{2}$$

where I is the output current (A), U is the input voltage (V) and K_a is amplification coefficient (A/V).

2.3. Mathematical Model of Hydraulic Power Mechanism

The hydraulic power mechanism of HAGC system is mainly realized by controlling the motion of the hydraulic cylinder with the electro-hydraulic servo valve. Its structural principle is displayed in Figure 2. In order to improve the response performance of the system, the servo valve is generally used to control the rodless chamber of the hydraulic cylinder, and the rod chamber of the hydraulic cylinder is supplied with oil at a constant pressure.

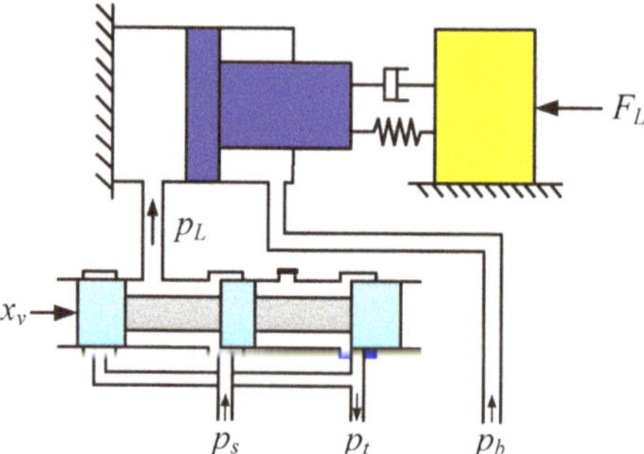

Figure 2. Schematic diagram of the servo valve control hydraulic cylinder.

When the servo valve works in the right position, the high pressure oil directly enters into the rodless chamber of the hydraulic cylinder. At this time, the piston rod of the cylinder drives the load to realize the pressing down action. When the servo valve operates in the left position, the fast lifting action of the roll can be achieved. During the rolling process, oil at a constant pressure of 1 MPa is always passed through the rod chamber to increase the damping of the system.

2.3.1. Flow Equation of Electro-Hydraulic Servo Valve

The function of the servo valve is to control the movement of the valve spool with weak current signal to achieve the control of high power hydraulic energy. There are many advantages such as small volume, high power amplification, fast response and high dynamic performance.

According to the working principle of the servo valve, when the spool displacement x_v is used as the input and the load flow Q_L is taken as the output, the basic flow equation of the servo valve can be obtained:

$$Q_L = f(x_v, p_L) = \begin{cases} C_d W x_v \sqrt{\dfrac{2(p_s - p_L)}{\rho}} & x_v \geq 0 \\ C_d W x_v \sqrt{\dfrac{2(p_L - p_t)}{\rho}} & x_v < 0 \end{cases} \tag{3}$$

where C_d is the flow coefficient of valve port, W is the area gradient of valve port (m), x_v is the displacement of main spool (m), ρ is the hydraulic oil density (kg/m^3), p_s is the oil supply pressure (MPa), p_t is the return pressure (MPa) and p_L is the working pressure of rodless chamber of the hydraulic cylinder (MPa).

The relationship between spool displacement of the servo valve and input current can be expressed as:

$$G_v(s) = \frac{x_v}{I_c} = \frac{K_{sv}}{\frac{s^2}{\omega_{sv}} + \frac{2\xi_{sv}}{\omega_{sv}}s + 1}, \tag{4}$$

where I_c is the input current of the servo valve (A), K_{sv} is the amplification coefficient of the spool displacement on the input current (m/A), ω_{sv} is the natural angular frequency of the servo valve (rad/s) and ξ_{sv} is the damping coefficient of the servo valve (N·s/m).

The servo valve also has nonlinear saturation characteristics and its input current is limited by:

$$I_c = \left\{ \begin{array}{ll} I & I < I_N \\ I_N & I \geq I_N \end{array} \right., \tag{5}$$

where I_N is the rated current of the servo valve (A).

2.3.2. Basic Flow Equation of Hydraulic Cylinder

The flow from the servo valve into the hydraulic cylinder not only meets the flow required to push the piston, but also compensates for internal and external leakage in the cylinder, as well as the flow required to compensate for oil compression and chamber deformation.

The flow continuity equation for the rodless chamber of the hydraulic cylinder can be expressed by:

$$Q_L = A_p\dot{x}_1 + C_{ip}(p_L - p_b) + C_{ep}p_L + \frac{V_0 + A_p x_1}{\beta_e}\dot{p}_L \tag{6}$$

where A_p is the effective working area of the piston (m^2), x_1 is the displacement of the piston rod (mm), C_{ip} is internal leakage coefficient (m$^3 \cdot$ s$^{-1} \cdot$ Pa^{-1}), C_{ep} is external leakage coefficient (m$^3 \cdot$ s$^{-1} \cdot$ Pa^{-1}), p_b is the working pressure of the rod chamber (MPa), V_0 is the initial volume of the control chamber (including the oil inlet pipe and the rodless chamber) (m^3) and β_e is the bulk modulus of oil (MPa).

Since the change of piston displacement of the hydraulic cylinder is small when the hydraulic system is working stably, that is, $A_p x_1 \ll V_0$, then the total volume of the hydraulic cylinder control chamber is approximately equal to the initial volume. In addition, with regard to the actual system, the external leakage is small and can be ignored. Therefore, the continuous flow equation of the hydraulic cylinder control chamber can be further written as:

$$Q_L = A_p\dot{x}_1 + C_{ip}(p_L - p_b) + \frac{V_0}{\beta_e}\dot{p}_L. \tag{7}$$

2.4. Mathematical Model of Load

The external load of the HAGC system consists of several sets of rolls with symmetrical structure. The basic structure of the load roll system of the commonly used four-high mill is shown in Figure 3. In consideration of the load roll system of the six-high mill, the basic structure is similar, and there is a set of intermediate rolls between the support roll and the work roll.

At present, in order to facilitate the analysis, the load roll system is mainly divided according to the lumped model and distribution parameter model, into single degree of freedom (DOF) load model and multi-DOF mass distribution load model, respectively. Moreover, numerous research studies indicate that the stiffness of the upper and lower roll systems of the rolling mill is asymmetrical. The analysis for the HAGC system according to the two-DOF mass distribution load model is more consistent with the actual working conditions [44,45].

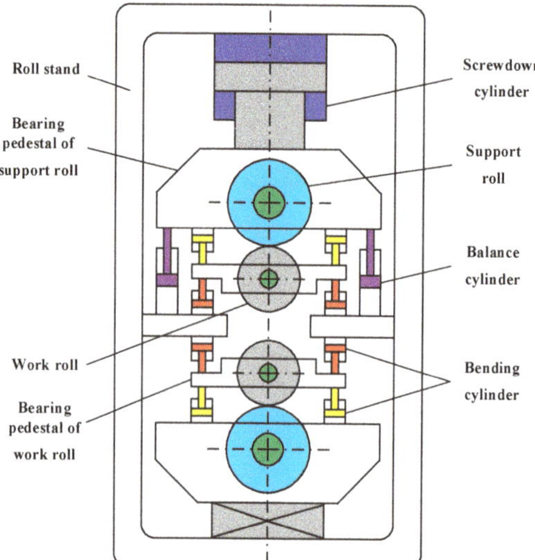

Figure 3. Structure diagram of four-high load roll system.

In order to get closer to the actual working conditions, the modeling method of the load roll system is studied based on the two-DOF asymmetric mass distribution model. The upper roll system is used as a mass system and the lower roll system is utilized as another mass system, then the two-DOF mechanical model of the load roll system is established, as illustrated in Figure 4.

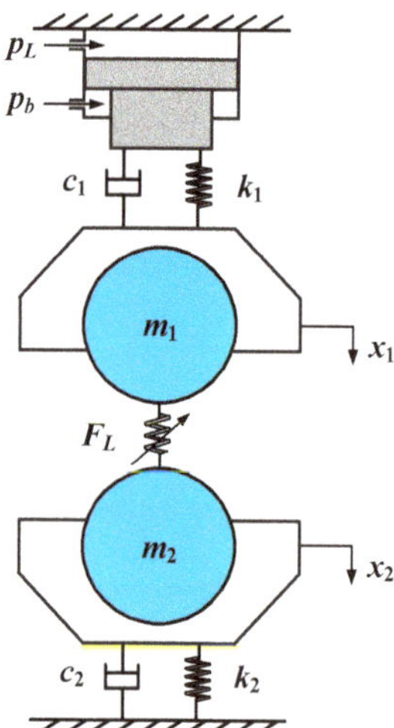

Figure 4. Two degrees of freedom mechanics model of the load roll system.

According to Newton's second law, the load force balance equation of the HAGC system can be expressed as:

$$p_L A_p - p_b A_b = m_1 \ddot{x}_1 + c_1 \dot{x}_1 + k_1 x_1 + F_L, \tag{8}$$

$$F_L = m_2 \ddot{x}_2 + c_2 \dot{x}_2 + k_2 x_2, \tag{9}$$

where m_1 is the equivalent mass of moving parts of the upper roll system (URS) (kg); m_2 is the equivalent mass of the moving parts of the lower roll system (LRS) (kg); c_1 is the linear damping coefficient of moving parts of URS (N·s/m); c_2 is the linear damping coefficient of moving parts of LRS (N·s/m); k_1 is the linear stiffness coefficient between the upper frame beam and the moving parts of URS (N/m); k_2 is the linear stiffness coefficient between the lower frame beam and the moving parts of LRS (N/m); x_1 is the displacement of URS (mm); x_2 is the displacement of LRS (mm); A_b is the effective working area of the rod chamber piston (m^2); and F_L is the load force acting on the roll system (N).

2.5. Mathematical Model of Sensor

The feedback component of the HAGC position closed-loop system is mainly the displacement sensor. In the actual working process, the response time of the sensor needs to be considered, so the sensor can be represented as an inertia link.

The transfer function of the displacement sensor is:

$$G_x(s) = \frac{K_x}{T_x s + 1}, \tag{10}$$

where K_x is the amplification coefficient of the displacement sensor (V/m) and T_x is the time constant of the displacement sensor.

3. Incremental Transfer Model of Position Closed-Loop System

3.1. Incremental Transfer Model of Hydraulic Transmission Part

When the system is in equilibrium at the working point A, according to the mathematical model and information transfer relationship established above, the equilibrium equations of the hydraulic transmission part of the HAGC system can be derived as:

$$Q_{LA} = f(x_{vA}, p_{LA}), \tag{11}$$

$$Q_{LA} = A_p \dot{x}_{1A} + C_{ip}(p_{LA} - p_b) + \frac{V_0}{\beta_e} \dot{p}_{LA}, \tag{12}$$

$$p_{LA} A_p - p_b A_b = m_1 \ddot{x}_{1A} + c_1 \dot{x}_{1A} + k_1 x_{1A} + F_{LA}, \tag{13}$$

where Q_{LA} is the value of the load flow Q_L at the working point A; x_{vA} is the value of spool displacement x_v at the working point A; p_{LA} is the value of working pressure p_L at the working point A; and x_{1A} is the value of piston rod displacement x_1 at the working point A.

When the system makes small disturbances near the working point A, all the variables of the system change around the equilibrium point, as follows:

$$Q_L = Q_{LA} + \Delta Q_L, \tag{14}$$

$$x_v = x_{vA} + \Delta x_v, \tag{15}$$

$$p_L = p_{LA} + \Delta p_L, \tag{16}$$

$$x_1 = x_{1A} + \Delta x, \tag{17}$$

where ΔQ_L is the disturbance quantity of the load flow Q_L at the working point A; Δx_v is the disturbance quantity of spool displacement x_v at the working point A; Δp_L is the disturbance quantity of working pressure p_L at the working point A; and Δx is the disturbance quantity of piston rod displacement x_1 at the working point A.

The load flow of the servo valve is expanded by Taylor series near the working point A, and the high-order minor terms are omitted, so:

$$Q_L = Q_{LA} + \frac{\partial Q_L}{\partial x_v}\Big|_A \Delta x_v + \frac{\partial Q_L}{\partial p_L}\Big|_A \Delta p_L. \tag{18}$$

Then, the approximate equation of disturbance flow can be deduced when the system makes a small disturbance motion near the working point A.

$$\begin{aligned}\Delta Q_L = Q_L - Q_{LA} &= \frac{\partial Q_L}{\partial x_v}\Big|_A \Delta x_v + \frac{\partial Q_L}{\partial p_L}\Big|_A \Delta p_L \\ &= K_q \Delta x_v - K_c \Delta p_L \end{aligned} \tag{19}$$

where K_q is the flow gain, $K_q = \frac{\partial Q_L}{\partial x_v}$; and K_c is the flow–pressure coefficient, $K_c = -\frac{\partial Q_L}{\partial p_L}$.

When the system makes small disturbance motion near the working point A, the flow continuity equation of the hydraulic cylinder can be expressed as:

$$Q_{LA} + \Delta Q_L = A_p(\dot{x}_{1A} + \Delta\dot{x}) + C_{ip}[(p_{LA} + \Delta p_L) - p_b] + \frac{V_0}{\beta_e}(\dot{p}_{LA} + \Delta\dot{p}_L). \tag{20}$$

In combination with Equations (12) and (20), there is:

$$\Delta Q_L = A_p\Delta\dot{x} + C_{ip}\Delta p_L + \frac{V_0}{\beta_e}\Delta\dot{p}_L. \tag{21}$$

When the system makes small disturbance motion near the working point A, the load force balance equation can be expressed as:

$$(p_{LA} + \Delta p_L)A_p - p_b A_b = m_1(\ddot{x}_{1A} + \Delta\ddot{x}) + c_1(\dot{x}_{1A} + \Delta\dot{x}) + k_1(x_{1A} + \Delta x) + F_{LA}. \tag{22}$$

In combination with Equations (13) and (22), there is:

$$\Delta p_L A_p = m_1\Delta\ddot{x} + c_1\Delta\dot{x} + k_1\Delta x. \tag{23}$$

In combination with Equations (19), (21) and (23), the incremental equations of the hydraulic transmission part can be deduced when the system makes small disturbance motion near the working point A.

$$\begin{cases} \Delta Q_L = K_q\Delta x_v - K_c\Delta p_L \\ \Delta Q_L = A_p\Delta\dot{x} + C_{ip}\Delta p_L + \frac{V_0}{\beta_e}\Delta\dot{p}_L \\ \Delta p_L = (m_1\Delta\ddot{x} + c_1\Delta\dot{x} + k_1\Delta x)/A_p \end{cases} \tag{24}$$

The incremental Equation (24) is further organized as follows:

$$\begin{aligned} K_q\Delta x_v = {} & \frac{V_0 m_1}{\beta_e A_p}\Delta\ddot{x} + \left[\left(\frac{V_0 c_1}{\beta_e A_p} + \frac{(C_{ip}+K_c)m_1}{A_p}\right)\right]\Delta\ddot{x} \\ & + \left[\left(\frac{V_0 k_1}{\beta_e A_p} + \frac{(C_{ip}+K_c)c_1}{A_p} + A_p\right)\right]\Delta\dot{x} + \frac{(C_{ip}+K_c)k_1}{A_p}\Delta x \end{aligned} \tag{25}$$

By performing Laplace transformation on Equation (25), the relationship between the load displacement disturbance Δx and the spool displacement disturbance Δx_v can be derived.

$$\Delta x = \frac{A_p}{s\left[\frac{V_0 m_1}{\beta_e}s^2 + (K_{ce}m_1 + \frac{V_0 c_1}{\beta_e})s + (K_{ce}c_1 + \frac{V_0 k_1}{\beta_e} + A_p^2)\right] + k_1 K_{ce}}K_q\Delta x_v \tag{26}$$

where K_{ce} is total flow–pressure coefficient ($\mathrm{m^3 \cdot s^{-1} \cdot Pa^{-1}}$), $K_{ce} = C_{ip} + K_c$.

Suppose that:

$$G_1(s) = \frac{A_p}{s\left[\frac{V_0 m_1}{\beta_e}s^2 + (K_{ce}m_1 + \frac{V_0 c_1}{\beta_e})s + (K_{ce}c_1 + \frac{V_0 k_1}{\beta_e} + A_p^2)\right] + k_1 K_{ce}}. \tag{27}$$

In addition, according to the aforementioned theoretical formula given as Equation (3), there is:

$$K_q = \frac{\partial Q_L}{\partial x_v} = \begin{cases} C_d W \sqrt{\frac{2(p_s - p_L)}{\rho}} & x_v \geq 0 \\ C_d W \sqrt{\frac{2(p_L - p_t)}{\rho}} & x_v < 0 \end{cases}. \tag{28}$$

From Equations (26)–(28), the information transfer relationship between the displacement disturbance Δx of the load and the displacement disturbance Δx_v of the servo valve spool can be identified, which is transmitted by the transfer function $G_1(s)$ and the nonlinear mathematical expression K_q.

3.2. Incremental Transfer Model of the Feedback and Control Part

When the HAGC system adopts the position closed loop, based on the mathematical model of displacement feedback and control, the relationship between spool displacement disturbance Δx_v and load displacement disturbance Δx can be deduced.

$$\begin{aligned} \Delta x_v &= G_c(s)K_a G_v(s)G_x(s)\Delta x \\ &= \frac{K_p(1+\frac{1}{T_i s}+T_d s)K_a K_x K_{sv}}{(T_x s+1)(\frac{s^2}{\omega_{sv}^2}+\frac{2\xi_{sv}}{\omega_{sv}}s+1)}\Delta x \end{aligned} \tag{29}$$

Assume that:

$$G_3(s) = \frac{K_p(1 + \frac{1}{T_i s} + T_d s)K_a K_x K_{sv}}{(T_x s + 1)(\frac{s^2}{\omega_{sv}^2} + \frac{2\xi_{sv}}{\omega_{sv}}s + 1)}. \tag{30}$$

It can be seen from Equations (29) and (30) that the information relationship between the spool displacement disturbance Δx_v and the load displacement disturbance Δx is transmitted by the transfer function $G_3(s)$. In addition, according to the input current limitation condition expression (Equation (5)) of the servo valve, it can be found that $G_3(s)$ possesses a nonlinear saturation characteristic and is a nonlinear transfer function.

4. Absolute Stability Condition for Position Closed-Loop System

On the basis of the aforementioned derived transfer relationship, the transfer block diagram of the disturbance of the position closed-loop system is established, as shown in Figure 5. For purpose of researching the absolute stability of system, the transfer block diagram of the disturbance is the mathematical model which uses the frequency method.

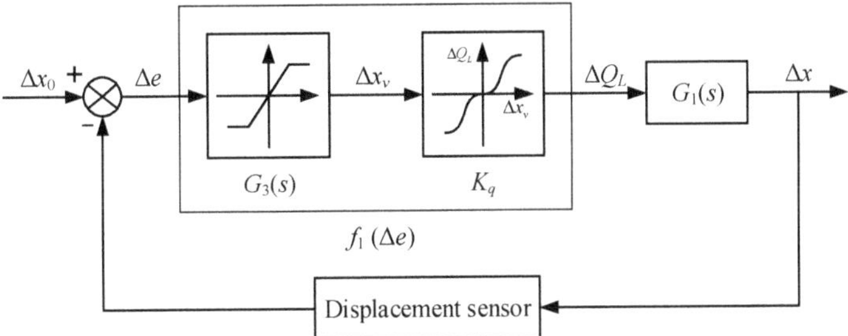

Figure 5. Transfer block diagram of the disturbance of the position closed-loop system.

In this work, the Popov frequency criterion is introduced to determine the absolute stability of the position closed-loop control of the HAGC system. For this, in the transfer function $G_1(s)$, suppose that $s = i\omega$, then the frequency characteristic is obtained:

$$G_1(i\omega) = \mathrm{Re}_1(\omega) + i\mathrm{Im}_1(\omega). \tag{31}$$

The expression (Equation (27)) of $G_1(s)$ is substituted into Equation (31), then the real frequency and imaginary frequency characteristics can be acquired:

$$\begin{aligned}
\mathrm{Re}_1(\omega) \;=\; & A_p[k_1 K_{ce} - (K_{ce}m_1 + \tfrac{V_0 c_1}{\beta_e})\omega^2] \\
& \times \left\{ [k_1 K_{ce} - (K_{ce}m_1 + \tfrac{V_0 c_1}{\beta_e})\omega^2]^2 + [(K_{ce}c_1 + \tfrac{V_0 k_1}{\beta_e} + A_p^2)\omega - \tfrac{V_0 m_1}{\beta_e}\omega^3]^2 \right\}^{-1}
\end{aligned} \tag{32}$$

$$\begin{aligned}
\mathrm{Im}_1(\omega) \;=\; & -A_p[(K_{ce}c_1 + \tfrac{V_0 k_1}{\beta_e} + A_p^2)\omega - \tfrac{V_0 m_1}{\beta_e}\omega^3] \\
& \times \left\{ [k_1 K_{ce} - (K_{ce}m_1 + \tfrac{V_0 c_1}{\beta_e})\omega^2]^2 + [(K_{ce}c_1 + \tfrac{V_0 k_1}{\beta_e} + A_p^2)\omega - \tfrac{V_0 m_1}{\beta_e}\omega^3]^2 \right\}^{-1}
\end{aligned} \tag{33}$$

The expression of corrected frequency characteristic $G_1^*(i\omega)$ is defined as:

$$G_1^*(i\omega) = X_1(\omega) + iY_1(\omega), \tag{34}$$

$$X_1(\omega) = \mathrm{Re}_1(\omega), \quad Y_1(\omega) = \omega\mathrm{Im}_1(\omega). \tag{35}$$

Then, according to Equations (32), (33) and (35), the corrected real frequency and imaginary frequency characteristics can be obtained:

$$\begin{aligned}
X_1(\omega) \;=\; & A_p[k_1 K_{ce} - (K_{ce}m_1 + \tfrac{V_0 c_1}{\beta_e})\omega^2] \\
& \times \left\{ [k_1 K_{ce} - (K_{ce}m_1 + \tfrac{V_0 c_1}{\beta_e})\omega^2]^2 + [(K_{ce}c_1 + \tfrac{V_0 k_1}{\beta_e} + A_p^2)\omega - \tfrac{V_0 m_1}{\beta_e}\omega^3]^2 \right\}^{-1}
\end{aligned} \tag{36}$$

$$\begin{aligned}
Y_1(\omega) \;=\; & -A_p\omega[(K_{ce}c_1 + \tfrac{V_0 k_1}{\beta_e} + A_p^2)\omega - \tfrac{V_0 m_1}{\beta_e}\omega^3] \\
& \times \left\{ [k_1 K_{ce} - (K_{ce}m_1 + \tfrac{V_0 c_1}{\beta_e})\omega^2]^2 + [(K_{ce}c_1 + \tfrac{V_0 k_1}{\beta_e} + A_p^2)\omega - \tfrac{V_0 m_1}{\beta_e}\omega^3]^2 \right\}^{-1}
\end{aligned} \tag{37}$$

The intersection between $G_1^*(i\omega)$ and the real axis is the critical point of the Popov frequency criterion. The coordinate is defined as $(-P_1^{-1}, 0)$. The abscissa value of the critical point can be obtained by using Equations (36) and (37):

$$X_1(\omega^*) = -\frac{A_p V_0 m_1 \beta_e}{\beta_e(K_{ce}m_1\beta_e + V_0 c_1)(K_{ce}c_1 + A_p^2) + V_0^2 k_1 c_1}. \tag{38}$$

Then by the definition of Popov line, we can know that:

$$P_1 = -\frac{1}{X_1(\omega^*)} = \frac{\beta_e(K_{ce}m_1\beta_e + V_0 c_1)(K_{ce}c_1 + A_p^2) + V_0^2 k_1 c_1}{A_p V_0 m_1 \beta_e}. \tag{39}$$

According to Popov's theorem [46,47], if the nonlinear characteristic function $f_1(\Delta e) = G_3(s)K_q\Delta e$ of the position closed-loop system satisfies Equation (40), the equilibrium point of the system is absolutely stable, that is:

$$f(0) = 0, \; 0 < \frac{f_1(\Delta e)}{\Delta e} \leq P_1. \tag{40}$$

From Equation (40), it can be concluded that if the characteristic curve of the nonlinear transfer function $G_3(s)K_q$ is located in the sector region, the position closed-loop system is globally asymptotically stable. The sector region is composed of the horizontal axis and the Popov line l_1 which passes through the origin with a slope P_1, as shown in Figure 6a. Conversely, if the characteristic curve of $G_3(s)K_q$ exceeds the sector region (as illustrated in Figure 6b), the position closed-loop system is unstable. At this time, complex nonlinear dynamic behavior is likely to occur when the system parameters change.

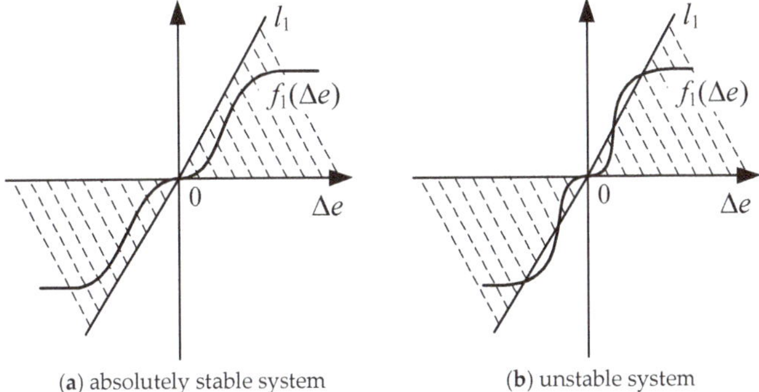

(**a**) absolutely stable system   (**b**) unstable system

Figure 6. Relation between the nonlinear characteristic curve of the position closed-loop system and l_1.

From the above analysis, the absolute stability conditions of the position closed-loop system can be derived:

$$G_3(s)K_q \leq \frac{\beta_e(K_{ce}m_1\beta_e + V_0c_1)(K_{ce}c_1 + A_p^2) + V_0^2k_1c_1}{A_pV_0m_1\beta_e}. \tag{41}$$

The expression of $G_3(s)$ and K_q are substituted into Equation (41), then the absolute stability condition of the position closed-loop system when the spool displacement is positive ($x_v \geq 0$) can be obtained as:

$$\frac{\beta_e(K_{ce}m_1\beta_e + V_0c_1)(K_{ce}c_1 + A_p^2) + V_0^2k_1c_1}{A_pV_0m_1\beta_e} \geq \frac{K_p(1 + \frac{1}{T_is} + T_ds)K_aK_xK_{sv}}{(T_xs + 1)(\frac{s^2}{\omega_{sv}} + \frac{2\xi_{sv}}{\omega_{sv}}s + 1)}C_dW\sqrt{\frac{2(p_s - p_L)}{\rho}}. \tag{42}$$

When the spool displacement is negative ($x_v < 0$), the absolute stability condition of the position closed-loop system can be acquired as:

$$\frac{\beta_e(K_{ce}m_1\beta_e + V_0c_1)(K_{ce}c_1 + A_p^2) + V_0^2k_1c_1}{A_pV_0m_1\beta_e} \geq \frac{K_p(1 + \frac{1}{T_is} + T_ds)K_aK_xK_{sv}}{(T_xs + 1)(\frac{s^2}{\omega_{sv}} + \frac{2\xi_{sv}}{\omega_{sv}}s + 1)}C_dW\sqrt{\frac{2(p_L - p_t)}{\rho}}. \tag{43}$$

5. Conclusions

In this paper, the function of key position closed-loop system in HAGC was introduced in detail. Based on the theoretical analysis, the mathematical model of each component was established. According to the connection relationship of each component element, the incremental transfer model of the position closed-loop system was derived. Moreover, according to the derived information transfer relationship, the transfer block diagram of the disturbance of the system was established. Furthermore, the Popov frequency criterion method was introduced to derive the absolute stability condition. The absolute stability conditions of the system are acquired in the following two conditions: when the spool displacement of the servo valve is positive or negative.

The obtained results lay a theoretical foundation for the study of the instability mechanism of the HAGC system. This research can provide a significant basis for the further investigation on the vibration traceability and control of the HAGC system.

Author Contributions: Conceptualization, Y.Z. and W.J.; Methodology, S.T.; Investigation, Y.Z. and S.T.; Writing-Original Draft Preparation, Y.Z.; Writing-Review & Editing, J.Z. and G.L.; Supervision, C.W.

Funding: This research was funded by National Natural Science Foundation of China (No. 51805214, 51875498), China Postdoctoral Science Foundation (No. 2019M651722), Natural Science Foundation of Hebei Province (No. E2018203339), Nature Science Foundation for Excellent Young Scholars of Jiangsu Province (No. BK20190101), Open Foundation of National Research Center of Pumps, Jiangsu University (No. NRCP201604) and Open Foundation of the State Key Laboratory of Fluid Power and Mechatronic Systems (No. GZKF-201714).

Conflicts of Interest: The authors declare no conflict of interest.

Nomenclature

HAGC	hydraulic automatic gauge control
PID	Proportion-integration-differentiation
DOF	degree of freedom
K_p	proportionality coefficient
T_i	integral time constant
T_d	differential time constant
s	Laplace operator
I	output current
U	input voltage
K_a	amplification coefficient
Q_L	load flow
x_v	spool displacement
C_d	flow coefficient of valve port
W	area gradient of valve port
ρ	hydraulic oil density
p_s	oil supply pressure
p_t	return pressure
p_L	working pressure of rodless chamber of hydraulic cylinder
I_c	input current of servo valve
K_{sv}	amplification coefficient of the spool displacement on the input current
ω_{sv}	natural angular frequency of servo valve
ξ_{sv}	damping coefficient of servo valve
I_N	rated current of servo valve
A_p	effective working area of piston
x_1	displacement of piston rod
C_{ip}	internal leakage coefficient
C_{ep}	external leakage coefficient
p_b	working pressure of the rod chamber
V_0	initial volume of the control chamber
β_e	bulk modulus of oil
m_1	equivalent mass of moving parts of the upper roll system (URS)
m_2	equivalent mass of the moving parts of the lower roll system (LRS)
c_1	linear damping coefficient of moving parts of URS
c_2	linear damping coefficient of moving parts of LRS
k_1	linear stiffness coefficient between upper frame beam and moving parts of URS
k_2	linear stiffness coefficient between lower frame beam and moving parts of LRS
x_1	displacement of URS
x_2	displacement of LRS
A_b	effective working area of rod chamber piston
F_L	load force acting on roll system
K_x	amplification coefficient of the displacement sensor
T_x	time constant of the displacement sensor
Q_{LA}	the value of load flow at the working point A
x_{vA}	the value of spool displacement at the working point A

p_{LA}	the value of working pressure at the working point A
x_{1A}	the value of piston rod displacement at the working point A
ΔQ_L	disturbance quantity of load flow at the working point A
Δx_v	disturbance quantity of spool displacement at the working point A
Δp_L	disturbance quantity of working pressure at the working point A
Δx	disturbance quantity of piston rod displacement at the working point A
K_q	flow gain
K_c	flow–pressure coefficient
K_{ce}	total flow–pressure coefficient

References

1. Tang, S.N.; Zhu, Y.; Li, W.; Cai, J.X. Status and prospect of research in preprocessing methods for measured signals in mechanical systems. *J. Drain. Irrig. Mach. Eng.* **2019**, *37*, 822–828.
2. Tang, B.; Jiang, H.; Gong, X. Optimal design of variable assist characteristics of electronically controlled hydraulic power steering system based on simulated annealing particle swarm optimisation algorithm. *Int. J. Veh. Des.* **2017**, *73*, 189–207. [CrossRef]
3. He, R.; Liu, X.; Liu, C. Brake performance analysis of ABS for eddy current and electrohydraulic hybrid brake system. *Math. Probl. Eng.* **2013**, *2013*, 979384. [CrossRef]
4. Yu, Y.; Zhang, C.; Han, X.J.; Bi, Q.S. Dynamical behavior analysis and bifurcation mechanism of a new 3- D nonlinear periodic switching system. *Nonlinear Dyn.* **2013**, *73*, 1873–1881. [CrossRef]
5. Roman, N.; Ceanga, E.; Bivol, I.; Caraman, S. Adaptive automatic gauge control of a cold strip rolling process. *Adv. Electr. Comput. Eng.* **2010**, *10*, 7–17. [CrossRef]
6. Hu, Y.J.; Sun, J.; Wang, Q.L.; Yin, F.C.; Zhang, D.H. Characteristic analysis and optimal control of the thickness and tension system on tandem cold rolling. *Int. J. Adv. Manuf. Technol.* **2019**, *101*, 2297–2312. [CrossRef]
7. Sun, J.L.; Peng, Y.; Liu, H.M. Dynamic characteristics of cold rolling mill and strip based on flatness and thickness control in rolling process. *J. Cent. South Univ.* **2014**, *21*, 567–576. [CrossRef]
8. Prinz, K.; Steinboeck, A.; Muller, M.; Ettl, A.; Kugi, A. Automatic gauge control under laterally asymmetric rolling conditions combined with feedforward. *IEEE Trans. Ind. Appl.* **2017**, *53*, 2560–2568. [CrossRef]
9. Prinz, K.; Steinboeck, A.; Kugi, A. Optimization-based feedforward control of the strip thickness profile in hot strip rolling. *J. Process Control* **2018**, *64*, 100–111. [CrossRef]
10. Kovari, A. Influence of internal leakage in hydraulic capsules on dynamic behavior of hydraulic gap control system. *Mater. Sci. Forum* **2015**, *812*, 119–124. [CrossRef]
11. Li, J.X.; Fang, Y.M.; Shi, S.L. Robust output-feedback control for hydraulic servo-position system of cold-strip rolling mill. *Control Theory Appl.* **2012**, *29*, 331–336.
12. Sun, W.Q.; Shao, J.; Song, Y.; Guan, J.L. Research and development of automatic control system for high precision cold strip rolling mill. *Adv. Mater. Res.* **2014**, *952*, 283–286. [CrossRef]
13. Yi, J.G. Modelling and analysis of step response test for hydraulic automatic gauge control. *J. Mech. Eng.* **2015**, *61*, 115–122. [CrossRef]
14. Liu, H.S.; Zhang, J.; Mi, K.F.; Gao, J.X. Simulation on hydraulic-mechanical coupling vibration of cold strip rolling mill vertical system. *Adv. Mater. Res.* **2013**, *694*, 407–414. [CrossRef]
15. Wang, J.; Sun, B.; Huang, Q.; Li, H. Research on the position-pressure master-slave control for rolling shear hydraulic servo system. *Stroj. Vestn.* **2015**, *61*, 265–272.
16. Hua, C.C.; Yu, C.X. Controller design for cold rolling mill HAGC system with measurement delay perturbation. *J. Mech. Eng.* **2014**, *50*, 46–53. [CrossRef]
17. Zhang, B.; Wei, W.; Qian, P.; Jiang, Z.; Li, J.; Han, J.; Mujtaba, M. Research on the control strategy of hydraulic shaking table based on the structural flexibility. *IEEE Access* **2019**, *7*, 43063–43075. [CrossRef]
18. Wang, C.; Hu, B.; Zhu, Y.; Wang, X.; Luo, C.; Cheng, L. Numerical study on the gas-water two-phase flow in the self-priming process of self-priming centrifugal pump. *Processes* **2019**, *7*, 330. [CrossRef]
19. Wang, C.; Shi, W.; Wang, X.; Jiang, X.; Yang, Y.; Li, W.; Zhou, L. Optimal design of multistage centrifugal pump based on the combined energy loss model and computational fluid dynamics. *Appl. Energy* **2017**, *187*, 10–26. [CrossRef]

20. Qian, J.Y.; Gao, Z.X.; Liu, B.Z.; Jin, Z.J. Parametric study on fluid dynamics of pilot-control angle globe valve. *ASME J. Fluids Eng.* **2018**, *140*, 111103. [CrossRef]

21. Qian, J.Y.; Chen, M.R.; Liu, X.L.; Jin, Z.J. A numerical investigation of the flow of nanofluids through a micro Tesla valve. *J. Zhejiang Univ. Sci. A* **2019**, *20*, 50–60. [CrossRef]

22. Hou, C.W.; Qian, J.Y.; Chen, F.Q.; Jiang, W.K.; Jin, Z.J. Parametric analysis on throttling components of multi-stage high pressure reducing valve. *Appl. Therm. Eng.* **2018**, *128*, 1238–1248. [CrossRef]

23. Wang, C.; He, X.; Zhang, D.; Hu, B.; Shi, W. Numerical and experimental study of the self-priming process of a multistage self-priming centrifugal pump. *Int. J. Energy Res.* **2019**, *43*, 4074–4092. [CrossRef]

24. Wang, C.; He, X.; Shi, W.; Wang, X.; Wang, X.; Qiu, N. Numerical study on pressure fluctuation of a multistage centrifugal pump based on whole flow field. *AIP Adv.* **2019**, *9*, 035118. [CrossRef]

25. He, X.; Jiao, W.; Wang, C.; Cao, W. Influence of surface roughness on the pump performance based on Computational Fluid Dynamics. *IEEE Access* **2019**, *7*, 105331–105341. [CrossRef]

26. Wang, C.; Chen, X.X.; Qiu, N.; Zhu, Y.; Shi, W.D. Numerical and experimental study on the pressure fluctuation, vibration, and noise of multistage pump with radial diffuser. *J. Braz. Soc. Mech. Sci. Eng.* **2018**, *40*, 481. [CrossRef]

27. Hu, B.; Li, X.; Fu, Y.; Zhang, F.; Gu, C.; Ren, X.; Wang, C. Experimental investigation on the flow and flow-rotor heat transfer in a rotor-stator spinning disk reactor. *Appl. Therm. Eng.* **2019**, *162*, 114316. [CrossRef]

28. Ye, S.G.; Zhang, J.H.; Xu, B.; Zhu, S.Q. Theoretical investigation of the contributions of the excitation forces to the vibration of an axial piston pump. *Mech. Syst. Signal Process.* **2019**, *129*, 201–217. [CrossRef]

29. Zhang, J.H.; Xia, S.; Ye, S.; Xu, B.; Song, W.; Zhu, S.; Xiang, J. Experimental investigation on the noise reduction of an axial piston pump using free-layer damping material treatment. *Appl. Acoust.* **2018**, *139*, 1–7. [CrossRef]

30. Bai, L.; Zhou, L.; Jiang, X.P.; Pang, Q.L.; Ye, D.X. Vibration in a multistage centrifugal pump under varied conditions. *Shock Vib.* **2019**, *2019*, 2057031. [CrossRef]

31. Bai, L.; Zhou, L.; Han, C.; Zhu, Y.; Shi, W.D. Numerical study of pressure fluctuation and unsteady flow in a centrifugal pump. *Processes* **2019**, *7*, 354. [CrossRef]

32. Wang, L.; Liu, H.L.; Wang, K.; Zhou, L.; Jiang, X.P.; Li, Y. Numerical simulation of the sound field of a five-stage centrifugal pump with different turbulence models. *Water* **2019**, *11*, 1777. [CrossRef]

33. Ding, S.; Zheng, W.X. Controller design for nonlinear affine systems by control Lyapunov functions. *Syst. Control Lett.* **2013**, *62*, 930–936. [CrossRef]

34. Liu, L.; Ding, S.H.; Ma, L.; Sun, H.B. A novel second-order sliding mode control based on the Lyapunov method. *Trans. Inst. Meas. Control* **2018**, *41*, 014233121878324. [CrossRef]

35. Zhang, J. Integral barrier Lyapunov functions-based neural control for strict-feedback nonlinear systems with multi-constraint. *Int. J. Control Autom. Syst.* **2018**, *16*, 2002–2010. [CrossRef]

36. Zhang, J.; Li, G.S.; Li, Y.H.; Dai, X.K. Barrier Lyapunov functions-based localized adaptive neural control for nonlinear systems with state and asymmetric control constraints. *Trans. Inst. Meas. Control* **2019**, *41*, 1656–1664. [CrossRef]

37. Yang, C.; Zhang, Q.; Zhou, L. Strongly absolute stability of Lur'e descriptor systems: Popov-type criteria. *Int. J. Robust Nonlinear Control* **2009**, *19*, 786–806. [CrossRef]

38. Saeki, M.; Wada, N.; Satoh, S. Stability analysis of feedback systems with dead-zone nonlinearities by circle and Popov criteria. *Automatica* **2016**, *66*, 96–100. [CrossRef]

39. Xia, L.; Jiang, H. An electronically controlled hydraulic power steering system for heavy vehicles. *Adv. Mech. Eng.* **2016**, *8*, 1687814016679566. [CrossRef]

40. Zhang, R.; Wang, Y.; Zhang, Z.D.; Bi, Q.S. Nonlinear behaviors as well as the bifurcation mechanism in switched dynamical systems. *Nonlinear Dyn.* **2015**, *79*, 465–471. [CrossRef]

41. Bi, Q.S.; Li, S.L.; Kurths, J.; Zhang, Z.D. The mechanism of bursting oscillations with different codimensional bifurcations and nonlinear structures. *Nonlinear Dyn.* **2016**, *85*, 993–1005. [CrossRef]

42. Xue, Z.H.; Cao, X.; Wang, T.Z. Vibration test and analysis on the centrifugal pump. *J. Drain. Irrig. Mach. Eng.* **2018**, *36*, 472–477.

43. Zhu, Y.; Tang, S.N.; Quan, L.X.; Jiang, W.L.; Zhou, L. Extraction method for signal effective component based on extreme-point symmetric mode decomposition and Kullback-Leibler divergence. *J. Braz. Soc. Mech. Sci. Eng.* **2019**, *41*, 100. [CrossRef]

44. Zhu, Y.; Qian, P.F.; Tang, S.N.; Jiang, W.L.; Li, W.; Zhao, J.H. Amplitude-frequency characteristics analysis for vertical vibration of hydraulic AGC system under nonlinear action. *AIP Adv.* **2019**, *9*, 035019. [CrossRef]
45. Zhu, Y.; Tang, S.; Wang, C.; Jiang, W.; Yuan, X.; Lei, Y. Bifurcation characteristic research on the load vertical vibration of a hydraulic automatic gauge control system. *Processes* **2019**, *7*, 718. [CrossRef]
46. Liu, Y.Z.; Chen, L.Q. *Nonlinear Vibration*; Higher Education Press: Beijing, China, 2001; pp. 57–123.
47. Ding, W.J. *Self-Excited Vibration*; Tsinghua University Press: Beijing, China, 2009; Volume 84, pp. 238–242.

Article

Effects of a Dynamic Injection Flow Rate on Slug Generation in a Cross-Junction Square Microchannel

Jin-yuan Qian [1,2,3], Min-rui Chen [1], Zan Wu [2], Zhi-jiang Jin [1] and Bengt Sunden [2,*]

[1] Institute of Process Equipment, Zhejiang University, Hangzhou 310027, China; qianjy@zju.edu.cn (J.-y.Q.); chenminrui@zju.edu.cn (M.-r.C.); jzj@zju.edu.cn (Z.-j.J.)
[2] Department of Energy Sciences, Lund University, P.O. Box 118, SE-22100 Lund, Sweden; Zan.wu@energy.lth.se
[3] State Key Laboratory of Fluid Power and Mechatronic Systems, Zhejiang University, Hangzhou 310027, China
* Correspondence: bengt.sunden@energy.lth.se; Tel.: +86-0571-87951216

Received: 3 September 2019; Accepted: 16 October 2019; Published: 18 October 2019

Abstract: The injection flow rates of two liquid phases play a decisive role in the slug generation of the liquid-liquid slug flow. However, most injection flow rates so far have been constant. In order to investigate the effects of dynamic injection flow rates on the slug generation, including the slug size, separation distance and slug generation cycle time, a transient numerical model of a cross-junction square microchannel is established. The Volume of Fluid method is adopted to simulate the interface between two phases, i.e., butanol and water. The model is validated by experiments at a constant injection flow rate. Three different types of dynamic injection flow rates are applied for butanol, which are triangle, rectangular and sine wave flow rates. The dynamic injection flow rate cycles, which are related to the constant slug generation cycle time t_0, are investigated. Results show that when the cycle of the disperse phase flow rate is larger than t_0, the slug generation changes periodically, and the period is influenced by the cycle of the disperse phase flow rate. Among the three kinds of dynamic disperse flow rate, the rectangular wave influences the slug size most significantly, while the triangle wave influences the separation distance and the slug generation time more prominently.

Keywords: liquid-liquid slug flow; slug generation; dynamic injection flow rate; microchannel; Computational Fluid Dynamics

---

1. Introduction

Due to the excellent uniform size and dynamic flow characteristics, slug flow inside microchannels has been widely used in a variety of fields [1,2]. A lot of scholars have contributed to this research topic. Many researchers pay attention to the gas-liquid two-phase slug flow in microchannels. For instance, Ryo et al. [3] conducted experiments to measure the void fraction and pressure drop of gas–liquid slug flow in circular microchannels with different diameters. They proposed a model to predict the distributions very well. Liu et al. [4] investigated the effects of capillary number and flow rate ratio on the slug–bubble flow by conducting experiments and numerical simulations. They developed a correlation to describe the relationship between bubble length, capillary number and flow rate ratio. Patel et al. [5] conducted experiments in a microchannel to observe the effects of operating conditions and the channel size on the liquid film thickness of bubbles in an air–water slug flow regime. It was found that the liquid film thickness in the corner of the square channel cross section decreased with the increase of the capillary number. Additionally, a correlation for the calculation of the film thickness was proposed with an error within an acceptable range.

Rocha et al. [6] conducted numerical simulations to investigate the Taylor bubbles in circular milli and microchannels for a wide range of Reynolds and Capillary numbers, and the shape and velocity of isolated Taylor bubbles, as well the liquid films surrounding the bubbles, were analyzed. They found different recirculation patterns of liquid slugs and different velocity characteristics of bubbles, for low and high Capillary numbers. Meanwhile, the thickness of films surrounding the bubbles were found almost stagnant at low Capillary and Reynolds numbers, and the bubble velocity could be estimated extremely accurately by the stagnant film hypothesis, especially at low Capillary numbers. Kingston et al. [7] created a novel experimental facility to control the hydrodynamic quantification and heat transfer of the microchannel slug flow boiling transport. Svetlov et al. [8] carried out experiments with a coaxial-spherical micro mixer and optimized the nozzle positions. The dependence of the Taylor flow regime upon liquid superficial velocity was weakened, and equations used to calculate the bubbles and slugs were obtained. Yao et al. [9–11] conducted a series of experiments to investigate the gas-liquid mass transfer in the slug flow regime by an online photographic method to measure the mass transfer coefficient of CO_2 bubbles. Yin et al. [12] also adopted CO_2 as the gas phase, and observed slug-bubbly flow, slug flow and slug-annular flow, where the void fraction and frictional pressure drop of the slug flow regime were mainly investigated. Liu et al. [13] carried out a numerical study to investigate the transition from the slug flow to the annular flow of convective boiling with high heat flux in a microchannel, and it was found that the transition of the flow regime had a positive impact on bubble evaporation. Magnini and Thome [14,15] investigated the effects of the primary flow parameters on the boiling heat transfer performance of a slug flow by means of numerical simulation. It was concluded that the heat transfer performance was enhanced by flow conditions, making the liquid film of the bubble thinner, and they proposed a new physical model to update the three-zone model, which had been widely used for slug flow boiling in microchannels. Li et al. [16] investigated the two-phase flow of R32 in a parallel-port microchannel, where the two-phase flow was generated by heating the subcooled refrigerant, and proposed the flow pattern map of R32.

In addition, many scholars have paid attention to the research of liquid-liquid slug flow. Garstecki et al. [17] conducted a pioneering investigation on the process of the formation of droplets and bubbles in a T-junction microchannel, and threw light on the mechanism of the break-up of the dispersed phase fluid at a low capillary number. Sattari-Najafabadi et al. [18,19] conducted experiments in cross-junction rectangular microchannels to investigate the effects of the aspect ratio and the channel size on the liquid-liquid mass transfer within the slug regime. They found that a smaller square microchannel was conducive to mass transfer. Kovalev et al. [20] investigated the flow hydrodynamics of liquid-liquid flow with a low viscosity ratio in a T-junction rectangular microchannel by presenting an experimental study. They observed six flow patterns, including a specific slug flow, where micro-sized droplets were generated. Carneiro et al. [21] studied the droplet formation regimes of immiscible fluids with low viscosities, which contained surfactant, by experimental and numerical methods. The effects of surfactant mass transport limitations, and of the interface rheology, are analyzed. A numerical method, which was based upon a coupled level-set and volume of fluid method, estimated the droplet size accurately. Zhang et al. [22] carried out experiments in a T-junction microchannel to investigate the effect of the elasticity of the PEO-glycerol solution on the dynamics of droplet generation. They observed four flow regimes, and researched the stages of droplet generation in detail. Qian et al. [23] investigated the slug flow pattern in a T-junction microchannels. The effects of the flow rates of two phases and the geometry properties of microchannel on the generation of slugs were analyzed in detail. A new flow stage, called lag stage merges, was observed. Man et al. [24] used a step surface to break liquid-liquid slugs into droplets by experiments, and found a critical slug length for the breakup. Wehking et al. [25] conducted an experimental study in a T-junction microchannel to investigate viscosity, interfacial tension and flow geometry on droplet formation. They developed empirical correlations to predict the transition regions for the slug and dripping regimes. Zhang et al. [26] conducted experiments in circular microchannels with different diameters at different temperatures to investigate the hydrodynamics and mass transfer characteristics of liquid-liquid slug flow; results revealed the roles of inertia and viscous force in the transition of liquid-liquid flow patterns, as well as the effects of temperature on mass transfer.

Song et al. [27] investigated the hydrodynamics and mass transfer performance of a liquid-liquid slug flow regime in a microreactor, where the chemical oxidative polymerization of aniline was happening. They found that the internal recirculation in dispersed phase slugs was beneficial for the mass transfer in the polymerization, and obtained a combination of residence time and temperature, with which a high level polyaniline yield was reached. Qian et al. [28] investigated the mixing effeciency on droplet formation process in microchannels with three shapes, aided by the CFD method. The mixing performance in different microchannels were compared, and found that the cross-shaped T-junction was the best. Van Loo et al. [29] studied the droplet formation in a cross-junction microchannel by experimental methods. They mainly focused on two steps of slug generation, and several scaling laws were proposed to relate the droplet volume and generation frequency. Zhao et al. [30] investigated the behavior of slug flow to prepare cryogel beads in a rectangular microchannel, and correlations of the liquid-liquid slug flow parameters were obtained for the bead diameter estimation. Malekzadeh et al. [31] utilized the OpenFOAM code to simulate the droplet formation in a T-junction microchannel, and found that when the contact angle increased, the droplet detachment time increased in the slug flow regime, while it decreased in the dripping regime. Gupta et al. [32] analyzed the effects of viscoelasticity on the liquid threads breakup in a cross-junction microchannel, by mesoscale lattice Boltzmann methods. It was found that the break-up point of the liquid thread was affected more pronouncedly with matrix viscoelasticity. Chen et al. [33] proposed a 3D physical model to describe the slug generation process in a cross-junction microchannel. Li et al. [34] studied the mechanism of liquid-liquid droplet formation in a T-junction microchannel numerically and analytically. They found that the surface force and viscous force played significant roles in the droplet regime. Ładosz et al. [35] proposed two models to calculate the pressure drop of liquid-liquid slug flow in square microchannels, and found that the models were reliable in predicting the pressure drop of liquids with similar viscosities. Qian et al. [36] conducted numerical investigation in serpentine microchannels to analyze the effects of bend radius of the microchannels on the mixing efficiency in liquid-liquid two-phase flow, the pressure drop in the microchannel was researched as well.

Our research group has also paid attention to the microchannel topic, Qian et al. [37] conducted numerical simulations on the nanofluids flowing through a micro Tesla valve. Based on the investigations mentioned above, the fluid injection flow rate is commonly constant. However, a dynamic injection flow rate may precisely control the slug size and the distance between two neighboring slugs. For instance, when a fixed slug size is required, and the continuous phase is expensive, the dynamic injection flow rate can control the slug size and the separation distance between slugs to reduce the cost. In this paper, firstly slugs generated in a cross-junction microchannel at constant flow rates are studied experimentally and numerically. The experimental and numerical data are compared to validate the reliability of the numerical method. Then, as the experimental conditions are limited, the numerical simulations of slug generation are extended to handle the condition of dynamic dispersed phase injection flow rates. The slug size, the separation distance and the generation time are analyzed to discover the effects of the dynamic dispersed phase injection flow rate on the slug generation process.

2. Experimental Setup

Figure 1 presents the experimental setup. It was built up by syringe pumps (New Era, NE-4000, New Era Pump Systems, 138 Toledo St, Farmingdale, NY 11735, US), a cross-junction microchannel chip, a microscope (Motic, SMZ-171, Motic Microscopes, Roanoke, VA, US), a digital camera (Olympus OM-DE-M1, Olympus Cameras, Shinjuku, Tokyo, Japan), and a laser light source. In this setup, the microchannel chip was fabricated by borosilicate glass. A cross-junction microchannel had been carved onto the chip, and the thermal bonding technique, which had excellent chemical resistance, was adopted to seal the channel. The microchannel geometrical parameters are shown in Figure 2. The cross section of the microchannel was a square, and the channel width W was 0.6 mm. To avoid the entrance effect, the length of the entrance sections was 10 *W*. In addition, the length of the section downstream from the cross-junction was 50 *W*, so that the slug flowing pattern could be fully developed. The continuous phase (water) was injected from the two side channels, and the dispersed phase (butanol) was injected from the central channel. The slug size was represented by its maximum length *d*, and the separation distance was represented by its minimum length *l*, as shown in Figure 2.

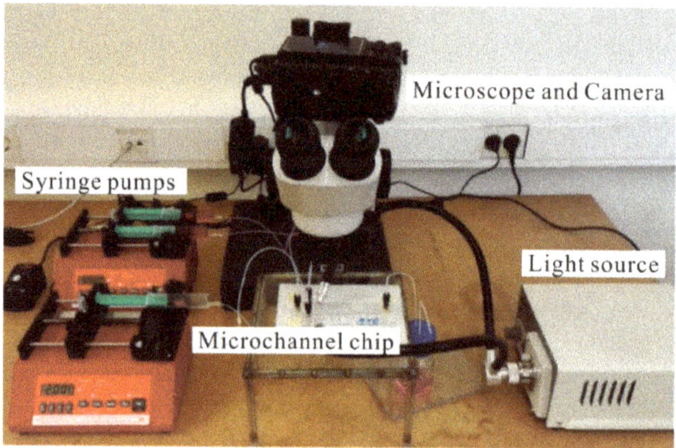

Figure 1. Experimental setup.

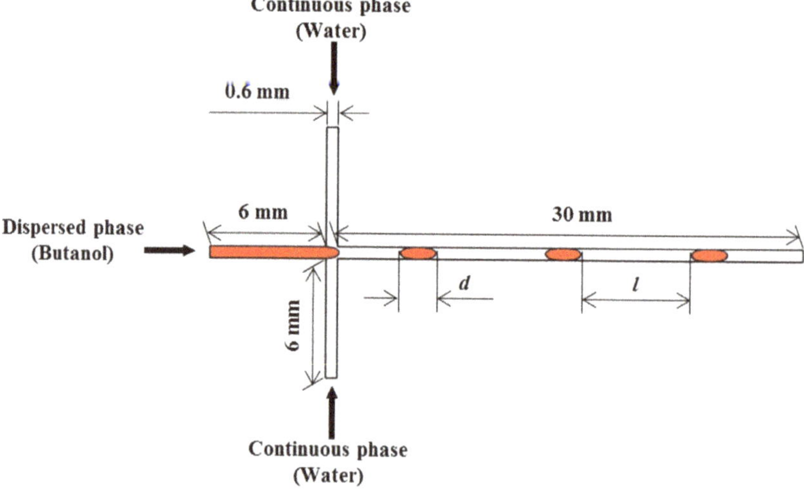

Figure 2. Geometry parameters of the microchannel.

In this experiment, syringe pumps were used to control the injection flow rate, which was kept constant, similar to the common situations. The slugs inside the microchannel were observed by a microscope, and the images were captured by a digital camera. A laser light source was employed to provide sufficient lightness for the image shoot. Finally, a bottle was connected to the outlet of the microchannel to collect waste liquids.

During the experiment, the water phase was firstly introduced into the microchannel for several minutes before injecting the butanol phase in order to get stable slugs. The slug length and slug velocity were extracted and averaged from 10 snapshots and videos. The experimental setup had been used in previous studies, and the experimental uncertainty is within ± 5%, as mentioned in reference [38].

3. Numerical Method

3.1. Mathematical Model

The CFD simulation of flow in a micro device has been studied by many researchers, especially for the discussion on the liquid film capture [39]. It should be said that Volume of Fluid (VOF) model is suitable and can be adopted in this investigation. In this model, the volume fraction of the continuous phase α, which ranges from 0 to 1, is used to judge whether the control volume contains the interface of the two phases or not. If α is equal to 0, the control volume is full of the dispersed phase. If α is equal to 1, the control volume is full of the continuous phase. If α is larger than 0 but less than 1, the control volume is mixed by the two phases, and contains the interface between the two phases. In a control volume, the physical properties of the fluid, such as density and viscosity, are calculated by the volume fractions of the two phases.

When the VOF model is adopted, the tracking of the interface(s) between the phases is accomplished by the solution of a continuity equation for the volume fraction of one (or more) of the phases. In the present investigation, there are two phases. Taking the continuous phase as an example, this equation can be expressed as the following form:

$$\frac{1}{\rho_c}\left[\frac{\partial}{\partial t}(\alpha_c\rho_c) + \nabla \cdot \left(\alpha_c\rho_c\vec{v}_c\right)\right] = \frac{1}{\rho_c}\sum_{p=1}^{n}\left(\dot{m}_{dc} - \dot{m}_{cd}\right) \tag{1}$$

where, ρ is the density, α is the fraction of continuous phase, the subscripts c and d represent the continuous and disperse phase, respectively. The symbol $\dot{m}_{dc}$ is the mass transfer from the dispersed phase to the continuous phase, and the symbol $\dot{m}_{cd}$ is the mass transfer from the continuous phase to the dispersed phase.

In the VOF model, a single momentum equation is solved throughout the domain, and the resulting velocity field is shared among the phases. The momentum equation, expressed as Equation (2), is dependent upon the volume fractions of all phases through the properties ρ and μ.

$$\frac{\partial}{\partial t}\left(\rho\vec{v}\right) + \nabla \cdot \left(\rho\vec{v}\vec{v}\right) = -\nabla p + \nabla \cdot \left[\mu\left(\nabla\vec{v} + \nabla\vec{v}^{T}\right)\right] + \rho\vec{g} + \vec{F} \tag{2}$$

where, μ is the viscosity. For the continuous and dispersed phases in this study, the properties ρ and μ are calculated by following equations:

$$\rho = \rho_c\alpha + \rho_d(1-\alpha) \tag{3}$$

$$\mu = \mu_c\alpha + \mu_d(1-\alpha) \tag{4}$$

Taking the continuous phase as an example, the source term is written as follows:

$$F = 2\gamma\frac{\rho\kappa_c\nabla\alpha}{\rho_c + \rho_d} \tag{5}$$

where, γ represents the surface tension between the two phases, and κ_c is the curvature of the interface which is determined as follows:

$$\kappa_c = \nabla \cdot \hat{n} \tag{6}$$

For the control volumes far away from the wall and the cells close to the wall, the unit normal vectors are derived from Equation (5) and (6), respectively:

$$\hat{n} = \frac{n}{|n|}, n = \nabla \cdot \alpha \tag{7}$$

$$\hat{n} = \hat{n}_c \cos\theta + \hat{t}_c \sin\theta \tag{8}$$

where, θ is the contact angle, $\hat{n}_c$ and $\hat{t}_c$ represent the normal and tangential vectors, respectively.

3.2. Mesh and Boundary Conditions

A 3D physical model of the cross-junction microchannel corresponding to the experimental setup is established. A hexahedral mesh is used to discretize the flow channel and a mesh independence check is conducted. Aided by the software Fluent 17.2, the numerical simulations are conducted. With constant continuous and disperse phase flow rate of 0.001 m/s, the slug size is 1,183 μm when the mesh size is 40 μm, and the slug size is 1,172 μm when the mesh size is 35 μm. The relative difference is within 0.95%. Therefore, meshes with size of 40 μm will provide numerical results with satisfactory accuracy, and the corresponding number of control volumes is 271,800.

Consistent with the experiments, water is set as the continuous phase and butanol is set as the dispersed phase. The fluid physical properties are listed in Table 1. At room temperature, about 7.7 g of butanol can be dissolved per 100 g of water, which indicates that butanol is slightly soluble in water. The images obtained from the experiments, in which the interfaces of water and butanol were clear, proved that the slight solubility affected little on the generation of slugs. The boundary conditions are set as follows: The three inlets are set as velocity inlets. With the advantage of the user defined functions (UDFs), the dispersed phase injection flow rate can be set as a dynamic flow rate by a specified function. The outlet is set as an outflow. Then, for all other surfaces, the wall condition is adopted. The wettability of a liquid is determined by the liquid and solid properties. The contact angle of butanol is set as 150°. The Pressure-Implicit with Splitting of Operators (PISO) scheme is adopted to couple the pressure and velocity. Concerning the spatial discretization, the Green-Gauss Cell-Based algorithm is used for gradients, the PRESTO! algorithm is used for pressure, a second order upwind scheme is adopted for momentum, and the Geo-Reconstruct is employed for the volume fraction.

Table 1. Physical properties of two fluids.

Fluid	Density (kg.m^{-3})	Viscosity (mPa.s)	Surface Tension (N.m^{-1})
Water	998.2	1.003	0.0018
Butanol	810	2.95	

4. Results and Discussion

4.1. Numerical Method Validation

The flowing pattern obtained from the numerical simulation is compared with the experimental data for the validation purpose. These experiments and simulations were conducted for the condition of a constant injection flow rate.

4.1.1. Experimental Results

In the experiments, the constant injection flow rates both for the continuous and dispersed phase were set as 0.001 m/s, 0.002 m/s, 0.003 m/s and 0.004 m/s, respectively. The images captured by the digital camera were used to measure the slug size and separation distance. According to the characteristics of similar figures, the ratio between the channel width (600 μm) and the real slug size equals the ratio between the values measured from the pictures. As shown in Figure 3, x represents the measured slug maximum length, and y represents the measured channel width.

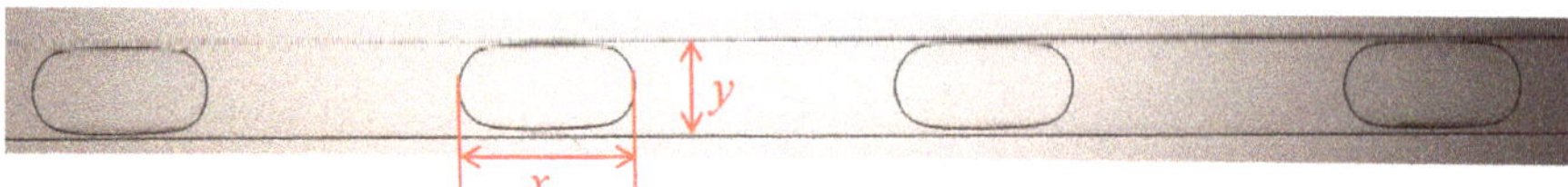

Figure 3. Schematic diagram of measurement mode.

Thus the slug size d can be calculated by the Equation $d = 600 \times \frac{x}{y}$. With the same method, the separation distance was measured.

In order to make the slug size and separation distance more accurate, 10 values were measured for each operating condition. The measured slug sizes are listed in Table 2, and the measured separation distances are listed in Table 3. These data show that the slugs were generated uniformly for the constant injection flow rates. Comparing every measured slug size and separated distance with their average values, the maximum related differences are less than ±0.5%, which indicates that the measurements of slug size and separation distance are stable, and that the slug size and separation distance can be represented by their average values.

Table 2. The slug size of experiments.

u_c (m/s)	u_d (m/s)	Sizes of 10 Slugs (μm)					Average Slug Size (μm)	The Maximum Related Difference (%)
0.001	0.001	1182	1185	1188	1180	1183	1183	0.42
		1186	1183	1183	1181	1182		
0.002	0.002	1065	1069	1068	1064	1067	1066	−0.38
		1062	1066	1065	1067	1066		
0.003	0.003	1029	1026	1024	1027	1026	1026	0.39
		1030	1025	1026	1028	1024		
0.004	0.004	928	926	923	924	926	926	0.32
		926	925	929	926	927		

Table 3. The separation distance of experiments.

u_c (m/s)	u_d (m/s)	Lengths of 10 Separation Distances (μm)					Average Separation Distance (μm)	The Maximum Related Difference (%)
0.001	0.001	1951	1950	1948	1953	1948	1950	0.15
		1952	1950	1949	1950	1947		
0.002	0.002	1518	1516	1519	1520	1515	1518	−0.19
		1520	1518	1517	1516	1519		
0.003	0.003	1274	1278	1280	1279	1274	1278	−0.31
		1278	1279	1281	1278	1276		
0.004	0.004	1299	1302	1295	1300	1299	1299	−0.31
		1298	1300	1299	1299	1301		

4.1.2. Validation of Numerical Method

Numerical simulations for the operating conditions of constant injection flow rates were conducted. The injection flow rates corresponded to the experiments. The flow patterns obtained from the numerical simulations were compared with the experiments. As shown in Figure 4, the simulated flow patterns agree well with the experiments. In Figure 4, the symbols u_c and u_d represent the continuous and disperse phase flow rate, respectively.

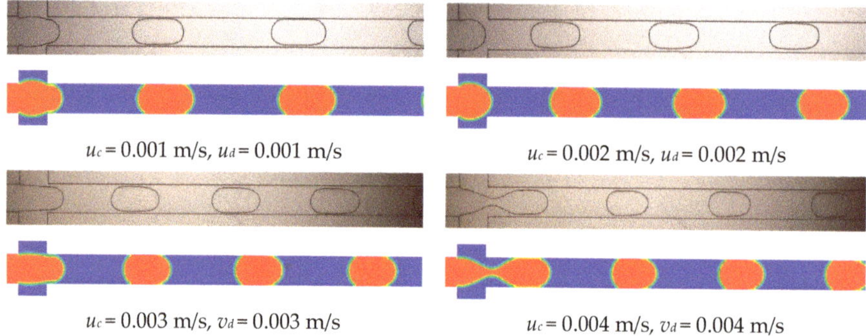

Figure 4. Flowing pattern comparison of experimental results (up) and simulated results (down).

The simulated slug size, which is the maximum slug length on the symmetry plane of the microchannel, and the simulated separation distance, which is its minimum length on the symmetry plane of the microchannel, are both compared with the experimental data. The results are shown in Table 4. The relative errors are within ± 8%. In fact, there are errors in both experiments and numerical simulations. Affected by the accuracy of the instruments, the injection flow rates are not always kept fixed, which will make the slug size change in the experiments. On the other hand, the pictures obtained from the experiments and the phase fraction contours obtained from the simulations are evaluated by image processing software, in which errors in the determination of the droplet boundary may occur. Accordingly, the maximum relative error, which is less than 8%, is deemed acceptable. The satisfactory comparison between the simulations and the experiments indicates that the numerical method is reliable.

Table 4. Slug size comparison between experimental and simulated results.

Injection Flow Rate		Slug Length (d/μm)			Separation Distance (l/μm)		
u_c (m/s)	u_d (m/s)	Experiment	Numerical Simulation	Relative Error (%)	Experiment	Numerical Simulation	Relative Error (%)
0.001	0.001	1183	1236.01	4.5	1950	1812.05	−7.1
0.002	0.002	1066	1104.23	3.6	1518	1533.74	1.0
0.003	0.003	1026	986.18	−3.9	1278	1288.89	0.9
0.004	0.004	926	934.19	0.9	1299	1243.61	−4.3

4.2. Numerical Simulation Results of Dynamic Dispersed Injection Flow Rate

In this section, the slug generation with a constant injection flow rate is numerically simulated at first. Then, the slug generation with a dynamic dispersed phase injection flow rate is analyzed. The dispersed phase dynamic injection flow rate follows three common wave types, namely, a triangle wave dynamic injection flow rate u_{dt}, a rectangular wave dynamic injection flow rate u_{dr} and a sine wave dynamic injection flow rate u_{ds}, as depicted in Figure 5. The effects of the dynamic dispersed phase injection flow rate for different types and different cycles on the slug generation are also discussed.

4.2.1. Comparison of Constant and Dynamic Injection Flow Rates

In this section, the slug generation process under conditions of constant as well as dynamic injection flow rates is investigated.

Firstly, numerical simulations for a constant injection flow rate are conducted. The continuous phase injection flow rate at the central inlet u_c is 0.001 m/s, and the dispersed phase injection flow rate at the side inlets u_d is 0.001 m/s, corresponding to the experiment as shown in the first line of Table 2. It is found that the slugs are generated uniformly. The slug generation characteristics, including the slug size, the separation distance and the generation time, are almost kept constant, i.e., independent of time. According to Table 2, the slug size d_0 is 1,236 μm, the separation distance l_0 is 1,812 μm. In addition, the slug generation time t_0 is 1.002 s.

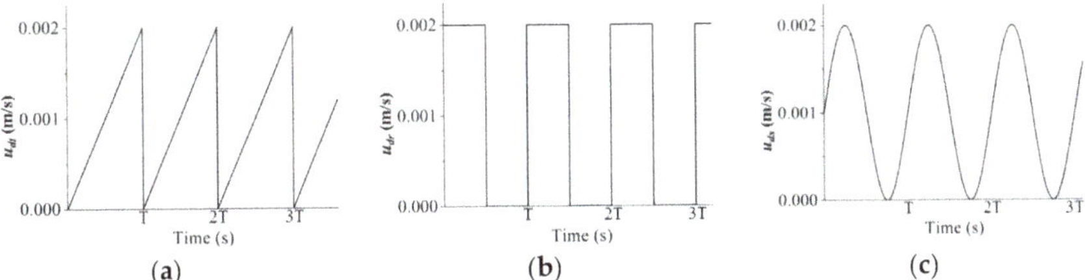

Figure 5. Dynamic dispersed injection flow rate of different types. (**a**) Triangle wave dynamic injection flow rate (u_{dt}); (**b**) Rectangular wave dynamic injection flow rate (u_{dr}); (**c**) Sine wave dynamic injection flow rate (u_{ds}).

Then, the slug generation under the condition of the triangle wave dynamic dispersed phase injection flow rate is numerically simulated. The cycles of the dynamic dispersed phase injection flow rate, which are related to the slug generation time t_0, are set as 0.5 t_0, t_0 and 2 t_0, respectively. Figure 6 presents the slug flow patterns for the triangle wave dynamic dispersed phase injection flow rate with different cycles. When the cycles of the triangle wave dynamic dispersed injection flow rate are $0.5t_0$ and t_0, the slugs are generated uniformly. The slug sizes are 1,235.36 μm and 1,235.16 μm, and the separation distances are 1,811.88 μm and 1,810.45 μm, respectively. Compared with the values obtained from a constant injection flow rate, the relative differences of slug sizes and separation distances are both within ± 0.1%, which reveals that the triangle wave dynamic dispersed phase injection flow rate with cycles of 0.5 t_0 and t_0 has a marginal effect on the slug generation. When the cycle of triangle wave dynamic dispersed phase injection flow rate is $2t_0$, however, the slug generation is not uniform anymore. The slug size and the separation distance change periodically.

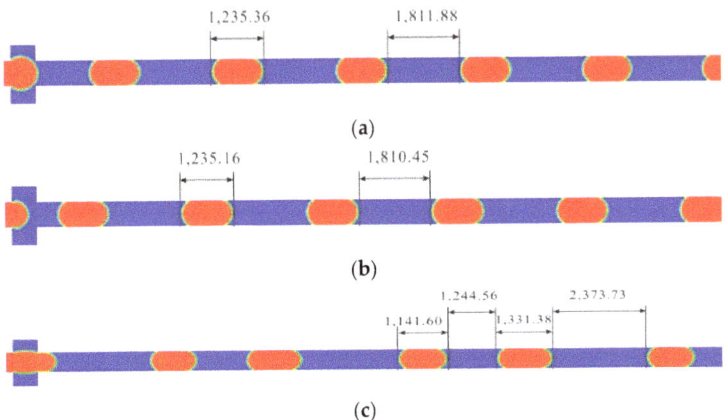

Figure 6. Flow types obtained from numerical simulations for triangle wave dynamic dispersed phase injection flow rate (μm). (**a**) The flow type for the condition of triangle wave dynamic dispersed phase injection flow rate with a cycle of 0.5 t_0; (**b**) The flow type for the condition of triangle wave dynamic dispersed phase injection flow rate with a cycle of t_0; (**c**) The flow type for the condition of triangle wave dynamic dispersed phase injection flow rate with a cycle of 2 t_0.

The times spent in generating each of the slugs are depicted in Figure 7, to further investigate the effects of the dynamic dispersed phase injection flow rate on the slug generation. When the cycles of the dynamic dispersed phase injection flow rate are 0.5 t_0 and t_0, the slug generation time is equal to t_0, after the flow has become stable. However, the slug generation time changes periodically, when the cycle of the dynamic dispersed phase injection flow rate is 2 t_0.

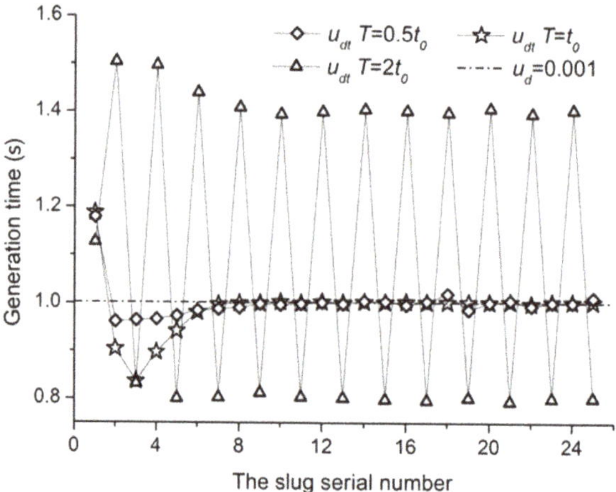

Figure 7. Slug generation time obtained from numerical simulations for a triangle wave dynamic dispersed phase injection flow rate.

The slug generation process includes three parts. At first, the dispersed phase fills the intersection of the microchannel. Then the dispersed phase stretches in the confluence channel, and at the same time, it is necked by the force from the continuous phase, which is blocked by the dispersed phase. There is a pressure drop between the intersection of the microchannel and the front end of the dispersed phase. When the pressure drop, which is affected by the continuous and dispersed phases, is larger than the tension of the two phases, the dispersed phase is broken, and a slug is generated. With a constant injection flow rate, the dispersed phase stretches in the confluent channel to a certain length, and combined with the force of the continuous phase, the critical pressure drop is achieved, at which a slug is generated, so the slug size is constant. When the cycles of the dynamic dispersed phase flow rate are 0.5 t_0 and t_0, the length of dispersed phase stretching in the confluence channel equals the constant injection flow rate in the period time of t_0. By combining the continuous and dispersed phases, the slug is generated at the critical pressure drop, so the slug size equals the constant dispersed injection flow rate. When the cycle of the dynamic dispersed phase flow rate is 2 t_0, the dispersed phase stretching in the confluence channel is not equal to the constant injection flow rate. A new combination of continuous and dispersed phases results in the critical pressure drop. When the dispersed phase is broken, the length over which it is stretching in the confluence channel is different from the case of the constant injection flow rate, so the slug flow type will change.

In summary, the dynamic dispersed phase injection flow rate can influence the slug generation, especially when the cycle is larger than t_0. In the following sections, the effects of the cycle and wave type on the slug generation are discussed.

4.2.2. Cycle of the Dynamic Injection Flow Rate

In this section, the effects of the triangle wave dynamic dispersed phase injection flow rates with different cycles on the slug generation are investigated. According to the above description, when the cycle is larger than t_0, the dynamic dispersed phase injection flow rate will affect the slug generation obviously. Here the cycles of the dynamic dispersed phase injection flow rate are set as 2 t_0, 3 t_0 and 4 t_0.

With different cycles of the dynamic dispersed phase injection flow rate, the flow pattern, the slug size, the separation distance and the slug generation time, are analyzed. Figure 8 reveals that the slug generation changes periodically, when the cycles are 2 t_0, 3 t_0 and 4 t_0. The slug size, the separation distance and the slug generation time oscillate near the values obtained under the condition of a constant injection flow rate. When the cycle of the dynamic dispersed phase injection flow rate is 2 t_0, there are two points in one slug generation cycle. When the cycle of the dynamic dispersed phase injection flow rate is 3 t_0, there are three points in one slug generation cycle. Similarly, when the cycle of the dynamic dispersed phase injection flow rate is $4t_0$, there are four points in one slug generation cycle. Under the conditions of a dynamic dispersed phase injection flow rate with the three cycles investigated in this section, the slug generation is periodical, and the number of slugs generated in one generation cycle is related to the flow rate cycle. As mentioned in Section 4.1, the critical pressure drop, resulting from the combination of the continuous and dispersed phases, causes the slug generation. The number of points in one cycle of the slug generation reflects how many times the critical pressure drop is achieved.

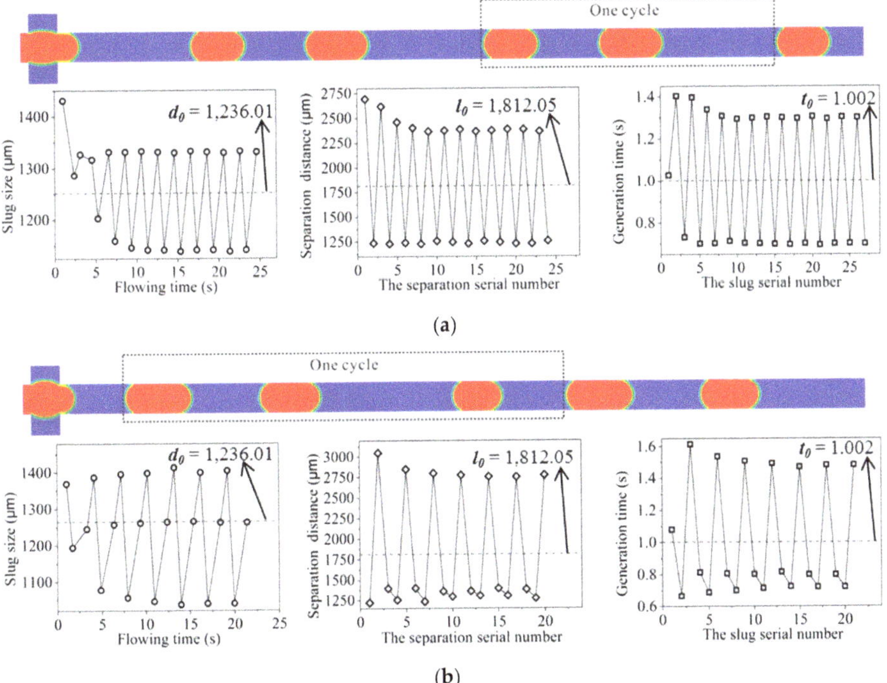

Figure 8. *Cont.*

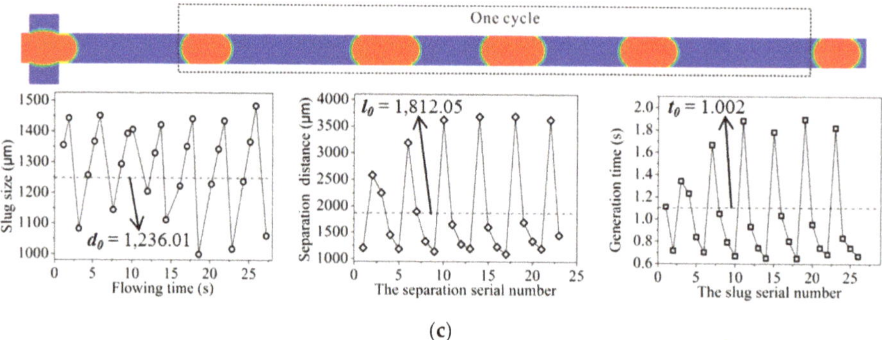

(**c**)

Figure 8. Slug generation obtained from numerical simulations for a triangle wave dynamic dispersed injection flow rate. (**a**) The slug generation for the condition of a triangle wave dynamic dispersed injection flow rate with a cycle of 2 t_0; (**b**) The slug generation for the condition of a triangle wave dynamic dispersed injection flow rate with a cycle of 3 t_0; (**c**) The slug generation for the condition of a triangle wave dynamic dispersed injection flow rate with a cycle of 4 t_0.

The case with a dynamic dispersed phase injection flow rate cycle 3 t_0 is taken as an example for further investigation. As mentioned above, there are three points in one slug generation cycle, when the cycle of the dispersed phase injection flow rate is 3 t_0. Within one slug generation cycle, the sum of the slug sizes is about 3,708 μm, the sum of separation distances is about 5,436 μm, and the sum of the generation times is about 3.006 s. These values are about three times those under the condition of a constant injection flow rate 0.001 m/s. As shown in Figure 9, the dynamic triangle wave injection flow rate meets the following characteristic:

$$\int_t^{t+T} u_{dt}dt = \int_t^{t+T} 0.001dt \qquad (9)$$

where, *t* is any instant of time, and *T* is the cycle of the dynamic injection flow rate. Equation (7) reveals that for the cycles investigated, the amounts of dispersed phase flow into the microchannel are equal, when the flow time is as long as a flow rate cycle, no matter that the dispersed phase injection flow rate is constant or dynamic. Thus when the cycle of the dynamic injection flow rate is $3t_0$, the sums of slug generation characteristics, including the slug size, the separation distance, and the slug generation time, are three times the values obtained under the condition of a constant injection flow rate.

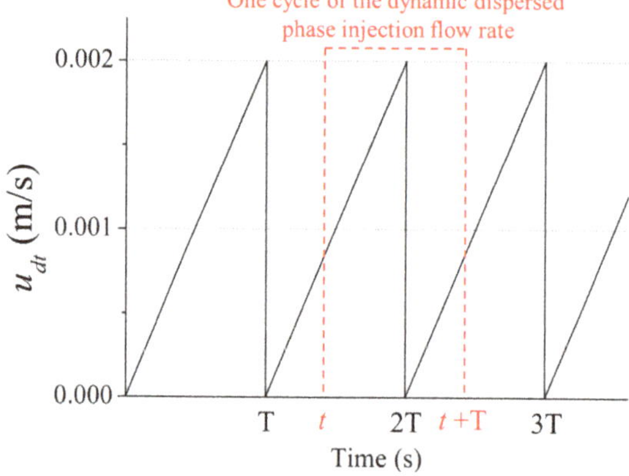

Figure 9. Integration of a triangle wave dynamic phase injection flow rate during one cycle.

4.2.3. Different Types of the Dynamic Injection Flow Rate

The type of the dynamic dispersed phase injection flow rate also plays an important role on the slug generation. With the same cycle of 2 t_0, the effects of the dynamic dispersed injection flow rate type, including a triangle wave, a rectangular wave and a sine wave, are studied. Figure 10 shows the slug size under the condition of dynamic dispersed phase injection flow rates of different types. The slug size changes periodically for the three kinds of dynamic dispersed phase injection flow rates, and it oscillates near 1236.01 µm, which is the slug size of the constant injection flow rate. In terms of the slug size amplitude, the triangle wave dynamic dispersed phase injection flow rate has the minimum amplitude, which is about 93 µm, while the rectangular wave dynamic dispersed phase injection flow rate has the maximum amplitude, which is about 250 µm.

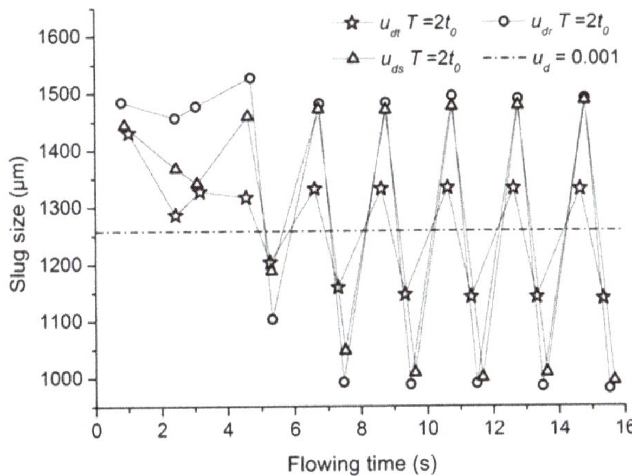

Figure 10. Slug size obtained from numerical simulations for dynamic dispersed injection flow rate of different types.

The separation distances for dynamic dispersed phase injection flow rates of different type are shown in Figure 11. For the three types of the dynamic dispersed injection flow rate, the separation distance changes periodically, and it oscillates near 1,812.05 µm, which is the separation distance of the constant injection flow rate. Concerning the amplitude of the separation distance, the sine wave dynamic dispersed phase injection flow rate has the minimum amplitude, which is about 230 µm, while the amplitude of the separation distance for the triangle wave dynamic dispersed phase injection flow rate is the maximum, which is about 569 µm.

Figure 12 presents the slug generation time for the different types of the dynamic dispersed phase injection flow rate. It is found that for these three types of the dynamic dispersed phase injection flow rate, the slug generation time changes periodically. When the dynamic dispersed phase injection flow rate is a sine wave, the amplitude of the slug generation time is the minimum, which is about 0.12 s. However, when the dynamic dispersed phase injection flow rate is a triangle wave, the amplitude of the slug generation time is the maximum, which is about 0.30 s. It is indicated that for the different types of the dynamic dispersed phase injection flow rate, the slug size, the separation distance and the slug generation time change periodically, and they oscillate near the values obtained under the condition of a constant injection flow rate. The types of the dynamic dispersed phase injection flow rate affect the vibration amplitude obviously.

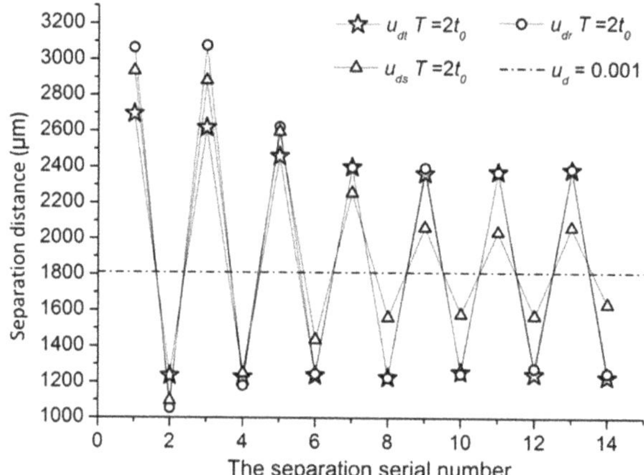

Figure 11. Separation distance obtained from numerical simulations for dynamic dispersed injection flow rate of different types.

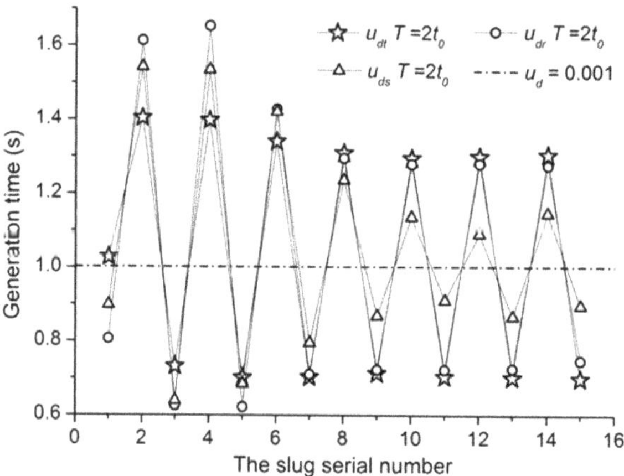

Figure 12. Slug generation time obtained from numerical simulations for dynamic dispersed injection flow rate of different types.

5. Conclusions

This work investigated the effects of a dynamic dispersed phase injection flow rate on the slug generation, in which the different cycles and different types of the dispersed phase injection flow rate were simulated numerically. The results show that the dynamic dispersed phase injection flow rate influences the slug generation observably. Especially when the cycle is larger than t_0, the slug generation changes periodically. For the same flux of the dispersed phase in one flow rate cycle, the sums of slug generation characteristics, including the slug size, the separation distance and the slug generation time, are equal, no matter whether the dispersed phase injection flow rate is constant or dynamic. Finally, for a fixed cycle, the dynamic dispersed phase injection flow rates with different types affect the oscillating amplitude of the slug size, separation distance and slug generation time obviously. Among these, the rectangular wave dynamic dispersed phase injection flow rate has the greatest influence on the slug size oscillation amplitude, while the triangle wave dispersed phase injection affects the separation distance and the slug generation time most significantly.

Author Contributions: Data curation, J.-y.Q., M.-r.C. and Z.W.; Formal analysis, M.-r.C.; Funding acquisition, J.-y.Q.; Investigation, M.-r.C.; Methodology, Z.W.; Project administration, Z.-j.J. and B.S.; Resources, B.S.; Supervision, J.-y.Q., Z.-j.J. and B.S.; Writing—Original Draft, J.-y.Q. and M.-r.C.; Writing—Review & Editing, B.S.

Funding: This research was funded by National Natural Science Foundation of China through, grant number 51,805,470; the Fundamental Research Funds for the Central Universities, grant number 2018QNA4013, and the Youth Funds of the State Key Laboratory of Fluid Power and Mechatronic Systems (Zhejiang University), grant number SKLoFP-QN-1801; the Fundamental Research Funds for the Central Universities, grant number 2018FZA213.

Conflicts of Interest: The authors declare no conflict of interest.

References

1. Maan, A.A.; Nazir, A.; Khan, M.K.I.; Boom, R.; Schroën, K. Microfluidic emulsification in food processing. *J. Food Eng.* **2015**, *147*, 1–7. [CrossRef]
2. Qian, J.; Li, X.; Wu, Z.; Bengt, S. A comprehensive review on liquid–liquid two-phase flow in microchannel: Flow pattern and mass transfer. *Microfluid. Nanofluidics* **2019**, *23*, 116. [CrossRef]
3. Kurimoto, R.; Nakazawa, K.; Minagawa, H.; Yasuda, T. Prediction models of void fraction and pressure drop for gas-liquid slug flow in microchannels. *Exp. Therm. Fluid Sci.* **2017**, *88*, 124–133. [CrossRef]
4. Liu, D.; Ling, X.; Peng, H. Experimental and numerical studies on microbubble generation and flow behavior in a microchannel with double flow junctions. *Ind. Eng. Chem. Res.* **2016**, *55*, 12276–12286. [CrossRef]
5. Patel, R.S.; Weibel, J.A.; Garimella, S.V. Characterization of liquid film thickness in slug-regime microchannel flows. *Int. J. Heat Mass Transf.* **2017**, *115*, 1137–1143. [CrossRef]
6. Rocha, L.A.M.; Miranda, J.M.; Campos, J.B.L.M. Wide range simulation study of taylor bubbles in circular milli and microchannels. *Micromachines* **2017**, *8*, 154. [CrossRef]
7. Kingston, T.A.; Weibel, J.A.; Garimella, S.V. An experimental method for controlled generation and characterization of microchannel slug flow boiling. *Int. J. Heat Mass Transf.* **2017**, *106*, 619–628. [CrossRef]
8. Svetlov, S.D.; Abiev, R.S. Formation mechanisms and lengths of the bubbles and liquid slugs in a coaxial-spherical micro mixer in Taylor flow regime. *Chem. Eng. J.* **2018**, *354*, 269–284. [CrossRef]
9. Yao, C.; Dong, Z.; Zhao, Y.; Chen, G. An online method to measure mass transfer of slug flow in a microchannel. *Chem. Eng. Sci.* **2014**, *112*, 15–24. [CrossRef]
10. Yao, C.; Dong, Z.; Zhao, Y.; Chen, G. The effect of system pressure on gas-liquid slug flow in a microchannel. *AIChE J.* **2014**, *60*, 1132–1142. [CrossRef]
11. Yao, C.; Liu, Y.; Zhao, S.; Dong, Z.; Chen, G. Bubble/droplet formation and mass transfer during gas–liquid–liquid segmented flow with soluble gas in a microchannel. *AIChE J.* **2017**, *63*, 1727–1739. [CrossRef]
12. Yin, Y.; Zhu, C.; Guo, R.; Fu, T.; Ma, Y. Gas-liquid two-phase flow in a square microchannel with chemical mass transfer: Flow pattern, void fraction and frictional pressure drop. *Int. J. Heat Mass. Transf.* **2018**, *127*, 484–496. [CrossRef]
13. Liu, Q.; Wang, W.; Palm, B. A numerical study of the transition from slug to annular flow in micro-channel convective boiling. *Appl. Therm. Eng.* **2017**, *112*, 73–81. [CrossRef]
14. Magnini, M.; Thome, J.R. A CFD study of the parameters influencing heat transfer in microchannel slug flow boiling. *Int. J. Therm. Sci.* **2016**, *110*, 119–136. [CrossRef]
15. Magnini, M.; Thome, J.R. An updated three-zone heat transfer model for slug flow boiling in microchannels. *Int. J. Multiphase Flow* **2017**, *91*, 296–314. [CrossRef]
16. Li, H.; Hrnjak, P. Flow visualization of R32 in parallel-port microchannel tube. *Int. J. Heat Mass Transf.* **2019**, *128*, 1–11. [CrossRef]
17. Garstecki, P.; Fuerstman, M.J.; Stonec, H.A.; Whitesides, G.M. Formation of droplets and bubbles in a microfluidic T-junction—Scaling and mechanism of break-up. *Lab. Chip* **2006**, *6*, 437–446. [CrossRef]
18. Sattari-Najafabadi, M.; Esfahany, M.N.; Wu, Z.; Sundén, B. Hydrodynamics and mass transfer in liquid-liquid non-circular microchannels: Comparison of two aspect ratios and three junction structures. *Chem. Eng. J.* **2017**, *322*, 328–338. [CrossRef]
19. Sattari-Najafabadi, M.; Nasr Esfahany, M.; Wu, Z.; Sundén, B. The effect of the size of square microchannels on hydrodynamics and mass transfer during liquid-liquid slug flow. *AIChE J.* **2017**, *63*, 5019–5028. [CrossRef]

20. Kovalev, A.V.; Yagodnitsyna, A.A.; Bilsky, A.V. Flow hydrodynamics of immiscible liquids with low viscosity ratio in a rectangular microchannel with T-junction. *Chem. Eng. J.* **2018**, *352*, 120–132. [CrossRef]

21. Carneiro, J.; Campos, J.B.L.M.; Miranda, J.M. High viscosity polymeric fluid droplet formation in a flow focusing microfluidic device—Experimental and numerical study. *Chem. Eng. Sci.* **2019**, *195*, 442–454. [CrossRef]

22. Zhang, Q.; Zhu, C.; Du, W.; Liu, C.; Fu, T.; Ma, Y.; Li, H.Z. Formation dynamics of elastic droplets in a microfluidic T-junction. *Chem. Eng. Res. Des.* **2018**, *139*, 188–196. [CrossRef]

23. Qian, J.; Li, X.; Wu, Z.; Jin, Z.; Zhang, J.; Bengt, S. Slug formation analysis of liquid–liquid two-phase flow in T-junction microchannels. *J. Therm. Sci. Eng. Appl.* **2019**, *11*. [CrossRef]

24. Man, J.; Li, Z.; Li, J.; Chen, H. Phase inversion of slug flow on step surface to form high viscosity droplets in microchannel. *Appl. Phys. Lett.* **2017**, *110*, 181601. [CrossRef]

25. Wehking, J.D.; Gabany, M.; Chew, L.; Kumar, R. Effects of viscosity, interfacial tension, and flow geometry on droplet formation in a microfluidic T-junction. *Microfluid. Nanofluid.* **2014**, *16*, 441–453. [CrossRef]

26. Zhang, Q.; Liu, H.; Zhao, S.; Yao, C.; Chen, G. Hydrodynamics and mass transfer characteristics of liquid–liquid slug flow in microchannels: The effects of temperature, fluid properties and channel size. *Chem. Eng. J.* **2019**, *358*, 794–805. [CrossRef]

27. Song, Y.; Song, J.; Shang, M.; Xu, W.; Liu, S.; Wang, B.; Lu, Q.; Su, Y. Hydrodynamics and mass transfer performance during the chemical oxidative polymerization of aniline in microreactors. *Chem. Eng. J.* **2018**, *353*, 769–780. [CrossRef]

28. Qian, J.; Li, X.; Gao, Z.; Jin, Z. Mixing efficiency analysis on droplet formation process in microchannels by numerical methods. *Processes* **2019**, *7*, 33. [CrossRef]

29. Van Loo, S.; Stoukatch, S.; Kraft, M.; Gilet, T. Droplet formation by squeezing in a microfluidic cross-junction. *Microfluid. Nanofluid.* **2016**, *20*, 146. [CrossRef]

30. Zhao, W.; Zhang, S.; Lu, M.; Shen, S.; Yun, J.; Yao, K. Immiscible liquid-liquid slug flow characteristics in the generation of aqueous drops within a rectangular microchannel for preparation of poly(2-hydroxyethylmethacrylate) cryogel beads. *Chem. Eng. Res. Des.* **2014**, *92*, 2182–2190. [CrossRef]

31. Malekzadeh, S.; Roohi, E. Investigation of different droplet formation regimes in a T-junction microchannel using the VOF technique in openfoam. *Microgravity Sci. Technol.* **2015**, *27*, 231–243. [CrossRef]

32. Gupta, A.; Sbragaglia, M. A lattice Boltzmann study of the effects of viscoelasticity on droplet formation in microfluidic cross-junctions. *Eur. Phys. J. E* **2016**, *39*, 1–15. [CrossRef] [PubMed]

33. Chen, X.; Glawdel, T.; Cui, N.; Ren, C.L. Model of droplet generation in flow focusing generators operating in the squeezing regime. *Microfluid. Nanofluid.* **2015**, *18*, 1341–1353. [CrossRef]

34. Li, X.B.; Li, F.C.; Yang, J.C.; Kinoshita, H.; Oishi, M.; Oshima, M. Study on the mechanism of droplet formation in T-junction microchannel. *Chem. Eng. Sci.* **2012**, *69*, 340–351. [CrossRef]

35. Ładosz, A.; von Rohr, P.R. Pressure drop of two-phase liquid-liquid slug flow in square microchannels. *Chem. Eng. Sci.* **2018**, *191*, 398–409. [CrossRef]

36. Qian, J.; Li, X.; Gao, Z.; Jin, Z. Mixing efficiency and pressure drop analysis of liquid-liquid two phases flow in serpentine microchannels. *J. Flow Chem.* **2019**, *9*, 187–197. [CrossRef]

37. Qian, J.; Chen, M.; Liu, X.; Jin, Z. A numerical investigation of the flow of nanofluids through a micro Tesla valve. *J. Zhejiang Univ. Sci. A* **2018**, *20*, 50–60. [CrossRef]

38. Cao, Z.; Wu, Z.; Sundén, B. Dimensionless analysis on liquid-liquid flow patterns and scaling law on slug hydrodynamics in cross-junction microchannels. *Chem. Eng. J.* **2018**, *344*, 604–615. [CrossRef]

39. Fletcher, D.F.; Haynes, B.S. CFD simulation of Taylor flow: Should the liquid film be captured or not? *Chem. Eng. Sci.* **2017**, *167*, 334–335. [CrossRef]